D1454080

SYMMETRY
IN BONDING AND SPECTRA

AN INTRODUCTION

SYMMETRY IN BONDING AND SPECTRA

AN INTRODUCTION

BODIE E. DOUGLAS
and
CHARLES A. HOLLINGSWORTH
Department of Chemistry
University of Pittsburgh
Pittsburgh, Pennsylvania

1985

ACADEMIC PRESS, INC.

(Harcourt Brace Jovanovich, Publishers)

Orlando San Diego New York London
Toronto Montreal Sydney Tokyo

ACADEMIC PRESS, INC.
Orlando, Florida 32887

United Kingdom Edition published by
ACADEMIC PRESS INC. (LONDON) LTD.
24–28 Oval Road, London NW1 7DX

Library of Congress Cataloging in Publication Data

Douglas, Bodie Eugene, Date
Symmetry in bonding and spectra.

Includes index.
1. Chemical bonds. 2. Spectrum analysis. 3. Symmetry (Physics) I. Hollingsworth, Charles A. II. Title.
QD461.D64 1985 541.2'24 84-20350
ISBN 0-12-221340-8 (alk. paper)

PRINTED IN THE UNITED STATES OF AMERICA

85 86 87 88 9 8 7 6 5 4 3 2 1

Contents

Preface

Symmetry is of such great importance in chemical applications generally that it is an essential part of the training for advanced undergraduate or beginning graduate students. The necessary mathematical background for the study of symmetry includes vectors, matrix algebra, and group theory. In this book we have included all of the essential mathematics for the benefit of those students who may need it. Vectors are reviewed briefly, and the essentials of matrix algebra are introduced. Group theory is an extensive branch of mathematics, but for the applications treated here, the essentials can be covered in little space. Since our objective is to enable students to handle applications, particularly applications to chemical bonding and spectroscopy, we have discussed only the required parts of group theory. In general, we have given proofs only where needed. In practice, many applications are handled using character tables and routine procedures. For example, the symmetry species of normal vibrations are generated by simple procedures in Chapter 10, but only after the multidimensional vector problem is illustrated.

Many courses and books dealing with the material treated here are called "Applications of Group Theory." It is symmetry, however, that is of primary importance in the applications. Group theory is needed to realize the full power of symmetry, but in fact, the amount of group theory required is not very great. In this volume we have emphasized the central role of symmetry.

The authors have long felt a need for the mathematical tools used here and for their chemical applications. One of us (CAH) published a book on the mathematical tools (*Vectors, Matrices, and Group Theory for Scientists and Engineers*, 1967). The other author (BD) has taught a first year graduate course on applications of symmetry for some time. This text is the result of a long perceived need for the development of extensive supplements for the teaching of this course.

Whereas some readers can skip the sections on vectors and matrices, we believe it is important that the essential background for the applications be introduced or reviewed briefly for those students with gaps in their background or in need of review. Chemistry students have had simple molecular orbital theory, ligand field theory, and treatments of molecular shapes. Instructors might omit these sections, but some readers will benefit from the concise reviews. Some quantum mechanics, providing the basis for later treatments, is reviewed in Chapter 4 so that the reader need not seek the necessary material from other treatises.

The usual procedures are given for deriving spectral terms (not those limited in practice to one- or two-electron cases) and electronic energy-level correlation diagrams. However, McDaniel's spin factoring is presented as a more useful approach to all but the most simple cases. Spin factoring allows us to generate with little difficulty electronic energy-level correlation diagrams for any electron configuration and any realistic symmetry.

Many of the applications are inorganic, reflecting our interests and the course that led to this text. The treatment of bonding and vibrational spectra includes organic molecules, and the approaches are generally applicable. The examples of symmetry-controlled reactions are mostly organic, reflecting the development of this topic.

In a typical one-term course it may not be possible to cover all of the material in the book. Considerable time can be saved for well-prepared students by omitting background material. A course with inorganic emphasis probably requires slighting Chapter 11, and organic emphasis can be achieved by omitting the first part of Chapter 5 (spectral terms) and part of Chapter 9.

Solved examples in the text illustrate theory and applications or introduce special points. Extensive problem sets cover the important methods and applications, and the answers are given in Appendix 4. Character tables (Appendix 1) include functions for identification of p, d, and f orbitals. Appendix 2 gives the chemically important double-group character tables, and Appendix 3 presents important correlation tables.

The authors wish to express their appreciation to Andrew R. Chopnak for drawing all of the original illustrations. Colleagues and students have made valuable contributions to the development of the manuscript.

1

The Importance of Symmetry in Chemistry

In this introductory chapter we draw on your familiarity with symmetry to examine elements of symmetry. With just this background and your knowledge of chemistry we can see how symmetry applies to orbitals in atoms and molecules. The same symmetry considerations apply to energy states of atoms and molecules and therefore to electronic and vibrational spectroscopy. The detailed treatment of an example can be followed pictorially. Do not be distracted by the terminology and notation used in the example; definitions and explanations will follow in later chapters.

1.1 Symmetry in Nature

We find the following dictionary definitions of symmetry: (1) similarity of form or arrangement on either side of a dividing line or plane; (2) excellence or beauty of form or proportion as a result of the similarity of form or arrangement. The word symmetry is derived from Greek and Latin words meaning "measure together." Thus we see that symmetry referred specifically to the similarity relative to a *mirror* plane. As an example, one side of a vase reflected through a mirror (a mirror plane or a plane of symmetry) might be identical to the other side of the vase. We now use symmetry in a broader sense to include other symmetry properties or *symmetry elements*. Thus the vase has rotational symmetry about a line down its center (the rotational axis) if it appears the same after being rotated about this line. Synonyms for symmetry include *harmony* and *balance*.

We associate symmetry with beauty and harmony in nature. The symmetry element shared by most animals, birds, and fish is the mirror plane. Features falling in the mirror plane (cut by the plane) occur singly; those off this plane occur in pairs. Your hands look alike, but they do not match up because of the difference between the front and back of the hand and the thumb occurs on one side. They are nonsuperimposable mirror images. In constructing a model of a dinosaur it is only necessary to recreate one side and then the other side can be duplicated through symmetry. Most leaves have mirror planes. However, as is usually true in nature, the symmetry is not perfect. Leaves from an oak tree differ from one another, but the best description of an oak leaf is the idealized symmetrical leaf.

Many flowers have mirror planes. Some flowers are radially symmetrical. The morning glory has a single petal with rotational symmetry such that it appears the same after rotation by any angle, except, of course, for small ridges and imperfections. Trillium has three petals arranged such that the blossom looks the same after rotation by 120°. Rotation by 360°/3 is called a threefold rotation or C_3. There are three mirror planes. Flowers with rotational symmetry and four identical petals (C_4 or 90° rotation), five identical petals (C_5 or $360°/5 = 72°$ rotation), and six identical petals (C_6 or 60° rotation) are common. In some flowers with six petals they occur in sets of three, with three in front and rotated by 60° relative to the other set of three. Such a flower has only C_3 rotational symmetry. Daisies and sunflowers have 12, 20, or more petals radiating from a center disk. Some have very-high-order rotational symmetry except for imperfections. We could describe a flower with radial symmetry by describing one petal and the order of rotational symmetry.

Although snowflakes differ considerably in appearance, they have the same hexagonal symmetry—a sixfold rotational axis and six mirror planes. Even exquisitely well-formed crystals have imperfections. In fact, without slight dislocations of the three-dimensional "mosaic" pattern of small units of the crystal, we could not obtain reasonable intensities in an x-ray diffraction pattern. The symmetry of a water molecule differs from that of an ice crystal, but the overall arrangement must lead to the symmetry of the crystal. From the detailed study of the x-ray diffraction pattern of the crystal, we can obtain the structure of the water molecule.

A cubic crystal has mirror planes bisecting each edge. They divide the crystal successively into halves, quarters, and eighths. If an x-ray crystallographic study is used to establish the structure of one-eight of the unit cell, the asymmetric unit, then the rest of the cell can be generated by repeated reflections through the mirror planes.

1.2 Electron Clouds in Atoms and Molecules

This size of an isolated atom has little meaning. The probability of finding an electron beyond the region of maximum density decreases with increasing distance from the nucleus but only approaches zero at infinity. Atomic size has meaning as the distance of closest approach by another atom.

An isolated atom has an "amorphous" electron cloud. The atomic orbitals become defined and take the shape dictated by the symmetry imposed by neighboring atoms. We use as our description of atomic orbitals the wave functions of the one-electron system, the *s*, *p*, *d*, and *f* orbitals for the various quantum shells of hydrogen. Yet, these have meaning only relative to some defined direction. Half-filled or filled *p*, *d*, or *f* orbitals are spherical. If we introduce a probe to try to determine the "shape" of an electron cloud, then we define a direction and an "orbital" will appear in some preferred orientation relative to the probe. With one direction defined, the *p* orbitals consist of one familiar dumbbell-shaped orbital ($m_l = 0$) along z (chosen as the unique direction) and a large donut in the xy plane for the coalesced p_x and p_y ($m_l = \pm 1$) orbitals. If we define a second direction, say x, then y is defined as well, and we obtain the usual descriptions of p_x, p_y, and p_z.

Atomic *s*, *p*, *d*, and *f* orbitals are distinguishable by their symmetry, and there are parallel distinctions for molecular orbitals. In fact, the σ, π, and δ notation was introduced to apply specifically to diatomic and linear molecules for which there is a close analogy to the atomic orbitals. We now use the notation more generally based on the orbital symmetry. An *s* orbital is spherical, a *p* orbital has two lobes with opposite signs of the amplitude of the wave function, separated by a nodal plane. A *d* orbital has two nodal planes. If we reflect from one point through the origin (the center of inversion) to an equivalent point, then the signs of the wave functions are the same for *s* and *d* but opposite for *p* and *f*. We say the *s* and *d* orbitals have even parity, *g* (*gerade*), and the *p* and *f* orbitals have odd parity, *u* (*ungerade*). Molecular orbitals are labeled similarly. The σ bonding orbitals, with no nodal planes, are *g*. If we "collapse" a diatomic molecule so that the nuclei coalesce, then the σ orbital would become spherical. Pi bonding orbitals and σ^* antibonding orbitals, with one nodal plane, are *u*. The nodal plane is through the bonded atoms for π and between them for σ^*. In each case the orbital for a collapsed atom would look like a *p* orbital. There are two nodal planes for a δ bonding orbital (see Fig. 6.2) and for a π^* antibonding orbital. The parity is *g*. For the δ orbital both nodal planes pass through the bonded atoms, but for π^*, one is through the bonded atoms and one is between them. These symmetry characteristics are summarized in Table 1.1.

We can tell a great deal about the relative energies of atomic and molecular orbitals by the number of nodal planes and/or radial nodes. There are no radial nodes for an orbital type (*s*, *p*, *d*, or *f*) the first time it is introduced, and the number of radial nodes increases by one with each quantum shell after that. The energies of the orbitals in Fig. 1.1 increase from top to bottom and from left to right as the number of nodes increases. Nature prefers to delocalize charge as much as possible. Each node causes more localization and increases

TABLE 1.1
Symmetry Properties of Atomic and Molecular Orbitals

Atomic orbitals	Molecular orbitals	Parity	Number of nodal planes	Orientation of nodal plane(s)
s	σ	g	0	—
p	π	u	1	Through bonded atoms
	σ^*	u	1	Between bonded atoms
d	δ	g	2	Through bonded atoms
	π^*	g	2	One through bonded atoms and one between them

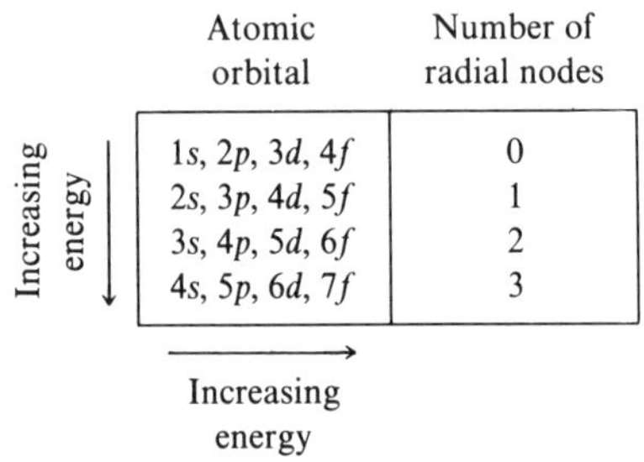

Atomic orbital	Number of radial nodes
$1s, 2p, 3d, 4f$	0
$2s, 3p, 4d, 5f$	1
$3s, 4p, 5d, 6f$	2
$4s, 5p, 6d, 7f$	3

Increasing energy ↓ (down the rows); Increasing energy → (across each row)

Fig. 1.1 Number of radial nodes and relative energies of atomic orbitals.

the energy. This is a useful and reliable guide in obtaining the relative energies of molecular orbitals representing various linear combinations of atomic orbitals. Symmetry serves as a guide for obtaining these combinations, but it gives us little information about energies.

Electron clouds representing molecular orbitals can be described by their symmetry properties, and these same symmetry descriptions apply to the mathematical expressions for the wave functions involved. For large molecules with many electrons, high-order secular equations are involved. Symmetry can be used to factor these equations into lower-order equations, simplifying their solution.

1.3 Spectroscopy

Orbitals can be described by their symmetry properties. The configurations representing the occupancy of the orbitals give rise to energy states that also are described by symmetry properties. Selection rules depend upon the symmetries of the ground state, the excited state, and an operator for the particular type of transition involved.

Symmetry selection rules for spectroscopy are remarkably similar for various kinds of spectra. They differ in the operator involved. The symmetry of the operator is determined by the nature of the displacement of electrons or atoms. We can determine which transitions are "allowed" or "forbidden" by simple examination of the symmetries involved.

Vibrational spectra involve transitions from the vibrational ground state, with the full symmetry of the molecule involved, to any of various vibrationally excited states. An animated representation of a vibrating molecule looks as though the motion is random and erratic. Actually the motion can be resolved into vibrational modes that must be consistent with the symmetry of the molecule itself. Stringent symmetry selection rules apply differently to vibrational transitions appearing in infrared and Raman spectra.

Nuclear magnetic resonance spectroscopy depends on coupling among nuclei and distinctions between chemically equivalent and nonequivalent nuclei. For a large and perhaps floppy molecule, it is sometimes difficult to examine sketches or models to determine which groups or atoms are really equivalent. The task is often simplified using symmetry. Atoms or groups interchanged by any symmetry element *must* be equivalent.

1.4 Symmetry Rules for Chemical Reactions

Bonding combinations of orbitals are determined by symmetry based on the shape and parity of orbitals involved. Only atomic orbitals that have *identical* symmetry properties relative to the symmetry elements of the molecule can combine. Since two reacting molecules must come together, break some bonds, and form new bonds, it seems evident that symmetry restrictions apply to reactions that occur through continuous interaction from initiation to completion. These are the Symmetry Rules or Woodward–Hoffmann Rules for chemical reactions.

1.5 Group Theory

Our discussion has dealt with symmetry, and that is the central theme of this book. We want to examine those aspects of chemistry—structure, bonding, spectroscopy, and reactions—where symmetry is of particularly great importance. Symmetry itself can become complex. The way to organize the symmetry properties of a molecule is to recognize that all of the symmetry elements taken together must form a mathematical group. This places restrictions on symmetry properties, and it provides a very powerful framework for dealing with and applying symmetry. The framework is *group theory*, a large, elegant part of mathematics. We need only a very small part of group theory. For instance, we rely on the orthogonality of symmetry species of a symmetry group. This is just one of the far reaching implications of the "great orthogonality theorem."

Group theory is essential to achieve the full power of symmetry, but group theory is a small part of this book. To a great extent we use routine procedures based upon group theory. No apology is needed. We are interested in gaining insight into problems in chemistry. Group theory is a useful tool; we use only that part needed. We are not concerned with the development of group theory except as necessary for the understanding of its chemical applications. The important relationships we examine are based upon *symmetry*, group theory just helps us apply it.

Ordinarily we deal with chemical isomers using a trial and error procedure. When the number of isomers becomes very large, symmetry and group theory provide the means for determining the number of isomers and for generating the isomers.

1.6 Symmetry Groups and Symmetry Species

We can examine a simple case to see how some properties of a molecule are characterized by symmetry. Proper definitions and derivations or justifications will come later. The square planar molecule XeF_4 is shown in Fig. 1.2. The molecule is unchanged in appearance by rotation by 90° about the z axis perpendicular to the plane of the molecule. This is called a fourfold (360°/4) rotational axis, C_4. We can also rotate by 180° about the z axis without change, so there is a C_2 axis (360°/2) coincident with C_4. There is also a twofold axis along x and one along y. Since these axes are interchanged with one another by fourfold rotation about z, they are equivalent to one another but different from C_2 coincident with C_4. We distinguish these rotations as C_2'. There are also two equivalent twofold axes bisecting edges, but since they are not interchanged with the C_2' axes, we label them C_2''. There is a mirror plane, labeled σ_{xy}, slicing through each atom. There are two mirror planes through diagonally opposite F atoms (xz and yz planes) and two more between adjacent F atoms. We distinguish these as σ' and σ'', respectively, as shown in Fig. 1.2. Since reflection of any point through the origin gives an equivalent point, there is an inversion center, labeled i, at the origin. These are most, but not quite all, of the collection of symmetry elements for a symmetry group called $\mathbf{D}_{4h}$.

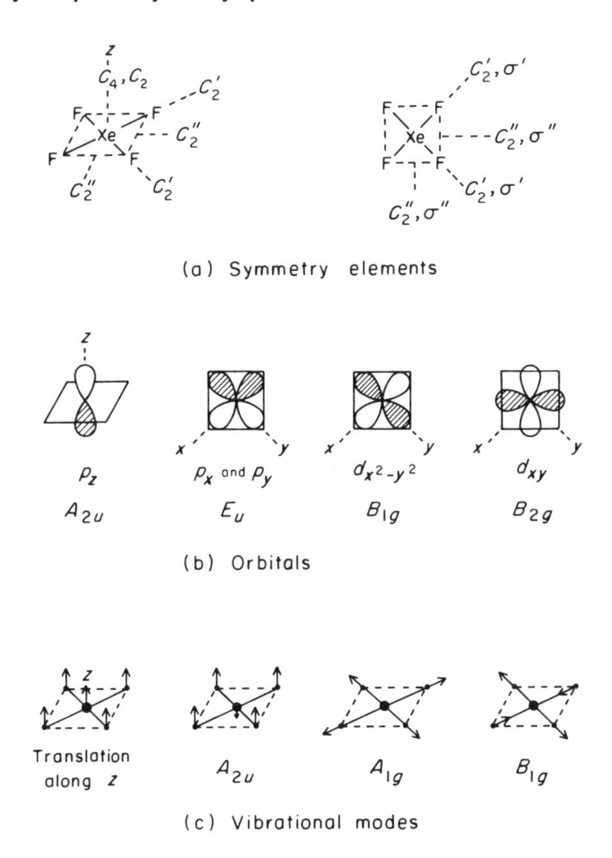

Fig. 1.2 (a) Some symmetry elements of XeF_4, the $\mathbf{D}_{4h}$ group; (b) symmetry properties of some orbitals in $\mathbf{D}_{4h}$; (c) symmetry properties of some vibrational modes of XeF_4. Translation along z is shown also.

The molecule is unchanged in appearance after carrying out the operation (rotation or reflection) associated with every one of the symmetry elements of the group. We could say that the molecule is described by, or belongs to, the totally symmetric *symmetry species* labeled A_{1g}. An s orbital is unchanged by any of the operations so it also belongs to A_{1g}. The d_{z^2} orbital with the positive lobes along z and the negative donut in the xy plane is also A_{1g}. The p_z orbital is unchanged by C_4 and C_2 rotations about the z axis and by the mirror planes, σ' and σ'', slicing through it. The signs of the lobes are reversed by C'_2 and C''_2 rotations, by reflection through the σ_{xy} plane and inversion through the origin. The symmetry species is A_{2u}. The symbol is A if rotation by 90° (C_4) causes no change, and it is B if the signs are reversed by the C_4 operation. The subscript is 1 if the signs are not changed by the C'_2 rotation, and it is 2 if they are changed by this rotation. The subscript is g if inversion causes no sign change and it is u if the sign is changed. The $d_{x^2-y^2}$ orbital has the signs of its lobes reversed by rotation by 90° (C_4) about z, but not by rotation by 180° (C_2') about x or y. There is no change in sign upon inversion through the origin. The symmetry species is B_{1g}. The d_{xy} orbital has the signs of its lobes reversed by 90° (C_4) rotation about z and by 180° rotation about the C_2' axes but not by inversion (i). The symmetry species is B_{2g}. The p orbitals change signs by the inversion operation, so they are all u. The p_x and p_y orbitals are interchanged by rotation by 90° (C_4) about z, so they must be treated as a pair. Together they belong to the E_u symmetry species.

We have not examined all of the symmetry species of the $\mathbf{D}_{4h}$ groups. The number of symmetry species of a group is determined using group theory. All of the atomic orbitals of Xe and the molecular orbitals of XeF_4 belong to symmetry species of the $\mathbf{D}_{4h}$ group. In fact, only combinations of F orbitals belonging to the *same symmetry species* as the Xe orbitals can combine to form molecular orbitals. All electronic energy states can be characterized by symmetry species of the molecule. For another molecule to react with XeF_4 by a concerted mechanism, there must be net positive overlap between the interacting orbitals of the two reacting molecules. This is another way of saying *they must belong to the same symmetry species.*

We can represent molecular vibrations by vectors showing the direction of motion of the atoms. If all atoms have along the z direction, then this is translation along z, not a vibration. The whole molecule moves. If the Xe atom moves in the opposite direction along z, compared to the motion of the four F atoms so that the center of gravity is unmoved, then the set of vectors is unchanged by C_4, but the signs (directions) are reversed by C_2', C_2'', σ_{xy}, and i. These properties are the same as for p_z or the A_{2u} symmetry species. If all F atoms move away from the origin in the xy plane and then in toward the origin, the vibrational mode is totally symmetric, belonging to the A_{1g} species. If two diagonally opposite F atoms move out while the other two move in, as shown in Fig. 1.2, then this is the pattern corresponding to the signs of the lobes of $d_{x^2-y^2}$ or the B_{1g} symmetry species.

All vibrational modes are characterized by symmetry species of the symmetry group. Using character tables that summarize all of the symmetry elements and symmetry species of a group, it is easy to determine which vibrations can be expected to appear in infrared spectra and which should appear in Raman spectra. For the $\mathbf{D}_{4h}$ group, vibrations might be "active" in infrared *or* Raman spectra but not both, and some vibrations are not active in either.

2

Properties of Groups

The algebraic theory of groups is a vast subject with applications in many fields. Our interest in group theory is as a tool for using symmetry to predict properties of chemical systems. In this chapter we give some of the elementary properties of groups that will be used in subsequent chapters. We show how the terminology of group theory is used to describe the symmetry of molecules and (briefly) crystals.

2.1 Definition

A mathematical group is an algebraic structure consisting of a collection of elements and a law for combining any two of the elements to obtain another element. The structure is a group if, and only if, the following four conditions are satisfied.

(1) *Closure.* The result of combining any two elements of the group is an element of the group.

(2) *Associativity.* The associative law of algebra holds. When the combination of elements is represented (as will usually be the case) by a multiplication symbolism, the associative law is then that for multiplication; that is, if A, B, and C are any three elements of the group, their combination satisfies

$$(AB)C - A(BC).$$

(3) *Identity.* The group has a *unique* element E such that $EA - AE = A$, where A is any element of the group.

(4) *Inverse.* Every element A has in the group one, and only one, inverse A^{-1} such that $AA^{-1} = A^{-1}A = E$.

In general the combination of two elements may or may not be commutative, that is, AB need not be the same as BA. The identity E is the only element that must commute with all the other elements of the group. Also, an element must always commute with its inverse. Groups in which all elements commute are called *Abelian* groups.

The number of elements in a group is called the *order* of the group. The order can be finite or infinite. A group within a group is called a *subgroup* of the larger group.

EXAMPLE 2.1 (a) Show that the set of four elements 1, -1, i, $-i$ (where $i = \sqrt{-1}$) with ordinary multiplication as the law of combination is an Abelian group.

(b) Does the set 1, -1 form a group under ordinary multiplication? How about the set i, $-i$?

(c) How about the single element 1?

SOLUTION (a) (1) Closure. The product of any two elements is an element of the set.

(2) Associativity. Ordinary multiplication is associative.

(3) Identity. The identity E is 1.

(4) The inverse of 1 is 1 (E is always its own inverse); -1 is the inverse of -1; $-i$ is the inverse of i and i is the inverse of $-i$.

The set is a group of order four. Ordinary multiplication is commutative. Therefore, the group is Abelian.

(b) The set 1, -1 is a group. The set i, $-i$ is not, since closure does not hold; also, there is no identity.

(c) The element 1 forms a group by itself. This is true of the identity element of any group.

EXAMPLE 2.2 Consider the set of all the positive integers (except zero) together with all the fractions made by taking the ratios of any two integers. Does this set form a group under ordinary multiplication if we assume that all fractions of the form n/n represent the integer 1?

SOLUTION (1) Closure. Yes, the multiplication of any two fractions is another fraction.

(2) Associativity. Yes.

(3) Identity. The integer 1 (or n/n) is E. It is unique by assumption.

(4) Inverse. n_2/n_1 is the inverse of n_1/n_2. The set is a group of infinite order.

2.2 Group Multiplication Tables

All of the abstract (algebraic) properties of a group are contained in its *multiplication table*. There is only one abstract group of order three. Its multiplication table can be written

	E	A	B
E	E	A	B
A	A	B	E
B	B	E	A

(2.1)

All group multiplication tables are *Latin squares*. That is, an element occurs once, and only once, in any row or column. For non-Abelian groups, our multiplication tables are constructed as follows.

The member in the top row is the right member in the product, and the member in the first (leftmost) column is the left member of the product, thus

	$\cdots$	y
$\vdots$		$\vdots$
x	$\cdots$	xy

The multiplication table for the group $(1, -1, i, -i)$ of Example 2.1 can be written

	1	-1	i	$-i$
1	1	-1	i	$-i$
-1	-1	1	$-i$	i
i	i	$-i$	-1	1
$-i$	$-i$	i	1	-1

(2.2)

The group (E, A, B, C) with the multiplication table

	E	A	B	C
E	E	A	B	C
A	A	E	C	B
B	B	C	A	E
C	C	B	E	A

(2.3)

represents the same abstract group as $(1, -1, i, -i)$ because they have the same multiplication table, and, thus, the same algebraic structure. We can go from one to the other by relabeling the elements: $1 \leftrightarrow E$, $-1 \leftrightarrow A$, $i \leftrightarrow B$, $-i \leftrightarrow C$.

Groups that have the same multiplication table (perhaps by relabeling) and therefore have the same algebraic structure are said to be *isomorphic*. If two groups G_1 and G_2 are isomorphic, we write $G_1 \sim G_2$ or $G_2 \sim G_1$.

The groups with the multiplication table

	E	A	B	C
E	E	A	B	C
A	A	E	C	B
B	B	C	E	A
C	C	B	A	E

(2.4)

are not isomorphic to those with multiplication table (2.3), because each element in table (2.4) is its own inverse, and that is not true for the elements in table (2.3). It follows that no rearrangement (or relabeling) of the elements can change table (2.4) to table (2.3). Tables (2.3) and (2.4) are the multiplication tables for the only two abstract groups of order four. All Latin squares of order four are isomorphic to one of these two.

Note that all groups of order three [table (2.1)] and all groups of order four [table (2.3) or (2.4)] are Abelian.

Consider a wheel with three spokes (Fig. 2.1). Referring to Fig. 2.1a, let the symbol C_3 represent a rotation through 120° (clockwise), taking the wheel from position (1) to position (2). Let the symbol C_3^2 represent a rotation through 240°, taking the wheel from position (1) to position (3). Let E represent no rotation, leaving the wheel in position (1).

Referring to Fig. 2.1b, let C_2 represent a rotation through 180° about the axis labeled C_2, taking the wheel from position (1) to position (4). Similarily we take C_2' and C_2'' to represent rotations through 180° about axes C_2' and C_2'', respectively. The open circle in the center is to indicate that the wheel has been turned over.

Referring to Fig. 2.1c, let σ_v represent an imaginary reflection through the plane labeled σ_v, which is perpendicular to the plane of the wheel. This imaginary process would take the wheel from position (1) to the imaginary position (7), where the spokes are oriented as in position (4), but the wheel has not been turned over as it has been in (4). Likewise, we use σ_v' and σ_v'' to represent reflections through the planes σ_v' and σ_v'', respectively. (It is conventional to call the direction of a rotation axis such as that in Fig. 2.1a vertical. The planes in Fig. 2.1c are then vertical—thus, the subscript v.)

For our rule of combining the operations in Fig. 2.1 we choose to consider the axes C_2, C_2', C_2'' and the planes σ_v, σ_v', σ_v'' to be fixed in space and the corresponding operations defined with respect to these fixed positions. Perhaps the rule of combination that we are using can best be explained by illustrations.

The combination of C_3 followed by C_3 is $C_3C_3 = C_3^2$. C_3^2 *followed* by C_3 is $C_3C_3^2 = E$. Also, $C_3^2C_3^2 = C_3$. More difficult to determine is C_3 *followed* by C_2

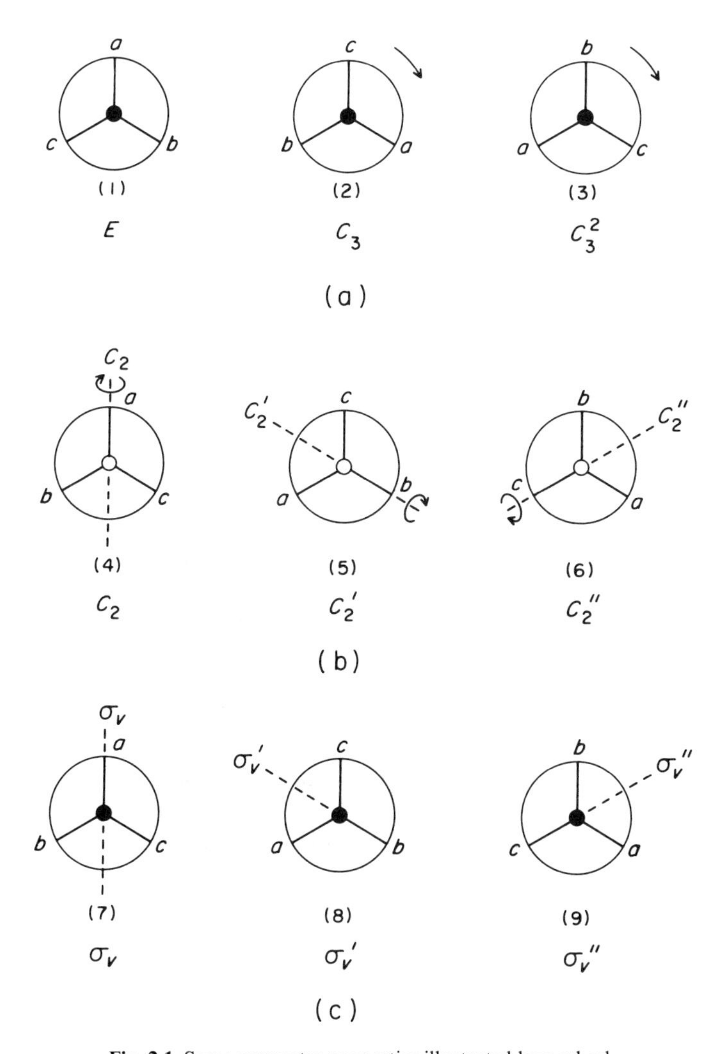

Fig. 2.1 Some symmetry properties illustrated by a wheel.

(about axis C_2, Fig. 2.1), that is, C_2C_3. We can obtain this product by use of the diagram

$$\overset{a}{c \curlywedge b} \xrightarrow{C_3} \overset{c}{b \curlywedge a} \xrightarrow{C_2} \overset{c}{a \curlywedge b} = C'_2.$$

Thus, $C_2C_3 = C'_2$. For C_2 *followed* by C_3, which is written C_3C_2, we have

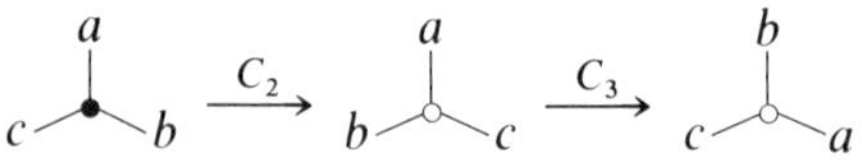

$$\overset{a}{c \;\bullet\; b} \xrightarrow{C_2} \overset{a}{b \;\circ\; c} \xrightarrow{C_3} \overset{b}{c \;\circ\; a} = C''_2.$$

Thus the operations C_3 and C_2 do not commute, i.e., $C_3C_2 \neq C_2C_3$. The products C_2C_2 and $\sigma_v\sigma_v$ are both E, that is, C_2 and σ_v are their own inverses. For the product $\sigma_v\sigma'_v$ we have

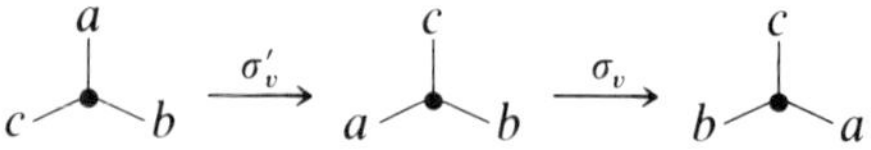

$$\overset{a}{c \;\bullet\; b} \xrightarrow{\sigma'_v} \overset{c}{a \;\bullet\; b} \xrightarrow{\sigma_v} \overset{c}{b \;\bullet\; a} = C_3.$$

EXAMPLE 2.3 Using the elements defined by Fig. 2.1 and the law of combination which we have illustrated, work the following exercises.

(a) Show that the elements $(E, \sigma_v, \sigma'_v, \sigma''_v)$ do not form a group.

(b) Using the product σ_vC_2 show that the nine elements in Fig. 2.1 do not form a group.

(c) Obtain the mutiplication table for the elements in Fig. 2.1a and b together. Is the group Abelian? What are its subgroups?

(d) Do the same as for Example 2.3c but for the elements in Fig. 2.1a and c together.

(e) What new elements must be added to the set of Fig. 2.1 so that all the elements in the figure will belong to one group? What is the physical (or imaginary) significance of the new elements?

SOLUTION (a) The set is not closed. We have seen that $\sigma_v\sigma'_v = C_3$.

(b) For σ_vC_2 we have

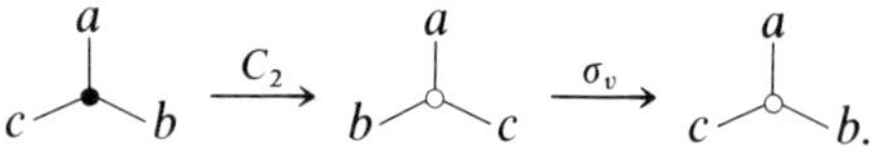

$$\overset{a}{c \;\bullet\; b} \xrightarrow{C_2} \overset{a}{b \;\circ\; c} \xrightarrow{\sigma_v} \overset{a}{c \;\circ\; b}.$$

This result is not one of the elements in Fig. 2.1 (it is similar to E, but has the open center). Thus the closure condition does not hold.

(c) The multiplication table is

	E	C_3	C_3^2	C_2	C'_2	C''_2
E	E	C_3	C_3^2	C_2	C'_2	C''_2
C_3	C_3	C_3^2	E	C''_2	C_2	C'_2
C_3^2	C_3^2	E	C_3	C'_2	C''_2	C_2
C_2	C_2	C'_2	C''_2	E	C_3	C_3^2
C'_2	C'_2	C''_2	C_2	C_3^2	E	C_3
C''_2	C''_2	C_2	C'_2	C_3	C_3^2	E

The group is not Abelian. For example, $C_3C_2 \neq C_2C_3$. The subgroups are (E); (E, C_3, C_3^2); (E, C_2); (E, C'_2); and (E, C''_2).

(d) The table is the same as that for $(E, C_3, C_3^2, C_2, C_2', C_2'')$ with σ_v, σ_v', and σ_v'' replacing C_2, C_2', and C_2'', respectively. The subgroups are obtained by making the same replacements. The groups $(E, C_3, C_3^2, \sigma_v, \sigma_v', \sigma_v'')$ and $(E, C_3, C_3^2, C_2, C_2', C_2'')$ are isomorphic.

(e) We found in Example 2.3b that one new element needed is that obtained from $\sigma_v C_2$, which is represented by

a

c b.

This element represents an imaginary reflection through the plane of the wheel (the horizontal plane). Its symbol is σ_h. We must add all the elements obtained by combining σ_h with those in Fig. 2.1. These are the elements obtained by reflecting those in Fig. 2.1 through the horizontal plane. The total number of *new* elements is three, and the order of the new group is twelve (see Problem 2.2).

REMARKS CONCERNING EXAMPLE 2.3 Although we leave most of our discussion of symmetry groups for later sections, we can use this example to introduce some of the concepts involved. So far we have made no assumptions about the symmetry of the wheel. Let us now consider wheels with certain symmetries.

(i) Suppose the wheel has circular symmetry so that one cannot distinguish positions (1), (2) and (3), but suppose the two sides of the wheel (i.e., above and below the page) are different so that positions in Fig. 2.1a are distinguishable from those in 2.1b. Suppose also that there are three equivalent arrows, one at the end of each spoke and pointing in the clockwise direction. These arrows will be reversed in direction by the imaginary operations of Fig. 2.1c. The only indistinguishable positions are those of Fig. 2.1a, which form the group (E, C_3, C_3^2). This group is called the *symmetry* group $\mathbf{C}_3$. (Symmetry groups in which one point—in this case the center of the wheel—remains fixed are also called *symmetry point* groups, or just *point* groups. The notation that we use for the operations and group designations is called the *Schoenflies* notation.)

(ii) Suppose now we remove the arrows, but make no other changes. The two sides of the wheel are still different. Now the imaginary positions of Fig. 2.1c are indistinguishable from one another and those of 2.1a. The symmetry of this wheel is described by the group $(E, C_3, C_3^2, \sigma_v, \sigma_v', \sigma_v'')$, which is designated $\mathbf{C}_{3v}$.

(iii) Suppose two sides of the wheel are the same, except the top side (above the page) has arrows in the clockwise direction, and the bottom side (below the page) has arrows in the counterclockwise direction (see Fig. 2.2). Now the positions in Fig. 2.1a and b will be indistinguishable from one another, but those in Fig. 2.1c will have their arrow opposite to those in 2.1a and b. The symmetry of this wheel is described by the group $(E, C_3, C_3^2, C_2, C_2', C_2'')$, which is given the symbol $\mathbf{D}_3$.

(iv) Finally, suppose that the arrows are absent and the two sides of the wheel are the same. Now all the positions in Fig. 2.1 are indistinguishable, but these do

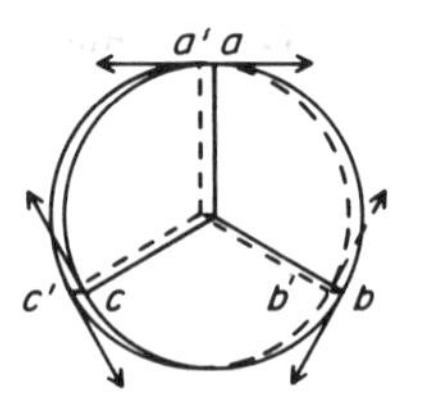

Fig. 2.2 Some symmetry properties of a wheel. The top side (from the front of the page and labeled $a\ b\ c$) and the bottom side (through the back of the page and labeled $a'\ b'\ c'$) are the same except for the arrows attached in opposite directions.

not form the complete set of symmetry operations of the wheel. A new element is the imaginary reflection through the plane of the wheel, i.e., σ_h. Some additional elements are obtained by combining σ_h with all the elements in Fig. 2.1. The order of the symmetry group thus obtained is 12. The group is designated $\mathbf{D}_{3h}$ (see Problem 2.2).

2.3 Group Generators and Their Relations

If $(a, b, c, \ldots)$ is a subset of the elements of a group and if every element of the group can be expressed as a product involving only elements of this set and their inverses, then this set is a set of *generators* of the group. Thus, the group $\mathbf{C}_3$ (Example 2.3) needs only a single generator. It can be C_3, since $C_3^2 = C_3C_3$, and $E = C_3C_3C_3 = C_3^3$. It can be C_3^2, since $C_3^2C_3^2 = C_3$, and $C_3^2C_3^2C_3^2 = E$. For the group $\mathbf{D}_3$ the generators can be (C_3, C_2) or (C_3^2, C_2), etc.

A group can be specified mathematically in terms of a set of generators and their *relations*. For the group $\mathbf{C}_3$ we have the single relation

$$C_3C_3C_3 = C_3^3 = E. \tag{2.5}$$

This is the *cyclic* group of order three. A cyclic group of order n has one generator A and a single defining relation

$$A^n = E. \tag{2.6}$$

Its elements are $[A, A^2, \ldots, A^n(=E)]$ (see Problem 2.5).

It should be mentioned that the expression $AA^{-1} = A^{-1}A = E$ is not a defining relation, since it applies to all groups. Also, relation (2.6) means that n is the smallest exponent of A that gives E; n is called the *period* of A. In any finite group, every element must have a period, and the powers of any element must form a cyclic subgroup or the group itself (see Problem 2.9).

In the abstract group to which $\mathbf{D}_3$ belongs, there are two generators A and B, and their relations can be expressed as

$$A^3 = E, \qquad B^2 = E, \qquad ABAB = E. \tag{2.7}$$

In applying these relations to the point group $\mathbf{D}_3$ we can pick C_3 to be A and C_2 to be B. (Other choices are possible. See Problems 2.4 and 2.14.)

The third relation in (2.7) can be expressed in many different ways. For example, from

$$ABAB = E$$

we have

$$B(ABAB)B^{-1} = BEB^{-1},$$
$$BABAE = E,$$
$$BABA = E,$$

and from this result we have

$$B(BABA) = BE,$$
$$B^2ABA = B,$$
$$EABA = B,$$
$$ABA = B.$$

2.4 Permutation Groups

Let us consider the six permutations of three letters abc:

$$E = \left(\frac{abc}{abc}\right), \qquad A = \left(\frac{abc}{cab}\right), \qquad B = \left(\frac{abc}{bca}\right),$$
$$a' = \left(\frac{abc}{acb}\right), \qquad b' = \left(\frac{abc}{cba}\right), \qquad c' = \left(\frac{abc}{bac}\right),$$

where the bottom row in the parentheses is the result of the operation on the top row. We have used the capital letters for the permutation in which all three letters have been moved and lower case letters for those in which only two letters have been moved. It is convenient to write the permutation operations in a *cyclic* form. For example, $A = (acb)$, which is to be read "a is replaced by c, c is replaced by b, and b is replaced by a" (see Problem 2.8).

In the cyclic notation we have

$$E, \qquad A = (acb), \qquad B = (abc),$$
$$a' = (bc), \qquad b' = (ac), \qquad c' = (ab).$$

Now we can illustrate the law of combination of two permutations. Consider the product Bc', which means "c' followed by B"; c' leaves the order bac; B then replaces a by b, b by c, and c by a; this leaves the order cba, which is the same result as the single permutation $(ac) = b'$.

If we return to Example 2.3 and think of the effect of the operations there as permutations of the spokes a, b, and c, then we recognize that the operators of Fig. 2.1(a) correspond in some way to the permutations A and B. Those of Fig. 2.1b (and also of c) correspond to a', b', and c'. We suspect that the group of permutations on three letters is isomorphic to the group $\mathbf{D}_3$ (and to the group $\mathbf{C}_{3v}$; see Example 2.3d). The multiplication table for the permutation group should correspond in some way to the table for $\mathbf{D}_3$ found in Example 2.3. In fact, the table for the permutation group can be obtained by substituting (A, B, b', a', c'), *respectively*, for $(C_3, C_3^2, C_2, C_2', C_2'')$ in the table for $\mathbf{D}_3$.

The total group of permutations on n letters is of order $n!$. This group is called the *symmetric* group of order $n!$ and of degree n. A permutation is even (odd) if it involves an even (odd) number of interchanges of letters. The set of even permutations in a symmetric group of degree n is a subgroup and is called the *alternating* group of degree n. The set of odd permutations does not form a group because it is not closed—the combination of two odd permutations is an even permutation.

In the symmetric group of degree three, the elements E, A, and B are the even permutations and make up the alternating group of degree three. It is isomorphic to the point group $\mathbf{C}_3$.

2.5 Conjugate Classes

An element A is said to be conjugate to an element B if there is an element X (and, therefore, an element $Y = X^{-1}$) in the group that transforms A into B according to

$$X^{-1}AX = YAY^{-1} = B. \tag{2.8}$$

The transformation (2.8) is called a *similarity transformation*. X is said to transform A, and B is called the transform of A by X.

If we calculate the transforms of the elements of $\mathbf{D}_3$ by all the elements of $\mathbf{D}_3$, we obtain the following table of transforms $X^{-1}AX$ (see Problem 2.10).

$A \backslash X$	E	C_3	C_3^2	C_2	C_2'	C_2''
E	E	E	E	E	E	E
C_3	C_3	C_3	C_3	C_3^2	C_3^2	C_3^2
C_3^2	C_3^2	C_3^2	C_3^2	C_3	C_3	C_3
C_2	C_2	C_2''	C_2'	C_2	C_2''	C_2'
C_2'	C_2'	C_2	C_2''	C_2''	C_2'	C_2
C_2''	C_2''	C_2'	C_2	C_2'	C_2	C_2''

(2.9)

We see that E is transformed into itself by all the elements of the group. The *sets* (C_3, C_3^2) and (C_2, C_2', C_2'') are each transformed into themselves by all the elements of the group. Sets that are transformed into themselves by all the elements of the group are called *conjugate classes*, or, simply, *classes*. Because of the isomorphism of $\mathbf{D}_3$, $\mathbf{C}_{3v}$ (see Example 2.3d), and the symmetric group of degree three, we know that the classes of these latter two groups will be (E), (C_3, C_3^2), and $(\sigma_v, \sigma_v', \sigma_v'')$ in $\mathbf{C}_{3v}$, and (E), (A, B), and (a', b', c') in the symmetric group.

Some properties of transforms and classes are the following (see Problem 2.12).

(1) No element transforms two different elements into the same element. That is, the same element does not occur more than once in any column of the table of transforms.

(2) No element occurs in more than one class.

(3) The number of elements in a class is a factor of the order of the group. Actually, if h is the order of the group, and the group has a class with m elements, then each element of the class is transformed into itself and every element of the class exactly h/m times.

(4) In an Abelian group each element is a class by itself ($X^{-1}AX = AX^{-1}X = A$).

2.6 Subgroups

2.6.1 SOME GENERAL PROPERTIES

The subgroups of the point group $\mathbf{D}_{3h}$ are $\mathbf{D}_3$, $\mathbf{C}_{3v}$, a group designated $\mathbf{C}_{3h}$ (consisting of the elements $E, C_3, C_3^2, \sigma_h, \sigma_h C_3, \sigma_h C_3^2$), $\mathbf{C}_3$, four groups with the symbol $\mathbf{C}_s$ (consisting of E and a reflection), three $\mathbf{C}_2$ groups (consisting of E and a twofold rotation), and the group $\mathbf{C}_1$ (consisting of the single element E).

$\mathbf{C}_1$ is a subgroup of every point group. Every element and its powers form a cyclic subgroup (see Problem 2.9). The cyclic subgroups of $\mathbf{D}_{3h}$ are $\mathbf{C}_3$, the four $\mathbf{C}_s$ groups and the three $\mathbf{C}_2$ groups, and (trivially) $\mathbf{C}_1$.

The order of any subgroup is a divisor of the order of the group (Lagrange's theorem). The order of $\mathbf{D}_{3h}$ is 12, and it has subgroups of orders 6, 3, 2, and 1.

2.6.2 CONJUGATE SUBGROUPS AND NORMAL SUBGROUPS

If H $(H_1, \ldots, H_g)$ is a subgroup of G, and X is an element of G not in H, then the transforms of the elements of H by X, that is, the set $X^{-1}HX$ $(X^{-1}H_1X, \ldots, X^{-1}H_gX)$ is a subgroup of G (see Problem 2.13). The subgroups H and $X^{-1}HX$ are called *conjugate* subgroups.

EXAMPLE 2.4 Find the subgroups in $\mathbf{D}_3$ that are conjugate to the subgroup (E, C_2).

SOLUTION We can use the table of transforms (2.9),

$$E^{-1}(E, C_2)E = (E, C_2),$$
$$C_3^{-1}(E, C_2)C_3 = (E, C_2''),$$
$$(C_3^2)^{-1}(E, C_2)C_3^2 = (E, C_2'),$$
$$C_2^{-1}(E, C_2)C_2 = (E, C_2),$$
$$(C_2')^{-1}(E, C_2)C_2' = (E, C_2''),$$
$$(C_2'')^{-1}(E, C_2)C_2'' = (E, C_2').$$

Thus the subgroups (E, C_2), (E, C_2'), and (E, C_2'') are conjugates.

Subgroups that transform into themselves (not necessarily each element into itself) by *all* the elements of G are called *normal* (also, *self-conjugate*, or *invariant*) subgroups. The subgroup $\mathbf{C}_3$ is a normal subgroup of $\mathbf{D}_3$ and of $\mathbf{C}_{3v}$ (see Example 2.5). $\mathbf{C}_1$ is a normal subgroup of every group.

EXAMPLE 2.5 Show that $\mathbf{C}_3$ is a normal subgroup of $\mathbf{D}_3$.

SOLUTION From the table of transforms (2.9) we have

$$E^{-1}(E, C_3, C_3^2)E = (E, C_3, C_3^2),$$
$$C_3^{-1}(E, C_3, C_3^2)C_3 = (E, C_3, C_3^2),$$
$$(C_3^2)^{-1}(E, C_3, C_3^2)C_3^2 = (E, C_3, C_3^2),$$
$$C_2^{-1}(E, C_3, C_3^2)C_2 = (E, C_3^2, C_3),$$
$$(C_2')^{-1}(E, C_3, C_3^2)C_2' = (E, C_3^2, C_3),$$
$$(C_2'')^{-1}(E. C_3, C_3^2)C_2'' = (E, C_3^2, C_3).$$

Thus, the subgroup $\mathbf{C}_3$ is transformed into itself by every element of $\mathbf{D}_3$ and so is *normal*.

2.6.3 DIRECT AND SEMIDIRECT PRODUCTS OF SUBGROUPS

A group G is the *direct product* of the subgroups $H_1, H_2, \ldots,$ and H_m and is written

$$G = H_1 \times H_2 \times \cdots \times H_m \tag{2.10}$$

provided the following two conditions hold.

(1) Every element of each subgroup commutes with every element of all the other subgroups.
(2) Each element g of G can be expressed in a unique way in the form

$$g = h_1 h_2 \cdots h_m, \tag{2.11}$$

where h_i is an element of H_i.

The subgroups in Eq. (2.10) and the elements in Eq. (2.11) can be taken in any order.

Consider the group $\mathbf{C}_{3h}$ $(E, C_3, C_3^2, \sigma_h, \sigma_h C_3, \sigma_h C_3^2)$ and the subgroups $\mathbf{C}_3$ (E, C_3, C_3^2), and $\mathbf{C}_s$ (E, σ_h). It is clear from physical consideration that σ_h commutes with the operations in $\mathbf{C}_3$. Also, condition (2) holds, since we can write the elements of $\mathbf{C}_{3h}$ as $(E_1, E_2, E_2 C_3, E_2 C_3, E_2 C_3^2, E_1 \sigma_h, \sigma_h C_3^2, \sigma_h C_3^2)$, where we have used the E_1 and E_2 for the identity elements in subgroups $\mathbf{C}_3$ and $\mathbf{C}_s$, respectively. It follows that $C_{3h} = C_3 \times C_s$.

If the elements of the subgroup H_1 do not all commute with those of subgroup H_2 but all the elements of G are products of these two subgroups, then we call G the *semidirect* product of H_1 and H_2, and we write[1]

$$G = H_1 \wedge H_2. \tag{2.12}$$

For example, the group $\mathbf{C}_{3v}$ is the semidirect product of $\mathbf{C}_3$ (E, C_3, C_3^2) and $\mathbf{C}_s$ (E, σ_v). $\mathbf{C}_3 \wedge \mathbf{C}_s$ and $\mathbf{C}_s \wedge \mathbf{C}_3$ both give the elements of $\mathbf{C}_{3v}$ but in different orders.

2.7 Point Groups

2.7.1 SYMMETRY ELEMENTS AND OPERATIONS

So far we have met examples of the following *symmetry elements* and corresponding *symmetry operations*: (1) the n-fold rotation axis C_n and the corresponding operation C_n (consisting of a rotation through an angle of

[1] R. L. Flurry, Jr., *Symmetry Groups*, Prentice-Hall, Englewood Cliffs, New Jersey, 1980.

$2\pi/n$) and the multiples C_n^k ($k = 2, 3, \ldots, n$); (2) and the reflection plane and the corresponding reflection operator.

If a system has more than one symmetry axis, and one of these has a larger value of n than any of the others, then that axis is called the *principal rotation* axis. If there is only one rotation axis, then that axis is the principal axis.

Reflection planes containing the principal rotation axis are designated σ_v (we will meet an exception to this notation shortly). Our use of the symbol σ_h to represent a reflection plane normal to the principal axis is standard.

There are other symmetry elements and corresponding symmetry operations. We shall define them by example. Consider Fig. 2.3, which represents two equilateral triangles, one directly above the other, in the staggered position. This could represent an ethane molecule in the staggered configuration, the vertices representing the six hydrogens. There is a C_3 axis normal to the planes of the triangles. There are three C_2 axes parallel to the planes of the triangles. There are three reflection planes through the C_3 axis. Although these are vertical planes, they are designated σ_d to indicate that they bisect a dihedral angle formed by two planes containing the C_3 axis and adjacent C_2 axes. In addition to these symmetry elements there are two new types illustrated in Fig. 2.3.

There is an *inversion center* midway between the centers of the two triangles. The operation of inversion interchanges the vertices 1 and 6, 2 and 4, and 3 and 5. The inversion operation is designated i.

A rotation through $2\pi/6$ (C_6) followed, or preceded, by a reflection through a horizontal plane midway between the two triangles is a symmetry operation. It puts 1 in the place of 5, 5 in the place of 2, etc. The conventional symbol for this combined operation is S_6, and the vertical axis is called an S_6 axis. An S_n axis is called an n-fold *improper* rotation axis. An improper rotation can be

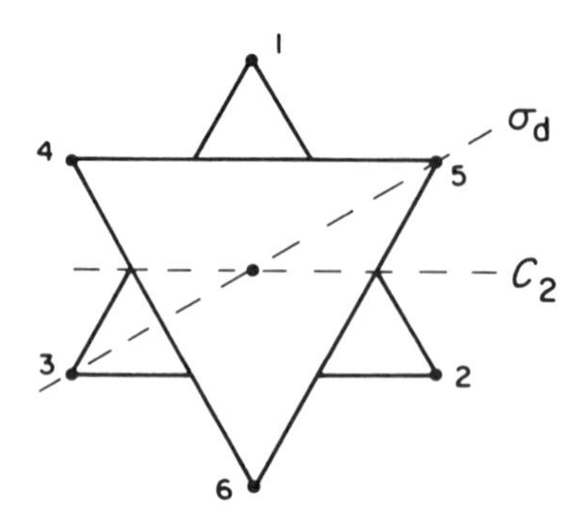

Fig. 2.3 An illustration of $\mathbf{D}_{3d}$ symmetry. Although the vertices of the triangles are numbered, they are equivalent.

defined in two ways. First, as we have illustrated above, it is a rotation followed, or preceded, by reflection through a plane normal to the rotation axis; that is,

$$S_n = \sigma_h C_n = C_n \sigma_h. \tag{2.13}$$

As in the example above, the σ_h and C_n elements need not be present individually. They may both be present, however; for example, the point group $\mathbf{C}_{3h}$ contains the operations C_3 and σ_h individually, and their combination gives S_3.

EXAMPLE 2.6 (a) Express the operations of $\mathbf{C}_{3h}$ in terms of the operations E, C_3, σ_h, S_3, and their powers.
(b) Prove that $\mathbf{C}_{3h}$ is a cyclic group of order six.
(c) Prove that $\mathbf{C}_{3h}$ and $\mathbf{C}_6$ are isomorphic.

SOLUTION (a) In Section 2.6.3 we found that $\mathbf{C}_{3h}$ is the direct product of the subgroups $\mathbf{C}_3$ and $\mathbf{C}_s$. Its elements were expressed there as (E, C_3, C_3^2, σ_h, $\sigma_h C_3$, $\sigma_h C_3^2$). The operator $\sigma_h C_3$ is S_3. Let us consider the powers of S_3.

$$S_3 = \sigma_h C_3,$$
$$S_3^2 = \sigma_h^2 C_3^2 = C_3^2,$$
$$S_3^3 = \sigma_h^3 C_3^3 = \sigma_h,$$
$$S_3^4 = \sigma_h^4 C_3^4 = C_3,$$
$$S_3^5 = \sigma_h^5 C_3^5 = \sigma_h C_3^2,$$
$$S_3^6 = \sigma_h^6 C_3^6 = E.$$

The operators in $\mathbf{C}_{3h}$ are E, C_3, C_3^2, σ_h, S_3, S_3^5.
(b) From part (a) we see that the powers S_3^k ($k = 1, 2, \ldots, 6$), give the whole group, and so the group is cyclic of order six.
(c) All cyclic groups of order six have the one generator A and the elements A, $A^2, A^3, A^4, A^5, A^6 = E$ and so must be isomorphic (see the list of isomorphisms in Section 2.7.3; also see Problem 2.17).

A second definition of improper rotations is in terms of a rotation followed, or preceded, by inversion i. The angle of rotation involved in this second definition is *not* $2\pi/n$ for S_n, where n has the value appropriate for the first definition, Eq. (2.13). For example, we have, by Eq. (2.13),

$$S_1 = \sigma_h C_1 = \sigma_h, \tag{2.14}$$

and it is not difficult to convince oneself that

$$\sigma_h = iC_2. \tag{2.15}$$

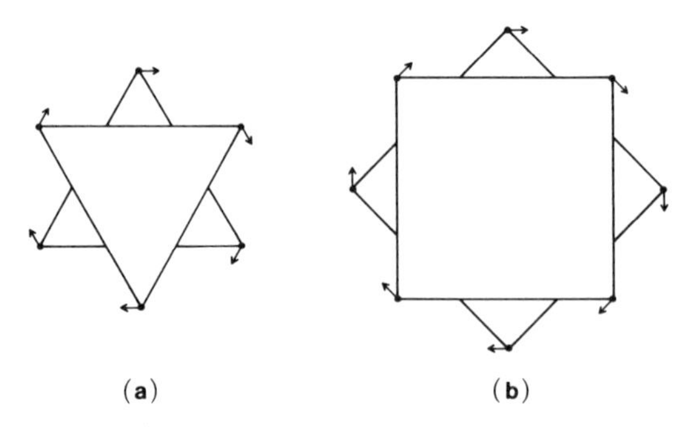

Fig. 2.4 Illustration of (a) $\mathbf{S}_6$ and (b) $\mathbf{S}_8$ symmetry.

It follows that

$$S_1 = iC_2. \tag{2.16}$$

Also, from Eqs. (2.13) and (2.15) we obtain (see Problem 2.15)

$$S_2 = i. \tag{2.17}$$

In the example illustrated by Fig. 2.3 the operations C_3 and i are both present, so, therefore, is the combination iC_3, which corresponds to the S_6 element. [The operation S_6 given by Eq. (2.13) is actually iC_3^2; see Problem 2.16.]

The first definition is conceptually the simpler of the two, and the value of n that is used in the conventional symbol S_n is that which agrees with Eq. (2.13). For the mathematical representation of an improper rotation about an arbitrary axis the second definition is sometimes more convenient (see e.g., Problem 3.5).

The point group giving the symmetry described in Fig. 2.3 is of order 12 and has the symbol $\mathbf{D}_{3d}$. It is the direct product of the group $\mathbf{D}_3$ and the inversion group $\mathbf{C}_i$ (E, i); $\mathbf{D}_{3d} = \mathbf{C}_i \times \mathbf{D}_3$. (The inversion operation commutes with all symmetry operations.)

The group $\mathbf{C}_{3h}$, which contains S_3, is a direct product. There is a type of group that contains the operator S_n when n is even. The symbol for this type of group is $\mathbf{S}_n$. We must consider two cases: when n is not divisible by 4, and when n is divisible by 4. When n is not divisible by 4, $\mathbf{S}_n$ is the direct product $\mathbf{C}_i \times \mathbf{C}_{n/2}$. This type of group can be illustrated with $\mathbf{S}_6$ as an example by the use of Fig. 2.4a, which is similar to Fig. 2.3, except that equivalent clockwise arrows have been placed at each vertex. Because of these arrows the C_2 and σ_d symmetry elements have been lost. The C_3 axis and the center of inversion remain. That an S_6 is present can be seen schematically as follows.

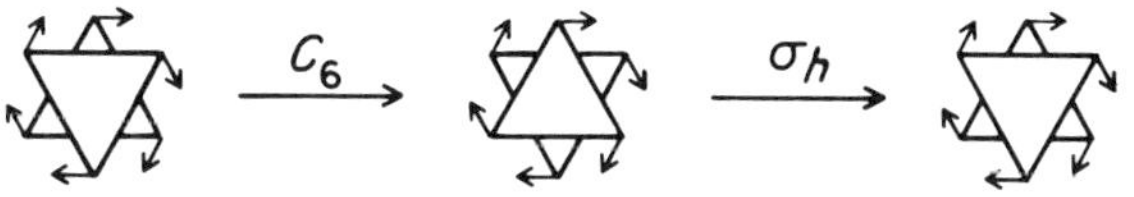

C_6 $\longrightarrow$ σ_h $\longrightarrow$

The case in which n is divisible by 4 is illustrated in Fig. 2.4b. The symmetry of the figure contains a C_4 and S_8, but there is no center of inversion.

For convenience, we now summarize the various symmetry operations.

- C_n rotation through $2\pi/n$
- σ reflection through a plane
- σ_v reflection through a vertical plane, i.e., through a plane containing the principal rotation axis
- σ_d a special σ_v, the reflection plane bisects the dihedral angle made by the principal axis and two adjacent C_2 axes that are perpendicular to the principal axis
- σ_h reflection through a plane normal to the principal rotation axis
- i inversion through a point
- S_n improper rotation through $2\pi/n$; it can be defined by Eq. (2.13)

2.7.2 PICTORIAL APPROACH TO GROUPING SYMMETRY OPERATIONS BY CLASSES

The rigorous method for determining which operations are in the same class is by the similarity transformation (see Sect. 2.5). However, since this can be tedious for large groups, a simpler approach is useful. First, we recognise that in the Abelian groups ($\mathbf{C}_n$, $\mathbf{C}_{nh}$, and $\mathbf{S}_n$) every operation is in its own class. The $\mathbf{S}_n$ (and $\mathbf{C}_{nh} = \mathbf{S}_n$ for odd n) groups are cyclic groups using the S_n generator. Second we identify the useful distinctions to be made: Are the following in the same or separate classes?

(1) All vertical planes
(2) C_n and C_n^{n-1} $(=C_n^{-1})$
(3) S_n and S_n^{n-1} $(= S_n^{-1})$

A rule that works is that symmetry elements that are not interchanged by any other operation of the group (this excludes the cyclic groups) are in separate classes. It is always true that all σ_v planes are interchanged by C_n operations and all σ_d planes are interchanged by C_n, but σ_v is never interchanged with σ_d.

For the cyclic $\mathbf{C}_3$ group, C_3 and C_3^2 are in separate classes, and in the $\mathbf{C}_{3h}$ group (E, C_3, C_3^2, σ_h, S_3, S_3^5) each element is still in a separate class. Of course, this is the cyclic $\mathbf{S}_3$ group. Here reflection of a curved arrow representing the

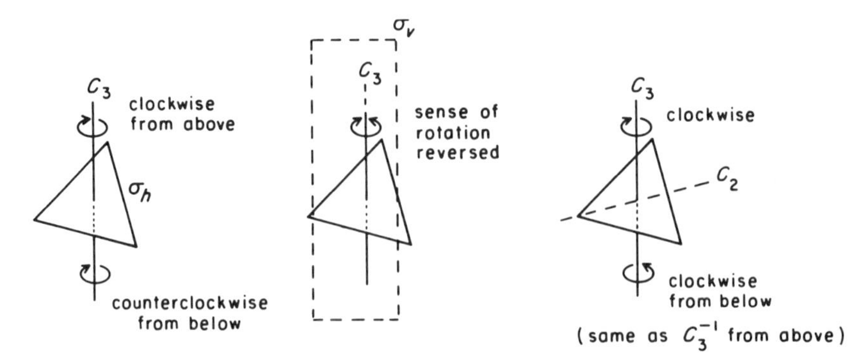

Fig. 2.5 Effect of σ_h, σ_v, and C_2 operations on the sense of rotation about a rotation axis.

clockwise sense of rotation from above gives a curved arrow in the counterclockwise sense when viewed from below the plane (see Fig. 2.5). This is not equivalent to C_3^2 (C_3^{-1}) from above, and they are in separate classes. In the $\mathbf{C}_{3v}$ group (E, $2C_3$, $3\sigma_v$) reflection of the curved arrow representing clockwise C_3 rotation through a vertical plane (σ_v) reverses the sense of the arrow, giving C_3^{-1} or C_3^2, so C_3 and C_3^2 are in the same class. The C_n and C_n^{n-1} operations are also in the same class for $\mathbf{D}_n$ (and $\mathbf{D}_{nh}$ and $\mathbf{D}_{nd}$). Any of the C_2 operations takes the clockwise C_3 axis from above into a clockwise C_3 axis from below and this is equivalent to C_n^{-1} (C_n^{n-1}) from above. Likewise C_2 axes perpendicular to S_n and vertical planes (σ_v or σ_d) place S_n and S_n^{n-1} in the same class.

These are operationally valid rules that can be useful. In fact, as we have seen (Example 2.3c) and as we shall see using matrices in Chapter 3, $C_3C_2 \neq C_3^{-1}$, we need the full similarity transformation $C_2^{-1}C_3C_2 = C_3^{-1}$ [see the table of transforms (2.9) and Problem 3.17].

2.7.3 THE POINT GROUPS

Adding a symmetry operation to a simple point group (order n) generates new operations and a larger group of an order higher by an integral factor. We have examined the axial groups with one principal axis C_n. These include $\mathbf{C}_n$, $\mathbf{C}_{nv}$, $\mathbf{C}_{nh}$, $\mathbf{D}_n$, $\mathbf{D}_{nh}$, $\mathbf{D}_{nd}$, and $\mathbf{S}_n$ ($\mathbf{C}_s = \mathbf{S}_1$ and $\mathbf{C}_i = \mathbf{S}_2$). There are also groups involving multiple axes of high order, representing the five Platonic solids (Fig. 2.6). The cubic groups include the cube and octahedron ($\mathbf{O}_h$, $6C_4$, $8C_3$) and the tetrahedron ($\mathbf{T}_d$, $8C_3$). The rotational groups are $\mathbf{O}$ and $\mathbf{T}$. Adding a center of symmetry to $\mathbf{O}$ gives $\mathbf{O}_h$, and $\mathbf{T} \times \mathbf{C}_i = \mathbf{T}_h$. The icosahedral groups include the regular icosahedron and dodecahedron ($\mathbf{I}_h$, $24C_5$, $20C_3$) and the rotational subgroup $\mathbf{I}$.

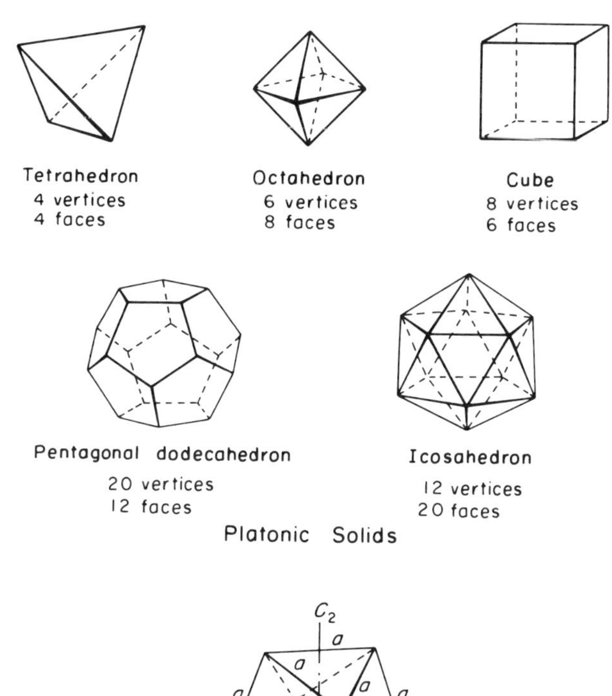

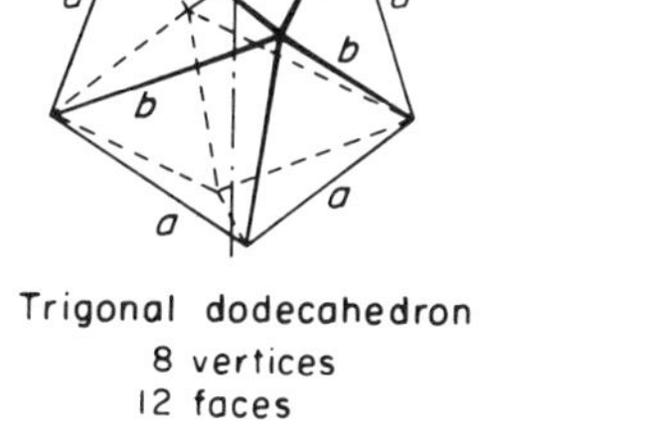

Fig. 2.6 The five regular polyhedra (Platonic solids) and the trigonal dodecahedron (not a regular polyhedron since b edges are longer than a).

In Table 2.1 we list the Schoenflies symbols, a set of generators, and other information for the point groups. The symmetry operations, grouped by class, are tabulated in the character tables for the point groups (Appendix 1).

The following isomorphisms exist.

For all n	$\mathbf{C}_{nv} \sim \mathbf{D}_n$; $\mathbf{D}_{2n} \sim \mathbf{D}_{nd}$
For even n	$\mathbf{C}_n \sim \mathbf{S}_n$
For odd n	$\mathbf{D}_{nh} \sim \mathbf{D}_{nd}$; $\mathbf{C}_{nh} \sim \mathbf{C}_{2n}$
Special	$\mathbf{C}_2 \sim \mathbf{C}_s \sim \mathbf{C}_i$; $\mathbf{D}_2 \sim \mathbf{C}_{2h}$; $\mathbf{O} \sim \mathbf{T}_d$

TABLE 2.1
Point Groups

Type	Schoenflies symbol	A set of generators	Order	Comments
Nonaxial	$\mathbf{C}_1$	E	1	No symmetry
	$\mathbf{C}_s$	σ	2	$\mathbf{C}_s = \mathbf{C}_{1h} = \mathbf{C}_{1v} = \mathbf{S}_1$
	$\mathbf{C}_i$	i	2	$\mathbf{C}_i = \mathbf{S}_2$
Axial	$\mathbf{C}_n$	C_n	n	
	$\mathbf{S}_{2n}$	S_{2n}	$2n$	When n is divisible by 4, $\mathbf{S}_n = \mathbf{C}_{n/2} \times \mathbf{C}_i$ When n is odd the group is called $\mathbf{C}_{nh}$
	$\mathbf{C}_{nh}$	C_n, σ_h	$2n$	$\mathbf{C}_{nh} = \mathbf{C}_n \times \mathbf{C}_i$ (n even); $\mathbf{C}_{nh} = \mathbf{C}_n \times \mathbf{C}_s$ (any n)
	$\mathbf{C}_{nv}$	C_n, σ_v	$2n$	
Dihedral	$\mathbf{D}_n$	C_n, C_2	$2n$	Angle between C_n and C_2 is 90°
	$\mathbf{D}_{nh}$	C_n, C_2, σ_h	$4n$	$\mathbf{D}_{nh} = \mathbf{D}_n \times \mathbf{C}_i$ (n even); $\mathbf{D}_{nh} = \mathbf{D}_n \times \mathbf{C}_s$ (any n)
	$\mathbf{D}_{nd}$	C_n, C_2, σ_d	$4n$	$\mathbf{D}_{nd} = \mathbf{D}_n \times \mathbf{C}_i$ (n odd)
Linear	$\mathbf{C}_\infty$	C_∞	∞	All finite and infinitesimal proper rotations about a fixed axis
	$\mathbf{C}_{\infty v}$	C_∞, σ_v	∞	
	$\mathbf{D}_{\infty h}$	C_∞, C_2, σ_h	∞	Angle between C_∞ and C_2 is 90°
Cubic	$\mathbf{T}$	C_3, C_2	12	The proper rotations that take a regular tetrahedron into itself; angle between C_3 and C_2 is 54.74°
	$\mathbf{T}_h$	C_3, C_2, i	24	$\mathbf{T}_h = \mathbf{T} \times \mathbf{C}_i$
	$\mathbf{T}_d$	C_3, S_4	24	All operations, including reflections, that take a regular tetrahedron into itself
	$\mathbf{O}$	C_3, C_4	24	The proper rotations that take a cube or a regular octahedron into itself; angle between C_3 and C_4 is 54.74°
	$\mathbf{O}_h$	C_3, C_4, i	48	All operations, including reflections and the inversion, that take a cube or regular octahedron into itself $\mathbf{O}_h = \mathbf{O} \times \mathbf{C}_i$
Icosahedral	$\mathbf{I}$	C_3, C_5	60	The proper rotations that take a regular isocahedron or dodecahedron into itself; angle between C_3 and C_5 is 37.38°
	$\mathbf{I}_h$	C_3, C_5, i	120	All the operations, including reflections and inversion, that take a regular isocahedron or dodecahedron into itself $\mathbf{I}_h = \mathbf{I} \times \mathbf{C}_i$

TABLE 2.1 (*Continued*)

Type	Schoenflies symbol	A set of generators	Order	Comments
Spherical	**K**	$C(\phi)$, $C'(\phi)$	∞	All proper rotations in space, i.e., all finite and infinitesimal proper rotations about any axis; $C(\phi)$ and $C'(\phi)$ are any rotations about any two different axis
	$\mathbf{K}_h$	$C(\phi)$, $C'(\phi)$, i	∞	All proper and improper rotations in space $\mathbf{K}_h = \mathbf{K} \times \mathbf{C}_i$

2.8 Assigning Molecules to Point Groups

2.8.1 LINEAR MOLECULES

Linear molecules can be assigned to point groups following a flow chart, but the presence of a C_∞ axis is strikingly apparent. Unsymmetrical linear molecules, such as CO, HCI, H—C≡N, O═C═S, and N═N═O, with a C_∞ axis and an infinite number of vertical planes belong to $\mathbf{C}_{\infty v}$. Symmetrical molecules or ions such as H_2, Cl_2, O═C═O, $N{=}N{=}N^-$, $F{-}H{-}F^-$, and H—C≡C—H have an infinite number of C_2 axes perpendicular to C_∞, a center of symmetry, and σ_h, corresponding to $\mathbf{D}_{\infty h}$. Identification of σ_h or i is sufficient to make the distinction between $\mathbf{D}_{\infty h}$ and $\mathbf{C}_{\infty v}$.

2.8.2 MOLECULES WITH ONE AXIS OF ROTATION

Molecules with no symmetry belong to the $\mathbf{C}_1$ point group (see Fig. 2.7). If a molecule has no rotational symmetry but has a reflection plane, such as HOCl, then the group is $\mathbf{C}_s$; if there is an inversion center, it is $\mathbf{C}_i$, as illustrated by the Pt complex with fixed chelate ring conformations. We can use a flow chart such as shown in Fig. 2.8 to aid in assigning molecules to point groups. If there is only one axis of rotation, the group is $\mathbf{C}_n$, $\mathbf{C}_{nv}$, $\mathbf{C}_{nh}$, or $\mathbf{S}_n$. If the C_n axis is the only symmetry element, the group is $\mathbf{C}_n$ (see Fig. 27). For N_2H_4 ($\mathbf{C}_2$), the conformation is considered as shown to preserve $\mathbf{C}_2$ symmetry. The ethylenediamine chelate ring of *cis*-$[CoCl_2(NH_2C_2H_4NH_2)_2]^+$ is in a gauche (puckered) conformation, but the barrier toward inversion is low and inversion occurs rapidly at room temperature. The chelate ring is considered a symmetrical group spanning an edge of the octahedron. The glycinate ion

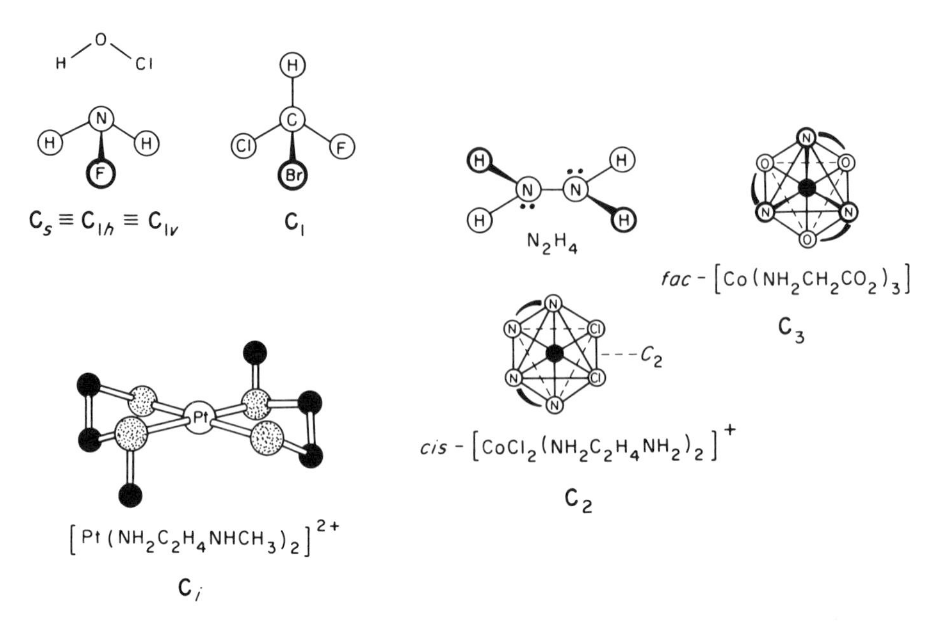

Fig. 2.7 Molecules and ions with $\mathbf{C}_s$, $\mathbf{C}_i$, and $\mathbf{C}_n$ symmetry.

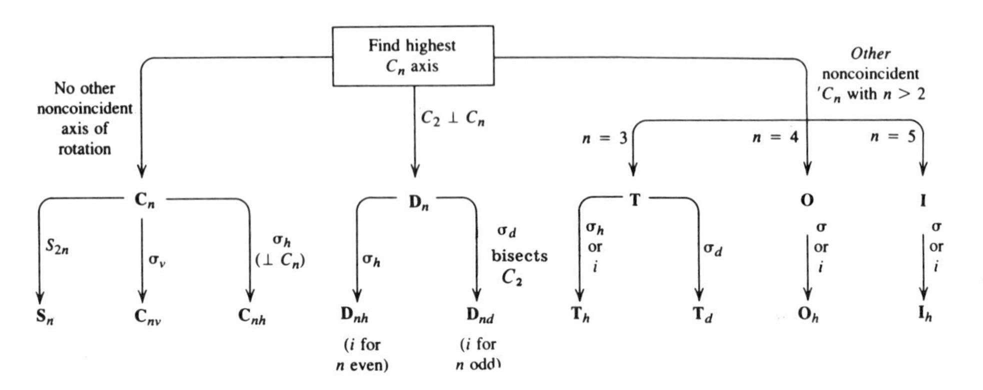

Fig. 2.8 Flow chart for assignment of molecules to point groups. Special nomenclature: $\mathbf{S}_1 \equiv \mathbf{C}_{1v} \equiv \mathbf{C}_{1h} \equiv \mathbf{C}_s$, $\mathbf{S}_n \equiv \mathbf{C}_{nh}$ for odd n, $\mathbf{S}_2 \equiv \mathbf{C}_i$.. (From B. Douglas, D. H. McDaniel, and J. J. Alexander, *Concepts and Models of Inorganic Chemistry*, 2nd ed., Wiley, New York, 1983, p. 101. © 1983 John Wiley & Sons, Inc.; reprinted by permission.)

($NH_2CH_2CO_2^-$) chelate ring is considered symmetrical except for different ends.

Finding one vertical plane requires $n\ \sigma_v$ for the point group $\mathbf{C}_{nv}$ (see Fig. 2.9). In the ammonia complexes there is free rotation about the metal–nitrogen

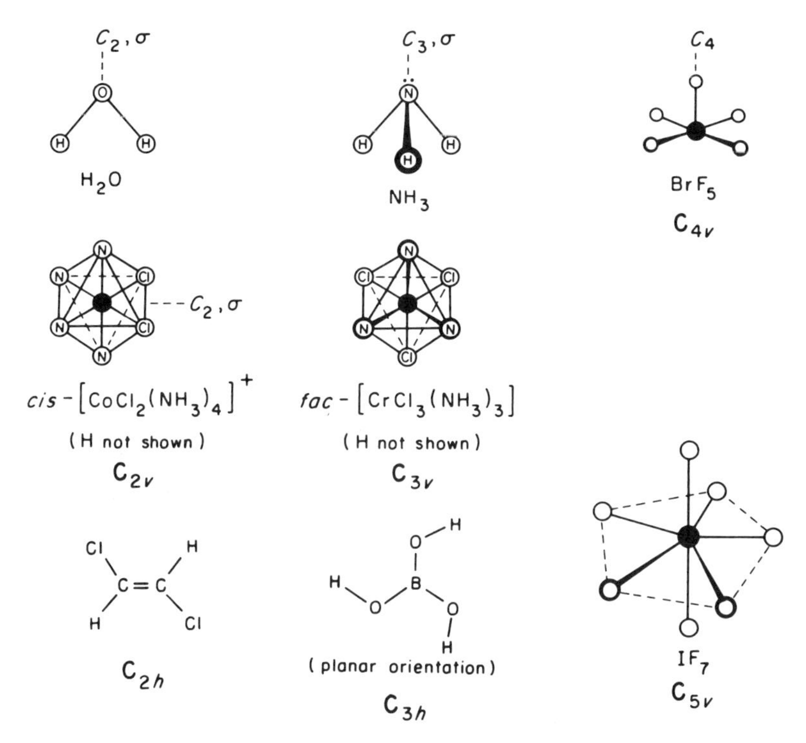

Fig. 2.9 Molecules and ions with $\mathbf{C}_{nv}$ and $\mathbf{C}_{nh}$ symmetry.

bond, giving the NH_3 effectively (time-averaged) cylindrical symmetry about the M–N bond. In determining the symmetry of the complex, the NH_3 is treated as though it had the (spherical) symmetry of an atom. If there is a σ_h; the group is $\mathbf{C}_{nh}$. The *trans*-substituted ethylene ($\mathbf{C}_{2h}$) is planar because of the double bond. The orientation of OH groups about the B—O bond of $B(OH)_3$ is shown fixed to preserve the $\mathbf{C}_{3h}$ symmetry. Of course, isolated $B(OH)_3$ molecules are not encountered. In aqueous solution or in the solid, the orientations of the OH groups is determined by hydrogen-bonding interactions.

Improper rotations (S_n) occur in many point groups ($\mathbf{C}_{nh}$, $\mathbf{D}_{nh}$, $\mathbf{D}_{nd}$, $\mathbf{T}_h$, $\mathbf{T}_d$, $\mathbf{O}_h$, $\mathbf{I}_h$) as well as in the $\mathbf{S}_n$ groups. Molecules belonging to $\mathbf{S}_n$ (n even) point groups have $C_{n/2}$ and S_n axes but no reflection planes. If for $C_{n/2}$, $(n/2)$ is odd, there is an inversion center. The $\mathbf{S}_2$ group is called $\mathbf{C}_i$. Examples of the uncommon groups $\mathbf{S}_4$, $\mathbf{S}_6$, and $\mathbf{S}_8$ are shown in Fig. 2.10. The symmetry of the Nd complex deviates slightly from $\mathbf{S}_8$. Note that one pyridine ring is tipped more than others.

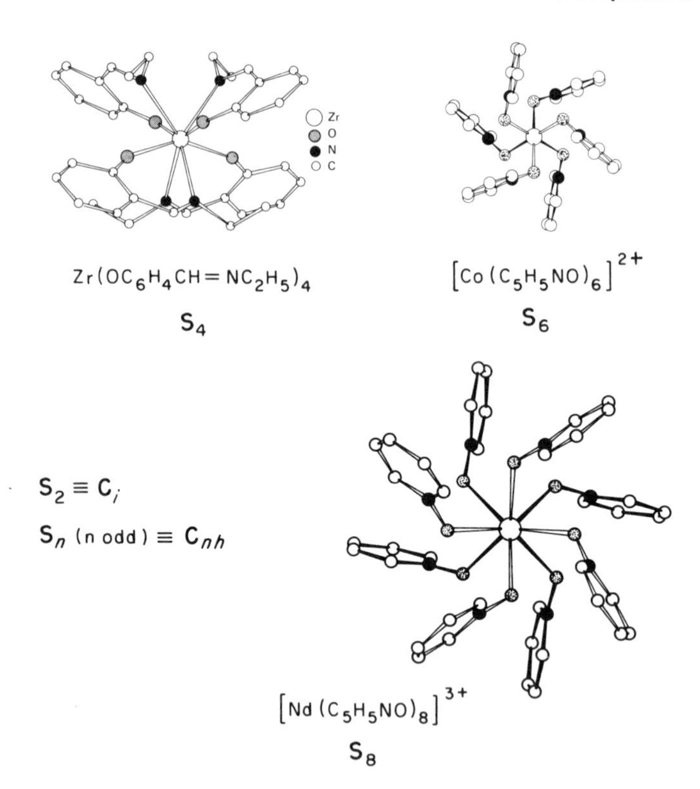

Fig. 2.10 Examples of $\mathbf{S}_n$ point groups $[Zr(OC_6H_4CH{=}NC_2H_5)_4]$ from D. C. Bradley, M. B. Hursthouse, and I. F. Rendall, *Chem. Commun.* **1970,** 368. $[Co(C_5H_5NO)_6]^{2+}$ from T. J. Bergendahl and J. S. Wood, *Inorg. Chem.* 1975, **14,** 338; © 1975, American Chemical Society. $[Nd(C_5H_5NO)_8]^{3+}$ from A. R. Al-Karaghouli and J. S. Wood, *Inorg. Chem.* 1979, **18,** 1177; © 1979, American Chemical Society, reproduced with permission.

2.8.3 DIHEDRAL GROUPS

Molecules belonging to dihedral groups have n C_2 axes perpendicular to C_n. Examples of the pure rotational $\mathbf{D}_n$ groups are shown in Fig. 2.11. $\mathbf{D}_2$ is unique since there are three C_2 axes, and the choice of the "principal" axis is arbitrary. Note that in the $[Pt(NH_2C_2H_4NH_2)_2]^{2+}$ complex ($\mathbf{D}_2$), the two ethylenediamine rings have the same fixed conformation (they are interchanged by the vertical C_2 axis). In the example used with $\mathbf{C}_i$ symmetry (Fig. 2.7), the rings have opposite conformations.

The $\mathbf{D}_{nh}$ group, incorporating σ_h and the elements of $\mathbf{D}_n$, is rather common. It is interesting that $\mathbf{D}_{3h}$ includes trigonal planar (CO_3^{2-}) and trigonal

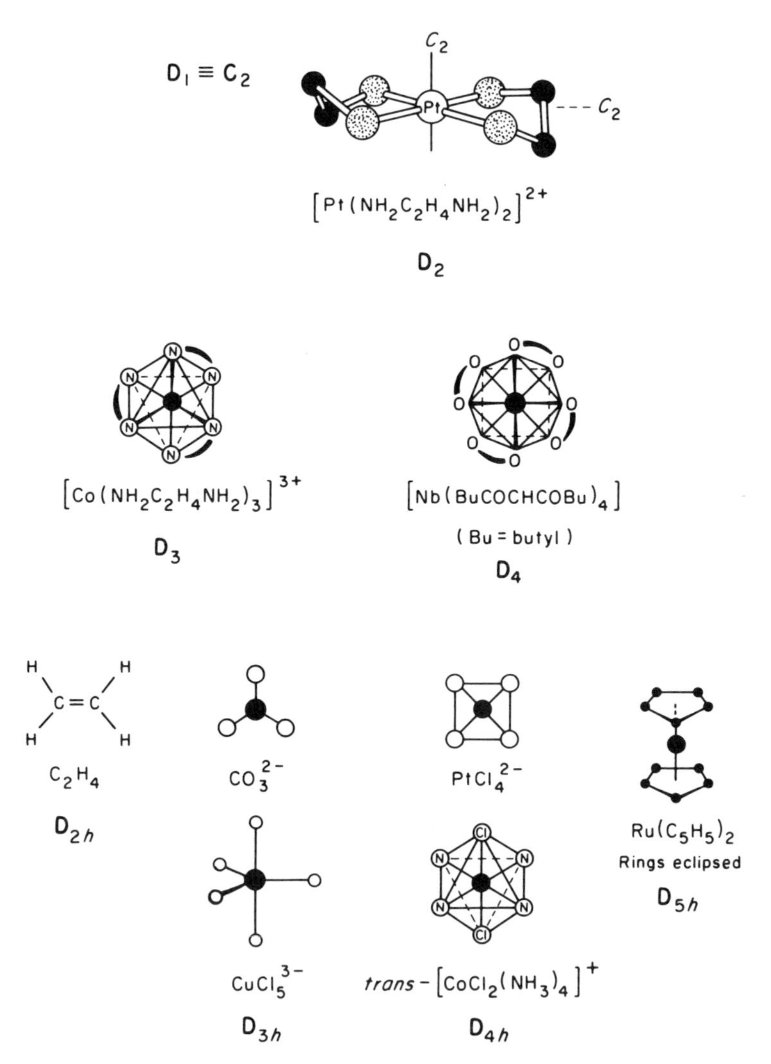

Fig. 2.11 Molecules or ions with $\mathbf{D}_n$ and $\mathbf{D}_{nh}$ symmetry.

bipyramidal ($CuCl_5^{3-}$ and PCl_5) arrangements. Similarly $\mathbf{D}_{4h}$ includes square planar and *trans*-disubstituted octahedral complexes. Both the planar C_5H_5 ion and the eclipsed sandwich compound $Ru(C_5H_5)_2$ are $\mathbf{D}_{5h}$.

Adding n vertical reflection planes (σ_d) bisecting the angles formed by the C_2 axes of $\mathbf{D}_n$ gives the group $\mathbf{D}_{nd}$. The $\mathbf{D}_{nd}$ groups have S_{2n} axes. In the examples shown in Fig. 2.12, the reflection planes are readily apparent. Look for the C_2 axes bisecting the angles formed by those σ_d planes.

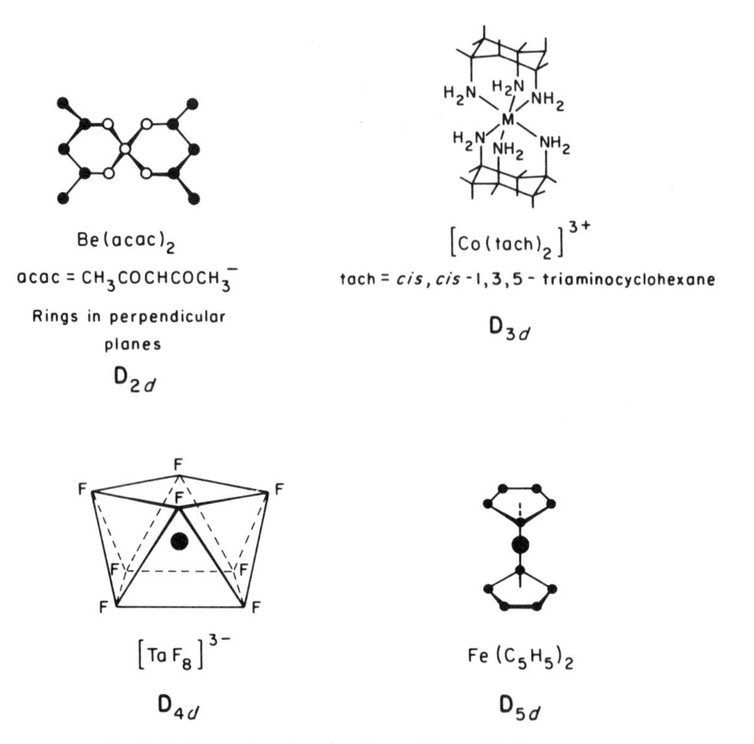

Fig. 2.12 Examples of molecules and ions with $\mathbf{D}_{nd}$ symmetry.

2.8.4 CUBIC AND ICOSAHEDRAL GROUPS

Multiple C_n axes of high order make it easy to identify molecules as belonging to one of the cubic groups, but distinction among the groups requires closer examination. Molecules with multiple C_3 axes and, in addition, only C_2 axes (no σ, i, or S_n) belong to $\mathbf{T}$ (the pure rotational group). Most tetrahedral molecules, such as CH_4, NH_4^+, BF_4^-, MnO_4^-, and P_4, have S_4 axes (and σ_d planes) (see Fig. 2.13), making the group $\mathbf{T}_d$. There is also a centrosymmetric group, $\mathbf{T}_h$. An example of this uncommon group is shown in Fig. 2.13. The $\mathbf{T}_h$ complex has some features of an octahedron, including an inversion center.

Octahedral symmetry is obvious because of the multiple C_4 axes. The pure rotational group $\mathbf{O}$ (order 24) also has multiple C_3 and C_2 axes. Addition of σ_h (or i) generates the additional symmetry of the full $\mathbf{O}_h$ group (order 48).

Multiple C_5 axes (and multiple C_3 and C_2 axes) identify the icosahedral rotational group $\mathbf{I}$ (order 60). Adding an inversion center gives the full $\mathbf{I}_h$

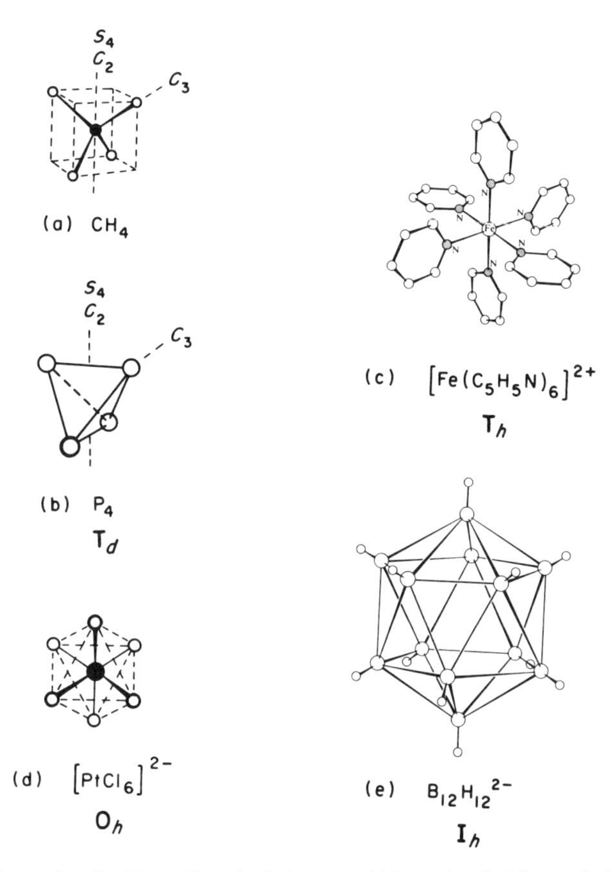

Fig. 2.13 Examples of cubic and icosahedral groups. (c) is reprinted with permission from R. J. Doedens and L. E. Dahl, *J. Am. Chem. Soc.* 1966, **88,** 4847, © 1966, American Chemical Society; (e) is from E. L. Muetterties, *The Chemistry of Boron and Its Compounds*, Wiley, New York, 1967, p. 233, © 1967 John Wiley & Sons, Inc.; reprinted by permission.

symmetry (order 120). The $B_{12}H_{12}^{2-}$ ion ($\mathbf{I}_h$, Fig. 2.13) is stable, even in aqueous solution (unlike other boron hydrides).

EXAMPLE 2.7 Assign the rotational group and the full symmetry group to each of the following:

(a) Biphenyl
 (i) with the phenyl rings coplanar,
 (ii) with the rings in perpendicular planes,
 (iii) with the rings in planes making 45°.

(b) $[Co_6(CO)_{14}]^{4-}$ with an octahedral Co_6 cluster, six terminal CO's, and one CO over each octahedral face.

(c) The polyhedra in Fig. 2.6.
(d) The S_8 ring in crown configuration.
(e) $Fe_2(cot)(CO)_6$ (cot = cyclooctatetraene).
The structure is as follows.

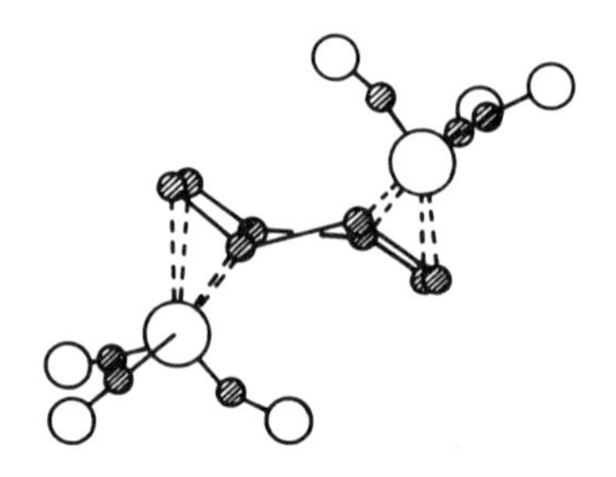

SOLUTION (a) (i) C_2 perpendicular to plane of rings and two axes in that plane. Rotational group $\mathbf{D}_2$. There are three σ_h and i, making the full group $\mathbf{D}_{2h}$.
(ii) C_2 along the long axis of the molecule plus two C_2 bisecting the planes of the rings is $\mathbf{D}_2$. There are two σ_d in the planes of the rings, giving $\mathbf{D}_{2d}$,
(iii) The same C_2 axes exist as for (ii), but there are no σ, so the group is $\mathbf{D}_2$.
(b) Multiple C_4 axes make the rotational group $\mathbf{O}$. There are also the reflection planes and the inversion center of the $\mathbf{O}_h$ group.
(c) Tetrahedron—multiple C_3 axes make it $\mathbf{T}$. The σ_d make it $\mathbf{T}_d$. Octahedron and Cube—multiple C_4 axes make them $\mathbf{O}$. The σ and i make them $\mathbf{O}_h$. Pentagonal dodecahedron and icosahedron—multiple C_5 axes make them $\mathbf{I}$. The σ and i make them $\mathbf{I}_h$: Trigonal dodecahedron—there is a C_2 axis vertical and C_2 axes through the edges labeled b (perpendicular to the vertical C_2), making it $\mathbf{D}_2$. There are σ_d through the a edges across the top of the figure (approximately in the plane of the paper) and perpendicular to this plane, making it $\mathbf{D}_{2d}$.
(d) There is a C_4 axis and four C_2 axes perpendicular to C_4 giving $\mathbf{D}_4$. There are four σ_d giving $\mathbf{D}_{4d}$. As a check we see that there is also an S_8 axis.
(e) There is a C_2 axis perpendicular to the plane of the paper and no other proper rotation. The rotational group is $\mathbf{C}_2$. There is a σ plane in the plane of the paper, giving $\mathbf{C}_{2h}$. We can use the presence of i as a check.

2.8.5 SYMMETRY RESTRICTIONS FOR OPTICAL ACTIVITY

A substance is optically active if its molecules are not superimposable on their mirror images. This condition is met for molecules having only proper rotations, that is, those belonging to $\mathbf{C}_n$, $\mathbf{D}_n$, $\mathbf{T}$, $\mathbf{O}$, or $\mathbf{I}$. We commonly look for an inversion center or a reflection plane, since molecules possessing i or σ cannot be optically active. This is a necessary, but not sufficient limitation. A molecule cannot be optically active if it has *any* S_n axis. This includes σ ($\mathbf{S}_1$) and i ($\mathbf{S}_2$). Examples of the unusual groups $\mathbf{S}_4$, $\mathbf{S}_8$ (no i or σ) and $\mathbf{S}_6$ (includes i) are shown in Fig. 2.10. The zinc complex with $\mathbf{S}_4$ symmetry is an example of a molecule that lacks σ or i, but cannot be optically active because of the S_4 operation. Optically active compounds such as CHBrClF have no symmetry ($\mathbf{C}_1$) and are properly described as *asymmetric*. Many other optically active species, such as $[Co(en)_3]^{3+}$ ($\mathbf{D}_3$), have rather high rotational symmetry. They are described as *dissymmetric*.

2.9 Crystal Symmetry

Let us begin our consideration of crystal symmetry with the lattice as an imaginary construct consisting of an infinite number of geometric points in space arranged such that the environment of a point (i.e., the arrangement of points about this point) is the same for every point of the lattice. The environment about any lattice point has a symmetry that corresponds to one of the point groups. It can be shown (see Section 3.14) that only certain point groups are compatible with a lattice structure.

In addition to the point symmetry (i.e., the rotational symmetry) just discussed, a lattice has a *translational* symmetry; it is clear that because all lattice points are equivalent it must be possible to displace the lattice upon itself in such a way that there has been no change in the lattice. Such displacements are called *primitive translations.* It is clear that these primitive translations can be combined with one another and that (as operations) they satisfy the group postulates.

Some structures are invariant under the combination of a rotation and a *nonprimitive* translation. That is, the translation alone does not take the lattice into itself, but combined with the rotation it does. If the rotation is proper, such a combined operation is called a *screw* operation. If the rotation is improper (actually, a reflection), the combined operation is called a glide operation. An idea of why there can be such operations can be obtained from Figs. 2.14 and 2.15.

As a result, there are a total of seven different lattice systems (called *crystal systems*) classified according to the point symmetry of its lattice points. Some of these systems are compatible with more than one lattice (see Table 2.2). Altogether there are 14 possible lattices. These are called *Bravais lattices*. We can illustrate how different lattices can belong to the same system by considering the cubic system, which can have three different types of lattice. The point symmetry of the cubic system is $\mathbf{O}_h$. The simplest cubic lattice is simple cubic, consisting of cubes with a lattice point on each corner. However, it is clear that the center of each cube has $\mathbf{O}_h$ symmetry also. A lattice point could be added at the center of each cube. This gives the body-centered cubic

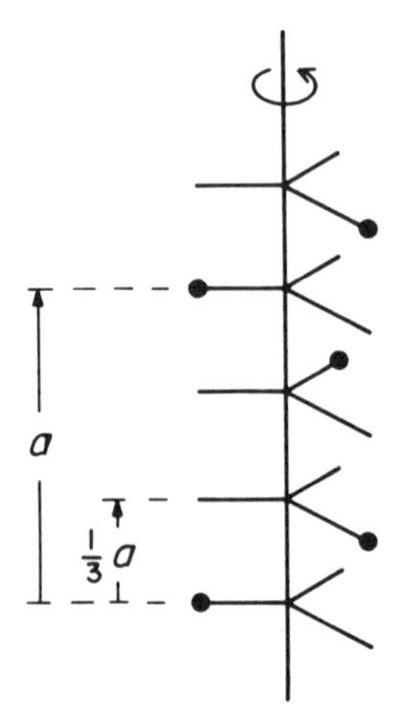

Fig. 2.14 Diagram for a threefold screw axis.

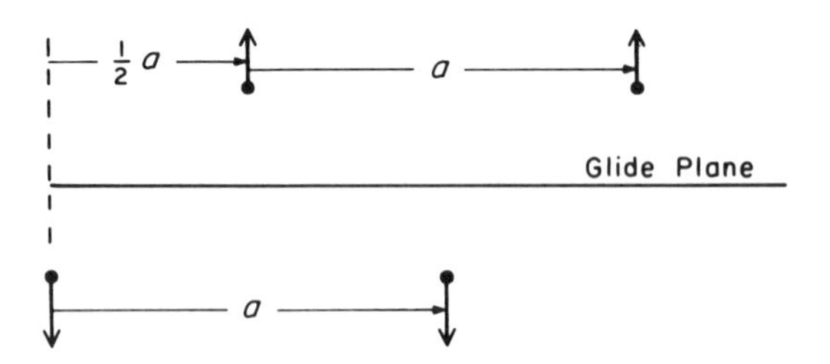

Fig. 2.15 Diagram for a glide plane.

lattice. But the center of each cube face also has $\mathbf{O}_h$ symmetry. Placing a lattice point in the center of each face gives the face-centered cubic lattice. However, placing lattice points at both the cube center and face centers does not give a new lattice, because the result is merely simple cubic with a smaller unit cell.

So far we have considered imaginary lattices consisting of equivalent geometric points. To get a *crystal* we must have some objects (atoms, ions, or molecules) on the lattice. We call these objects the *basis* of the crystal. A very important fact is that the basis might not have the whole symmetry of the lattice site. If the basis in free space is spherical, such as an atom, then it will take the symmetry of the site when introduced into the crystal. However, if it is a molecule, or a nonspherical ion such as H_2O or CO_3^{2-}, it may have less symmetry than the site. The basis cannot have more symmetry than the site, because the crystal forces will always reduce the symmetry until it is no greater

TABLE 2.2
Crystal Systems

Systems	Classes[a]	Bravais lattices[b]	Laue groups
Cubic	$\mathbf{O}_h, \mathbf{T}_h, \mathbf{T}_d, \mathbf{O}, \mathbf{T}$	P, I, F	$\mathbf{O}_h, \mathbf{T}_h$
Hexagonal	$\mathbf{D}_{6h}, \mathbf{C}_{6h}, \mathbf{D}_{3h}, \mathbf{C}_{6v}, \mathbf{D}_6, \mathbf{C}_{3h}, \mathbf{C}_6$	P	$\mathbf{D}_{6h}, \mathbf{C}_{6h}$
Trigonal	$\mathbf{D}_{3d}, \mathbf{D}_3, \mathbf{C}_{3v}, \mathbf{S}_6, \mathbf{C}_3$	P	$\mathbf{D}_{3d}, \mathbf{S}_6$
Tetragonal	$\mathbf{D}_{4h}, \mathbf{C}_{4h}, \mathbf{D}_{2d}, \mathbf{C}_{4v}, \mathbf{D}_4, \mathbf{S}_4, \mathbf{C}_4$	P, I	$\mathbf{D}_{4h}, \mathbf{C}_{4h}$
Orthorhombic	$\mathbf{D}_{2h}, \mathbf{C}_{2v}, \mathbf{D}_2$	P, I, C, F	$\mathbf{D}_{2h}$
Monoclinic	$\mathbf{C}_{2h}, \mathbf{C}_s, \mathbf{C}_2$	P, C	$\mathbf{C}_{2h}$
Triclinic	$\mathbf{C}_i, \mathbf{C}_1$	P	$\mathbf{C}_i$

[a] The highest symmetry (the first class listed in each case) is the symmetry of the lattice points.

[b] P = simple, I = body-centered, F = face-centered, C = end-centered (two opposite faces centered).

than that of the site. Only certain basis symmetries are compatible with the symmetry of any given crystal system. These various possible symmetries are called the *crystal classes* (see Table 2.2). Altogether there are 32 different crystal classes. The symmetry of a class is given by a subgroup (or the whole group) of the symmetry of the crystal *system* to which it belongs. Also, a class will belong to the *system* of *lowest* symmetry for which the class is a subgroup. Thus, the class $\mathbf{C}_{4v}$ belongs to the tetragonal system and not the cubic system, although $\mathbf{C}_{4v}$ is a subgroup of both.[2] The point groups that correspond to the 32 crystal classes are called the crystallographic point groups.

In crystals that contain more than one type of object (more than one type of atom, ion, or molecule) in the unit cell, sites other than the points of the Bravais lattice may be occupied. The symmetry of such a site depends upon its position in the unit cell. For example, in a simple cubic lattice a general point on the edge of a unit cell (say the x axis) is invariant under operations that leave all points on the edge (x axis) fixed. It therefore has $\mathbf{C}_{4v}$ symmetry. A general point on the cube diagonal has $\mathbf{C}_{3v}$ symmetry. An object at any site cannot have symmetry greater than the site. An object, which when free is spherical, will take on the symmetry of its site in the crystal.

The *space group* of a crystal consists of all the rotations (proper and improper), translations, and combinations of rotations and translations that leave the crystal invariant. If there are no glide planes or screw axes, the group is called *symmorphic*. If a glide plane or screw axis is present, the group is *nonsymmorphic*. There are 73 symmorphic and 157 nonsymmorphic space groups, making a total of 230.

[2] For an explanation of this, see G. Weinreich, *Solids: Elementary Theory for Advanced Students*, Wiley, New York, 1965.

The x-ray diffraction patterns normally possess centrosymmetric symmetry, whether or not the crystal has a center of symmetry. For this reason, the centrosymmetric crystallographic point groups have special significance and are called the Laue groups (see Table 2.2).

There are extensive tables giving various properties of crystallographic structures.[3] Our purpose should be served if we give an example of the type of information available and some brief explanation. In Section 10.10 we shall use the information in the form given by Fateley *et al.*[4] Therefore, we take our example from their table of site symmetries.

One crystal form of TiO_2 (anatase) belongs to the tetragonal system, class $\mathbf{D}_{4h}$. Its space group is designated $\mathbf{D}_{4h}^{19}$. It is the 19th space group in class $\mathbf{D}_{4h}$ (there are 20 space groups in that class). Another symbol for that group is $\mathbf{I}4_1/amd$. The letter **I** means the cell is body-centered. 4_1, a, m, and d signify, respectively, a fourfold screw axis, a glide plane, a mirror plane, and a second glide plane. The solidus (/) indicates that 4_1 and a are perpendicular. The site symmetries are the following: $2\mathbf{D}_{2d}(2)$, $2\mathbf{C}_{2h}(4)$, $\mathbf{C}_{2v}(4)$, $2\mathbf{C}_2(8)$, $\mathbf{C}_s(8)$, $\infty\mathbf{C}_1(16)$. The number in parentheses indicates the number of equivalent sites with that symmetry. (Sites are equivalent if an operation of the group takes one into the other.) The number in front of the symbol indicates the number of nonequivalent sets of these sites. In TiO_2 (anatase) there are, per unit cell, two titaniums in equivalent $\mathbf{D}_{2h}$ sites and four oxygens in equivalent $\mathbf{C}_{2v}$ sites.

Additional Readings

F. A. Cotton, *Chemical Applications of Group Theory*, 2nd ed. Wiley (Interscience), New York, 1971.

R. L. Flurry, Jr., *Symmetry Groups*, Prentice-Hall, Englewood Cliffs, New Jersey, 1980.

L. H. Hall, *Group Theory and Symmetry in Chemistry*, McGraw-Hill, New York, 1969.

D. C. Harris and M. D. Bertolucci, *Symmetry and Spectroscopy*, Oxford Univ. Press, London and New York, 1978.

R. McWeeny, *Symmetry: An Introduction to Group Theory and Its Applications*, Macmillan, New York, 1963.

M. Orchin and H. H. Jaffe, *Symmetry, Orbitals, and Spectra*, Wiley (Interscience), New York, 1970.

J. A. Salthouse and M. J. Ware, *Point Group Character Tables and Related Data*, Cambridge Univ. Press, London and New York, 1972, a useful compilation.

M. Tinkham, *Group Theory and Quantum Mechanics*, McGraw-Hill New York, 1964.

[3] N. F. M. Henry and K. Lonsdale, eds. *International Tables for X-Ray Crystallography*, vol. 1, 2nd ed., Kynock, Birmingham, England, 1965.

[4] W. G. Fateley, F. R. Dollish, N. T. McDevitt, and F. F. Bentley, *Infrared and Raman Selection Rules for Molecular and Lattice Vibrations*, Wiley (Interscience), New York, 1972.

Problems

2.1 Show that the set of positive and negative integers (including zero) with ordinary addition as the law of combination forms an Abelian group.

2.2 Refer to remark (iv) of Example 2.3. Find the elements of the group $\mathbf{D}_{3h}$, formed by adding σ_h to $\mathbf{D}_3$, that are not given by Fig. 2.1. What is the order of $\mathbf{D}_{3h}$?

2.3 Using the symbol $\sigma_h C_3 = C_3 \sigma_h = S_3$, evaluate the following products in $\mathbf{D}_{3h}$ (see Problem 2.2): (a) $\sigma'_v C_2$, (b) $C_2 \sigma'_v$, (c) $C'_2 \sigma'_v$, (d) $S_3 S_3$, (e) S_3^3, (f) $S_3 C_2$, (g) $C_2 S_3$.

2.4 Use the relation (2.7) to obtain the multiplication table for $\mathbf{D}_3$. (*Hint*: The independent elements can be taken to be E, A, A^2, B, AB, $A^2 B$, and these can be correlated with the elements of $\mathbf{D}_3$.)

2.5 Find the generators and relations for the group (1, -1, i, $-i$) (Example 2.1).

2.6 Find the possible sets of generators of the group $\mathbf{D}_{3h}$.

2.7 What are the possible generators and their relations for the symmetric group of degree three?

2.8 Write the permutation

$$\begin{pmatrix} abcdefgh \\ bdehfcga \end{pmatrix}$$

in cycle form. What is the period of the permutation?

2.9 Prove that every element of a finite group has a period and the element is the generator of a cyclic subgroup of order equal to the period; or that the whole group is cyclic and the element is the generator of the whole group.

2.10 Evaluate the following transforms for $\mathbf{D}_3$: (a) C_3 by C_3, (b) C_3 by C_2, (c) C_2 by C'_2, (d) C_2 by C_2.

2.11 Prove that the inverse of the product $ABC \cdots XY$ is $(ABC \cdots XY)^{-1} = Y^{-1}X^{-1} \cdots C^{-1}B^{-1}A^{-1}$.

2.12 Prove the properties (1), (2), and (3) of transforms and classes stated at the end of Section 2.5.

2.13 Prove that if X is not an element of the subgroup H, then the set $X^{-1}HX$ (with X fixed) is a subgroup or the whole group G.

2.14 The point group $\mathbf{D}_2$ is a group of order four with two generators. Find the generator relations and construct the multiplication table.

2.15 Prove that $S_2 = i$.

2.16 Prove that S_6 given by Eq. (2.13) is the same as iC_3^2.

2.17 Prove the following: (a) $\mathbf{S}_n \sim \mathbf{C}_n$ for even n. (b) $\mathbf{C}_{nh} \sim \mathbf{C}_{2n}$ for odd n.

2.18 Identify subgroups of $\mathbf{D}_{4h}$.

2.19 For which of the following groups are C_n (major axis) and C_n^{n-1}, S_n and S_n^{n-1} in the same class? (a) $\mathbf{C}_n$, (b) $\mathbf{S}_n$, (c) $\mathbf{C}_{nv}$, (d) $\mathbf{C}_{nh}$, (e) $\mathbf{D}_n$, (f) $\mathbf{T}_h$, (g) $\mathbf{T}_d$.

2.20 What group is obtained by adding the symmetry operation indicated? (a) $\mathbf{C}_3$ plus S_3, (b) $\mathbf{C}_3$ plus S_6, (c) $\mathbf{C}_3$ plus i, (d) $\mathbf{C}_4$ plus i, (e) $\mathbf{S}_4$ plus i.

2.21 What group is obtained by deleting the symmetry operation indicated? (a) $\mathbf{O}_h$ minus i, (b) $\mathbf{D}_{3d}$ minus i, (c) $\mathbf{D}_{4d}$ minus S_8, (d) $\mathbf{D}_{4h}$ minus i, (e) $\mathbf{D}_{4h}$ minus S_4.

2.22 Solid models and ball and stick models are useful for identifying symmetry elements and point groups. Assign point groups to figures made from cutouts given on pages 130–131 in B. E. Douglas, D. H. McDaniel, and J. J. Alexander, *Concepts and Models of Inorganic Chemistry*, 2nd ed., Wiley, New York, 1983.

2.23 Construct models for the $\mathbf{T}$, $\mathbf{T}_d$, $\mathbf{T}_h$, $\mathbf{O}$, and $\mathbf{O}_h$ point groups as instructed by A. F. Gremillion, *J. Chem. Educ.* 1982, **59,** 195.

2.24 Assign the point groups to the following: (a) N═N═O (linear), (b) NO_2 (angular), (c) NO_2^+ (linear), (d) NO_3^- (planar), (e) PF_3 (pyramidal), (f) ClF_3 (Table 6.1), (g) SF_4 (Table 6.1), (h) ICl_2^- (linear), (i) XeF_4 (planar), (j) IF_7 (Table 6.1).

2.25 Assign the point groups to the following: (a) C_2H_6 (staggered), (b) C_2H_6 (elcipsed), (c) $H_2C{=}C{=}CH_2$,
(d) spiropentane

H₂C   CH₂
  C
H₂C   CH₂

(e) cubane, C_8H_8, (f) *trans*-$C_2H_2Cl_2$, (g) *cis*-$C_2H_2Cl_2$, (h) 1,1-$C_2H_2Cl_2$, (i) C_6H_{12} (boat), (j) C_6H_{12} (chair), (k) cyclobutadiene (planar, delocalized π bonding), (l) cyclopropane.

2.26 For 1,3,5-trinitrobenzene assign the point group assuming
(a) all NO_2 in the plane of the ring,
(b) all NO_2 in planes perpendicular to the plane of the ring,
(c) all NO_2 in planes skewed in the same direction to the same extent,
(d) free rotation about the C—N bonds.

2.27 Assign the point group to the following:
(a) a new roller brush for a car wash with the bristles sticking straight out,
(b) the same brush after use, with the bristles bent back uniformly,
(c) a symmetrical Christmas tree (idealized).

2.28 Assign point groups to the following:

(a) $Mo(S_2C_2H_2)_3$ (planar bidentate ligands coordinated along the tetragonal edges of a trigonal prism),

(b) $Cr(C_6H_6)_2$ (sandwich compound, rings eclipsed),

(c) $Ni(C_4H_4)_2$ (sandwich compound, rings staggered),

(d) $U(C_8H_8)_2$ (sandwich compound, rings eclipsed),

(e) $Be(acac)_2$ (acac, planar, bidentate 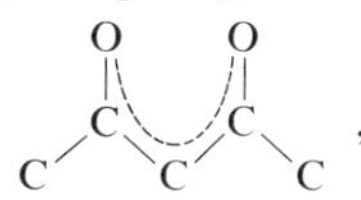, complex tetrahedral,

(f) $Cu(acac)_2$ (complex planar).

2.29 Assign the point groups for the following "octahedral" complexes.

(a) $[Co(en)_3]^{3+}$ (en is a symmetrical bidentate ligand)

(b) *cis*-$[CoCl_2(NH_3)_4]^+$

(c) *trans*-$[CoCl_2(NH_3)_4]^+$

(d) *cis*-$[CoCl_2(en)_2]^+$

(e) *trans*-$[CoCl_2(en)_2]^+$

(f) *trans*-$[CoCl_2(trien)]^+$
(trien = $NH_2C_2H_4NHC_2H_4NHC_2H_4NH_2$)

(g) *symmetrical-cis*-$[CoCl_2(trien)]^+$

(h) *unsymmetrical-cis*-$[CoCl_2(trien)]^+$

(i) *facial*-$[CrCl_3(NH_3)_3]$

(j) *meridional*-$[CrCl_3(NH_3)_3]$

(k) *fac*-$[CrCl_3(dien)]$
(both isomers, dien = $NH_2C_2H_4NHC_2H_4NH_2$)

(l) *mer*-$[CrCl_3(dien)]$

(m) *fac*-$[Co(NH_2CH_2CO_2)_3]$

(n) *mer*-$[Co(NH_2CH_2CO_2)_3]$

(o) *trans*-$[CoCl_2(H_2O)_4]$ (all water molecules skewed by the same angle)

2.30 Assign point groups to the following polynuclear compounds.

(a) P_4 (tetrahedral)

(b) P_4O_6 (add one O along each edge)

(c) P_4O_{10} (add one more terminal O/P)

(d) B_2H_6

(e) B_4H_{10} (two BH with B—B bond, two BH_2, and four bridging *H*)

(f) B_5H_9 (five BH, borons form a square pyramid with four bridging H in base)

(g) S_4N_4

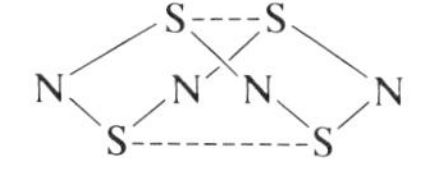

(h) $[Cu_2(CH_3CO_2)_4(H_2O)_2]$ (Each Cu "octahedral", four bridging bidentate acetates, one axial H_2O/Cu, plus Cu—Cu bond)
(i) $Re_2Cl_8^{2-}$ (Re—Re bond, two planar $ReCl_4^-$ eclipsed)
(j) $Re_3Cl_{12}^{3-}$ (Re_3 triangular, three bridging Cl in plane, one in-plane equatorial Cl/Re, and two "axial" Cl/Re)
(k) Te_6^{4+} (trigonal prism)
(l) $Mo_6Cl_8^{4+}$ (Mo_6 octahedron with one Cl above each face)
(m) $Ta_6Cl_{12}^{2+}$ (Ta_6 octahedron with one Cl along each edge)
(n) $W_2Cl_9^{3-}$ (two octahedra sharing a face)
(o) $Mn_2(CO)_{10}$ (two octahedra staggered by 45° sharing an apex)
(p) $Ru_3(CO)_{12}$ (Ru_3 triangular, each Ru has two axial and two equatorial CO)
(q) $Ir_4(CO)_{12}$ (Ir_4 tetrahedron with three terminal CO/Ir)

2.31 What is the site symmetry (consider only near neighbors) of each ion in the following crystals? Use models or consult inorganic texts. (a) NaCl, (b) CsCl, (c) ZnS (wurtzite, hcp), (d) ZnS (zinc blende, ccp), (e) CaF_2, (f) NiAs, (g) PtS, (h) Cu (ccp) (i) Mg (hcp).

2.32 In a body-centered cubic structure a metal has eight nearest neighbors and six more at only slightly (~15%) greater distance. What is the site symmetry for nearest neighbors and for all 14 neighbors?

2.33 Metals with cubic close-packed (ccp) structures are generally more malleable and ductile than metals with other structures.
(a) For the ccp and hexagonal close packed (hcp) structures, the close-packed layers are perpendicular to what rotation axis?
(b) How many such axes are there for ccp and hcp structures?
(c) Account for the greater malleability of ccp structures on this basis.

3

Vectors, Matrices, and Group Representations

The purpose of this chapter is to introduce the reader to the representation theory of groups. This is the aspect of group theory that is central to most applications in chemistry. Because the mathematical tools of representation theory are vectors and matrices, we begin with a brief introduction to these.

3.1 Introduction

In the next several sections we give some necessary mathematical tools, presented without proofs or discussion of the reasons for introducing various mathematical entities. Their usefulness will become apparent when we proceed to applications.

3.2 Cartesian Vectors

In three dimensions, a vector can be represented in terms of its components in a Cartesian coordinate system.[1] If we pick the origin as the starting point of the vector, then the components are the coordinates of the end x, y, z. The

[1] A Cartesian system is one in which the axes are mutually perpendicular (orthogonal), and the unit of length is the same along all axes.

vector is written as a row (x, y, z) or a column

$$\begin{bmatrix} x \\ y \\ z \end{bmatrix}.$$

In Cartesian systems these two forms are equivalent.

The rules of vector algebra that we need are the following.

(1) *Multiplication by a scalar s*. Each component is multiplied by the scalar (a scalar is a number, or any variable, that obeys all the laws of ordinary arithmetic):

$$s\mathbf{v} = (sx, sy, sz). \tag{3.1}$$

Division by a scalar s corresponds to multiplication by $1/s$.

(2) *Addition* (or *substraction*) of two vectors. The corresponding components are added (or subtracted):

$$\mathbf{v}_1 + \mathbf{v}_2 = (x_1 + x_2, y_1 + y_2, z_1 + z_2). \tag{3.2}$$

(3) *Scalar product* (also called *dot product*) of two vectors. The result is a scalar (not a vector) defined as the sum of the products of the corresponding components:

$$\mathbf{v}_1 \cdot \mathbf{v}_2 = x_1 x_2 + y_1 y_2 + z_1 z_2. \tag{3.3}$$

(4) The *magnitude* (or length) of the vector $\mathbf{v}$ is defined by

$$|\mathbf{v}| = \sqrt{\mathbf{v} \cdot \mathbf{v}} = \sqrt{x^2 + y^2 + z^2}. \tag{3.4}$$

(5) Two vectors are *orthogonal* (perpendicular) when their scalar product is equal to zero,

$$\mathbf{v}_1 \cdot \mathbf{v}_2 = 0. \tag{3.5}$$

EXAMPLE 3.1 Given the vectors $\mathbf{v}_1 = (1, 2, 1)$ and $\mathbf{v}_2 = (2, 0, 3)$, calculate the following:

(a) $2\mathbf{v}_1 + \frac{1}{2}\mathbf{v}_2$,
(b) $\mathbf{v}_1 \cdot \mathbf{v}_2$,
(c) $|\mathbf{v}_1|$,
(d) $\mathbf{v}_2/|\mathbf{v}_1|$.

SOLUTION (a) $2\mathbf{v}_1 + \frac{1}{2}\mathbf{v}_2 = (2 + 1, 4 + 0, 2 + \frac{3}{2}) = (3, 4, \frac{7}{2})$
(b) $\mathbf{v}_1 \cdot \mathbf{v}_2 = 2 + 0 + 3 = 5$,
(c) $|\mathbf{v}_1| = \sqrt{1 + 4 + 1} = \sqrt{6}$,
(d) $\mathbf{v}_2/|\mathbf{v}_1| = (1/\sqrt{6})(2, 0, 3) = (2/\sqrt{6}, 0, 3/\sqrt{6})$.

EXAMPLE 3.2 Test the following pairs of vectors for orthogonality: (a) $(1,1,1)$ and $(-1,1,-1)$, (b) $(1,0,0)$ and $(0,1,0)$, (c) $(1,0,0)$ and $(0,0,1)$, (d) $(0,1,0)$ and $(0,0,1)$, (e) $(2,1,1)$ and $(-2,1,1)$, (f) $(1,2,1)$ and $(1,-1,1)$.

SOLUTION The scalar products of all pairs except (a) and (e) are zero. Therefore, all pairs are orthogonal except pairs (a) and (e).

EXAMPLE 3.3 Find a unit vector that is orthogonal to both of the unit vectors $(1/\sqrt{3})(1,1,1)$ and $(1/\sqrt{2})(1,0,-1)$.

SOLUTION Let the vector be $\mathbf{u} = (x, y, z)$. We have

$$\mathbf{u} \cdot \frac{1}{\sqrt{3}}(1,1,1) = \frac{1}{\sqrt{3}}(x + y + z) = 0,$$

$$\mathbf{u} \cdot \frac{1}{\sqrt{2}}(1,0,-1) = \frac{1}{\sqrt{2}}(x - z) = 0,$$

$$\mathbf{u} \cdot \mathbf{u} = x^2 + y^2 + z^2 = 1.$$

Simultaneous solution of these three equations gives

$$\mathbf{u} = \frac{1}{\sqrt{6}}(1, -2, 1).$$

In our work it will often happen that the components of a vector are complex numbers. In such cases the ordinary scalar product must be replaced by the *Hermitian scalar product*, in which $\mathbf{v}_i^* \cdot \mathbf{v}_j$ is used instead of $\mathbf{v}_i \cdot \mathbf{v}_j$. The asterisk indicates complex conjugation of the components of that vector. This modification of the scalar product prevents nonzero vectors from having zero length. The complex conjugation is often not indicated explicitly, but it is understood that $\mathbf{v}_i \cdot \mathbf{v}_j$ means $\mathbf{v}_i^* \cdot \mathbf{v}_j$ when complex vectors are involved. (Since in all cases of concern to us $\mathbf{v}_i^* \cdot \mathbf{v}_j$ is real, there is no difference between $\mathbf{v}_i^* \cdot \mathbf{v}_j$ and $\mathbf{v}_i \cdot \mathbf{v}_j^*$.)

EXAMPLE 3.4 Find the length of the vector $(i, 1, 0)$ where $i = \sqrt{-1}$.

SOLUTION The ordinary scalar product would give

$$\sqrt{(i,1,0) \cdot (i,1,0)} = \sqrt{-1 + 1 + 0} = 0.$$

We must use the Hermitian scalar product

$$\sqrt{(-i,1,0) \cdot (i,1,0)} = \sqrt{(1 + 1 + 0)} = \sqrt{2}.$$

The concept of n-dimensional Cartesian vectors is obtained directly by extending the definitions given above to rows or columns with n components.

If a set of vectors $\mathbf{v}_i$, each of unit length, are all mutually orthogonal, then we call the set *orthonormal* and write

$$\mathbf{v}_i^* \cdot \mathbf{v}_j = \delta_{ij}, \tag{3.6}$$

where δ_{ij} is the Kronecker *delta* defined by

$$\delta_{ij} = \begin{cases} 1 & \text{when} \quad i = j, \\ 0 & \text{when} \quad i \neq j. \end{cases} \tag{3.7}$$

In n dimensions, not more than n different nonzero vectors can be mutually orthogonal. In an n-dimensional space the orthonormal set

$$\begin{gathered} (1, 0, 0, \ldots, 0), \\ (0, 1, 0, \ldots, 0), \\ \vdots \\ (0, 0, 0, \ldots, 1), \end{gathered} \tag{3.8}$$

represents a Cartesian coordinate system in that space.

In three dimensions the scalar product has the following meaning:

$$\mathbf{v} \cdot \mathbf{u} = |\mathbf{v}|\,|\mathbf{u}| \cos\theta, \tag{3.9}$$

where θ is the angle between the vectors $\mathbf{v}$ and $\mathbf{u}$. If $\mathbf{u}$ is a unit vector (i.e., if $|\mathbf{u}| = 1$), then $\mathbf{v} \cdot \mathbf{u}$ is the projection of $\mathbf{v}$ in the direction of $\mathbf{u}$. We will find it convenient to extend this concept to n dimensions.

EXAMPLE 3.5 Given a rectangular parallelepiped with edges of lengths a, b, and c, calculate the angles between its diagonal and its three edges.

SOLUTION Place a corner at the origin and let the three edges be along Cartesian axes. The diagonal is in the direction of the vector (a, b, c). The angles are, by Eq. (3.9),

$$\cos\theta_a = \frac{(1, 0, 0) \cdot (a, b, c)}{|(a, b, c)|} = \frac{a}{\sqrt{a^2 + b^2 + c^2}}$$

$$\theta_a = \arccos \frac{a}{\sqrt{a^2 + b^2 + c^2}},$$

$$\theta_b = \arccos \frac{b}{\sqrt{a^2 + b^2 + c^2}},$$

$$\theta_c = \arccos \frac{c}{\sqrt{a^2 + b^2 + c^2}}.$$

EXAMPLE 3.6 Find the projection of the vector $(3, 1, 2, 0)$ in the direction of each of the following vectors: (a) $(1, 1, 1, 1)$, (b) $(0, 0, 1, 0)$, (c) $(1, 0, -1, 0)$, (d) $(1, -1, -1, 1)$.

SOLUTION (a) $(3, 1, 2, 0) \cdot (1, 1, 1, 1)/|(1, 1, 1, 1)| = 3$, (b) 2, (c) $1/\sqrt{2}$, (d) 0.

3.3 Determinants

The nth-order determinant is written as an $n \times n$ array with n rows and n columns,

$$|A| = \begin{vmatrix} a_{11} & a_{12} & \cdots & a_{1n} \\ a_{21} & a_{22} & \cdots & a_{2n} \\ \vdots & \vdots & \ddots & \vdots \\ a_{n1} & a_{n2} & \cdots & a_{nn} \end{vmatrix},$$

and is defined by

$$|A| = \sum (-1)^h a_{1i_1} a_{2i_2} \cdots a_{ni_n}, \tag{3.11}$$

where $i_1, i_2, \ldots, i_n$ represent the subscripts $1, 2, \ldots, n$ in some order. The summation is over all of the $n!$ permutations of the subscripts $i_1, i_2, \ldots, i_n$, and h is the number of interchanges required to get the term to the form $a_{j_1 1} a_{j_2 2} \cdots a_{j_n n}$, where $j_1, j_2, \ldots, j_n$ represent the subscripts $1, 2, \ldots, n$ in some order.

It follows directly from the definition that $|A|$ can also be defined by

$$|A| = \sum (-1)^h a_{i_1 1} a_{i_2 2} \cdots a_{i_n n}, \tag{3.12}$$

and so the value of $|A|$ is not changed by interchanging all rows with corresponding columns. That is, the *transpose* of a determinant has the same value as the determinant.

In principle we can evaluate a determinant by writing down all the $n!$ terms with the proper sign. However, there are recipes that make this task easier. For second- and third-order determinants, the following diagrams are useful:

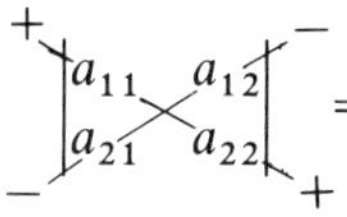

$$\begin{vmatrix} a_{11} & a_{12} \\ a_{21} & a_{22} \end{vmatrix} = a_{11}a_{22} - a_{21}a_{22}, \tag{3.13}$$

$$\begin{vmatrix} a_{11} & a_{12} & a_{13} \\ a_{21} & a_{22} & a_{23} \\ a_{31} & a_{32} & a_{33} \end{vmatrix} \begin{matrix} a_{11} & a_{12} \\ a_{21} & a_{22} \\ a_{31} & a_{32} \end{matrix}$$

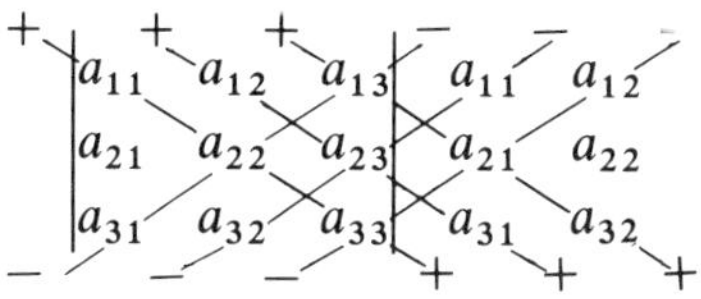

$$= a_{11}a_{22}a_{33} + a_{12}a_{23}a_{31} + a_{13}a_{21}a_{32} - a_{13}a_{22}a_{31} - a_{11}a_{23}a_{32} - a_{12}a_{21}a_{32}. \tag{3.14}$$

There are no simple diagramatic schemes for the evaluation of higher-order determinants, but there are algebraic recipes. We shall describe only one of these—the expansion by cofactors.

The determinant remaining after the ith row and jth column have been deleted is called the *minor* of the element a_{ij}. This minor multiplied by $(-1)^{i+j}$ is called the *cofactor* of a_{ij}, and we shall designate it by A_{ij}. Expansions of $|A|$ are then given by

$$|A| = \sum_{i=1}^{n} a_{ik} A_{ik} \tag{3.15}$$

or

$$|A| = \sum_{i=1}^{n} a_{ki} A_{ki}. \tag{3.16}$$

Equation (3.15) is the expansion in terms of the kth column, and Eq. (3.16) is the expansion in terms of the kth row. For example,

$$\begin{vmatrix} a_{11} & a_{12} & a_{13} \\ a_{21} & a_{22} & a_{23} \\ a_{31} & a_{32} & a_{33} \end{vmatrix} = a_{11}A_{11} + a_{12}A_{12} + a_{13}A_{13}$$

$$= a_{11}\begin{vmatrix} a_{22} & a_{23} \\ a_{32} & a_{33} \end{vmatrix} - a_{12}\begin{vmatrix} a_{21} & a_{23} \\ a_{31} & a_{33} \end{vmatrix} + a_{13}\begin{vmatrix} a_{21} & a_{22} \\ a_{31} & a_{33} \end{vmatrix} \tag{3.17}$$

is the expansion in terms of the first row.

It is clear that by a repetition of this process an nth-order determinant can be put into a form which contains only third- (or second-) order determinants.

EXAMPLE 3.7 Evaluate the determinant

$$\begin{vmatrix} 2 & 5 & 4 & 0 \\ 0 & 1 & 2 & 1 \\ 0 & 1 & 3 & 1 \\ 3 & 6 & 0 & 1 \end{vmatrix}.$$

SOLUTION Expansion in terms of the cofactors of the first column gives

$$D = 2(-1)^{(1+1)}\begin{vmatrix} 1 & 2 & 1 \\ 1 & 3 & 1 \\ 6 & 0 & 1 \end{vmatrix} + 3(-1)^{(1+4)}\begin{vmatrix} 5 & 4 & 0 \\ 1 & 2 & 1 \\ 1 & 3 & 1 \end{vmatrix}.$$

The third-order determinants can be evaluated by the use of Eq. (3.14) (see Problem 3.1), but we shall expand them and use Eq. (3.13). Expansion of the first determinant in terms of the third row and the second determinant in terms of the first row gives

$$\begin{aligned} D &= 2\left\{6(-1)^{(1+3)}\begin{vmatrix}2 & 1\\3 & 1\end{vmatrix} + 1(-1)^{(3+3)}\begin{vmatrix}1 & 2\\1 & 3\end{vmatrix}\right\} \\ &\quad -3\left\{5(-1)^{(1+1)}\begin{vmatrix}2 & 1\\3 & 1\end{vmatrix} + 4(-1)^{(1+2)}\begin{vmatrix}1 & 1\\1 & 1\end{vmatrix}\right\} \\ &= 2\{6(2-3) + (3-2)\} - 3\{5(2-3) - 4(1-1)\} \\ &= 5. \end{aligned}$$

There are many properties of determinants that allow operations that often greatly simplify the evaluation of a determinant. A list of some of these properties and operations are given in Appendix 3A (see Problems 3.2 and 3.3).

3.4 Matrices

An $m \times n$ matrix is an array with m rows and n columns,

$$\mathbf{A} = \begin{bmatrix} a_{11} & a_{12} & \cdots & a_{1n} \\ a_{21} & a_{22} & \cdots & a_{2n} \\ \vdots & \vdots & & \vdots \\ a_{m1} & a_{m2} & \cdots & a_{mn} \end{bmatrix}. \tag{3.18}$$

It does not have a value but rather represents a mathematical operator. Our meaning of "operator" will be clarified as we proceed. The matrix can be thought of as being made up of n m-dimensional column vectors or of m n-dimensional row vectors.

The $n \times m$ matrix obtained by interchanging all rows with corresponding columns is called the *transpose* of $\mathbf{A}$ (written $\mathbf{A}^{\mathrm{T}}$)

$$\mathbf{A}^{\mathrm{T}} = \begin{bmatrix} a_{11} & a_{21} & \cdots & a_{m1} \\ a_{12} & a_{22} & \cdots & a_{m2} \\ \vdots & \vdots & & \vdots \\ a_{1n} & a_{2n} & \cdots & a_{mn} \end{bmatrix}. \tag{3.19}$$

$\mathbf{A}$ and $\mathbf{A}^{\mathrm{T}}$ do not in general represent the same operator, although in special cases they do.

We shall often deal with square (i.e., $n \times n$) matrices. In this case the elements $a_{11}, a_{22}, \ldots, a_{nn}$ make up what is called the *principal diagonal*, and these elements are called *diagonal* elements. The others are *off-diagonal* elements. The sum of the diagonal elements is called the *trace* (also *spur*) in matrix algebra, and it is designated tr **A**. We will see that a matrix sometimes represents a group operation, and in that case its trace is called the *character* of the representation of the operation. Associated with any square matrix is a determinant $|\mathbf{A}|$, the determinant of **A**.

The *rank* of a matrix can be defined as follows. Consider an $m \times n$ matrix and delete any $m - r$ rows and any $n - r$ columns. The remaining r rows and r columns form an $r \times r$ matrix called a *minor* of **A**. Consider all the possible minors of **A** and their determinants. The order of the largest nonzero determinant thus obtained is called the *rank* of the matrix.

A *unit* matrix **I** (also called the *identity* matrix) is any square matrix that has all off-diagonal elements zero and all diagonal elements unity.

The following algebraic operations with matrices are defined.

(1) *Multiplication by a scalar*. Each element of the matrix is multiplied by s,

$$s\mathbf{A} = \begin{bmatrix} sa_{11} & sa_{12} & \cdots & sa_{1n} \\ sa_{21} & & & \vdots \\ \vdots & & & \\ sa_{m1} & \cdots & & sa_{mn} \end{bmatrix}. \tag{3.20}$$

(2) *Addition and subtraction* of two $m \times n$ matrices. Corresponding elements are added or subtracted,

$$\mathbf{A} \pm \mathbf{B} = \begin{bmatrix} a_{11} & a_{12} & \cdots & a_{1n} \\ a_{21} & & & \vdots \\ \vdots & & & \\ a_{m1} & & \cdots & a_{mn} \end{bmatrix} \pm \begin{bmatrix} b_{11} & b_{12} & \cdots & b_{1n} \\ b_{21} & & & \vdots \\ \vdots & & & \\ b_{m1} & & \cdots & b_{mn} \end{bmatrix}$$

$$= \begin{bmatrix} a_{11} \pm b_{11} & a_{12} \pm b_{12} & \cdots & a_{1n} \pm b_{1n} \\ a_{21} \pm b_{21} & & & \vdots \\ \vdots & & & \\ a_{m1} \pm b_{m1} & \cdots & & a_{mn} \pm b_{mn} \end{bmatrix} \tag{3.21}$$

(3) *Multiplication* of an $m_1 \times n$ matrix **A** by an $n \times m_2$ matrix **B** (same n, but possibly different m values) gives an $m_1 \times m_2$ matrix **AB**. The matrix **A** is considered to be an array of m_1 n-dimensional row vectors, and **B** is considered to be an array of m_2 n-dimensional column vectors. The ij element of the product **AB** is the scalar product of ith row vector of **A** with the jth

column vector of **B**. That is, we have

$$\begin{bmatrix} a_{11} & a_{12} & \cdots & a_{1n} \\ \vdots & \vdots & & \vdots \\ a_{i1} & a_{i2} & \cdots & a_{in} \\ \vdots & \vdots & & \vdots \\ a_{m_1 1} & a_{m_1 2} & \cdots & a_{m_1 n} \end{bmatrix} \begin{bmatrix} b_{11} & b_{12} & \cdots & b_{1j} & \cdots & b_{1m_2} \\ & & & b_{2j} & & \\ \vdots & \vdots & & \vdots & & \vdots \\ b_{n1} & b_{n2} & & b_{nj} & \cdots & b_{nm_2} \end{bmatrix} = \begin{bmatrix} & \vdots & \\ \cdots & (AB)_{ij} & \cdots \\ & \vdots & \end{bmatrix} \tag{3.22}$$

where

$$(AB)_{ij} = a_{i1}b_{1j} + a_{i2}b_{2j} + \cdots + a_{in}b_{nj}. \tag{3.23}$$

As an example, take **A** to be a 4 × 3 matrix and **B** to be a 3 × 2 matrix. Let us represent the ith row vector of **A** by the symbol $\mathbf{a}_{i*}$ and the jth column vector of **B** by $\mathbf{b}_{*j}$; then we can write

$$\begin{bmatrix} a_{11} & a_{12} & a_{13} \\ a_{21} & a_{22} & a_{23} \\ a_{31} & a_{32} & a_{33} \\ a_{41} & a_{42} & a_{43} \end{bmatrix} \begin{bmatrix} b_{11} & b_{12} \\ b_{21} & b_{22} \\ b_{31} & b_{32} \end{bmatrix} = \begin{bmatrix} \mathbf{a}_{1*}\cdot\mathbf{b}_{*1} & \mathbf{a}_{1*}\cdot\mathbf{b}_{*2} \\ \mathbf{a}_{2*}\cdot\mathbf{b}_{*1} & \mathbf{a}_{2*}\cdot\mathbf{b}_{*2} \\ \mathbf{a}_{3*}\cdot\mathbf{b}_{*1} & \mathbf{a}_{3*}\cdot\mathbf{b}_{*2} \\ \mathbf{a}_{4*}\cdot\mathbf{b}_{*1} & \mathbf{a}_{4*}\cdot\mathbf{b}_{*2} \end{bmatrix} \tag{3.24}$$

$$\begin{aligned} \mathbf{a}_{1*}\cdot\mathbf{b}_{*1} &= a_{11}b_{11} + a_{12}b_{21} + a_{13}b_{31}, \\ \mathbf{a}_{1*}\cdot\mathbf{b}_{*2} &= a_{11}b_{12} + a_{12}b_{22} + a_{13}b_{32}, \\ \mathbf{a}_{2*}\cdot\mathbf{b}_{*1} &= a_{21}b_{11} + a_{22}b_{21} + a_{23}b_{31}, \end{aligned} \tag{3.25}$$

For square matrices both products **AB** and **BA** have meaning. However, since the rule is *row from the left matrix–dot–column from the right matrix*, these two products are not in general the same matrix. When they are the same we say **A** and **B** *commute*. It is important, however, that whether or not **A** and **B** commute,

$$\mathrm{tr}(\mathbf{AB}) = \mathrm{tr}(\mathbf{BA}). \tag{3.26}$$

EXAMPLE 3.8 Given the two matrices

$$\mathbf{A} = \begin{bmatrix} 5x & 4 & 3 \\ 4 & 3 & 2y \end{bmatrix}, \qquad \mathbf{B} = \begin{bmatrix} x & 0 & 1 & y \\ 4 & 0 & 2 & 1 \\ 2 & 0 & 1 & 1 \end{bmatrix},$$

calculate the products (a) **AB** and (b) **BA**.

SOLUTION

(a)

$$\mathbf{AB} = \begin{bmatrix} 5x^2 + 16 + 6 & 0 & 5x + 8 + 3 & 5xy + 4 + 3 \\ 4x + 12 + 4y & 0 & 4 + 6 + 2y & 4y + 3 + 2y \end{bmatrix}$$

$$= \begin{bmatrix} 5x^2 + 22 & 0 & 5x + 11 & 5xy + 7 \\ 4x + 4y + 12 & 0 & 2y + 10 & 6y + 3 \end{bmatrix}.$$

(b) **BA** is not defined since the number of columns of **B** is not the same as the number of rows of **A**.

EXAMPLE 3.9 Show that

$$\text{tr } \mathbf{AB} = \text{tr } \mathbf{BA}$$

where

$$\mathbf{A} = \begin{bmatrix} a_{11} & a_{12} & a_{13} \\ a_{21} & a_{22} & a_{23} \\ a_{31} & a_{32} & a_{33} \end{bmatrix}, \qquad \mathbf{B} = \begin{bmatrix} b_{11} & b_{12} & b_{13} \\ b_{21} & b_{22} & b_{23} \\ b_{31} & b_{32} & b_{33} \end{bmatrix}$$

SOLUTION

$$\begin{aligned} \text{tr } \mathbf{AB} &= (a_{11}, a_{12}, a_{13}) \cdot (b_{11}, b_{21}, b_{31}) \\ &\quad + (a_{21}, a_{22}, a_{23}) \cdot (b_{12}, b_{22}, b_{32}) \\ &\quad + (a_{31}, a_{32}, a_{33}) \cdot (b_{13}, b_{23}, b_{33}) \\ &= a_{11}b_{11} + a_{12}b_{21} + a_{13}b_{31} \\ &\quad + a_{21}b_{12} + a_{22}b_{22} + a_{23}b_{32} \\ &\quad + a_{31}b_{13} + a_{32}b_{23} + a_{33}b_{33}, \\ \text{tr } \mathbf{BA} &= (b_{11}, b_{12}, b_{13}) \cdot (a_{11}, a_{21}, a_{31}) \\ &\quad + (b_{21}, b_{22}, b_{23}) \cdot (a_{12}, a_{22}, a_{32}) \\ &\quad + (b_{31}, b_{32}, b_{33}) \cdot (a_{13}, a_{23}, a_{33}) \\ &= b_{11}a_{11} + b_{12}a_{21} + b_{13}a_{31} \\ &\quad + b_{21}a_{12} + b_{22}a_{22} + b_{23}a_{32} \\ &\quad + b_{31}a_{13} + b_{32}a_{23} + b_{33}a_{33}. \end{aligned}$$

The two results contain the same terms in different order.

We will usually be concerned with either the products of two $n \times n$ matrices, which gives an $n \times n$ matrix, or the product of a square matrix and column vector, which gives another column vector.

A square matrix $\mathbf{A}$ may have an inverse $\mathbf{A}^{-1}$ that is defined by

$$\mathbf{A}\mathbf{A}^{-1} = \mathbf{A}^{-1}\mathbf{A} = \mathbf{I}, \tag{3.27}$$

where $\mathbf{I}$ is the unit matrix. It can be shown that

$$\mathbf{A}^{-1} = \frac{1}{|\mathbf{A}|}\begin{bmatrix} A_{11} & A_{21} & \cdots & A_n \\ A_{12} & & \ddots & \vdots \\ \vdots & & & \\ A_{1n} & & \cdots & A_{nn} \end{bmatrix}, \tag{3.28}$$

where A_{kl} is the cofactor of a_{kl} in $|\mathbf{A}|$. (Note the transpose of positions in Eq. (3.28). A_{ji} in $\mathbf{A}^{-1}$ takes the position of a_{ij} in $\mathbf{A}$.) It is clear that for $\mathbf{A}^{-1}$ to exist, we must have $|\mathbf{A}| \neq 0$, that is, the rank of $\mathbf{A}$ must be n.

There is a second type of matrix multiplication that we will encounter in applications. It is called the *direct product* of two matrices and is written $\mathbf{A} \times \mathbf{B}$. We shall define this operation by example. Consider

$$\mathbf{A} = \begin{bmatrix} a_{11} & a_{12} \\ a_{21} & a_{22} \end{bmatrix} \tag{3.29}$$

and

$$\mathbf{B} = \begin{bmatrix} b_{11} & b_{12} & b_{13} \\ b_{21} & b_{22} & b_{23} \\ b_{31} & b_{32} & b_{33} \end{bmatrix}. \tag{3.30}$$

The direct product $\mathbf{A} \times \mathbf{B}$ is the 6×6 matrix given by

$$\mathbf{A} \times B = \begin{bmatrix} a_{11}\mathbf{B} & a_{12}\mathbf{B} \\ a_{21}\mathbf{B} & a_{22}\mathbf{B} \end{bmatrix}$$

$$= \begin{bmatrix} a_{11}b_{11} & a_{11}b_{12} & a_{11}b_{13} & a_{12}b_{11} & a_{12}b_{12} & a_{12}b_{13} \\ a_{11}b_{21} & a_{11}b_{22} & a_{11}b_{23} & a_{12}b_{21} & a_{12}b_{22} & a_{12}b_{23} \\ a_{11}b_{31} & a_{11}b_{32} & a_{11}b_{33} & a_{12}b_{31} & a_{12}b_{32} & a_{12}b_{33} \\ a_{21}b_{11} & a_{21}b_{12} & a_{21}b_{13} & a_{22}b_{11} & a_{22}b_{12} & a_{22}b_{13} \\ a_{21}b_{21} & a_{21}b_{22} & a_{21}b_{23} & a_{22}b_{21} & a_{22}b_{22} & a_{22}b_{23} \\ a_{21}b_{31} & a_{21}b_{32} & a_{21}b_{33} & a_{22}b_{31} & a_{22}b_{32} & a_{22}b_{33} \end{bmatrix} \tag{3.31}$$

The direct product $\mathbf{B} \times \mathbf{A}$ is the 6×6 matrix given by

$$\mathbf{B} \times \mathbf{A} = \begin{bmatrix} b_{11}\mathbf{A} & b_{12}\mathbf{A} & b_{13}\mathbf{A} \\ b_{21}\mathbf{A} & b_{22}\mathbf{A} & b_{23}\mathbf{A} \\ b_{31}\mathbf{A} & b_{32}\mathbf{A} & b_{33}\mathbf{A} \end{bmatrix} \tag{3.32}$$

In general, the direct product is not commutative. There are no restrictions on the number of rows or columns in either matrix. When $\mathbf{A}$ and $\mathbf{B}$ are both square (but not necessarily of the same order) the following important result applies:

$$\text{tr}(\mathbf{A} \times \mathbf{B}) = \text{tr}(\mathbf{B} \times \mathbf{A}) = (\text{tr } \mathbf{A})(\text{tr } \mathbf{B}) \tag{3.33}$$

That is, the trace of the direct product is equal to the product of the traces.

EXAMPLE 3.10 For the matrices

$$\mathbf{A} = \begin{bmatrix} 1 & 0 & 2 \\ 3 & 2 & 0 \\ 1 & 3 & 1 \end{bmatrix}, \qquad \mathbf{B} = \begin{bmatrix} a & b \\ c & d \end{bmatrix},$$

calculate the direct products (a) $\mathbf{A} \times \mathbf{B}$, (b) $\mathbf{B} \times \mathbf{A}$. (c) Show that Eq. (3.33) holds.

SOLUTION

(A)

$$\mathbf{A} \times \mathbf{B} = \begin{bmatrix} 1\mathbf{B} & 0\mathbf{B} & 2\mathbf{B} \\ 3\mathbf{B} & 2\mathbf{B} & 0\mathbf{B} \\ 1\mathbf{B} & 3\mathbf{B} & 1\mathbf{B} \end{bmatrix} = \begin{bmatrix} a & b & 0 & 0 & 2a & 2b \\ c & d & 0 & 0 & 2c & 2d \\ 3a & 3b & 2a & 2b & 0 & 0 \\ 3c & 3d & 2c & 2d & 0 & 0 \\ a & b & 3a & 3b & a & b \\ c & d & 3c & 3d & c & d \end{bmatrix}$$

(b)

$$\mathbf{B} \times \mathbf{A} = \begin{bmatrix} a\mathbf{A} & b\mathbf{A} \\ c\mathbf{A} & d\mathbf{A} \end{bmatrix} = \begin{bmatrix} a & 0 & 2a & b & 0 & 2b \\ 3a & 2a & 0 & 3b & 2b & 0 \\ a & 3a & a & b & 3b & b \\ c & 0 & 2c & d & 0 & 2d \\ 3c & 2c & 0 & 3d & 2d & 0 \\ c & 3c & c & d & 3d & d \end{bmatrix}.$$

(c)

$$\text{tr}(\mathbf{A} \times \mathbf{B}) = a + d + 2a + 2d + a + d = 4a + 4d,$$

$$\text{tr}(\mathbf{B} \times \mathbf{A}) = a + 2a + a + d + 2d + d = 4a + 4d.$$

Some elementary properties of matrices in general and the definitions of some special matrices and their special properties are given in Appendix 3B.

3.5 Linear Transformations

Consider a rotation about the z axis in three dimensions. If the original system is (x, y, z) and the rotated system is (x', y', z'), then we have (see Problem 3.5),

$$
\begin{aligned}
x' &= (\cos\theta)x + (\sin\theta)y,\\
y' &= -(\sin\theta)x + (\cos\theta)y,\\
z' &= z,
\end{aligned}
\tag{3.34}
$$

where θ is the angle of rotation. Equations (3.34) apply for *clockwise* rotation of a *body* in a *fixed coordinate system*, that is, from positive x axis toward negative y axis. They also apply for *counterclockwise* rotation of the *coordinate axes* with the *body fixed.*

Equation (3.34) can be written

$$\mathbf{x}' = \mathbf{R}_z\mathbf{x}, \tag{3.35}$$

where $\mathbf{x}$ and $\mathbf{x}'$ represent the column vectors

$$\mathbf{x} = \begin{bmatrix} x \\ y \\ z \end{bmatrix} \quad \text{and} \quad \mathbf{x}' = \begin{bmatrix} x' \\ y' \\ z' \end{bmatrix} \tag{3.36}$$

and $\mathbf{R}_z$ is the matrix,

$$\mathbf{R}_z = \begin{bmatrix} \cos\theta & \sin\theta & 0 \\ -\sin\theta & \cos\theta & 0 \\ 0 & 0 & 1 \end{bmatrix} \tag{3.37}$$

Similarly, rotations about the x and y axes are represented by the rotation matrices

$$\mathbf{R}_x = \begin{bmatrix} 1 & 0 & 0 \\ 0 & \cos\theta & \sin\theta \\ 0 & -\sin\theta & \cos\theta \end{bmatrix}, \tag{3.38}$$

$$\mathbf{R}_y = \begin{bmatrix} \cos\theta & 0 & \sin\theta \\ 0 & 1 & 0 \\ -\sin\theta & 0 & \cos\theta \end{bmatrix}. \tag{3.39}$$

A reflection through the xy plane can be expressed as

$$x' = x, \qquad y' = y, \qquad z' = -z, \tag{3.40}$$

and the corresponding matrix is

$$\boldsymbol{\sigma}_{xy} = \begin{bmatrix} 1 & 0 & 0 \\ 0 & 1 & 0 \\ 0 & 0 & -1 \end{bmatrix}. \tag{3.41}$$

The matrix for inversion through the origin is the negative unit matrix

$$\mathbf{i} = \begin{bmatrix} -1 & 0 & 0 \\ 0 & -1 & 0 \\ 0 & 0 & -1 \end{bmatrix} = (-1)\begin{bmatrix} 1 & 0 & 0 \\ 0 & 1 & 0 \\ 0 & 0 & 1 \end{bmatrix}. \tag{3.42}$$

The operators $\boldsymbol{\sigma}_{xy}$ and $\mathbf{i}$ are simple examples of what are called *improper* rotations. Any improper rotation can be expressed in matrix form as a proper (ordinary) rotation followed, or preceded, by inversion (see Problems 3.4 and 3.5).

In general, any linear transformation can be expressed as

$$\mathbf{x}' = \mathbf{T}\mathbf{x}, \tag{3.43}$$

where $\mathbf{T}$ is a matrix and $\mathbf{x}$ and $\mathbf{x}'$ are column vectors. In n dimensions, $\mathbf{T}$ will be an $n \times n$ matrix.

Any transformation that changes one Cartesian system (i.e., a system with a set of orthonormal axes) to another Cartesian system must correspond to an *orthogonal* matrix. A necessary and sufficient condition that a matrix be orthogonal is that its columns form an orthonormal set, or equivalently, that its rows form an orthonormal set. Some other properties of orthogonal matrices are given in Appendix 3B.

If two vectors $\mathbf{v}_1$ and $\mathbf{v}_2$ undergo the same orthogonal transformation to $\mathbf{v}'_1$ and $\mathbf{v}'_2$, then their scalar product remains unchanged,

$$\mathbf{v}'_1 \cdot \mathbf{v}'_2 = \mathbf{v}_1 \cdot \mathbf{v}_2, \tag{3.44}$$

that is, the scalar product is invariant under orthogonal transformation. As a special case of this, the length of a vector is invariant under orthogonal transformation.

Suppose we carry out two transformations consecutively, first

$$\mathbf{x}' = \mathbf{T}_1\mathbf{x} \tag{3.45}$$

followed by

$$\mathbf{x}'' = \mathbf{T}_2\mathbf{x}'. \tag{3.46}$$

The combined transformation can be expressed as

$$\mathbf{x}'' = \mathbf{T}\mathbf{x}, \tag{3.47}$$

where

$$\mathbf{T} = \mathbf{T}_2\mathbf{T}_1 \tag{3.48}$$

and the multiplication in $\mathbf{T}_2\mathbf{T}_1$ is matrix multiplication. Since matrices (even orthogonal ones) do not necessarily commute, the order in which the transformations are performed may be significant.

Suppose we have a matrix equation

$$\mathbf{y} = \mathbf{A}\mathbf{x}, \tag{3.49}$$

where $\mathbf{x}$ and $\mathbf{y}$ are vectors and $\mathbf{A}$ is a matrix. Suppose now that both $\mathbf{x}$ and $\mathbf{y}$ undergo the same linear transformation $\mathbf{T}$,

$$\mathbf{x}' = \mathbf{T}\mathbf{x}, \tag{3.50}$$

$$\mathbf{y}' = \mathbf{T}\mathbf{y}. \tag{3.51}$$

In the new system, Eq. (3.49) becomes

$$\mathbf{y}' = \mathbf{A}'\mathbf{x}', \tag{3.52}$$

where

$$\mathbf{A}' = \mathbf{T}\mathbf{A}\mathbf{T}^{-1}. \tag{3.53a}$$

Equation (3.53a) is often expressed in terms of the inverse of $\mathbf{T}$. If $\mathbf{t}$ is $\mathbf{T}^{-1}$, we obtain

$$\mathbf{A}' = \mathbf{t}^{-1}\mathbf{A}\mathbf{t}. \tag{3.53b}$$

This type of transformation of the matrix $\mathbf{A}$ is called a *similarity* transformation. A very important theorem states that

$$\operatorname{tr} \mathbf{A}' = \operatorname{tr} \mathbf{A}. \tag{3.54}$$

That is, the trace of a matrix is invariant under a similarity transformation.

Suppose we have two spaces, for example, a two-dimensional space (x_1, x_2) and a three-dimensional space (y_1, y_2, y_3). Let us define a six-dimensional space by combining these two sets in dictionary order to obtain $(z_1, z_2, \ldots, z_6)$, where

$$\begin{aligned} z_1 &= x_1 y_1, \quad & z_2 &= x_1 y_2, \quad & z_3 &= x_1 y_3, \\ z_4 &= x_2 y_1, \quad & z_5 &= x_2 y_2, \quad & z_6 &= x_2 y_3. \end{aligned} \tag{3.55}$$

If the two original spaces undergo the transformations

$$\mathbf{x}' = \mathbf{A}\mathbf{x}, \tag{3.56}$$

$$\mathbf{y}' = \mathbf{B}\mathbf{y}, \tag{3.57}$$

then the combined space undergoes the transformation

$$\mathbf{z}' = (\mathbf{A} \times \mathbf{B})\mathbf{z}, \tag{3.58}$$

where $(\mathbf{A} \times \mathbf{B})$ is the direct product (see Problem 3.6).

3.6 Eigenvalues and Eigenvectors of Matrices

Any square matrix $\mathbf{A}$ has certain characteristic directions such that any vectors (called *eigenvectors*) in these directions suffer only changes in length under the operation $\mathbf{A}$. The mathematical expression for such behavior can be written

$$\mathbf{A}\mathbf{v} = \lambda\mathbf{v}, \tag{3.59}$$

where $\mathbf{v}$ is an eigenvector of $\mathbf{A}$ and λ is a scalar called the *eigenvalue* corresponding to the eigenvector $\mathbf{v}$. All the $n \times n$ matrices of interest to us have n linearly independent eigenvectors. Their eigenvalues may, or may not, all be different. If two or more eigenvectors have the same eigenvalues, we say that the matrix is *degenerate*.

Equation (3.59) can be written

$$(\mathbf{A} - \lambda\mathbf{I})\mathbf{v} = \mathbf{0}, \tag{3.60}$$

where $\mathbf{I}$ is the unit matrix. Equation (3.60) represents n homogeneous linear equations with the n components of $\mathbf{v}$ as unknowns. This requires that (see Appendix 3A.12)

$$|\mathbf{A} - \lambda\mathbf{I}| = 0. \tag{3.61}$$

The determinant $|\mathbf{A} - \lambda\mathbf{I}|$ is called the *secular* determinant, and Eq. (3.61) is the *secular* equation. The secular determinant is a polynomial of degree n in λ. The n roots of secular equation (3.61) are the eigenvalues of $\mathbf{A}$. Substitution of an eigenvalue into Eq. (3.60) [or (3.59)] produces the set of equations whose solution is the components of the *eigendirection* corresponding to that eigenvalue. Here we have used the term "eigendirection" to emphasize the fact that the magnitude, and even the sign, of the eigenvectors are not fixed. It is often useful to have the eigenvectors normalized (see Examples 3.11 and 3.12).

Some important properties of eigenvalues and eigenvectors are listed in Appendix 3B. As a consequence of Appendix 3B.3(d) and 3B.5 the eigenvectors of a Hermitian matrix can form an orthonormal set. If a matrix is constructed with columns equal to the eigenvectors of such an orthonormal set (see Section 3.2) of eigenvectors of $\mathbf{A}$, the matrix will be *unitary* (or orthogonal). Some properties of unitary matrices are given in Appendix 3B.4. If this unitary matrix is used as the transformation $\mathbf{T}$ in Eq. (3.53a), the transformed matrix $\mathbf{A}'$ will be diagonal. The off-diagonal elements will be zero, and the diagonal elements will be eigenvalues of $\mathbf{A}$.

EXAMPLE 3.11 All matrices of the form $\left[\begin{smallmatrix} a & b \\ b & a \end{smallmatrix}\right]$, where a and b are arbitrary scalars, commute (see Problem 3.7). Calculate their eigenvalues and eigenvectors. Comment on the dependence of these eigenvalues and eigenvectors on the values of a and b. Are the eigenvectors orthogonal? What transformation will diagonalize these matrices? Carry out the diagonalization.

SOLUTION The eigenvalues are given by the roots of the secular equation

$$\begin{bmatrix} a-\lambda & b \\ b & a-\lambda \end{bmatrix} = 0 \qquad \text{or} \qquad \begin{aligned} (a-\lambda)^2 - b^2 &= 0, \\ (a-\lambda)^2 &= b^2, \\ \lambda &= a \pm b. \end{aligned}$$

The eigenvectors are given by

$$\begin{bmatrix} a-\lambda & b \\ b & a-\lambda \end{bmatrix}\begin{bmatrix} x \\ y \end{bmatrix} = 0 \qquad \text{or} \qquad (a-\lambda)x + by = 0, \qquad bx + (a-\lambda)y = 0.$$

For $\lambda = a + b$, these give $x = y$, and the eigenvector is in the direction of $\pm(1, 1)$. A normalized eigenvector is $\mathbf{v}_{a+b} = (1/\sqrt{2}, 1/\sqrt{2})$. For $\lambda = a - b$, we obtain $x = -y$, and the eigenvector is in the direction of $\pm(1, -1)$. A normalized eigenvector is $\mathbf{v}_{a-b} = (1/\sqrt{2}, -1/\sqrt{2})$.

The eigenvalues depend on both a and b. However, the eigenvectors are independent of both a and b, in agreement with the theorem that commuting matrices have the same eigenvectors (see Appendix 3B.7).

The eigenvectors are also orthogonal, in agreement with the theorem that for symmetric matrices, eigenvectors corresponding to different eigenvalues are orthogonal [see Appendix 3B.3(d)]. The exception is when $b = 0$. In that case both eigenvalues have the value a, and any vector in the xy plane is an eigenvector. Therefore, any two independent vectors in the xy plane, including $(1, 1)$ and $(1, -1)$, can be used as the eigenvectors (see Appendix 3B.5).

The transformation $\mathbf{T}$ [see Eq. (3.53a)] that diagonalized the matrices contains the *normalized* eigenvector as columns,

$$\mathbf{T} = \begin{bmatrix} 1/\sqrt{2} & 1/\sqrt{2} \\ 1/\sqrt{2} & -1/\sqrt{2} \end{bmatrix} = \frac{1}{\sqrt{2}}\begin{bmatrix} 1 & 1 \\ 1 & -1 \end{bmatrix}, \qquad \mathbf{T}^{-1} = \mathbf{T}^{\mathrm{T}} = \frac{1}{\sqrt{2}}\begin{bmatrix} 1 & 1 \\ 1 & -1 \end{bmatrix}.$$

(Note that in this particular case $\mathbf{T}^{-1} = \mathbf{T}^{\mathrm{T}} = \mathbf{T}$.)
Equation (4.53a) gives

$$\tfrac{1}{2}\begin{bmatrix} 1 & 1 \\ 1 & -1 \end{bmatrix}\begin{bmatrix} a & b \\ b & a \end{bmatrix}\begin{bmatrix} 1 & 1 \\ 1 & -1 \end{bmatrix} = \begin{bmatrix} a+b & 0 \\ 0 & a-b \end{bmatrix}.$$

EXAMPLE 3.12 Consider the matrix

$$\mathbf{A} = \begin{bmatrix} a & b & 0 \\ b & c & 0 \\ 0 & 0 & d \end{bmatrix}.$$

(a) Find its eigenvalues.

(b) Find the eigenvectors for the case $a = 1, b = 2, c = 1$, and d still arbitrary. Show that these eigenvectors are orthogonal to one another.

(c) Repeat part (b) for the two cases $d = 3$ and -1.

SOLUTION (a) The eigenvalues are the roots of

$$\begin{vmatrix} a-\lambda & b & 0 \\ b & c-\lambda & 0 \\ 0 & 0 & d-\lambda \end{vmatrix} = 0 \quad \text{or} \quad \begin{vmatrix} a-\lambda & b \\ b & c-\lambda \end{vmatrix}(d-\lambda) = 0,$$

$$[(a-\lambda)(c-\lambda) - b^2](d-\lambda) = 0,$$

$$[\lambda^2 - (a+c)\lambda + ac - b^2](d-\lambda) = 0,$$

$$\lambda^2 - (a+c)\lambda + ac - b^2 = 0,$$

and

$$\lambda = d.$$

The roots of the quadratic are

$$\lambda = \frac{a + c \pm \sqrt{(a+c)^2 - 4(ac - b^2)}}{2}$$

$$= \frac{a + c \pm \sqrt{(a-c)^2 + 4b^2}}{2}.$$

(b) With $a = 1$, $b = 2$, $c = 1$, and d still arbitrary, the eigenvalues become $\lambda = d$, 3, and -1. The eigenvectors are obtained by substituting these values into the matrix equation

$$\begin{bmatrix} 1-\lambda & 2 & 0 \\ 2 & 1-\lambda & 0 \\ 0 & 0 & d-\lambda \end{bmatrix} \begin{bmatrix} x \\ y \\ z \end{bmatrix} = 0$$

and solving for x, y, and z. For $\lambda = d$, we have

$$(1-d)x + 2y + 0z = 0,$$

$$2x + (1-d)y + 0z = 0,$$

$$0x + 0y + 0z = 0.$$

Except for $d = 3$ or -1, the only solutions are $x = 0$, $y = 0$, and z arbitrary. Thus, the eigenvector corresponding to the eigenvalue d is the z axis, that is, in the direction of the vector $(0, 0, 1)$, or equivalently, in the direction of $(0, 0, -1)$.

For $\lambda = 3$, we have

$$-2x + 2y = 0,$$

$$2x - 2y = 0,$$

$$(d-3)z = 0.$$

Except for $d = 3$, z must be zero, and $x = y$. The eigenvector is, therefore, in the direction of the vectors $\pm(1, 1, 0)$.

For $\lambda = -1$,

$$2x + 2y = 0,$$

$$2x + 2y = 0,$$

$$(d + 1)z = 0.$$

Except for $d = -1$, z must be zero, and $x = -y$. The eigenvector is in the direction of the vectors $\pm(1, -1, 0)$.

These vectors are orthogonal to one another because their scalar products are all zero.

(c) For each of the cases $d = 3$ and $d = -1$, the matrix has two equal eigenvalues and so is degenerate. For $d = 3$, any linear combination of (0, 0, 1) and (1, 1,0) can be an eigenvector. This can be checked easily. For $\lambda = 3$ and $d = 3$, we have the requirement $x = y$, but z can have any value. Thus,

$$(\alpha, \alpha, \beta) = \alpha(1, 1, 0) + \beta(0, 0, 1),$$

where α and β are arbitrary, is an eigenvector.

Likewise, for $\lambda = -1$ and $d = -1$, any linear combination of $(1, -1, 0)$ and $(0, 0, 1)$ is an eigenvector because the requirement is $x = -y$, but z can have any value.

3.7 Matrix Representations of Symmetry Groups

We have seen that symmetry operations (i.e., rotations, reflections, and inversion) can be expressed in terms of orthogonal matrices. Also, successive operations are expressed by matrix multiplication. The corresponding matrices obey the group multiplication table. It follows that any symmetry group can be represented in terms of matrices. Actually, as we shall see, there is an infinite number of different matrix representations for any given group, but only a finite number of these are what are called nonequivalent *irreducible* representations.

Let us consider the group $\mathbf{C}_{2v}$. We can obtain a three-dimensional representation by the method used to obtain Eqs. (3.37)–(3.42). The representation so obtained is

$$\underset{E}{\begin{bmatrix} 1 & 0 & 0 \\ 0 & 1 & 0 \\ 0 & 0 & 1 \end{bmatrix}} \underset{C_2(z)}{\begin{bmatrix} -1 & 0 & 0 \\ 0 & -1 & 0 \\ 0 & 0 & 1 \end{bmatrix}} \underset{\sigma_{xz}}{\begin{bmatrix} 1 & 0 & 0 \\ 0 & -1 & 0 \\ 0 & 0 & 1 \end{bmatrix}} \underset{\sigma_{yz}}{\begin{bmatrix} -1 & 0 & 0 \\ 0 & 1 & 0 \\ 0 & 0 & 1 \end{bmatrix}} \tag{3.62}$$

The Cartesian coordinate system (x, y, z) that was used to obtain this representation is called the *basis* of the representation.

If we had picked the C_2 axis to be the x axis, for example, we would have obtained a different set of matrices as our representation. However, this new set can be transformed to the old set (and vice versa) by an orthogonal similarity transformation, that is, a rotation. The same transformation is required for each matrix. For this reason these different representations are said to be *equivalent*.

We can obtain a six-dimensional representation by use of Fig. 3.1. (Fig. 10.2 in Chapter 10 is used to obtain a nine-dimensional representation.) If we take the coordinates in the six-dimensional space in the order $(x_1, y_1, z_1, x_2, y_2, z_2)$ as a basis, we have, for example,

$$\begin{bmatrix} x_1 \\ y_1 \\ z_1 \\ x_2 \\ y_2 \\ z_2 \end{bmatrix} \xrightarrow{C_2} \begin{bmatrix} -x_2 \\ -y_2 \\ z_2 \\ -x_1 \\ -y_1 \\ z_1 \end{bmatrix} = \begin{bmatrix} x'_1 \\ y'_1 \\ z'_1 \\ x'_2 \\ y'_2 \\ z'_0 \end{bmatrix}. \tag{3.63}$$

Written out, Eq. (3.63) is

$$\begin{aligned} x'_1 &= 0x_1 + 0y_1 + 0z_1 - x_2 + 0y_2 + 0z_2, \\ y'_1 &= 0x_1 + 0y_1 + 0z_1 + 0x_2 - y_2 + 0z_2, \\ z'_1 &= 0x_1 + 0y_1 + 0z_1 + 0x_2 + 0y_2 + z_2, \\ x'_2 &= -x_1 + 0y_1 + 0z_1 + 0x_2 + 0y_2 + 0z_2, \\ y'_2 &= 0x_1 - y_1 + 0z_1 + 0x_2 + 0y_2 + 0z_2, \\ z'_2 &= 0x_1 + 0y_1 + z_1 + 0x_2 + 0y_2 + 0z_2. \end{aligned} \tag{3.64}$$

In matrix notation this is

$$\mathbf{x}' = \mathbf{C}_2\mathbf{x}, \tag{3.65}$$

where

$$\mathbf{C}_2 = \begin{bmatrix} 0 & 0 & 0 & -1 & 0 & 0 \\ 0 & 0 & 0 & 0 & -1 & 0 \\ 0 & 0 & 0 & 0 & 0 & 1 \\ -1 & 0 & 0 & 0 & 0 & 0 \\ 0 & -1 & 0 & 0 & 0 & 0 \\ 0 & 0 & 1 & 0 & 0 & 0 \end{bmatrix}. \tag{3.66}$$

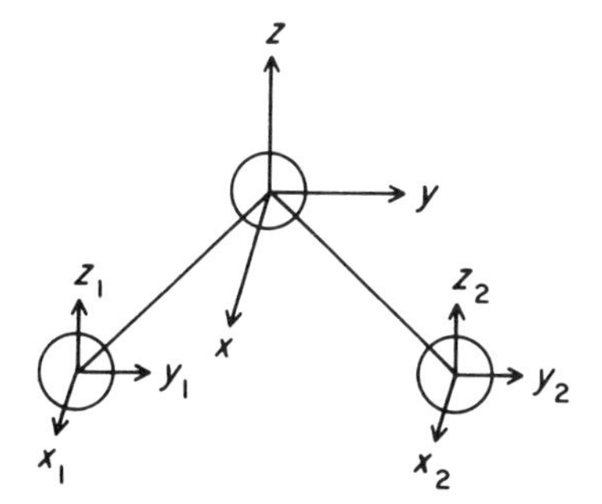

Fig. 3.1 Bent triatomic molecule ($\mathbf{C}_{2v}$ symmetry). The coordinate system for the molecule and the central atom is (x, y, z). The coordinate systems for the other two atoms are (x_1, y_1, z_1) and (x_2, y_2, z_2).

Likewise we obtain

$$\boldsymbol{\sigma}_{xz} = \begin{bmatrix} 0 & 0 & 0 & 1 & 0 & 0 \\ 0 & 0 & 0 & 0 & -1 & 0 \\ 0 & 0 & 0 & 0 & 0 & 1 \\ 1 & 0 & 0 & 0 & 0 & 0 \\ 0 & -1 & 0 & 0 & 0 & 0 \\ 0 & 0 & 1 & 0 & 0 & 0 \end{bmatrix}, \tag{3.67}$$

$$\boldsymbol{\sigma}_{yz} = \begin{bmatrix} -1 & 0 & 0 & 0 & 0 & 0 \\ 0 & 1 & 0 & 0 & 0 & 0 \\ 0 & 0 & 1 & 0 & 0 & 0 \\ 0 & 0 & 0 & -1 & 0 & 0 \\ 0 & 0 & 0 & 0 & 1 & 0 \\ 0 & 0 & 0 & 0 & 0 & 1 \end{bmatrix}. \tag{3.68}$$

If we had picked the coordinates in a different order, we would have obtained different matrices. However, the two representations would be equivalent.

These examples illustrate how various matrix representations of a group are possible. These are all what we call *faithful* representations because in these representations each group operation has a different matrix. A mathematically trivial, but very important, unfaithful representation that exists for every group is the one in which every group element is represented by the number 1. That is, every group element is represented by the one-dimensional unit matrix. It is clear the number 1 satisfies every multiplication table. This representation is called the *completely symmetric* or the *identity* representation. A representation in which the n-dimensional $\mathbf{I}$ represents every element of the group is also possible, but it has no special significance. The one-dimensional identity representation is an *irreducible* representation, while the representation consisting of the n-dimensional $\mathbf{I}$ is *reducible*.

Now let us examine the difference between reducible and irreducible representations. Suppose we have an n-dimensional representation, and the matrix representing the pth group operator is designated $\mathbf{G}(p)$, where p goes from 1 to h (h is the order of the group). If this representation is reducible, there will be a coordinate transformation, the same for all matrices of the representation, that will reduce simultaneously all of the h matrices to a blocked form

$$\mathbf{G}(p) = \begin{bmatrix} \mathbf{G}_1(p) & \mathbf{0} & \mathbf{0} & \cdots \\ \mathbf{0} & \mathbf{G}_2(p) & \mathbf{0} & \cdots \\ \mathbf{0} & \mathbf{0} & \mathbf{G}_3(p) & \cdots \\ \vdots & \vdots & \vdots & \ddots \end{bmatrix}, \qquad p = 1, \ldots, h. \tag{3.69}$$

[For the definition and some properties of matrices in blocked form see Appendix 3B.1(h).] Here the zeros represent submatrices, all of whose elements are zero. The matrices $\mathbf{G}_1(1), \mathbf{G}_1(2), \ldots, \mathbf{G}_1(h)$ are all of the same dimension. Likewise, $\mathbf{G}_2(1), \mathbf{G}_2(2), \ldots, \mathbf{G}_2(h)$ are all of the same dimension, etc. From the rule of matrix multiplication in blocked form it follows that $\mathbf{G}_1(1), \mathbf{G}_1(2), \ldots, \mathbf{G}_1(h)$ form a representation of the group. Also, $\mathbf{G}_2(1), \mathbf{G}_2(2), \ldots, \mathbf{G}_2(h)$ form a representation of the group, etc. Some of these representations may be the same or equivalent to one another. If further reduction is not possible, these representations are called *irreducible*.

The representation of $\mathbf{C}_{2v}$ given in (3.62) is already in blocked form—each block being one-dimensional. The corresponding irreducible representations are 1, -1, 1, -1; and 1, -1, -1, 1; and 1, 1, 1, 1. These one-dimensional representations are designated B_1, B_2, and A_1 (see Section 3.9 and the character table for $\mathbf{C}_{2v}$ in Appendix 1). If we designate the three-dimensional reducible representation by $\Gamma^{(3)}$, then we can write

$$\Gamma^{(3)} = A_1 + B_1 + B_2. \tag{3.70}$$

For irreducible representations that have higher dimensionality than one, there is the complication resulting from the fact that the representation can take an infinite number of equivalent forms, since any similarity transformation of a given representation leads to an equivalent representation. Fortunately, for almost all of our applications, we need only the characters of a representation (i.e., the traces of the matrices). Because of Eq. (3.54), the characters are the same for all equivalent representations; also, because operators in the same class are similarity transforms of one another, they have the same character.

3.8 Some Properties of Irreducible Representations

Let the matrix of the pth operator in the kth irreducible representation be $\mathbf{\Gamma}_k(p)$, and let

$$\mathbf{\Gamma}_k(p) = \begin{bmatrix} \Gamma_k(p)_{11} & \Gamma_k(p)_{12} & \cdots & \Gamma_k(p)_{1l_k} \\ \Gamma_k(p)_{21} & & & \vdots \\ \vdots & & & \\ \Gamma_k(p)_{l_k 1} & & \cdots & \Gamma_k(p)_{l_k l_k} \end{bmatrix}, \tag{3.71}$$

where we use the symbol

$$\Gamma_k(p)_{ij}(i,j = 1,2,\ldots,l_k;\ p = 1,2,\ldots,h;\ k = 1,2,\ldots,\kappa)$$

for the ij element of the matrix; l_k is the dimension of the representation; h is the order of the group; and κ is the number of nonequivalent irreducible representations of the group.

It can be shown that all irreducible representations can be put in unitary form.[1] Therefore, we assume that all the matrices $\mathbf{\Gamma}_k(p)$ are unitary. See Appendix 3B.4 for some properties of unitary matrices.

We define the h-dimensional vector

$$\mathbf{\Gamma}_{k,ij} = \sqrt{l_k/h}\,(\Gamma_k(1)_{ij}, \Gamma_k(2)_{ij}, \ldots, \Gamma_k(h)_{ij}). \tag{3.72}$$

There are l_k^2 such vectors for the kth representation and $\sum_{k=1}^{\kappa} l_k^2$ vectors in all.

Let $\chi_k(q)$ be the character of the operators in the qth class, that is,

$$\chi_k(q) = \sum_i \Gamma_k(q)_{ii}, \tag{3.73}$$

where q indicates any one of the operators in the qth class. Let the number of operators in the qth class be α_q. We define the κ-dimensional vector

$$\boldsymbol{\chi}_k = \sqrt{1/h}\,(\sqrt{\alpha_1}\,\chi_k(1), \sqrt{\alpha_2}\,\chi_k(2), \ldots, \sqrt{\alpha_\kappa}\,\chi_k(\kappa)). \tag{3.74}$$

[Here we have anticipated property (IR3) (given below) by taking the number of characters to be κ.]

In terms of these notations we state, without proof, the following important properties.

(IR1) The vectors $\mathbf{\Gamma}_{k,ij}$ are linearly independent and span an h-dimensional vector space. There are, therefore, h such vectors, and we have

$$\sum_{k=1}^{\kappa} l_k^2 = h. \tag{3.75}$$

[1] The reader can find the proof in any one of several standard texts. For example, M. Tinkham, *Group Theory and Quantum Mechanics*, McGraw-Hill, New York, 1964, p 103.

(IR2) "The great orthogonality theorem." The vectors $\boldsymbol{\Gamma}_{k,ij}$ form an orthonormal set, which means that

$$\boldsymbol{\Gamma}_{k,ij} \cdot \boldsymbol{\Gamma}_{k',i'j'} = \frac{\sqrt{l_k l'_k}}{h} \sum_p \Gamma_k(p)^*_{ij} \Gamma_{k'}(p)_{i'j'} = \delta_{kk'} \delta_{ii'} \delta_{jj'}. \tag{3.76}$$

(IR3) The number of nonequivalent irreducible representations is equal to the number of classes κ.

(IR4) Orthogonality theorem for the characters. The vectors $\boldsymbol{\chi}_k$ form an orthonormal set, which means that

$$\boldsymbol{\chi}_k \cdot \boldsymbol{\chi}_{k'} = \frac{1}{h} \sum_{q=1}^{\kappa} \alpha_q \chi_k^*(q) \chi_{k'}(q) = \frac{1}{h} \sum_{p=1}^{h} \chi_k^*(p) \chi_{k'}(p) = \delta_{kk'}. \tag{3.77}$$

[The first summation in Eq. (3.77) is taken over the classes (κ in number), and the second summation is taken over the operators (h in number). The two summations are equivalent.]

(IR5) The vectors $\boldsymbol{\chi}_k$ span the vector space of the characters of all the representations of the group. If $\boldsymbol{\chi}$ is the vector defined by Eq. (3.74) for a *reducible* representation, then

$$\boldsymbol{\chi} = \sum_k a_k \boldsymbol{\chi}_k \tag{3.78}$$

and

$$a_k = \boldsymbol{\chi} \cdot \boldsymbol{\chi}_k = \frac{1}{h} \sum_{q=1}^{\kappa} \alpha_q \chi(q) \chi_k(q) = \frac{1}{h} \sum_{p=1}^{h} \chi(p) \chi_k(p), \tag{3.79}$$

where a_k is the number of times the kth irreducible representation is contained in the reducible representation under consideration; in Eq. (3.70), $a_{A_1} = a_{B_1} = a_{B_2} = 1$.

We shall illustrate the use of these theorems by applying them to $\mathbf{C}_{3v}$. First, we determine the number of irreducible representations and their dimensions. The group has three classes (Section 2.5) and so has three irreducible representations (IR3). Equation (3.75) gives

$$l_1^2 + l_2^2 + l_3^2 = 6,$$

which is satisfied only by the dimensions 1, 1, and 2.

The identity representation is one of the one-dimensional representations. A second one-dimensional representation is obtained by taking $+1$ for the identity and rotations and -1 for the reflections. (It is clear that the multiplication table is satisfied.) The remaining irreducible representation must be two-dimensional. We can obtain a three-dimensional representation (which, of course, must be reducible) by using a three-dimensional Cartesian system as basis. We obtain (e.g., by the method used in Problem 3.5) the following reducible representation:

$$
\begin{matrix} E & C_3 & C_3^2 \\
\begin{bmatrix} 1 & 0 & 0 \\ 0 & 1 & 0 \\ 0 & 0 & 1 \end{bmatrix}
\begin{bmatrix} \frac{-1}{2} & \frac{\sqrt{3}}{2} & 0 \\ \frac{-\sqrt{3}}{2} & \frac{-1}{2} & 0 \\ 0 & 0 & 1 \end{bmatrix}
\begin{bmatrix} \frac{-1}{2} & \frac{-\sqrt{3}}{2} & 0 \\ \frac{\sqrt{3}}{2} & \frac{-1}{2} & 0 \\ 0 & 0 & 1 \end{bmatrix}
\end{matrix}
$$

$$
\begin{matrix} \sigma_v & \sigma_v' & \sigma_v'' \\
\begin{bmatrix} -1 & 0 & 0 \\ 0 & 1 & 0 \\ 0 & 0 & 1 \end{bmatrix}
\begin{bmatrix} \frac{1}{2} & \frac{-\sqrt{3}}{2} & 0 \\ \frac{-\sqrt{3}}{2} & \frac{-1}{2} & 0 \\ 0 & 0 & 1 \end{bmatrix}
\begin{bmatrix} \frac{1}{2} & \frac{\sqrt{3}}{2} & 0 \\ \frac{\sqrt{3}}{2} & \frac{-1}{2} & 0 \\ 0 & 0 & 1 \end{bmatrix}
\end{matrix}. \tag{3.80}
$$

It is clear that this three-dimensional representation contains the identity representation at least once. A two-dimensional representation is left. Is it irreducible? If it is reducible, it contains one (or both) of the one-dimensional representations. We can check this by using Eq. (3.79). The characters of the two one-dimensional representations and the two-dimensional representation remaining from (3.80) are as follows. For convenience we have used the conventional symbols for the representations (see Section 3.9). The characters for representation E are the traces of the 2×2 blocks in (3.80).

Classes		E	(C_3, C_3^2)	$(\sigma_v, \sigma_v', \sigma_v'')$
Identity representation	(A_1)	1	1	1
Second one-dimensional representation	(A_2)	1	1	-1
Two-dimensional representation	(E)	2	-1	0
α_q		1	2	3

(3.81)

Using Eq. (3.79) to determine whether the two-dimensional representation contains A_1 or A_2, we obtain

$$a_{A_1} = \tfrac{1}{6}|1 \cdot 2 \cdot 1 + 2 \cdot (-1) \cdot 1 + 3 \cdot 0 \cdot 1)| = 0,$$

$$a_{A_2} = \tfrac{1}{6}|1 \cdot 2 \cdot 1 + 2 \cdot (-1) \cdot 1 + 3 \cdot 0 \cdot (-1)| = 0.$$

Since E does not contain either A_1 or A_2, it must be irreducible, and is therefore equivalent to the representations conventionally labeled E.

We have obtained the following irreducible representations. (Of course, E can be put into any one of an infinite number of equivalent forms by the proper unitary transformation.)

	E	C_3	C_3^2	σ_v	σ'_v	σ''_v
A_1	1	1	1	1	1	1
A_2	1	1	1	-1	-1	-1
E	$\begin{bmatrix} 1 & 0 \\ 0 & 1 \end{bmatrix}$	$\begin{bmatrix} -1/2 & \sqrt{3}/2 \\ -\sqrt{3}/2 & -1/2 \end{bmatrix}$	$\begin{bmatrix} -1/2 & -\sqrt{3}/2 \\ \sqrt{3}/2 & -1/2 \end{bmatrix}$	$\begin{bmatrix} -1 & 0 \\ 0 & 1 \end{bmatrix}$	$\begin{bmatrix} 1/2 & -\sqrt{3}/2 \\ -\sqrt{3}/2 & -1/2 \end{bmatrix}$	$\begin{bmatrix} 1/2 & \sqrt{3}/2 \\ \sqrt{3}/2 & -1/2 \end{bmatrix}$

(3.82)

The six vectors $\boldsymbol{\Gamma}_{k,ij}$ are the following:

$$\begin{aligned} \boldsymbol{\Gamma}_{A_1} &= \sqrt{\tfrac{1}{6}}(1, 1, 1, 1, 1, 1), \\ \boldsymbol{\Gamma}_{A_2} &= \sqrt{\tfrac{1}{6}}(1, 1, 1, -1, -1, -1), \\ \boldsymbol{\Gamma}_{E,11} &= \sqrt{\tfrac{2}{6}}(1, -\tfrac{1}{2}, -\tfrac{1}{2}, -1, \tfrac{1}{2}, \tfrac{1}{2}), \\ \boldsymbol{\Gamma}_{E,12} &= \sqrt{\tfrac{2}{6}}(0, \sqrt{3}/2, -\sqrt{3}/2, 0, -\sqrt{3}/2, \sqrt{3}/2), \\ \boldsymbol{\Gamma}_{E,21} &= \sqrt{\tfrac{2}{6}}(0, -\sqrt{3}/2, \sqrt{3}/2, 0, -\sqrt{3}/2, \sqrt{3}/2), \\ \boldsymbol{\Gamma}_{E,22} &= \sqrt{\tfrac{2}{6}}(1, -\tfrac{1}{2}, -\tfrac{1}{2}, 1, -\tfrac{1}{2}, -\tfrac{1}{2}). \end{aligned} \tag{3.83}$$

It is easy to check that these form an orthonormal set (IR2). The matrix constructed by using these vectors as either rows or columns is an orthogonal matrix—both the rows and columns form orthogonal sets.

The three vectors $\boldsymbol{\chi}_k$ are

$$\begin{aligned} \boldsymbol{\chi}_{A_1} &= (1/\sqrt{6})(1, \sqrt{2}, \sqrt{3}), \\ \boldsymbol{\chi}_{A_2} &= (1/\sqrt{6})(1, \sqrt{2}, -\sqrt{3}), \\ \boldsymbol{\chi}_{E} &= (1/\sqrt{6})(2, -\sqrt{2}, 0). \end{aligned} \tag{3.84}$$

These form an orthonormal set (IR4), and the matrix (character matrix) formed by using these vectors as rows (or columns) is orthogonal.

Although it is obvious, we mention here for emphasis that the character of the identity operation E always has a value equal to the dimension of the representation, that is,

$$\chi_p(E) = l_k, \tag{3.85}$$

whether the representation is reducible or irreducible.

3.9 Character Tables

A brief discussion of the structure of character tables is in order. For reference, we give the table for $\mathbf{C}_{3v}$,

$\mathbf{C}_{3v}$	E	$2C_3$	3σ	Linear bases	Quadratic bases	Cubic bases
A_1	1	1	1	z	x^2+y^2, z^2	$z^3, x(x^2-3y^2), z(x^2+y^2)$
A_2	1	1	-1	R_z		$y(3x^2-y^2)$
E	2	-1	0	$(x, y)(R_x, R_y)$	$(x^2-y^2, xy)(xz, yz)$	$(xz^2, yz^2)[xyz, z(x^2-y^2)][x(x^2+y^2), y(x^2+y^2)]$

The first row labels each class by giving the number of operators in the class and a representative operator of the class. The first column contains the symbol for each irreducible representation. The number at the intersection of a column and a row is the character of that class in the corresponding representation. The characters of the identity E are the dimensions of the representation. The three columns on the right contain, respectively, linear, quadratic, and cubic bases of the representations. (Bases are discussed in Section 3.10.)

Point groups that are isomorphic have the same irreducible representations and characters, for example, $\mathbf{C}_{nv}$ and $\mathbf{D}_n$ (see Section 2.7.3 for the list of isomorphic point groups). Isomorphic point groups do not have all the same basis functions, however.

3.9.1 MULLIKEN SYMBOLS

The symbols for the representations are those suggested by Mulliken. A and B refer to one-dimensional representations. E and T refer, respectively, to two- and three-dimensional representations. The only finite point groups with representations of dimensions four and five are the icosahedral groups; these are labeled G and H, respectively. A few other rules that are convenient to remember are the following.

(1) The representations labeled B have bases that change sign under the operation C_n (see $\mathbf{C}_{4v}$).

(2) For groups that contain inversion as an operation, the subscripts g (gerade) and u (ungerade) indicate that the bases *do not* or *do* change sign under inversion.

(3) For A and B representations, the subscripts 1 and 2 indicate bases that *do not* and *do* change sign under a C_2 operation perpendicular to the principal rotation axes or under a σ_v operation when there is no perpendicular C_2. The single and double primes indicate bases that *do not* and *do* change sign under σ_h.

EXAMPLE 3.13 Use the character table to reduce the six-dimensional representation of $\mathbf{C}_{2v}$ given by Eqs. (3.66)–(3.68).

SOLUTION The characters of the irreducible representations of $\mathbf{C}_{2v}$ (obtained from the character table) and that of the six-dimensional representation $\Gamma^{(6)}$ (obtained by adding the diagonal elements) are as follows.

	E	C_2	σ_{xz}	σ_{yz}
A_1	1	1	1	1
A_2	1	1	-1	-1
B_1	1	-1	1	1
B_2	1	-1	-1	1
$\Gamma^{(6)}$	6	0	0	2

We wish to find the coefficients a_k,

$$\Gamma^{(6)} = a_{A_1}A_1 + a_{A_2}A_2 + a_{B_1}B_1 + a_{B_2}B_2.$$

We can use Eq. (3.79),

$$a_{A_1} = \tfrac{1}{4}(6 + 2) = 2,$$

$$a_{A_2} = \tfrac{1}{4}(6 - 2) = 1,$$

$$a_{B_1} = \tfrac{1}{4}(6 - 2) = 1,$$

$$a_{B_2} = \tfrac{1}{4}(6 + 2) = 2.$$

The result is

$$\Gamma^{(6)} = 2A_1 + A_2 + B_1 + 2B_2.$$

We could have obtained this result "by inspection," that is, by noting that the characters of $\Gamma^{(6)}$ can be obtained by taking in each column of the table twice the character of A_1, plus the character of A_2, plus the character of B_1, plus twice the character of B_2.

3.10 Bases for Representations

3.10.1 TRANSFORMATIONS OF FUNCTIONS

The components of the vector (x, y, z) form a basis for the three-dimensional representation of $\mathbf{C}_{2v}$ given in (3.62) and reduced in Eq. (3.70). The components of $(x_1, y_1, z_1, x_2, y_2, z_2)$ form the basis for the representation containing the matrices in Eqs. (3.66)–(3.68).

Often we are concerned with functions (instead of vector components) that are bases for representations. We must, therefore, define what we mean by the transformation of a function in terms of the transformation of the independent variables in the function. In this book we shall use the simple concept illustrated by the following example.[2]

Let the function be x^2y, and let the transformation of the variables be

$$x' = a_{11}x + a_{12}y, \qquad y' = a_{21}x + a_{22}y. \tag{3.86}$$

We define the transformation of x^2y to be

$$(x^2y)' = (x')^2(y') = (a_{11}x + a_{12}y)^2(a_{21}x + a_{22}y). \tag{3.87}$$

3.10.2 ORTHONORMAL SETS OF FUNCTIONS

An important concept in dealing with basis functions is that of orthogonality. Two functions f_1 and f_2 are *orthogonal* if

$$\int f_1^* \cdot f_2 \, d\tau = 0, \tag{3.88}$$

where $d\tau$ is the volume element corresponding to the variables in f_1 and f_2. A function f is said to be *normalized* if

$$\int f^* f \, d\tau = 1. \tag{3.89}$$

A set of functions $f_1, f_2, \ldots, f_n$ is said to form an *orthonormal* set if

$$\int f_i^* f_j \, d\tau = \delta_{ij}. \tag{3.90}$$

The reader will note the similarity in the definitions of orthonormal sets of functions and of vectors [see Eq. (3.6)]. For a very important principle involving the interplay of the concepts of orthonormal sets of functions and orthonormal sets of vectors, see Appendix 3B.4(g).

As examples of orthonormal sets, we have the normalized trigonometric functions $f_n(x)$ and $g_n(x)$. The functions

$$f_n(x) = \frac{1}{\sqrt{L}} \sin \frac{n\pi x}{L}, \qquad n = 1, 2, \ldots, \tag{3.91}$$

[2] This simple procedure will suffice for our applications, although there is a subtle difficulty associated with it. This difficulty could be eliminated by a slight, but inconvenient, modification of the definition. See, for example, M. Tinkham, *Group Theory and Quantum Mechanics*, McGraw-Hill, New York, 1964. For a more complete discussion and applications of the definition that we are using, see J. Mathews and R. L. Walker, *Mathematical Methods of Physics*, 2nd. ed., Benjamin, New York, 1970.

are orthonormal because

$$\frac{1}{L}\int_{-L}^{L} \sin\frac{m\pi x}{L}\sin\frac{n\pi x}{L}\,dx = \delta_{mn}, \tag{3.92}$$

and the functions

$$g_0 = \frac{1}{\sqrt{2L}}, \qquad n = 0, \tag{3.93}$$

$$g_n(x) = \frac{1}{\sqrt{L}}\cos\frac{n\pi x}{L}, \qquad n = 1, 2, \ldots, \tag{3.93}$$

are orthonormal because

$$\frac{1}{L}\int_{-L}^{L} \cos\frac{m\pi x}{L}\cos\frac{n\pi x}{L}\,dx = \delta_{mn}. \tag{3.94}$$

Also, $f_n(x)$ $(n = 1, 2, \ldots)$ is orthogonal to $g_m(x)$ $(m = 0, 1, 2, \ldots)$.

3.10.3 AN ILLUSTRATION OF BASIS FUNCTIONS

To illustrate the existence of basis functions, let us consider the normalized functions

$$\begin{aligned}
\Psi^{\mathrm{c}}_{nn'}(x, y) &= \cos\frac{n\pi x}{2}\cos\frac{n'\pi y}{2}, \\
\Psi^{\mathrm{s}}_{mm'}(x, y) &= \sin\frac{m\pi x}{2}\sin\frac{m'\pi y}{2}, \\
\Psi^{\mathrm{cs}}_{nm}(x, y) &= \cos\frac{n\pi x}{2}\sin\frac{m\pi y}{2}, \\
\Psi^{\mathrm{sc}}_{mn}(x, y) &= \sin\frac{m\pi x}{2}\cos\frac{n\pi y}{2},
\end{aligned} \tag{3.95}$$

where n and n' are odd integers, m and m' are even integers (nonzero), and $-1 \leq x \leq 1$ and $-1 \leq y \leq 1$. (These are the solutions for a quantum mechanical particle in a square well with the origin at the center; see Example 4.2.)

The symmetry of these functions can be described by the point group $\mathbf{C}_{4v}$. According to the character table, $\mathbf{C}_{4v}$ has four one-dimensional and one two-dimensional irreducible representations. The effects of the operations of $\mathbf{C}_{4v}$

on x and y are tabulated below. Only one operation from each class need be considered.

$$\begin{array}{c|ccccc} & E & C_4 & C_2 & \sigma_v & \sigma_d \\ \hline x' & x & -y & -x & -x & y \\ y' & y & x & -y & y & x \end{array} \tag{3.96}$$

By the use of this table we find the following behavior of the functions in (3.95).

(1) $\Psi^{c}_{nn}\,(n = n')$ is invariant under all the group operations and is therefore a basis for A_1.

(2) $\Psi^{s}_{mm}\,(m = m')$ is invariant under E, C_2, and σ_d but changes sign under C_4 and σ_v. It is a basis for B_2.

(3) $\Psi^{c}_{nn'}$ and $\Psi^{c}_{n'n}$ $(n \neq n')$ are each invariant under E, C_2, and σ_v but transform into one another under C_4 and σ_d. If we treat the pair $(\Psi^{c}_{nn'}, \Psi^{c}_{n'n})$ as the component of a two-dimensional vector, they form a two-dimensional representation $\Gamma^{c}_{nn'}$ with the following matrices and characters.

$$\begin{array}{cccccc} & E & C_4 & C_2 & \sigma_v & \sigma_d \\ & \begin{bmatrix} 1 & 0 \\ 0 & 1 \end{bmatrix} & \begin{bmatrix} 0 & 1 \\ 1 & 0 \end{bmatrix} & \begin{bmatrix} 1 & 0 \\ 0 & 1 \end{bmatrix} & \begin{bmatrix} 1 & 0 \\ 0 & 1 \end{bmatrix} & \begin{bmatrix} 0 & 1 \\ 1 & 0 \end{bmatrix} \\ \chi^{c}_{nn'} & 2 & 0 & 2 & 2 & 0 \end{array}$$

From these we obtain

$$\Gamma^{c}_{nn'} = A_1 + B_1.$$

By inspection, we can see that the linear combination

$$\phi^{c+}_{nn'} = \Psi^{c}_{nn'} + \Psi^{c}_{n'n} \tag{3.97}$$

is a basis for A_1 and that

$$\phi^{c-}_{nn'} = \Psi^{c}_{nn'} - \Psi^{c}_{n'n} \tag{3.98}$$

is a basis for B_1.

The functions $\Psi^{s}_{mm'}$ and $\Psi^{s}_{m'm}$ $(m \neq m')$ are each invariant under E and C_2. Each changes sign under σ_v. They change into the negative of one another under C_4. They change into one another under σ_d. The two-dimensional matrices and the characters are as follows.

$$\begin{array}{cccccc} & E & C_4 & C_2 & \sigma_v & \sigma_d \\ & \begin{bmatrix} 1 & 0 \\ 0 & 1 \end{bmatrix} & \begin{bmatrix} 0 & -1 \\ -1 & 0 \end{bmatrix} & \begin{bmatrix} 1 & 0 \\ 0 & 1 \end{bmatrix} & \begin{bmatrix} -1 & 0 \\ 0 & -1 \end{bmatrix} & \begin{bmatrix} 0 & 1 \\ 1 & 0 \end{bmatrix} \\ \chi^{s}_{mm'} & 2 & 0 & 2 & -2 & 0 \end{array}$$

These give

$$\Gamma^{s}_{mm'} = A_2 + B_2. \tag{3.99}$$

We find by inspection that the linear combination

$$\phi^{s+}_{mm'} = \Psi^{s}_{mm'} + \Psi^{s}_{m'm} \tag{3.100}$$

is a basis for B_2, and

$$\phi^{s-}_{mm'} = \Psi^{s}_{mm'} - \Psi^{s}_{m'm} \tag{3.101}$$

is a basis for A_2.

The functions Ψ^{sc}_{nm} and Ψ^{cs}_{mn} give the following matrices and characters.

$$\begin{array}{cccccc} & E & C_4 & C_2 & \sigma_v & \sigma_d \\ & \begin{bmatrix} 1 & 0 \\ 0 & 1 \end{bmatrix} & \begin{bmatrix} 0 & 1 \\ -1 & 0 \end{bmatrix} & \begin{bmatrix} -1 & 0 \\ 0 & -1 \end{bmatrix} & \begin{bmatrix} -1 & 0 \\ 0 & 1 \end{bmatrix} & \begin{bmatrix} 0 & 1 \\ 1 & 0 \end{bmatrix} \\ \chi_{nm} & 2 & 0 & -2 & 0 & 0 \end{array}$$

This is the irreducible representation E.

EXAMPLE 3.14 Use character tables (Appendix 1) to identify the symmetry species for p_z, d_{z^2}, $d_{x^2-y^2}$, d_{xz}, f_{z^3}, and f_{xyz} in $\mathbf{D}_3$, $\mathbf{D}_{3d}$, $\mathbf{D}_{4h}$, and $\mathbf{O}_h$.

SOLUTION

	p_z	d_{z^2}	$d_{x^2-y^2}$	d_{xz}	f_{z^3}	f_{xyz}
$\mathbf{D}_3$	A_2	A_1	E (with d_{xy})	E (with d_{yz})	A_2	B_1
$\mathbf{D}_{3d}$	A_{2u}	A_{1g}	E_g	E_g	A_{2u}	E_u (with $f_{z(x^2-y^2)}$)
$\mathbf{D}_{4h}$	A_{2u}	A_{1g}	B_{1g}	E_g	A_{2u}	B_{1u}
$\mathbf{O}_h$	$(p_x, p_y, p_z)T_{1u}$, $(d_{z^2}, d_{x^2-y^2})E_g$, $(d_{xz}, d_{yz}, d_{xy})T_{2g}$, $(f_{z^3}, f_{x^3}, f_{y^3})T_{1u}$, $f_{xyz}A_{1u}$					

EXAMPLE 3.15 By carrying out appropriate transformations, find the representations for the functions $x^2 - y^2$ (corresponding to $d_{x^2-y^2}$), xyz (corresponding to f_{xyz}), and $z(x^2 - y^2)$ (corresponding to $f_{z(x^2-y^2)}$) in $\mathbf{C}_{4v}$. Check the results with those given in the character table.

SOLUTION We can use the transformations of x and y in $\mathbf{C}_{4v}$ [table (3.96)], tabulating the effect on z (totally symmetric) and performing the algebraic operations on the result for each symmetry operation. Results are shown in Table 3.1.

We see that $d_{x^2-y^2}$ transforms as B_1, f_{xyz} transforms as B_2, and since z is totally symmetric, $z(x^2 - y^2)$ and $(x^2 - y^2)$ transform as B_1.

TABLE 3.1
Solution of Example 3.15

$\mathbf{C}_{4v}$	E	C_4	C_2	σ_v	σ_d	
z'	z	z	z	z	z	
x'^2	x^2	y^2	x^2	x^2	y^2	
y'^2	y^2	x^2	y^2	y^2	x^2	
$x'^2 - y'^2$	$x^2 - y^2$	$y^2 - x^2$	$x^2 - y^2$	$x^2 - y^2$	$y^2 - x^2$	
$\chi(x^2 - y^2)$	1	-1	1	1	-1	$\Gamma = B_1$
$x'y'z'$	xyz	$-xyz$	xyz	$-xyz$	xyz	
$\chi(xyz)$	1	-1	1	-1	-1	$\Gamma = B_2$
$\chi[z(x^2 - y^2]$	1	-1	1	1	-1	$\Gamma = B_1$

3.10.4 SOME PROPERTIES OF BASIC FUNCTIONS

The following theorems express important properties of basis functions of unitary representations of a group.

(B1) If the functions $f_1, \ldots, f_m$ and the functions $g_1, \ldots, g_n$ are bases for the (not necessarily irreducible) representations Γ_m and Γ_n, respectively, then the direct product set, $f_1 g_1, \ldots, f_1 g_n, f_2 g_1, \ldots, f_m g_n$, is a basis for the direct product representation $\Gamma_m \times \Gamma_n$, which may be reducible even when Γ_m and Γ_n are both irreducible. [The matrices of $\Gamma_m \times \Gamma_n$ are the direct products of the matrices of Γ_m and Γ_n, and by Eq. (3.33), the characters of $\Gamma_m \times \Gamma_n$ are the products of the characters of Γ_m and Γ_n.]

(B2) The set $f_1, \ldots, f_m$ and the set $g_1, \ldots, g_n$ of (B1) are orthogonal to each other unless the resolution of $\Gamma_m \times \Gamma_n$ in terms of its irreducible components contains the identity representation. If Γ_m and Γ_n are *irreducible*, then $\Gamma_m \times \Gamma_n$ will contain the identity representation only if Γ_m and Γ_n are *equivalent*. If Γ_m and Γ_n are the *same* irreducible representation, then f_i and g_i are orthogonal if they belong to different rows in the matrices of the representation.[3]

(B3) If the three sets of f_i, Q_i, and g_i (Q_i can be a set of functions or a set of operators) form bases for the irreducible representations Γ_n, Γ_m, and Γ_p, respectively, then the integral $\int f_i Q_j g_k \, d\tau$ is zero unless the direct $\Gamma_n \times \Gamma_m \times \Gamma_p$ contains the identity representation.

(B4) Suppose that the operator H is invariant under all the operations of the group (H is a basis for the identity representation). Suppose also that the sets of functions f_i and g_i are bases for irreducible representations. We have the following useful result: The integral $\int f_i^* H g_j \, d\tau$ is zero unless f_i and g_j belong to the same row ($i = j$) of the same irreducible representation.

[3] This holds if f_i and g_i are the same set of functions, so that f_i and f_j ($i \neq j$) are orthogonal.

(B5) If the set of functions f_i forms a basis for a unitary representation, then the function

$$f = f_1^* f_1 + f_2^* f_2 + \cdots + f_m^* f_m \tag{3.102}$$

is invariant under the operations of the group; that is, f is a basis for the identity representation.

The usefulness of theorems (B2)–(B4) is a result of the fact that they give conditions under which certain integrals *must* be zero. They do not rule out the possibility that an integral may be zero even when it is allowed to be nonzero by the theorems.

3.10.5 PROJECTION AND TRANSFER OPERATORS

A very important problem is that of constructing bases for irreducible representations. From a given function it is possible to obtain functions that belong to an irreducible representation provided the given function has a "component," that is, a nonzero "projection," in the "direction" of that irreducible representation. This can be done by the use of *projection* and *transfer* (*shift*) operators. These operators can be defined as follows.

Let $\Gamma_k(p)_{ij}$ represent the ij element of the matrix representing the pth operator O_p in the kth irreducible representation [see Eq. (3.71)]. We define the operator $O_{k,ij}$ as

$$O_{k,ij} = \frac{l_k}{h} \sum_p \Gamma_k(p)_{ij}^* O_p, \tag{3.103}$$

where h is the order of the group and l_k is the dimension of the representation. These operators, for which $i = j$, are *projection* operators $P_{k,ii}$, that is,

$$P_{k,ii} = O_{k,ii}. \tag{3.104}$$

The off-diagonal operators are *transfer*, or *shift*, operators

$$T_{k,ij} = O_{k,ij}, \qquad i \neq j. \tag{3.105}$$

A trace-projection operator is defined by

$$P_k = \sum_i P_{k,ii} = \sum_p \chi_k(p) O_p, \tag{3.106}$$

where $\chi_k(p)$ is the character for the pth operator.

When the representation is one-dimensional, P_k and $O_{k,ii}$ are the same, and $T_{k,ij}$ does not exist.

Suppose we start with a function f. We have

$$P_{k,ii} f = f_{k,ii}, \tag{3.107}$$

where $f_{k,ii}$ is a basis function corresponding to the ith row of the kth irreducible representation. If f has no component in the direction of the kth representation, then $P_{k,ii} f$ will be zero. The partner of $f_{k,ii}$ corrresponding to the jth row can be obtained by use of the transfer operator as follows:

$$T_{k,ij} f_{k,ii} = f_{k,ij}. \tag{3.108}$$

In this way, starting with the operator $P_{k,ii}$ and using all the transfer operators $T_{k,ij}$ for all $j \neq i$, we obtain a set of l_k partner functions.[4]

If we start with a different projection operator $P_{k,jj}$ $(j \neq i)$ and use the transfer operators $T_{k,jl}$ $(l \neq j)$, then we obtain a second set of functions (l_k in number) that form a basis for the kth irreducible representation. In this way we can obtain l_k sets of l_k basis functions. These sets may or may not all be different, and some of the sets may be zero, that is, some $P_{k,jj}$ may give zero.

The trace-projection operator defined by Eq. (3.106) gives the result

$$P_k f = \sum_p \chi_k(p) O_p f = \sum_i P_{k,ii} f = \sum_i f_{k,ii}, \tag{3.109}$$

which is a linear combination of basis functions, one function from each of the l_k different sets of partner functions. Because of the invariance of the trace under similarity transformation, the operator P_k holds for all the equivalent representations corresponding to the kth irreducible representation.

EXAMPLE 3.16 Use the projection and transfer operators to obtain bases for the irreducible representations of $\mathbf{C}_{3v}$ from the function

$$F(x, y, z) = ax + by + cz + d,$$

where a, b, c, and d are constants.

SOLUTION The representations are given in (3.82). The projection operators are

$$\begin{aligned}
P_{A_1} &= \tfrac{1}{6}[E + C_3 + C_3^2 + \sigma_v + \sigma_v' + \sigma_v''], \\
P_{A_2} &= \tfrac{1}{6}[E + C_3 + C_3^2 - \sigma_v - \sigma_v' - \sigma_v''], \\
P_{E,11} &= \tfrac{2}{6}[E - \tfrac{1}{2}C_3 - \tfrac{1}{2}C_3^2 - \sigma_v + \tfrac{1}{2}\sigma_v' + \tfrac{1}{2}\sigma_v''], \\
P_{E,22} &= \tfrac{2}{6}[E - \tfrac{1}{2}C_3 - \tfrac{1}{2}C_3^2 + \sigma_v - \tfrac{1}{2}\sigma_v' - \tfrac{1}{2}\sigma_v''], \\
P_E &= \tfrac{2}{6}[2E - C_3 - C_3^2 + 0\sigma_v + 0\sigma_v' + 0\sigma_v''] \\
&= \tfrac{1}{3}[2E - C_3 - C_3^2].
\end{aligned} \tag{3.110}$$

[4] We have used the set of partner functions as a column vector. If we had considered the set to be a row vector, as is often done, then the roles of i and j would have been interchanged.

The transfer operators are

$$T_{E,12} = \frac{2}{6}\left[0\mathrm{E} + \frac{\sqrt{3}}{2}C_3 - \frac{\sqrt{3}}{2}C_3^2 + 0\sigma_v - \frac{\sqrt{3}}{2}\sigma_v' + \frac{\sqrt{3}}{2}\sigma_v''\right]$$

$$= \frac{1}{\sqrt{3}}[C_3 - C_3^2 - \sigma_v' + \sigma_v''], \tag{3.111}$$

$$T_{E,21} = \frac{1}{\sqrt{3}}[-C_3 + C_3^2 - \sigma_v' + \sigma_v''].$$

We have, by use of the matrices of the representations as the operators,

$$\begin{aligned}
EF &= ax + by + cz + d,\\
C_3F &= (-a/2 - (\sqrt{3}/2)b)x + ((\sqrt{3}/2)a - b/2)y + cz + d,\\
C_3^2F &= (-a/2 + (\sqrt{3}/2)b)x + (-(\sqrt{3}/2)a - b/2)y + cz + d,\\
\sigma_vF &= -ax + by + cz + d,\\
\sigma_v'F &= (a/2 - (\sqrt{3}/2)b)x + (-(\sqrt{3}/2)a - b/2)y + cz + d,\\
\sigma_v''F &= (a/2 + (\sqrt{3}/2)b)x + ((\sqrt{3}/2)a - b/2)y + cz + d.
\end{aligned} \tag{3.112}$$

For the basis of A_1, summing for all operations, we have

$$P_{A_1}F = \tfrac{1}{6}[6cz + 6d] = cz + d.$$

The function F contains no basis for A_2, because

$$P_{A_2}E = \tfrac{1}{6}[0] = 0.$$

For the basis of E we can start with $P_{E,11}$,

$$P_{E,11}F = \tfrac{2}{6}[E - \tfrac{1}{2}C_3 - \tfrac{1}{2}C_3^2 - \sigma_v + \tfrac{1}{2}\sigma_v' + \tfrac{1}{2}\sigma_v'']F = ax. \tag{3.113}$$

This is one of the partner functions. We can get the second one by using the operator $T_{E,12}$

$$T_{E,12}ax = \frac{2}{6}\left[\frac{\sqrt{3}}{2}C_3 - \frac{\sqrt{3}}{2}C_3^2 - \frac{\sqrt{3}}{2}\sigma_v' + \frac{\sqrt{3}}{2}\sigma_v''\right]ax = ay. \tag{3.114}$$

If we use the operators $P_{E,22}$ and $T_{E,21}$, we obtain the pair bx and by (see Problem 3.8). When a and b are arbitrary constants this pair is equivalent to ax and ay. With $a = b = 1$, we have agreement with the linear pair (x, y) given in the character table.

If we apply the trace projection operator P_E, we obtain

$$P_EF = \tfrac{1}{3}(3ax + 3by) = ax + by,$$

which is a linear combination of functions—one from the $P_{E,11}$ set and one from the $P_{E,22}$ set.

When the matrices of an irreducible representation (and therefore the transfer operators $T_{k,ij}$) are not available, it is still possible to obtain basis functions. Also, in many applications it is convenient to use shortcuts that do not involve the transfer operators, even when these operators are available. The shortcuts are often simple but require some judgment based upon experience or intuition. It is best to describe them by example, and we shall do this in subsequent applications (Chapter 6).

3.10.6 THE TABULATED BASIS FUNCTIONS

The character tables give linear, quadratic, and cubic Cartesian basis functions. These apply for Cartesian bases with the z axis in the direction of the principal rotation axis.

The rotations R_x, R_y, and R_z are linear bases, but their transformation properties are different from those of ordinary coordinates. Their behavior under the symmetry operations is easy to visualize. We picture the rotation as a circular arrow about the axis of rotation and imagine what happens to the arrow when the symmetry operation is performed. Thus, R_z is invariant under any rotation about the z axis; it changes sign under any perpendicular two-fold rotation or any vertical reflection; R_x is transformed to $-R_y$ under a clockwise $90°$ rotation about the z axis; R_x, R_y, and R_z are all invariant under inversion (see Problem 3.17).

EXAMPLE 3.17 One of the four sp^3 hybrid orbitals is

$$\Psi = \tfrac{1}{2}(s + p_x + p_y + p_z).$$

Use the character table for $\mathbf{T}_d$ to obtain the parts of Ψ^2 that transform according to the various irreducible representations of that group. (The s orbital is invariant; p_x, p_y, and p_z transform as x, y, and z, respectively.)

SOLUTION We have

$$\Psi^2 = \tfrac{1}{4}(s^2 + p_x^2 + p_y^2 + p_z^2 + 2sp_x + 2sp_y + 2sp_z + 2p_xp_y + 2p_xp_z + 2p_yp_z).$$

We use the fact that s, s^2 are invariant; sp_x, sp_y, sp_z transform as x, y, z,; p_x^2, p_y^2, p_z^2 transform as x^2, y^2, z^2; p_xp_y, p_xp_z, p_yp_z transform as xy, xz, yz. It follows from the character table that the components are as follows:

A_1: $\tfrac{1}{4}(s^2 + p_x^2 + p_y^2 + p_z^2)$

A_2: none

E: none

T_1: none

T_2: $\tfrac{1}{2}[s(p_x + p_y + p_z) + (p_xp_y + p_xp_z + p_yp_z)]$

3.11 The C_n Groups

The rotation through an angle $2\pi/n$ about a fixed axis is an element of a cyclic group of order n with the elements $C_n, C_n^2, \ldots, C_n^{n-1}, C_n^n\,(=E)$. Since all rotations about a fixed axis commute, each element is in a class by itself. The group has n classes and n one-dimensional irreducible representations. If U is a basis for one of these representations and ε is the character of C_n in that representation, then we have

$$\begin{aligned} C_n U &= \varepsilon U, \\ C_n^2 U &= \varepsilon C_n U = \varepsilon^2 U, \\ &\vdots \\ C_n^{n-1} U &= \varepsilon^{n-1} U, \\ C_n^n U &= \varepsilon^n U = EU = U. \end{aligned} \tag{3.115}$$

Therefore,

$$\varepsilon^n = 1 \tag{3.116}$$

and so ε is one of the n nth roots of unity. The roots of unity are given by

$$\exp(ik2\pi/n) = \cos\frac{k2\pi}{n} + i\sin\frac{k2\pi}{n}, \qquad k = 0, 1, \ldots, n-1. \tag{3.117}$$

If we take

$$\varepsilon = \exp(i2\pi/n), \tag{3.118}$$

then the n roots are given by $1, \varepsilon, \varepsilon^2, \ldots, \varepsilon^{n-1}$. But they are also given, in some order, by $\varepsilon^s, \varepsilon^{s+1}, \varepsilon^{s+2}, \ldots, \varepsilon^{s+n-1}$ and also by $\varepsilon^s, (\varepsilon^s)^2, (\varepsilon^s)^{n-1}$, where s is any integer. It is clear from this that the n irreducible representations [which we designate here Γ_k $(k = 0, 1, \ldots, n-1)$] and their characters can be expressed as in Table 3.2, where $\varepsilon = \exp(i2\pi/n)$. Because of Eq. (3.116) these expressions can be simplified so that no exponent of ε is greater than $n-1$. Also, modification is possible because

$$(\varepsilon^k)^* = \varepsilon^{n-k}, \tag{3.119}$$

which allows us to group together in pairs the representations Γ_k and Γ_{n-k}, whose characters are just the complex conjugate of one another. For even n, there is a representation $\Gamma_{n/2}$ that is real and so, like the identity representation, does not occur as one of a pair. The final forms are illustrated by the cases $\mathbf{C}_5$ and $\mathbf{C}_6$, Tables 3.3 and 3.4.

TABLE 3.2
The Irreducible Representations and Their Characters of C

C_n	E	C_n	C_n^2	C_n^3	$\cdots$	C_n^{n-1}
Γ_0	1	1	1	1	$\cdots$	1
Γ_1	1	ε	ε^2	ε^3	$\cdots$	ε^{n-1}
Γ_2	1	ε^2	$(\varepsilon^2)^2$	$(\varepsilon^2)^3$	$\cdots$	$(\varepsilon^2)^{n-1}$
$\vdots$	$\vdots$					
Γ_{n-1}	1	ε^{n-1}	$(\varepsilon^{n-1})^2$	$(\varepsilon^{n-1})^3$	$\cdots$	$(\varepsilon^{n-1})^{n-1}$

TABLE 3.3
The C_5 Case [$\varepsilon = \exp(i2\pi/5)$]

C_5	E	C_5	C_5^2	C_5^3	C_5^4
Γ_0 A	1	1	1	1	1
Γ_1 E_1	1	ε	ε^2	$(\varepsilon^2)^*$	ε^*
Γ_4	1	ε^*	$(\varepsilon^2)^*$	ε^2	ε
Γ_2 E_2	1	ε^2	ε^*	ε	$(\varepsilon^2)^*$
Γ_3	1	$(\varepsilon^2)^*$	ε	ε^*	ε^2

TABLE 3.4
The C_6 Case [$\varepsilon = \exp(i2\pi/6)$]

C_6	E	C_6	C_3	C_2	C_3^2	C_6^5
Γ_0 A	1	1	1	1	1	1
Γ_3 B	1	-1	1	-1	1	-1
Γ_1 E_1	1	ε	$-\varepsilon^*$	-1	$-\varepsilon$	ε^*
Γ_5	1	ε^*	$-\varepsilon$	-1	$-\varepsilon^*$	ε
Γ_2 E_2	1	$-\varepsilon^*$	$-\varepsilon$	1	$-\varepsilon^*$	$-\varepsilon$
Γ_4	1	$-\varepsilon$	$-\varepsilon^*$	1	$-\varepsilon$	$-\varepsilon^*$

It is necessary to explain why the pairs Γ_k and Γ_{n-k} are classified as two-dimensional representations. This is because the quantum mechanical states that form the basis for these pairs are degenerate. When there is no *applied* magnetic field present, Hamiltonians are invariant under *time reversal*, and if the time-reversal operator is included, it interchanges the states belonging to Γ_k and Γ_{n-k}. Therefore, these representations (in this larger group) are no longer one-dimensional. It is then legitimate for quantum mechanical applications to consider the representation $\Gamma_k + \Gamma_{n-k}$ to be a two-dimensional irreducible representation. Its character will be equal to the sum of the characters of Γ_k and Γ_{n-k}, and they will be real.[5] Some authors give the tables in this combined form. For example, the table for $\mathbf{C}_5$ could be as follows (see Problem 3.11).

$\mathbf{C}_5$	E	C_5	C_5^2	C_5^3	C_5^4
A	1	1	1	1	1
E_1	2	$2\cos(2\pi/5)$	$2\cos(4\pi/5)$	$2\cos(4\pi/5)$	$2\cos(2\pi/5)$
E_2	2	$2\cos(4\pi/5)$	$2\cos(2\pi/5)$	$2\cos(2\pi/5)$	$2\cos(4\pi/5)$

(3.120)

It is not difficult to obtain basis functions for C_n. Consider the function

$$U_k(\theta) = \exp[-ik\theta], \qquad k = 0, \pm 1, \ldots, \pm(n-1), \tag{3.121}$$

where θ is the angle (measured in the counterclockwise direction) about the rotation axis. The function $U_k(\theta)$ is a basis function for an irreducible representation of C_n, since we have (for the clockwise direction of rotation)

$$C_n^q U_k(\theta) = \exp[-ik(\theta - 2\pi q/n)] = \exp[iqk2\pi/n]\, U_k(\theta) \tag{3.122}$$

and $\exp[iqk2\pi/n]$ is the character of C_n^q in Γ_k when k is positive and in $\Gamma_{|k|}^*$ $(=\Gamma_{n-|k|})$ when k is negative. The restriction of k to integers is required by the condition that $C_n^n U(\theta) = U(\theta)$.

3.12 Some Other Finite Point Groups

In this section we consider briefly some of the finite groups, giving particular attention to the ways in which the irreducible representations and characters of one group can be related to those of another.

The groups $\mathbf{C}_{nv}$ and $\mathbf{D}_n$ are isomorphic, and they therefore have equivalent irreducible representations and the same characters. (For a list of isomorphism of point groups, see Section 2.7.3.) It follows that the characters we obtained for $\mathbf{C}_{3v}$ in Section 3.8 also apply to $\mathbf{D}_3$. The class structure of $\mathbf{C}_{nv}$ and $\mathbf{D}_n$ depends upon whether n is even or odd. When n is odd the number of classes, and therefore the number of irreducible representations, is $\frac{1}{2}(n+3)$. When n is even this number is $\frac{1}{2}(n+6)$. The general forms of the character tables for $\mathbf{C}_{nv}$ and $\mathbf{D}_n$ are represented in Tables 3.5 and 3.6.

The groups $\mathbf{C}_{nh}$ are the direct product of the group $\mathbf{C}_s$ and the groups $\mathbf{C}_n$. The character matrix of $\mathbf{C}_{nh}$ is the direct product of the character matrices of

[5] This is because $\varepsilon^k + (\varepsilon^k)^* = 2\cos(2k\pi/n)$. See Problem 3.11.

TABLE 3.5
C_{nv} and D_n Character Table, n Odd[a]

C_{nv} (D_n)	E	$2C_n$	$2C_n^2$	$\cdots$	$2C_n^q$	$n\sigma_v$ (nC_2)
Γ_1	1	1	1	$\cdots$	1	1
Γ_2	1	1	1	$\cdots$	1	-1
Γ_3	2	$2\cos\alpha$	$2\cos 2\alpha$	$\cdots$	$2\cos q\alpha$	0
Γ_4	2	$2\cos 2\alpha$	$2\cos 4\alpha$	$\cdots$	$2\cos 2q\alpha$	0
$\vdots$	$\vdots$	$\vdots$	$\vdots$		$\vdots$	$\vdots$
Γ_{q+2}	2	$2\cos q\alpha$	$2\cos 2q\alpha$	$\cdots$	$2\cos q^2\alpha$	0

[a] $\alpha = 2\pi/n$, $q = (n-1)/2$.

TABLE 3.6
C_{nv} and D_n Character Table, n Even[a]

C_{nv} (D_n)	E	$2C_n$	$2C_n^2$	$\cdots$	$C_n^q = C_2$	$\frac{n}{2}\sigma_v'\left(\frac{n}{2}C_2'\right)$	$\frac{n}{2}\sigma_v''\left(\frac{n}{2}C_2''\right)$
Γ_1	1	1	1	$\cdots$	1	1	1
Γ_2	1	1	1	$\cdots$	1	-1	-1
Γ_3	1	-1	1	$\cdots$	$(-1)^q$	1	-1
Γ_4	1	-1	1	$\cdots$	$(-1)^q$	-1	1
Γ_5	2	$2\cos\alpha$	$2\cos 2\alpha$	$\cdots$	$2\cos q\alpha$	0	0
Γ_6	2	$2\cos 2\alpha$	$2\cos 4\alpha$	$\cdots$	$2\cos 2q\alpha$	0	0
$\vdots$	$\vdots$	$\vdots$	$\vdots$		$\vdots$	$\vdots$	$\vdots$
Γ_{q+3}	2	$2\cos(q-1)\alpha$	$2\cos 2(q-1)\alpha$	$\cdots$	$2\cos q(q-1)\alpha$	0	0

[a] $\alpha = 2\pi/n$, $q = n/2$.

$\mathbf{C}_s$ and $\mathbf{C}_n$. For example, the character matrix of $\mathbf{C}_{3h}$ is

$$\begin{matrix} E & \sigma_h \\ \end{matrix} \quad \begin{matrix} E & C_3 & C_3^2 \end{matrix} \quad \begin{matrix} E & C_3 & C_3^2 & \sigma_h & S_3 & S_3^5 \end{matrix}$$

$$\begin{bmatrix} 1 & 1 \\ 1 & -1 \end{bmatrix} \times \begin{bmatrix} 1 & 1 & 1 \\ 1 & \varepsilon & \varepsilon^* \\ 1 & \varepsilon^* & \varepsilon \end{bmatrix} = \begin{bmatrix} 1 & 1 & 1 & 1 & 1 & 1 \\ 1 & \varepsilon & \varepsilon^* & 1 & \varepsilon & \varepsilon^* \\ 1 & \varepsilon^* & \varepsilon & 1 & \varepsilon^* & \varepsilon \\ 1 & 1 & 1 & -1 & -1 & -1 \\ 1 & \varepsilon & \varepsilon^* & -1 & -\varepsilon & -\varepsilon^* \\ 1 & \varepsilon^* & \varepsilon & -1 & -\varepsilon^* & -\varepsilon \end{bmatrix}. \tag{3.123}$$

The reader should compare this with the character table for $\mathbf{C}_{3h}$. Those point groups that are direct products are indicated in Table 2.1. In all of these direct products one of the groups is either $\mathbf{C}_s$ or $\mathbf{C}_i$, and these two groups are isomorphic.

3.13 Groups of Infinite Order[6]

3.13.1 THE GROUPS $\mathbf{C}_\infty$, $\mathbf{C}_{\infty v}$, $\mathbf{C}_{\infty h}$, $\mathbf{D}_\infty$, AND $\mathbf{D}_{\infty h}$

Of the groups $\mathbf{C}_\infty$, $\mathbf{C}_{\infty v}$, $\mathbf{C}_{\infty h}$, $\mathbf{D}_\infty$, and $\mathbf{D}_{\infty h}$, only $\mathbf{C}_{\infty v}$ and $\mathbf{D}_{\infty h}$ (the symmetries of linear molecules) have direct applications in molecular symmetry. $\mathbf{C}_{\infty v}$ is also the symmetry of an applied electric field. $\mathbf{C}_{\infty h}$ is the symmetry of an applied magnetic field. $\mathbf{C}_\infty$ and $\mathbf{D}_\infty$ are of interest to us because of their subgroup relationships to the other groups.

The group $\mathbf{C}_\infty$ consists of all rotations about a fixed axis (including infinitesimal rotations). Since all rotations about a fixed axis commute, $\mathbf{C}_\infty$ is Abelian and has only one-dimensional irreducible representations. $\mathbf{C}_\infty$ is of infinite order and has an infinite number of irreducible representations. We can take as bases the functions

$$U_k(\theta) = \exp(-ik\theta), \qquad k = 0, \pm 1, \pm 2, \ldots, \pm\infty, \tag{3.124}$$

If C_∞^ϕ represents a rotation through an angle ϕ, then we have

$$C_\infty^\phi U_k(\theta) = \exp|-ik(\theta - \phi)| = \exp(ik\phi)U_k(\theta), \tag{3.125}$$

and it follows that the character of C_∞^ϕ in the kth representation is $\exp(ik\phi)$. The positive and negative values of any k give representations that are complex conjugates of one another. Time-reversal invariance allows us to combine these two representations into one two-dimensional representation, as we do for the groups $\mathbf{C}_n$.

The group $\mathbf{C}_{\infty v}$ is obtained from $\mathbf{C}_\infty$ by adding an operation σ_v as generator, which then generates an infinite number of such operations. All of the operations σ_v belong to the same class, since there is a rotation that will transform any vertical reflection into any other. $\mathbf{C}_{\infty v}$ is not Abelian. The operators σ_v do not commute with rotations. The operators σ_v mix the basis function U_k with the basis function U_{-k} $(= U_k^*)$. This gives a two-dimensional representation E_k with character $2\cos k\phi$ for the operator C_∞^ϕ and a character zero for the operators σ_v (see Problem 3.10).

There is a one-dimensional representation in which the σ_v operators have character -1, and all the other characters are $+1$. As indicated in the

[6] For infinite groups, equations involving summation over the group elements [e.g., Eq. (3.79)] must be modified. We shall not discuss this here, but we note that for some problems it is possible to use a finite subgroup in place of the infinite group. See D. P. Strommen and E. R. Lippincott, Comments on Infinite Point Groups, *J. Chem. Educ.*, 1972, **47**, 341. An illustration of the use of this method is given in Example 6.5. For an elementary discussion of Projection operators in infinite groups, see H. H. Jaffé and S. J. David, Projection Operators in Continuous Groups, *J. Chem. Educ.*, 1984, **61**, 503.

character table, a basis for this representation is rotation about the z axis, R_z. The character table shows that there are two elements in each class C_∞^ϕ in $\mathbf{C}_{\infty v}$. These two elements are C_∞^ϕ and its inverse $C_\infty^{-\phi}$. This is because there is a σ_v that will transform any C_∞^φ into its inverse.

If a horizontal plane of symmetry is added to the elements of $\mathbf{C}_\infty$, we obtain $\mathbf{C}_{\infty h}$. Since σ_h commutes with all of the rotations, the group is Abelian and has only one-dimensional irreducible representations, except for the doubling allowed by time inversion. $\mathbf{C}_{\infty h}$ is the direct product $\mathbf{C}_\infty \times \mathbf{C}_s$, and its irreducible representation can be obtained directly from those of $\mathbf{C}_\infty$ and $\mathbf{C}_s$.

If a C_2 axis perpendicular to the C_∞ axis is added to $\mathbf{C}_\infty$, we obtain $\mathbf{D}_\infty$. It is not the direct product of $\mathbf{C}_\infty$ and $\mathbf{C}_2$, because these operations do not all commute. Note that C_∞^ϕ and $C_\infty^{-\phi}$ will be in the same class because there is a C_2 operation that will transform C_∞^ϕ into $C_\infty^{-\phi}$.

$\mathbf{D}_{\infty h}$ is the direct product $\mathbf{D}_\infty \times \mathbf{C}_i$. Its irreducible representations can be obtained directly from those of $\mathbf{D}_\infty$ and $\mathbf{C}_i$.

3.13.2 THE GROUPS[7] $\mathbf{K}_h$ AND $\mathbf{K}$

The group corresponding to the symmetry of the sphere (and, therefore, of the isolated atom) is the rotation–inversion group $\mathbf{K}_h$, consisting of all proper and improper rotations. $\mathbf{K}$ is the subgroup of $\mathbf{K}_h$ that consists of all proper rotations. $\mathbf{K}_h$ is the direct product $\mathbf{K} \times \mathbf{C}_i$. *Parity* has meaning in $\mathbf{K}_h$ but not in $\mathbf{K}$. Functions that are unchanged by inversion have *even* parity, and those that change sign have *odd* parity. The representations of the *finite* subgroups of $\mathbf{K}_h$ that have even-parity basis functions have the subscript g attached to their symbol, and those with odd-parity basis functions have the subscript u. For the representations of $\mathbf{K}_h$ itself, the conventional symbols for even and odd parity are + and −, often as subscripts, but we shall use them as superscripts (see below). For the representations of $\mathbf{K}$ (and its subgroups), no parity designation is needed, since parity has no meaning there.

The spherical harmonics $Y_{lm}(\theta\phi)$ $(m = -l, \ldots, l)$ (see Table 3.7) are the angular parts of the eigenstates of the one-electron atom (neglecting spin), and they are also basis functions for the irreducible representation of dimension $2l + 1$ of $\mathbf{K}_h$. The symbols $\mathscr{D}^l$ are used for the irreducible representations when the basis functions are a standard form of the spherical harmonics. The parity is even when l is even and odd when l is odd, and so in this case the $\pm$ superscripts are not required.

Many-electron atomic functions (spectral terms) are also basis functions for the irreducible representations of $\mathbf{K}_h$. If spin is neglected, they are designated

[7] Other symbols for these two groups are $\theta(3)$ and $\theta^+(3)$; $R(3)$ and $R^+(3)$; or $R_h(3)$ and $R(3)$. The number (3) indicates that these are for three-dimensional physical space.

TABLE 3.7
Some Spherical Harmonics

Symbol	Polar	Cartesian	Normalization Constant
Y_{00}	1	1	$\frac{1}{2}(1/\pi)^{1/2}$
Y_{10}	$\cos\theta$	z/r	$\frac{1}{2}(3/\pi)^{1/2}$
$Y_{1\pm1}$	$\mp(\sin\theta)e^{\pm i\phi}$	$\mp(x \pm iy)/r$	$\frac{1}{2}(3/2\pi)^{1/2}$
Y_{20}	$(3\cos^2\theta - 1)$	$(3z^2 - r^2)/r^2$	$\frac{1}{4}(5/\pi)^{1/2}$
$Y_{2\pm1}$	$\mp(\sin\theta)(\cos\theta)e^{\pm i\phi}$	$\mp z(x \pm iy)r^2$	$\frac{1}{2}(15/2\pi)^{1/2}$
$Y_{2\pm2}$	$(\sin^2\theta)e^{\pm 2i\phi}$	$(x \pm iy)^2/r^2$	$\frac{1}{4}(15/2\pi)^{1/2}$
Y_{30}	$(5\cos^3\theta - 3\cos\theta)$	$z(5z^2 - 3r^2)/r^3$	$\frac{1}{4}(7/\pi)^{1/2}$
$Y_{3\pm1}$	$\mp\sin\theta(5\cos^2\theta - 1)e^{\pm i\phi}$	$\mp(x \pm iy)(5z^2 - r^2)/r^3$	$\frac{1}{8}(21/\pi)^{1/2}$
$Y_{3\pm2}$	$(\sin^2\theta)(\cos\theta)e^{\pm 2i\phi}$	$z(x \pm iy)^2/r^3$	$\frac{1}{4}(105/2\pi)^{1/2}$
$Y_{3\pm3}$	$\mp(\sin^3\theta)e^{\pm 3i\phi}$	$\mp(x \pm iy)^3/r^3$	$\frac{1}{8}(35/\pi)^{1/2}$

$\mathscr{D}^{L\pm}$ for the $(2L + 1)$-dimensional representation. The parity symbol $\pm$ is needed because the parity is *not* determined by whether L is even or odd. When electron spin is taken into account, the irreducible representation of dimension $2J + 1$ is designated $\mathscr{D}^{J\pm}$. In this case J can take on half-integral values as well as integral values; the corresponding irreducible representations do not belong to $\mathbf{K}_h$, but they are the so-called double-valued representations (see Section 9.9).

All proper rotations through a given angle ω belong to the same class, since there is a rotation in the group ($\mathbf{K}_h$ or $\mathbf{K}$) that will take any rotation axis into any other rotation axis. We can therefore calculate the characters for all proper rotations through an angle ω by considering the rotation about the z axis. We obtain (see Problem 3.12)

$$\begin{bmatrix} e^{-il\omega} & 0 & \cdots & 0 \\ 0 & e^{-i(l-1)\omega} & \cdots & \\ \vdots & & \ddots & \vdots \\ 0 & & & e^{il\omega} \end{bmatrix} \tag{3.126}$$

for the matrix representing a rotation through an angle ω about the z axis. The character is (see Problem 3.13)

$$\chi^{2l+1}(\omega) = \sum_{m=-l}^{l} e^{im\omega} = \frac{\sin[(l + \frac{1}{2})\omega]}{\sin(\frac{1}{2})\omega}. \tag{3.127}$$

Therefore, this is the character for all proper rotations through the angle ω. For $\mathscr{D}^{J-}(\omega)$, the character for the improper rotation is the negative of that given by Eq. (3.127)

Since all point groups are subgroups of $\mathbf{K}_h$, the irreducible representations $\mathscr{D}^{J+}$ and $\mathscr{D}^{J-}$ correspond to representations (often reducible) of the point groups. Both $\mathscr{D}^{J+}$ and $\mathscr{D}^{J-}$ correspond to $\mathscr{D}^{J}$ of K. The resolutions of $\mathscr{D}^{J\pm}$ in the point groups (and the resolutions of the point groups in their subgroups) have been tabulated.[8] Such tables are called *correlation* (or compatibility) tables. For finite subgroups, the resolutions can be obtained by the use of Eq. (3.79), as illustrated by Example 3.18. (See also Section 5.2.)

EXAMPLE 3.18 By the use of Eqs. (3.127) and (3.79), obtain the resolutions of (a) $\mathscr{D}^{2+}$ and $\mathscr{D}^{2-}$ in $\mathbf{C}_{2v}$, (b) $\mathscr{D}^{2+}$ and $\mathscr{D}^{2-}$ in $\mathbf{D}_2$, (c) $\mathscr{D}^{3+}$ and $\mathscr{D}^{3-}$ in $\mathbf{C}_{2v}$.

SOLUTION (a) For $\mathscr{D}^{2+}$ and $\mathscr{D}^{2-}$ we have from Eq. (3.127)

$$\chi(C_2) = \frac{\sin(5\pi/2)}{\sin(\pi/2)} = 1.$$

Since σ_v and σ_v' are improper two-fold rotations [i.e., $\sigma(xz) = iC_2(y)$ and $\sigma_v(yz) = iC_2(x)$], we have

$$\chi(\sigma_v) = \pm 1,$$

where the plus sign applies to $\mathscr{D}^{2+}$ and the minus sign to $\mathscr{D}^{2-}$. The characters for $\mathscr{D}^{2+}$ for the operations in $\mathbf{C}_{2v}$ are therefore as follows.

	$\chi(E)$	$\chi(C_2)$	$\chi(\sigma_v)$	$\chi(\sigma_v')$
$\mathscr{D}^{2+}$	5	1	1	1

Note that the value 5 for $\chi(E)$ follows directly from the dimensionality of the representation. This value can also be obtained by the use of Eq. (3.127), but not directly since

$$\frac{\sin[(l + (\frac{1}{2}))0]}{\sin(0/2)} = \frac{0}{0}.$$

However, we can use l'Hospital's rule as follows:

$$\lim_{x \to 0} \frac{\sin[2l + 1)x]}{\sin(x/2)} = \lim_{x \to 0} \left[\frac{d\{\sin(l + 1/2)x|\}}{dx} \bigg/ \frac{d\{\sin(x/2)\}}{dx} \right]$$

$$= \lim_{x \to 0} \frac{(2l + 1)\cos[(l + 1/2)x]}{\cos(x/2)}$$

$$= 2l + 1.$$

[8] G. F. Koster, J. O. Dimmock, R. G. Wheeler, and H. Statz, *Properties of the Thirty-Two Point Groups*, MIT Press, Cambridge, Massachusetts, 1963. E. B. Wilson, Jr., J. C. Decius, and P. C. Cross, *Molecular Vibrations*, McGraw-Hill, New York, 1955. J. A. Salthouse and M. J. Ware, *Point Group Character Tables and Related Data*, Cambridge Univ. Press, London and New York, 1972.

The use of Eq. (3.79) (or inspection) and the characters of $\mathbf{C}_{2v}$ give for $\mathscr{D}^{2+}$

$$a_{A_1} = \tfrac{1}{4}(5 + 1 + 1 + 1) = 2,$$
$$a_{A_2} = \tfrac{1}{4}(5 + 1 - 1 - 1) = 1,$$
$$a_{B_1} = \tfrac{1}{4}(5 - 1 + 1 - 1) = 1,$$
$$a_{B_2} = \tfrac{1}{4}(5 - 1 - 1 + 1) = 1.$$

Therefore, $\mathscr{D}^{2+} = 2A_1 + A_2 + B_1 + B_2$. For $\mathscr{D}^{2-}$ we have the following characters.

	$\chi(E)$	$\chi(C_2)$	$\chi(\sigma_v)$	$\chi\sigma'_v$
$\mathscr{D}^{2-}$	5	1	-1	-1

From these we obtain $\mathscr{D}^{2-} = A_1 + 2A_2 + B_1 + B_2$.

(b) The characters are as follows.

	E	$C_2(3)$	$C_2(y)$	$C_2(z)$
$\mathscr{D}^{2+}$	5	1	1	1
$\mathscr{D}^{2-}$	5	1	1	1

With the characters of $\mathbf{D}_2$ and Eq. (3.79), these give $\mathscr{D}^{2\pm} = 2A + B_1 + B_2 + B_3$. The fact that $\mathbf{D}_2$ has no improper rotations means that parity has no significance. Therefore, $\mathscr{D}^{2+}$ and $\mathscr{D}^{2-}$ of $\mathbf{K}_h$ have the same decomposition as $\mathscr{D}^2$ of $\mathbf{K}$.

(c) Equation (3.127) leads to

$$\chi(C_2) = \frac{\sin(7\pi/2)}{\sin(\pi/2)} = -1$$

for both $\mathscr{D}^{3+}$ and $\mathscr{D}^{3-}$. For the reflections, we obtain $\chi(\sigma_v) = -1$ for $\mathscr{D}^{3+}$ and $+1$ for $\mathscr{D}^{3-}$. The characters are therefore as follows.

	E	C_2	σ_v	σ'_v
$\mathscr{D}^{3+}$	7	-1	-1	-1
$\mathscr{D}^{3-}$	7	-1	1	1

The final results are

$$\mathscr{D}^{3+} = A_1 + 2A_2 + 2B_1 + 2B_2, \qquad \mathscr{D}^{3-} = 2A_1 + A_2 + 2B_1 + 2B_2.$$

REMARKS CONCERNING EXAMPLE 3.18 It does not matter in this example whether $\mathscr{D}^{2\pm}$ refers to $\mathscr{D}^{L\pm}$ or $\mathscr{D}^{J\pm}$. However, when J is half-integral, the analogous problem requires the use of the double-valued representations of the point groups (see Section 9.9).

3.14 Space Groups

With any lattice point as origin, it is possible to choose three independent (but not necessarily orthogonal) vectors τ_1, τ_2, and τ_3 and a set of coordinate axes in these directions so that any lattice point can be represented by three integers n_1, n_2, and n_3.[9] Also, any *primitive* lattice translation $\mathbf{t}$ can be expressed as

$$\mathbf{t} = n_1\tau_1 + n_2\tau_2 + n_3\tau_3. \tag{3.128}$$

The combination of two primitive translations $\mathbf{t}'$ and $\mathbf{t}''$ can be expressed as

$$\mathbf{t} = \mathbf{t}' + \mathbf{t}'' = (n_1' + n_1'')\tau_1 + (n_2' + n_2'')\tau_2 + (n_3' + n_3'')\tau_3. \tag{3.129}$$

The inverse of $\mathbf{t}$ is simply $-\mathbf{t}$. It is clear that the primitive translations form an Abelian subgroup of any space group.

Let us now consider the rotational symmetry operations of the lattice. The position of a lattice point can be represented by the vector (n_1, n_2, n_3). The effect of a rotation can be expressed by a rotation matrix $\boldsymbol{\Theta}$,

$$\mathbf{n}' = \boldsymbol{\Theta}\mathbf{n}. \tag{3.130}$$

Written out, Eq. (3.130) is

$$\begin{aligned} n_1' &= \theta_{11}n_1 + \theta_{12}n_2 + \theta_{13}n_3, \\ n_2' &= \theta_{21}n_1 + \theta_{22}n_2 + \theta_{23}n_3, \\ n_3' &= \theta_{31}n_1 + \theta_{32}n_2 + \theta_{33}n_3. \end{aligned} \tag{3.131}$$

Because Eq. (3.131) must hold for all integers n_1, n_2, and n_3 and because n_1', n_2', and n_3' must be integers, it follows that the matrix elements θ_{ij} must all be integers.

In particular,

$$\operatorname{tr}\boldsymbol{\Theta} = a_{11} + a_{22} + a_{33} = \pm\text{integer}. \tag{3.132}$$

But for any rotation (see Problem 3.5),

$$\operatorname{tr}\boldsymbol{\Theta} = 1 + 2\cos\theta, \tag{3.133}$$

where θ is the angle of rotation. It follows that

$$1 + 2\cos\theta = \pm\text{integer}. \tag{3.134}$$

[9] It is to be understood throughout this section that these integers can take on (independently) either positive or negative values.

The only possible values of $\cos\theta$ that can satisfy Eq. (3.134) are ± 1, $\pm\frac{1}{2}$, 0, and the possible values of θ are 60, 90, 120, 180, 240, 270, and 300°. Thus the crystallographic point groups are restricted to proper or improper rotations through these angles, which are integral multiples of 60 and 90°. By a similar but somewhat more complicated process, it can be shown that any rotation axis must be parallel to a primitive translation (see Problem 3.16).

A space-group operation can be represented by the symbol $[\boldsymbol{\Theta}, \mathbf{t}]$, which means a rotation $\boldsymbol{\Theta}$ *followed* by a translation $\mathbf{t}$. Its effect on a lattice site at position $\mathbf{r}$ is given by

$$[\boldsymbol{\Theta}, \mathbf{t}]\mathbf{r} = \boldsymbol{\Theta}\mathbf{r} + \mathbf{t}. \tag{3.135}$$

A pure rotation is represented by $[\boldsymbol{\Theta}, \mathbf{0}]$, and a pure translation is represented by $[\mathbf{I}, \mathbf{t}]$, where $\mathbf{0}$ is the zero vector and $\mathbf{I}$ is the identity matrix. An expression for the combination of two space-group operations can be found as follows:

$$\begin{aligned}
[\boldsymbol{\Theta}_2, \mathbf{t}_2][\boldsymbol{\Theta}_1, \mathbf{t}_1]r &= [\boldsymbol{\Theta}_2, \mathbf{t}_2](\boldsymbol{\Theta}_1 \mathbf{r} + \mathbf{t}_1) \\
&= \boldsymbol{\Theta}_2(\boldsymbol{\Theta}_1 \mathbf{r} + \mathbf{t}_1) + \mathbf{t}_2 \\
&= \boldsymbol{\Theta}_2\boldsymbol{\Theta}_1\mathbf{r} + \boldsymbol{\Theta}_2\mathbf{t}_1 + \mathbf{t}_2 \\
&= [\boldsymbol{\Theta}_2\boldsymbol{\Theta}_1, \boldsymbol{\Theta}_2\mathbf{t}_1 + \mathbf{t}_2]\mathbf{r}.
\end{aligned} \tag{3.136}$$

Thus, the combination is a rotation $\boldsymbol{\Theta}_2\boldsymbol{\Theta}_1$ followed by a translation $\boldsymbol{\Theta}_2\mathbf{t}_1 + \mathbf{t}_2$. From Eq. (3.136) it follows that the inverse of $[\boldsymbol{\Theta}, \mathbf{t}]$ is given by (see Problem 3.15)

$$[\boldsymbol{\Theta}, \mathbf{t}]^{-1} = [\boldsymbol{\Theta}^{-1}, -\boldsymbol{\Theta}^{-1}\mathbf{t}]. \tag{3.137}$$

That the pure translations $[\mathbf{I}, \mathbf{t}]$ form an invariant (normal) subgroup can be shown as follows: Let $[\mathbf{I}, \mathbf{t}]$ be any pure translation and $[\boldsymbol{\Theta}, \mathbf{t}']$ be any operator of the space group. Then we have

$$\begin{aligned}
[\boldsymbol{\Theta}, \mathbf{t}']^{-1}[\mathbf{I}, \mathbf{t}][\boldsymbol{\Theta}, \mathbf{t}'] &= [\boldsymbol{\Theta}, \mathbf{t}']^{-1}[\mathbf{I}\boldsymbol{\Theta}, \mathbf{I}\mathbf{t}' + \mathbf{t}] \\
&= [\boldsymbol{\Theta}^{-1}, -\boldsymbol{\Theta}\mathbf{t}'][\boldsymbol{\Theta}, \mathbf{t}' + \mathbf{t}] \\
&= [\mathbf{I}, \boldsymbol{\Theta}^{-1}(\mathbf{t}' + \mathbf{t}) - \boldsymbol{\Theta}\mathbf{t}'],
\end{aligned} \tag{3.138}$$

which is a pure translation.

More than a brief discussion of the irreducible representation of space groups would lead us too far astray. For a treatment of these, the reader is referred to Koster.[10] However, we are in a position to consider without difficulty a few elementary aspects of the theory.

[10] G. F. Koster, Space Groups and Their Representations, in *Solid State Physics*, vol. 5 (F. Seitz and D. Turnball, eds.). Academic Press, New York, 1957, p. 173.

Since translations commute, the irreducible representations of the translation group are all one-dimensional. Since

$$[\mathbf{I},\mathbf{t}_2][\mathbf{I},\mathbf{t}_1] = [\mathbf{I},\mathbf{t}_1 + \mathbf{t}_2], \tag{3.139}$$

the characters must satisfy

$$\chi(\mathbf{t}_2)\,\chi(\mathbf{t}_1) = \chi(\mathbf{t}_1 + \mathbf{t}_2). \tag{3.140}$$

The only function that satisfies Eq. (3.140) is the exponential function

$$\chi(\mathbf{t}) = \exp(\boldsymbol{\mu}\cdot\mathbf{t}), \tag{3.141}$$

where $\boldsymbol{\mu}$ is any constant vector. If $\chi(\mathbf{t})$ is to remain finite as $\mathbf{t}$ approaches positive or negative infinity, $\boldsymbol{\mu}$ must be an imaginary vector, that is,

$$\boldsymbol{\mu} = i\mathbf{k}, \tag{3.142}$$

where $i = \sqrt{-1}$ and $\mathbf{k}$ is a real vector. There is a representation for each value of $\mathbf{k}$, and since $\mathbf{k}$ is a vector, it corresponds to three-independent components k_1, k_2, and k_3. For a primitive translation, the character is[11]

$$\chi_k(\mathbf{t}) = \exp[i(k_1 n_1 \tau_1 + k_2 n_2 \tau_2 + k_3 n_3 \tau_3)]. \tag{3.143}$$

Note that the k representation is the direct product of the representation for the three separate translations $n_1\boldsymbol{\tau}_1$, $n_2\boldsymbol{\tau}_2$, and $n_3\boldsymbol{\tau}_3$, which have the characters

$$\chi_{k_j}(n_j\boldsymbol{\tau}_j) = \exp[ik_j n_j \tau_j], \qquad j = 1, 2, 3. \tag{3.144}$$

[Note carefully that in Eq. (3.143), $\mathbf{k}$ labels the irreducible representation and $\mathbf{t}$ designates the operator. The same applies to k_j and $n_j\boldsymbol{\tau}_j$, respectively, in Eq. (3.144).]

Because

$$\exp\{in_j[k_j + 2\pi(p_j/\tau_j)]\tau_j\} = \exp(in_j k_j \tau_j), \qquad j = 1, 2, 3, \quad p_j = \pm 1, \pm 2, \ldots, \tag{3.145}$$

we see that $k_j + 2\pi(p_j/\tau_j)$ and k_j correspond to the same representation. Therefore, we get all the different representations by taking

$$0 \le k_j \le 2\pi/\tau_j, \qquad j = 1, 2, 3. \tag{3.146}$$

Condition (3.146) still leaves (as it should) an infinite number of irreducible representations (one for each of the possible sets of values of set k_1, k_2, k_3) for a translation group, which is of infinite order for an infinite lattice. Most

[11] The symbol τ_i represents the magnitude $|\boldsymbol{\tau}_i|$.

workers avoid dealing with this group of infinite order by replacing the infinite lattice with a cyclic one;[12] that is, they assume that there is an integer N (assumed to be very large) such that the translations $N\boldsymbol{\tau}_j$ $(j = 1, 2, 3)$ each takes the lattice into itself. This means that

$$[\mathbf{I}, \boldsymbol{\tau}_j]^N = [\mathbf{I}, \mathbf{0}], \qquad j = 1, 2, 3, \tag{3.147}$$

and the corresponding characters satisfy

$$[\chi(\tau_j)]^N = 1, \qquad j = 1, 2, 3. \tag{3.148}$$

This situation is the same as that for the cyclic groups $\mathbf{C}_n$ [see Eq. (3.116)]. It follows that for the translation $\boldsymbol{\tau}_j$ in the q_jth representation, we can write

$$\chi_{q_j}(\tau_j) = \varepsilon^{q_j}, \tag{3.149}$$

and for the translation $n_j\boldsymbol{\tau}_j$,

$$\chi_{q_j}(n_j\tau_j) = \varepsilon^{n_j q_j}, \qquad n_j, q_j = 0, 1, 2, \ldots, N-1, \quad j = 1, 2, 3, \tag{3.150}$$

where ε is a primitive Nth root of unity,

$$\varepsilon = \exp(i2\pi/N). \tag{3.151}$$

Now the order of the group of primitive translations along $\boldsymbol{\tau}_j$ is N, since n_j takes on integral values from 0 to $N-1$, and the order of the group of primitive translations in three dimensions is N^3. The number of irreducible representations for translations along $\boldsymbol{\tau}_j$ is N, since q_j takes on integral values from 0 to $N-1$, and the total number of irreducible representations for translations in three dimensions is N^3.

The character of a three-dimensional primitive translation can be expressed as

$$\chi_q(\mathbf{t}) = \exp\left[\frac{i2\pi}{N}(q_1 n_1 + q_2 n_2 + q_3 n_3)\right]. \tag{3.152}$$

Equation (3.152) can be written

$$\chi_k(\mathbf{t}) = \exp[i\mathbf{k} \cdot \mathbf{t}] \tag{3.153}$$

if we take

$$\mathbf{k} = k_1\boldsymbol{\kappa}_1 + k_2\boldsymbol{\kappa}_2 + k_3\boldsymbol{\kappa}_3, \tag{3.154}$$

where the basis vectors $\boldsymbol{\kappa}_j$ form a set *reciprocal* to the set $\boldsymbol{\tau}_j$ so that they satisfy

$$\boldsymbol{\kappa}_j \cdot \boldsymbol{\tau}_{j'} = 2\pi\delta_{jj'}, \tag{3.155}$$

[12] For a justification of this cyclic model see G. Weinreich, *Solids: Elementary Theory for Advanced Students*, Wiley, New York, 1965.

and if we take

$$k_j = q_j/N. \tag{3.156}$$

The space spanned by the set $\boldsymbol{\kappa}_j$ is called the *reciprocal* lattice, and **k** is a *reciprocal* lattice vector. We see that the reciprocal lattice vector **k** identifies the irreducible representation, and the direct lattice vector **t** designates the operator. For each **t**, **k** ranges over the N^3 nonprimitive displacements $(q_1/N, q_2/N, q_3/N)$ of the reciprocal lattice, and for each **k**, **t** ranges over the N^3 primitive displacements (n_1, n_2, n_3) of the direct lattice. The reciprocal lattice belongs to the same crystal class as the direct lattice. The primitive cell in the reciprocal lattice is called the (first) *Brillouin* zone.

Basis functions for the irreducible representations of the translational subgroup can be expressed as

$$u_k(\mathbf{r}) = A_k(\mathbf{r})[\exp(-i\mathbf{k}\cdot\mathbf{r})], \tag{3.157}$$

where **r** is a position vector (not necessarily a lattice point) and $A_k(\mathbf{r})$ satisfies the periodic condition

$$A_k(\mathbf{r} + \mathbf{t}) = A_k(\mathbf{r}). \tag{3.158}$$

Because the translational subgroup is Abelian, its irreducible representations are one-dimensional, and the basis functions are not transformed into one another by translations. Let us briefly investigate what happens when rotations are added to obtain a space group. Which basis functions are transformed into one another to give representations of higher dimensions? To answer this question we shall use examples in which we assume that the group is symmorphic. (For a treatment of the more complicated nonsymmorphic cases we refer the reader to other texts.[13]) We proceed by looking at the effects the point group operations have on the various **k** vectors, which as we have seen, label the irreducible representations and the basis functions of the translational group.

Suppose the crystal class is $\mathbf{C}_{3v}$. Consider first a **k** in general position. Its symmetry will be $\mathbf{C}_1$. Only the identity in $\mathbf{C}_{3v}$ will leave **k**, and $u_k(\mathbf{r})$, unchanged. All other operations of $\mathbf{C}_{3v}$ will change the original **k** to another general position. The total number of different **k** vectors (and basis functions) thus obtained will be six. This set of six **k** vectors is called the *star* of **k**. The six basis functions labeled by these **k** vectors form a basis for a six-dimensional representation of the space group.

Next we start with a **k** with higher symmetry than $\mathbf{C}_1$. Take, for example, any **k** in one of the three reflection planes but not along the rotation axis. In

[13] M. Tinkham, *Group Theory and Quantum Mechanics*, McGraw-Hill, New York, 1964; M. Lax, *Symmetry Principles in Solid State and Molecular Physics*, Wiley, New York, 1974.

this case the star contains three vectors, and there will be a corresponding three-dimensional representation.

A **k** along the threefold axis will be invariant under all the point-group operations. There is only one **k** in the star, and only one-dimensional representations are involved. (This case includes $\mathbf{k} = \mathbf{0}$ as a special case.)

Additional Reading

G. F. Koster, *Space Groups and Their Representations*, Academic Press, New York, 1957.
See also the references cited at the end of Chapter 2.

Problems

3.1 Evaluate the determinant in Example 3.7 by using Eq. (3.14) to evaluate the third-order determinants obtained there.

3.2 By the use of properties 3A.5 and 3A.8 and an expansion in cofactors, show the following:

$$\begin{vmatrix} 1 & 1 & 2 & 1 \\ 5 & 5 & 10 & 10 \\ 3 & 1 & 4 & 2 \\ 1 & 4 & 1 & 1 \end{vmatrix} = 5\begin{vmatrix} 1 & 1 & 2 & 1 \\ 1 & 1 & 2 & 2 \\ 3 & 1 & 4 & 2 \\ 1 & 4 & 1 & 1 \end{vmatrix}$$

$$= 5\begin{vmatrix} 0 & 0 & 0 & -1 \\ 1 & 1 & 2 & 2 \\ 3 & 1 & 4 & 2 \\ 1 & 4 & 1 & 1 \end{vmatrix}$$

$$= 5\begin{vmatrix} 1 & 1 & 2 \\ 3 & 1 & 4 \\ 1 & 4 & 1 \end{vmatrix}$$

3.3 Prove the following:

$$\begin{vmatrix} a & b & 0 & 0 & 0 \\ 1 & 4 & p & q & r \\ 2 & 5 & l & m & n \\ x & y & 0 & 0 & 0 \\ 7 & 6 & u & v & w \end{vmatrix} = (-1)\begin{vmatrix} a & b \\ x & y \end{vmatrix} \cdot \begin{vmatrix} l & m & n \\ p & q & r \\ u & v & w \end{vmatrix}.$$

3.4 Show that the reflection σ_{xy} [Eq. (3.41)] is equivalent to a twofold rotation about the z axis followed (or preceded) by inversion.

3.5 A rotation through an angle θ about an axis in the direction of the unit vector $\mathbf{u} = (u_x, u_y, u_z)$ is given by the matrix $\mathbf{\Phi} = [\phi_{ij}]$, where

$$\phi_{xx} = (1 - \cos\theta)u_x^2 + \cos\theta, \qquad \phi_{zx} = (1 - \cos\theta)u_x u_z + u_y \sin\theta,$$

$$\phi_{xy} = (1 - \cos\theta)u_x u_y + u_z \sin\theta, \quad \phi_{zy} = (1 - \cos\theta)u_y u_z - u_x \sin\theta,$$

$$\phi_{xz} = (1 - \cos\theta)u_x u_z - u_y \sin\theta, \quad \phi_{zz} = (1 - \cos\theta)u_z^2 + \cos\theta.$$

$$\phi_{yx} = (1 - \cos\theta)u_x u_y - u_z \sin\theta,$$

$$\phi_{yy} = (1 - \cos\theta)u_y^2 + \cos\theta,$$

$$\phi_{yz} = (1 - \cos\theta)u_y u_z + u_x \sin\theta,$$

The matrix $(-1)\mathbf{\Phi}$ is the corresponding *improper* rotation.

(a) Prove that $\mathbf{u}$ is the axis of rotation by proving that $\mathbf{u}$ is invariant under $\mathbf{\Phi}$.

(b) Obtain the matrix for rotations about the z axis.

(c) Obtain the matrix for reflection through the plane perpendicular to $\mathbf{u}$.

(d) Obtain the matrix for reflection through the plane that contains the z axis and makes an angle ϕ with the xz plane.

(e) Show that the axis of rotation is an eigenvector of the rotation with an eigenvalue of $+1$ for the proper rotation and -1 for the improper rotation.

(f) Show that $\operatorname{tr} \mathbf{\Phi} = 1 + 2\cos\theta$.

3.6 Prove Eq. (3.58).

3.7 Show that all matrices of the form $\left[\begin{smallmatrix} a & b \\ b & a \end{smallmatrix}\right]$, where a and b are arbitrary scalars, commute.

3.8 Use the operators $P_{E,22}$ and $T_{E,21}$, as suggested in Example 3.14, to find the partners in the projection of $F = ax + by + cz + d$ in the direction of the irreducible representation E of $\mathbf{C}_{3v}$.

3.9 Let ϕ_k $(k = 1, \ldots, n)$ be an orthonormal set so that $\int \phi_k^* \phi_l \, d\tau = \delta_{kl}$. Let Ψ_i $(i = 1, \ldots, n)$ be a set of linear combinations

$$\Psi_i = \sum_j a_{ij} \phi_j.$$

Prove that the set Ψ_i is orthonormal (so that $\int \Psi_i^* \Psi_j \, d\tau = \delta_{ij}$) if, and only if, the matrix $[a_{ij}]$ is unitary (orthogonal if all a_{ij} are real.)

3.10 Prove that for $\mathbf{C}_{\infty v}$, the functions $u_k(\theta) = \exp(-ik\theta)$ and $u_k(\theta)^*$ are basis functions for the two-dimensional irreducible representation E_k.

3.11 Prove that for $\mathbf{C}_{nv}$, the characters in E_k are $2\cos(k2\pi/n)$ and zero for C_n^k and σ_v, respectively.

3.12 The spherical harmonics can be expressed as

$$Y_{lm}(\theta, \phi) = P_{lm}(\cos\theta)e^{im\phi}, \qquad m = 0, \pm 1, \ldots, \pm l,$$

where $P_{lm}(\cos\theta)$ is the Legendre function. Use $Y_{lm}(\theta, \phi)$ as basis functions and the fact that rotation about the z axis changes only ϕ to prove Eq. (3.126).

3.13 Prove Eq. (3.127).

3.14 Show that if two operators **A** and **B** have a set of n eigenvectors in common, then the $n \times n$ matrix representations of **A** and **B** with this set of eigenvectors as basis must commute.

3.15 Prove Eq. (3.137).

3.16 Prove that any rotation axis of a crystal must be parallel to a primitive translation.

3.17 If a rotation R_z about the z axis is represented by a curved arrow about the z axis, then the effect of a twofold rotation about an axis perpendicular to the z axis can be pictured as "turning the arrow over," which reverses the sense of the arrow, indicating the inverse R_z^{-1} is obtained. By use of matrix representations of the operators involved, prove that this process corresponds to a similarity transformation of R_z.

3.18 Obtain the representations for the p and d orbitals from the character tables and verify the results by examining the transformation properties of each orbital for $\mathbf{C}_{4v}$ and $\mathbf{D}_{3h}$.

3.19 Tabulate the effects of the operations of $\mathbf{D}_{4h}$ on the coordinates x, y, and z. From the results obtain the representations for d_{xy}, $d_{x^2-y^2}$, d_{xz}, f_{xyz}, and $f_{z(x^2-y^2)}$.

3.20 What symmetry operations cause x and y to belong to an E representation? What symmetry operations cause x, y, and z to belong to a T representation? What symmetry operations cause p_x and p_y to be degenerate, but not $d_{x^2-y^2}$ and d_{xy}?

Appendix: Some Additional Mathematical Background

3A PROPERTIES OF DETERMINANTS

The following are some useful properties of determinants.

3A.1 All columns may be interchanged with their corresponding rows without changing the value of the determinant, that is, a determinant and its transpose have the same value, $|d_{ij}| = |d_{ji}|$. For example,

$$\begin{vmatrix} d_{11} & d_{12} & d_{13} \\ d_{21} & d_{22} & d_{23} \\ d_{31} & d_{32} & d_{33} \end{vmatrix} = \begin{vmatrix} d_{11} & d_{21} & d_{31} \\ d_{12} & d_{22} & d_{32} \\ d_{13} & d_{23} & d_{33} \end{vmatrix}.$$

3A.2 Any property that can be expressed in terms of *columns* can be analogously expressed in terms of *rows*. Therefore, the following properties apply to rows as well as columns.

3A.3 Interchanging any two columns changes the sign of the determinant. For example,

$$\begin{vmatrix} d_{11} & d_{12} & d_{13} \\ d_{21} & d_{22} & d_{23} \\ d_{31} & d_{32} & d_{33} \end{vmatrix} = (-1)\begin{vmatrix} d_{13} & d_{12} & d_{11} \\ d_{23} & d_{22} & d_{21} \\ d_{33} & d_{32} & d_{31} \end{vmatrix}.$$

3A.4 If two columns are identical, the value of the determinant is zero. This follows directly from 3A.3.

3A.5 Multiplying all the elements of one column by a constant λ multiplies the value of the determinant by λ.

3A.6 Multiplying all the elements of the determinant by a constant λ multiplies the determinant by λ^n, where n is the order of the determinant.

3A.7 If all the elements of a column are zero, then the value of the determinant is zero.

3A.8 The value of a determinant is not changed if each element of one column is multiplied by a constant λ and added to the corresponding element of a second column. For example,

$$\begin{vmatrix} d_{11} & d_{12} & d_{13} \\ d_{21} & d_{22} & d_{23} \\ d_{31} & d_{32} & d_{33} \end{vmatrix} = \begin{vmatrix} d_{11} + \lambda d_{12} & d_{12} & d_{13} \\ d_{21} + \lambda d_{22} & d_{22} & d_{23} \\ d_{31} + \lambda d_{32} & d_{32} & d_{33} \end{vmatrix}.$$

3A.9 A determinant can be separated in terms of a column as follows:

$$\begin{vmatrix} d_{11} & \cdots & d_{1j} + d'_{1j} & \cdots & d_{1n} \\ \vdots & & \vdots & & \vdots \\ d_{n1} & \cdots & d_{nj} + d'_{nj} & \cdots & d_{nn} \end{vmatrix} = \begin{vmatrix} d_{11} & \cdots & d_{1j} & \cdots & d_{1n} \\ \vdots & & \vdots & & \vdots \\ d_{n1} & \cdots & d_{nj} & \cdots & d_{nn} \end{vmatrix} + \begin{vmatrix} d_{11} & \cdots & d'_{1j} & \cdots & d_{1n} \\ \vdots & & \vdots & & \vdots \\ d_{n1} & \cdots & d'_{nj} & \cdots & d_{nn} \end{vmatrix}.$$

3A.10 If $\mathbf{A}$ and $\mathbf{B}$ are square matrices of orders n_1 and n_2, respectively, $\mathbf{X}$ is an $n_2 \times n_1$ matrix, and $\mathbf{0}$ is the $n_1 \times n_2$ zero matrix, then the determinant of a compound matrix (in so-called block form) satisfies the equation

$$\begin{vmatrix} \mathbf{A} & \mathbf{0} \\ \mathbf{X} & \mathbf{B} \end{vmatrix} = |\mathbf{A}| \cdot |\mathbf{B}|.$$

More generally, if $\mathbf{A}$, $\mathbf{B}$, $\mathbf{C}$, etc., are square matrices, then

$$\begin{vmatrix} \mathbf{A} & \mathbf{0} & \mathbf{0} & \cdots \\ \mathbf{X} & \mathbf{B} & \mathbf{0} & \cdots \\ \mathbf{Y} & \mathbf{Z} & \mathbf{C} & \cdots \\ \vdots & \vdots & \vdots & \ddots \end{vmatrix} = |\mathbf{A}| \cdot |\mathbf{B}| \cdot |\mathbf{C}| \cdots.$$

(In applications the matrices $\mathbf{X}$, $\mathbf{Y}$, $\mathbf{Z}$, etc., are often zero matrices, so that the compound matrix is in diagonal block form.)

3A.11 The (row) · (column) multiplication defined for matrices works for determinents. However, because of property 3A.2 the multiplication of determinants can also be (row) · (row), (column) · (row), or (column) · (column). Because determinants have values, the value of the product of determinants is equal to the product of the values of the determinants. The multiplication of determinants is commutative.

3A.12 (Cramer's rule) If $\mathbf{A} = [a_{ij}]$ is a square matrix with $|\mathbf{A}| \neq 0$, then the system of linear equations

$$\begin{aligned} a_{11}x_1 + \cdots + a_{1n}x_n &= b_1, \\ &\vdots \\ a_{n1}x_1 + \cdots + a_{nn}x_n &= b_n, \end{aligned}$$

has the solution given by

$$x_1 = \frac{\begin{vmatrix} b_1 & a_{12} & \cdots & a_{1n} \\ \vdots & \vdots & & \vdots \\ b_n & a_{n2} & \cdots & a_{nn} \end{vmatrix}}{|\mathbf{A}|}$$

$$\vdots$$

$$x_n = \frac{\begin{vmatrix} a_{11} & \cdots & a_{1n-1} & b_1 \\ \vdots & & \vdots & \vdots \\ a_{n1} & \cdots & a_{nn-1} & b_n \end{vmatrix}}{|\mathbf{A}|}$$

Note that the numerator in x_i is the determinant obtained by replacing the ith column of $\mathbf{A}$ by the column of constant terms.

3A.13 If $\mathbf{A} = [a_{ij}]$ is a square matrix, then the system of linear homogeneous equations

$$\begin{aligned} a_{11}x_1 + \cdots + a_{1n}x_n &= 0 \\ &\vdots \\ a_{n1}x_1 + \cdots + a_{nn}x_n &= 0 \end{aligned}$$

has a solution other than $x_1 = x_2 = \cdots = x_n = 0$ if, and only if, $|\mathbf{A}| = 0$.

3B PROPERTIES OF MATRICES

We first list some properties of matrix multiplication.

3B.1 (a) $\mathbf{ABC} = \mathbf{A}(\mathbf{BC}) = (\mathbf{AB})\mathbf{C}$ (associative law).

(b) $\mathbf{A}(\mathbf{B} + \mathbf{C}) = \mathbf{AB} + \mathbf{AC}$, $(\mathbf{B} + \mathbf{C})\mathbf{A} = \mathbf{BA} + \mathbf{CA}$ (distributive law).

(c) $\mathbf{AA} \cdots \mathbf{A}$ (n times) $= \mathbf{A}^n$, $\mathbf{A}^n\mathbf{A}^m = \mathbf{A}^{n+m}$.

[In general, $(\mathbf{AB})^n \neq \mathbf{A}^n\mathbf{B}^n$. If A and B commute, then $(\mathbf{AB})^n = \mathbf{A}^n\mathbf{B}^n$.]

(d) $(\mathbf{ABC} \cdots \mathbf{Y})^{-1} = \mathbf{Y}^{-1} \cdots \mathbf{C}^{-1}\mathbf{B}^{-1}\mathbf{A}^{-1}$.

(e) $(\mathbf{ABC} \cdots \mathbf{Y})^{\mathrm{T}} = \mathbf{Y}^{\mathrm{T}} \cdots \mathbf{C}^{\mathrm{T}}\mathbf{B}^{\mathrm{T}}\mathbf{A}^{\mathrm{T}}$.

(f) It is possible that $\mathbf{AB} = \mathbf{0}$ when $\mathbf{A} \neq \mathbf{0}$ and $\mathbf{B} \neq \mathbf{0}$.

(g) In general, $(\mathbf{A} + \mathbf{B})^2 = \mathbf{A}^2 + \mathbf{AB} + \mathbf{BA} + \mathbf{B}^2$. If $\mathbf{A}$ and $\mathbf{B}$ commute, then $(\mathbf{A} + \mathbf{B})^2 = \mathbf{A}^2 + 2\mathbf{AB} + \mathbf{B}^2$.

(h) Suppose two matrices are partitioned (or blocked), for example, as follows:

$$\mathbf{A} = \begin{bmatrix} \mathbf{A}_{11} & \mathbf{A}_{12} & \mathbf{A}_{13} \\ \mathbf{A}_{21} & \mathbf{A}_{22} & \mathbf{A}_{23} \\ \mathbf{A}_{31} & \mathbf{A}_{32} & \mathbf{A}_{33} \end{bmatrix}, \qquad \mathbf{B} = \begin{bmatrix} \mathbf{B}_{11} & \mathbf{B}_{12} & \mathbf{B}_{13} \\ \mathbf{B}_{21} & \mathbf{B}_{22} & \mathbf{B}_{23} \\ \mathbf{B}_{31} & \mathbf{B}_{32} & \mathbf{B}_{33} \end{bmatrix},$$

where $\mathbf{A}_{ij}$ and $\mathbf{B}_{ij}$ are submatrices. If the number of columns in $\mathbf{A}_{11}$ is equal to the number of rows in $\mathbf{B}_{11}$, etc., then

$$\mathbf{AB} = \begin{bmatrix} (\mathbf{A}_{11}\mathbf{B}_{11} + \mathbf{A}_{12}\mathbf{B}_{21} + \mathbf{A}_{13}\mathbf{B}_{31}) & \cdots & (\mathbf{A}_{11}\mathbf{B}_{13} + \mathbf{A}_{12}\mathbf{B}_{23} + \mathbf{B}_{33}) \\ \vdots & & \vdots \\ (\mathbf{A}_{31}\mathbf{B}_{11} + \mathbf{A}_{32}\mathbf{B}_{21} + \mathbf{A}_{33}\mathbf{B}_{31}) & \cdots & (\mathbf{A}_{31}\mathbf{B}_{13} + \mathbf{A}_{32}\mathbf{B}_{23} + \mathbf{A}_{33}\mathbf{B}_{33}) \end{bmatrix},$$

where $\mathbf{A}_{11}\mathbf{B}_{11}$, etc., are the products of matrices.

3B.2 The following relationships apply to the direct (Kronecker) product:

(a) $\mathbf{A} \times \mathbf{B} \times \mathbf{C} = \mathbf{A} \times (\mathbf{B} \times \mathbf{C}) = (\mathbf{A} \times \mathbf{B}) \times \mathbf{C}$ (associative law),

(b) $\mathbf{A} \times (\mathbf{B} + \mathbf{C}) = \mathbf{A} \times \mathbf{B} + \mathbf{A} \times \mathbf{C}$,
$(\mathbf{B} + \mathbf{C}) \times \mathbf{A} = \mathbf{B} \times \mathbf{A} + \mathbf{C} \times \mathbf{A}$ (distributive law),

(c) $(\mathbf{A} \times \mathbf{B})(\mathbf{C} \times \mathbf{D}) = (\mathbf{AC}) \times (\mathbf{BD})$,
(d) $(\mathbf{A} \times \mathbf{B})^{-1} = \mathbf{A}^{-1} \times \mathbf{B}^{-1}$,
(e) $(\mathbf{A} \times \mathbf{B})^{\mathrm{T}} = \mathbf{A}^{\mathrm{T}} \times \mathbf{B}^{\mathrm{T}}$.

3B.3 *Hermitian* matrices have the following properties.

(a) If $\mathbf{H}$ is a Hermitian matrix, then $(\mathbf{H}^{\mathrm{T}})^* = \mathbf{H}$. A Hermitian matrix with all elements real is a *symmetric* matrix.
(b) Similarity transformations of a Hermitian matrix give Hermitian matrices.
(c) All the eigenvalues of a Hermitian are real.
(d) Eigenvectors belonging to different eigenvalues of a Hermitian matrix are orthogonal (see Examples 3.11 and 3.12).
(e) A Hermitian matrix can be diagonalized by a unitary transformation.

3B.4 *Unitary* matrices have the following properties.

(a) If $\mathbf{U}$ is a unitary matrix, then $(\mathbf{U}^{\mathrm{T}})^* = \mathbf{U}^{-1}$. An *orthogonal* matrix is a unitary matrix with all elements real.
(b) The eigenvalues of a unitary matrix are of the form $e^{i\alpha}$ $(-\pi \leq \alpha \leq \pi)$. The only real eigenvalues are ± 1.
(c) The eigenvectors of a unitary matrix are orthogonal.
(d) A unitary matrix can be diagonalized by a unitary transformation.
(e) The determinant of a unitary matrix satisfies $|\mathbf{U}| = e^{i\alpha}$ $(-\pi \leq \alpha \leq \pi)$. If $\mathbf{U}$ is an orthogonal matrix, then $|\mathbf{U}| = \pm 1$.
(f) A necessary and sufficient condition for a matrix to be unitary is that the rows (or the columns) form an orthonormal set.
(g) An orthonormal set of vectors, or of functions, remains orthonormal after a linear transformation if, and only if, the transformation is unitary. This can be stated in other terms as follows:
If f_i $(i = 1, 2, \ldots, n)$ is an orthonormal set of vectors, or functions, and we have the linear transformation

$$g_i = \sum_{j=1} a_{ij} f_j, \qquad i = 1, 2, \ldots, n,$$

then the set g_i is an orthonormal set of vectors, or functions, if, and only if, the set of row vectors,

$$\mathbf{r}_i = (a_{i1}, a_{i2}, \ldots, a_{in}), \qquad i = 1, 2, \ldots, n,$$

is orthonormal, or, equivalently, if, and only if, the set of column vectors

$$\mathbf{c}_i = (a_{1i}, a_{2i}, \ldots, a_{ni}), \qquad i = 1, 2, \ldots, n,$$

is orthonormal (see Problem 3.9).

3B.5 If a set of eigenvectors of a matrix have the same eigenvalue, any linear combination of these eigenvectors is an eigenvector with that same eigenvalue (see Examples 3.11 and 3.12). It is usually convenient to use a set of linear combinations that is orthonormal.

3B.6 If, and only if, two matrices commute, they have the same eigenvectors (see Problem 3.14).

3B.7 If, and only if, two matrices commute, they can be simultaneously diagonalized. That is, they will be diagonalized by the same similarity transformation.

3C THE GREAT ORTHOGONALITY THEOREM

Because we give without proof theorems that we need in dealing with irreducible representations, the importance of the orthogonality theorem [Eq. (3.76)] may not be apparent. The central role of this theorem justifies the following discussion and proof.

According to the orthogonality theorem the "normalized" matrix elements of the irreducible representations of a group of order h form the components of an orthonormal set of vectors $\mathbf{\Gamma}_{k,ij}$ [Eq. (3.72)]. These vectors (h in number) span an h-dimensional vector space. The l_k^2 vectors corresponding to an irreducible representation of dimension l_k span an l_k^2-dimensional subspace. Since the total space is the sum of its orthogonal subspaces, Eq. (3.75) follows directly. The orthonormal properties of the characters [Eq. (3.77)] are obtained by summation (over i, j, i' and j') of both sides of Eq. (3.76). The reduction equations (3.78) and (3.79) can be understood in terms of the concept of the scalar product as a projection—in the sense of a generalization of Eq. (3.9). That is, the scalar product $\chi \cdot \chi_k$ gives the "projection" ("component") of the reducible character vector χ in the "direction" of the character vector of the kth irreducible representation. Finally, the existence of the projection and transfer operators depends[14] upon Eq. (3.76).

The proof of the great orthogonality theorem, which we shall give, requires the following lemmas and corollaries.

LEMMA I If $\mathbf{M}$ is an $m \times n$ matrix with $m \neq n$, then the determinant $|\mathbf{M}\mathbf{M}^{*\mathrm{T}}|$ is zero.

Proof Construct the square matrix $\mathbf{M}_s$ by adding rows (or columns) of zeros to make up the deficiency in rows (or columns) in $\mathbf{M}$. We have

[14] See, for example, M. Tinkham, *Group Theory and Quantum Mechanics*, McGraw-Hill, New York, 1946, p. 40.

$\mathbf{MM}^{*\mathrm{T}} = \mathbf{M}_s\mathbf{M}_s^{*\mathrm{T}}$, and therefore, $|\mathbf{MM}^{*\mathrm{T}}| = |\mathbf{M}_s\mathbf{M}_s^{*\mathrm{T}}| = |\mathbf{M}_s||\mathbf{M}_s^{*\mathrm{T}}| = 0$, since $|\mathbf{M}_s| = 0$.

LEMMA II If $\mathbf{MM}^{*\mathrm{T}} = 0$, then[15] $\mathbf{M} = 0$.

Proof Let $\mathbf{MM}^{*\mathrm{T}} = \mathbf{X}$; then $\mathbf{X}_{ij} = 0$, and

$$\mathbf{X}_{ij} = \sum_k M_{ik} M_{kj}^{*\mathrm{T}} = \sum_k M_{ik} M_{jk}^{*} = 0.$$

In particular, for $i = j$ (and with $|M_{ij}|$ representing the absolute value of M_{ij}) we have

$$X_{ii} = \sum_k M_{ik} M_{ik}^{*} = \sum_k |M_{ik}|^2 = 0,$$

which means $M_{ik} = 0$ for all i and k.

LEMMA III If a diagonal matrix $\mathbf{A}_d$ commutes with a matrix $\mathbf{M}$, then $\mathbf{A}_d$ has repeated eigenvalues and can take the form

$$\mathbf{A}_d = \begin{bmatrix} \lambda_1 & 0 & \cdots & & & \\ 0 & \lambda_1 & & & & \\ \vdots & & \ddots & & & \\ & & & \lambda_1 & & \\ & & & & \lambda_2 & \\ & & & & & \ddots \end{bmatrix},$$

and $\mathbf{M}$ must take the diagonal block (reduced) form

$$\mathbf{M} = \begin{bmatrix} \mathbf{M}_1 & 0 & \cdots \\ 0 & \mathbf{M}_2 & \\ \vdots & & \ddots \end{bmatrix},$$

where $\mathbf{M}_1$, $\mathbf{M}_2$, etc., are submatrices of orders no greater than the multiplicatives g_1, g_2, etc., of λ_1, λ_2, etc.

Proof Let $\mathbf{X} = \mathbf{A}_d\mathbf{M} = \mathbf{M}\mathbf{A}_d$ and consider the element X_{ij}, where $i \le g_1$ and $j > g_1$. We have $X_{ij} = \lambda_1 M_{ij} = M_{ij}\lambda_2$, which gives $M_{ij} = 0$ when $\lambda_1 \ne \lambda_2$. For $i > g_1$ and $j \le g_1$, we have $\lambda_2 M_{ij} = M_{ij}\lambda_1$, and so $M_{ij} = 0$. For $i \le g_1$, M_{ij} can be different from zero only if $j \le g_1$. These give the submatrix $\mathbf{M}_1$. Likewise, we can show that for $g_1 < i \le g_2$, M_{ij} can be different from zero only if $g_1 < j \le g_2$. These give $\mathbf{M}_2$.

[15] In general the condition $\mathbf{AB} = \mathbf{0}$ does not require that either $\mathbf{A} = \mathbf{0}$ or $\mathbf{B} = \mathbf{0}$ [Appendix 3B.1(f)]. For example, consider the case

$$\mathbf{A} = \begin{bmatrix} 1 & 0 \\ 0 & 0 \end{bmatrix} \text{ and } \mathbf{B} = \begin{bmatrix} 0 & 0 \\ 0 & 1 \end{bmatrix} \text{ so } \mathbf{AB} = \begin{bmatrix} 0 & 0 \\ 0 & 0 \end{bmatrix}.$$

COROLLARY III.1 If $\mathbf{A}_d$ has no multiple eigenvalues, then $\mathbf{M}$ must be diagonal.

Proof In this case $g_1 = g_2 = \cdots = 1$, and $\mathbf{M}_1, \mathbf{M}_2$, etc., must be 1×1 matrices.

COROLLARY III.2 Any diagonal matrix $\mathbf{A}_d$ that commutes with all the matrices of an irreducible representation must be of the form $\lambda\mathbf{I}$, where λ is constant and $\mathbf{I}$ is the unit matrix ($\lambda\mathbf{I}$ is called a *constant*, on a *scalar* matrix).

Proof The matrices would be in reduced form unless $\lambda_1 = \lambda_2 = \cdots = \lambda$.

LEMMA IV If two commuting matrices undergo the same similarity transformation, then the transformed matrices commute.

Proof Let $\mathbf{AB} = \mathbf{BA}$ and $\mathbf{A}'$, $\mathbf{B}'$ be the transformed matrices. We have

$$\mathbf{TABT}^{-1} = \mathbf{TBAT}^{-1},$$

$$\mathbf{TAT}^{-1}\mathbf{TBT}^{-1} = \mathbf{TBT}^{-1}\mathbf{TAT}^{-1},$$

$$\mathbf{A}'\mathbf{B}' = \mathbf{B}'\mathbf{A}'.$$

LEMMA V The scalar matrix $\lambda\mathbf{I}$ is unchanged by similarity transformations.

Proof Since $\mathbf{I}$ commutes with all matrices, we have $\mathbf{T}(\lambda\mathbf{I})\mathbf{T}^{-1} = \lambda\mathbf{ITT}^{-1} = \lambda\mathbf{I}$.

LEMMA VI (Schur's) Any matrix $\mathbf{M}$ that commutes with all of the matrices of a unitary, irreducible, representation must be a scalar matrix $\lambda\mathbf{I}$.

Proof We shall show that $\mathbf{M}$ is equivalent to a diagonal matrix, and therefore, by Corollary III.2, $\mathbf{M} = \lambda\mathbf{I}$.

Let $\mathbf{\Gamma}(R)$ be the matrix representing the operator R. Because $\mathbf{M\Gamma}(R) = \mathbf{\Gamma}(R)\mathbf{M}$ holds for all the operators, we also have $\mathbf{M\Gamma}(R^{-1}) = \mathbf{\Gamma}(R^{-1})\mathbf{M}$. Taking the complex conjugate and transpose of both sides of the latter equation gives

$$[\mathbf{M\Gamma}(R^{-1})]^{*\mathrm{T}} = [\mathbf{\Gamma}(R^{-1})\mathbf{M}]^{*\mathrm{T}},$$

$$\mathbf{\Gamma}(R^{-1})^{*\mathrm{T}}\mathbf{M}^{*\mathrm{T}} = \mathbf{M}^{*\mathrm{T}}\mathbf{\Gamma}(R^{-1})^{*\mathrm{T}} \qquad \text{[Appendix 3B.1(e)]}.$$

Because Γ is unitary, $\mathbf{\Gamma}(R^{-1})^{*\mathrm{T}} = \mathbf{\Gamma}(R)$. Therefore, $\mathbf{\Gamma}(R)\mathbf{M}^{*\mathrm{T}} = \mathbf{M}^{*\mathrm{T}}\mathbf{\Gamma}(R)$, and we see that $\mathbf{M}^{*\mathrm{T}}$ commutes with $\mathbf{\Gamma}(R)$. It follows that the two Hermitian matrices $\mathbf{M}' = \mathbf{M} + \mathbf{M}^{*\mathrm{T}}$ and $\mathbf{M}'' = i(\mathbf{M} - \mathbf{M}^{*\mathrm{T}})$ each commute with $\mathbf{\Gamma}(R)$. Because $\mathbf{M}'$ and $\mathbf{M}''$ are Hermitian, they can be diagonalized by unitary transformations, and in the transformed systems they still commute with $\mathbf{\Gamma}(R)$ (Lemma IV). However, by Corollary III.2, $\mathbf{M}'$ and $\mathbf{M}''$ must be scalar matrices, and by Lemma V they must be scalar in all systems. Because $\mathbf{M} = \frac{1}{2}(\mathbf{M}' - i\mathbf{M}'')$, $\mathbf{M}$ must be a scalar matrix also.

LEMMA VII Let Γ_k, of dimension l_k, and $\Gamma_{k'}$, of dimension $l_{k'}$, be two unitary, irreducible, representations of the same group. Let there exist an $l_k \times l_{k'}$ matrix $\mathbf{M}$ such that $\mathbf{M}\boldsymbol{\Gamma}_k(R) = \boldsymbol{\Gamma}_{k'}(R)\mathbf{M}$ for every operator R of the group. Then

(1) $\mathbf{M}\mathbf{M}^{*\mathrm{T}} = \lambda\mathbf{I}$.
(2) If $l_k \neq l_{k'}$, then $\mathbf{M} = \mathbf{0}$.
(3) If $l_k = l_{k'}$, then either $\mathbf{M} = \mathbf{0}$ or $|\mathbf{M}| \neq 0$, in which case Γ_k and $\Gamma_{k'}$ are equivalant.

Proof Because $\mathbf{M}\boldsymbol{\Gamma}_k(R) = \boldsymbol{\Gamma}_{k'}(R)\mathbf{M}$ for all R, it is true that

$$\mathbf{M}\boldsymbol{\Gamma}_k(R^{-1}) = \boldsymbol{\Gamma}_{k'}(R^{-1})\mathbf{M},$$

and it follows (see proof of Lemma VI) that $\boldsymbol{\Gamma}_k(R)\mathbf{M}^{*\mathrm{T}} = \mathbf{M}^{*\mathrm{T}}\boldsymbol{\Gamma}_{k'}(R)$, and then $\mathbf{M}\boldsymbol{\Gamma}_k(R)\mathbf{M}^{*\mathrm{T}} = \mathbf{M}\mathbf{M}^{*\mathrm{T}}\boldsymbol{\Gamma}_{k'}(R)$. Substituting $\boldsymbol{\Gamma}_{k'}(R)\mathbf{M}$ for $\mathbf{M}\boldsymbol{\Gamma}_k(R)$ in this last equation gives

$$\boldsymbol{\Gamma}_{k'}(R)\mathbf{M}\mathbf{M}^{*\mathrm{T}} = \mathbf{M}\mathbf{M}^{*\mathrm{T}}\boldsymbol{\Gamma}_{k'}(R).$$

It follows from Lemma VI that $\mathbf{M}\mathbf{M}^{*\mathrm{T}} = \lambda\mathbf{I}$.

If $l_k \neq l_{k'}$ we have from Lemma I $|\mathbf{M}\mathbf{M}^{*\mathrm{T}}| = 0$, and therefore, $\lambda = 0$ and $\mathbf{M}\mathbf{M}^{*\mathrm{T}} = \mathbf{0}$. By Lemma II, $\mathbf{M} = \mathbf{0}$.

If $l_k = l_{k'}$, $\mathbf{M}$ can be zero, but it need not be. If $\mathbf{M} \neq \mathbf{0}$, then $\lambda \neq 0$ and $|\mathbf{M}| \neq 0$, in which case $\mathbf{M}$ has an inverse. It follows that $\mathbf{M}\boldsymbol{\Gamma}_k(R)\mathbf{M}^{-1} = \boldsymbol{\Gamma}_{k'}(R)\mathbf{M}\mathbf{M}^{-1} = \boldsymbol{\Gamma}_{k'}(R)$. Therefore, $\boldsymbol{\Gamma}_k(R)$ and $\boldsymbol{\Gamma}_{k'}(R)$ are related by a similarity transformation, meaning that Γ_k and $\Gamma_{k'}$ are equivalent.

We are now in a position to prove the orthogonality theorem, Eq. (3.76),

$$\boldsymbol{\Gamma}_{k,ij} \cdot \boldsymbol{\Gamma}_{k',i'j'} = \frac{\sqrt{l_k l_{k'}}}{h} \sum_p \Gamma_k(p)_{ij}^* \Gamma_{k'}(p)_{i'j'} = \delta_{kk'}\delta_{ii'}\delta_{jj'}. \tag{3.76}$$

The procedure is to construct first a matrix $\mathbf{M}$ that satisfies the condition of Lemma VII. We shall show that such a matrix is

$$\mathbf{M} = \sum_R \boldsymbol{\Gamma}_{k'}(R)\mathbf{A}\boldsymbol{\Gamma}_k(R^{-1}),$$

where $\mathbf{A}$ is an arbitrary $l_{k'} \times l_k$ matrix. We have

$$\begin{aligned}
\boldsymbol{\Gamma}_{k'}(S)\mathbf{M} &= \sum_R \boldsymbol{\Gamma}_{k'}(S)\boldsymbol{\Gamma}_{k'}(R)\mathbf{A}\boldsymbol{\Gamma}_k(R^{-1})\boldsymbol{\Gamma}_k(S^{-1})\boldsymbol{\Gamma}_k(S) \\
&= \sum_R \boldsymbol{\Gamma}_{k'}(SR)\mathbf{A}\boldsymbol{\Gamma}_k(R^{-1}S^{-1})\boldsymbol{\Gamma}_k(S) \\
&= \sum_R \boldsymbol{\Gamma}_{k'}(SR)\mathbf{A}\boldsymbol{\Gamma}_k((SR)^{-1})\boldsymbol{\Gamma}_k(S) \\
&= \left[\sum_R \boldsymbol{\Gamma}_{k'}(SR)\mathbf{A}\boldsymbol{\Gamma}_k((SR)^{-1})\right]\boldsymbol{\Gamma}_k(S).
\end{aligned}$$

But because the summation is over each group operator once, and only once, we have

$$\sum_R \mathbf{\Gamma}_{k'}(SR)\mathbf{A}\mathbf{\Gamma}_k((SR)^{-1}) = \sum_p \mathbf{\Gamma}_{k'}(p)\mathbf{A}\mathbf{\Gamma}_k(p^{-1}) = \mathbf{M},$$

where p represents the pth operator and p^{-1} represents its inverse. We thus obtain $\mathbf{\Gamma}_{k'}(S)\mathbf{M} = \mathbf{M}\mathbf{\Gamma}_k(S)$, which is the condition of Lemma VII. From that lemma we have that $\mathbf{M} = \mathbf{0}$ when $l_k \neq l_{k'}$. Also, $\mathbf{M} = \mathbf{0}$ when $l_k = l_{k'}$ unless Γ_k and $\Gamma_{k'}$ are equivalent (actually, in the context of the orthogonality theorem equivalent representations are the *same* representations). $\mathbf{M} = \mathbf{0}$ means

$$M_{i'i} = \sum_p \sum_\alpha \sum_\beta \Gamma_{k'}(p)_{i'\alpha} A_{\alpha\beta} \Gamma_k(p^{-1})_{\beta i} = 0, \qquad \text{all} \quad i', i.$$

Because $\mathbf{A}$ is arbitrary, we can take all $A_{\alpha\beta} = 0$ except $A_{j'j} = 1$, which gives

$$\sum_p \Gamma_{k'}(p)_{i'j'} \Gamma_k(p^{-1})_{ji} = 0,$$

or because Γ_k is unitary,

$$\sum_p \Gamma_{k'}(p)_{i'j'} \Gamma_k(p)^*_{ij} = 0.$$

This establishes the $\delta_{kk'}$ part of Eq. (3.76).

Now assume that Γ_k and $\Gamma_{k'}$ are the same representations (i.e., $k = k'$). We obtain, just as before, that the matrix

$$\mathbf{M} = \sum_p \mathbf{\Gamma}_k(p)\mathbf{A}\mathbf{\Gamma}_k(p^{-1})$$

satisfies the condition of Lemma VII, and in this case, the condition of Lemma VI. It follows that $\mathbf{M} = \lambda\mathbf{I}$. This means

$$M_{i'i} = \sum_p \sum_\alpha \sum_\beta \Gamma_k(p)_{i'\alpha} A_{\alpha\beta} \Gamma_k(p^{-1})_{\beta i} = \lambda\delta_{i'i}.$$

Choosing all $A_{\alpha\beta} = 0$ except $A_{j'j} = 1$, gives

$$\sum_p \Gamma_k(p)_{i'j'} \Gamma_k(p^{-1})_{ji} = \lambda(j'j)\delta_{i'i},$$

where we have indicated the dependence of λ upon the choice of $\mathbf{A}$. We can interchange the order of the factors on the left of this last result and sum over i, with $i = i'$, to obtain

$$\sum_p \sum_{i-1}^{l_k} \Gamma_k(p^{-1})_{ji} \Gamma_k(p)_{ij'} = \lambda(j'j) l_k.$$

But

$$\sum_{i=1}^{l_k} \Gamma_k(p^{-1})_{ji} \Gamma_k(p)_{ij'} = \Gamma_k(p^{-1}p)_{ij'} = \Gamma_k(E)_{jj'} = \delta_{jj'}.$$

We have, therefore,

$$\sum_{P} \delta_{jj'} = h\delta_{jj'} = \lambda(j'j)l_k$$

or

$$\lambda(j'j) = \frac{h\delta_{jj'}}{l_k}.$$

Substituting this result for λ into the expression for $M_{i'i}$ gives the $\delta_{ii'}\delta_{jj'}$ part of Eq. (3.76).

4

Symmetry Properties of Hamiltonians, Eigenfunctions, and Atomic Orbitals

The simplifying effects of symmetry in the quantum mechanical treatments of chemical systems have their origins in the symmetry of the Hamiltonians. To understand how this happens, we need to keep in mind certain elementary properties of Hamiltonians and of their exact and approximate eigenfunctions. We describe these properties briefly and then illustrate how symmetry can be used to predict certain properties of eigenfunctions and to simplify the problem of obtaining approximate eigenvalues and eigenfunctions.

4.1 Matrix Representations of the Hamiltonian Operator

The Schrödinger equation can be written

$$H\psi_m = E_m\psi_m, \qquad m = 1, 2, \ldots, n, \tag{4.1}$$

where H is the Hamiltonian operator and E_m is a constant equal in value to the energy of the mth quantum state, which is represented by the ψ function ψ_m. The constants E_m and the functions ψ_m, which satisfy Eq. (4.1), are called, respectively, the *eigenvalues* and the *eigenfunctions* of H. The number of states n is usually infinite. However, for our purpose, it will be satisfactory to think in terms of a finite number of states. When $E_m \neq E_{m'}$, the functions ψ_m and $\psi_{m'}$ are orthogonal (see Problem 4.1). When $E_m = E_{m'}$, we call ψ_m and $\psi_{m'}$ *degenerate*, and any linear combination of ψ_m and $\psi_{m'}$ is an eigenfunction with the eigenvalue E_m [see Problem 4.3(a)]. It is always possible to choose linear combinations that are mutually orthogonal. Since the functions can be normalized, it is possible to use eigenfunctions that form an orthonormal set, so that

$$\int \psi_m^* \psi_{m'} \, d\tau = \delta_{mm'}, \tag{4.2}$$

where $d\tau$ is the volume element corresponding to the variables of the ψ functions.

We define the mm'th element of the Hamiltonian matrix in the basis of its eigenvectors by

$$H_{mm'} = \int \psi_m^* H \psi_{m'} \, d\tau. \tag{4.3}$$

It is a principle of quantum mechanics that Hamiltonian operators are Hermitian ($H_{mm'}^* = H_{m'm}$). This ensures that all the eigenvalues (energies) are real (see Problem 4.2). From Eqs. (4.1) and (4.2) we obtain

$$H_{mm'} = E_m \, \delta_{mm'}, \tag{4.4}$$

which means the Hamiltonian is diagonal in the basis consisting of its orthonormal eigenfunctions [see Example 4.1(a)].

In applications, we usually have a Hamiltonian in the form of functions and differential operators, and we seek the eigenvalues and eigenfunctions. To obtain the Hamiltonian in matrix form we choose as basis a set of functions ϕ_i $(i = 1, 2, \ldots, p)$, and we assume that the eigenfunctions ψ_m can be expressed approximately as linear combinations of these,

$$\psi_m \simeq \sum_{i=1}^{p} a_{mi} \phi_i, \qquad m = 1, 2, \ldots, p, \tag{4.5}$$

where a_{mi} are constants. At the moment, we do *not* assume that the set ϕ_i is orthonormal, but it often is, and we shall later consider the simplifications resulting from orthogonality and normalization. The Hamiltonian matrix with the set ϕ_i as basis has the elements H_{ij}^{ϕ} given by,

$$H_{ij}^{\phi} = \int \phi_i^* H \phi_j \, d\tau. \tag{4.6}$$

The index ϕ on H_{ij}^{ϕ} is conventionally omitted, and we shall omit it in the future whenever there should be no confusion about which basis functions are being used. We use it here to emphasize the following important point. The finite set ϕ_i $(i = 1, 2, \ldots, p)$ will in general not form a *complete* set because it gives only approximations to the eigenfunctions of H, according to approximation (4.5). The $p \times p$ matrix H_{ij}^{ϕ} is an approximation of the $n \times n$ matrix given by Eq. (4.3). Ordinarily $p < n$ (often n is infinite).

Let us see how H_{ij}^{ϕ} can be used to obtain the approximations to the eigenvalues and eigenfunctions of H. Substituting from approximation (4.5) into Eq. (4.1) gives

$$H\left(\sum_{j=1}^{p} a_{mj}\phi_j\right) \simeq E_m\left(\sum_{j=1}^{p} a_{mj}\phi_j\right). \tag{4.7}$$

Multiplying by ϕ_i^* and integrating leads to

$$\sum_{j=1}^{p} a_{mj}H_{ij}^{\phi} \simeq E_m \sum_{j=1}^{p} a_{mj}S_{ij}, \tag{4.8}$$

where the *overlap integral* S_{ij} is defined by

$$S_{ij} = \int \phi_i^* \phi_j \, d\tau. \tag{4.9}$$

Approximation (4.8) can be written

$$\sum_{j=1}^{p} (H_{ij}^{\phi} - E_m S_{ij})a_{mj} \simeq 0. \tag{4.10}$$

In place of this approximation, we write

$$\sum_{j=1}^{p} (H_{ij}^{\phi} - \lambda_m S_{ij})a_{mj} = 0, \tag{4.11}$$

where λ_m is assumed to be an approximation to E_m. If in Eq. (4.11) we let i take on its values from 1 to p, then we obtain a set of linear homogeneous equations with the coefficients $a_{m1}, a_{m2}, \ldots, a_{mp}$ as unknowns. Such a set of equations has a set of nonzero solutions only if the determinant of its matrix is zero (see Chapter 3, Appendix 3A.13)

$$|H_{ij}^{\phi} - \lambda_m S_{ij}| = 0. \tag{4.12}$$

This equation is a polynomial equation of the pth degree in λ_m. Its solutions are approximations for p of the eigenvalues E_m of H. For each λ_m, there is a corresponding p-dimensional vector $(a_{m1}, a_{m2}, \ldots, a_{mp})$. The components of these vectors are the coefficients for approximation (4.5); p of the n states are thereby approximated. There is a *variation principle* that tells us something about which of the n states will be approximated [See Example 4.1(b) and Remark (2)].

In general, physically meaningful eigenfunctions must satisfy certain "auxiliary conditions." For example, the ψ functions for an atomic electron must be single-valued and must go to zero at the proper rate as the radius becomes infinite. It is to be expected that better approximations to ψ_m will be obtained if all the ϕ_i satisfy these auxiliary conditions (see Example 4.1). We can anticipate that (in some sense) the closer the functions ϕ_i are to the true functions ψ_m the better the approximations will be.

The problem exemplified by Eq. (4.12) is a generalized eigenvalue problem. When the functions ϕ_i are orthonormal so that

$$S_{ij} = \delta_{ij}, \tag{4.13}$$

Eq. (4.12) simplifies to

$$|H_{ij}^{\phi} - \lambda_m \delta_{ij}| = 0, \tag{4.14}$$

which is the ordinary eigenvalue problem. In the generalized problem (as well as the ordinary one), the values of λ_m are called the eigenvalues, the vectors $(a_{m1}, \ldots, a_{mp})$ are the eigenvectors, and the linear combinations in (4.5) are the eigenfunctions. Recall that eigenfunctions belonging to different eigenvalues are orthogonal (see Problem 4.1), and any linear combination of eigenfunctions with the same eigenvalue is an eigenfunction with that eigenvalue [see Problem 4.3(a)]. These combinations include those that produce mutually orthogonal functions. When the orthogonal eigenvectors and eigenfunctions are normalized, we have (see Problem 4.4)

$$\sum_{i=1}^{p} \sum_{j=1}^{p} a_{mi}^{*} a_{m'j} S_{ij} = \delta_{mm'}, \qquad m,m' = 1, 2, \ldots, p, \tag{4.15}$$

$$\int \psi_m^{*} \psi_{m'} \, d\tau = \delta_{mm'}, \qquad m,m' = 1, 2, \ldots, p, \tag{4.16}$$

where S_{ij} is given by Eq. (4.9). When the basis set ϕ_i is orthonormal, Eq. (4.15) reduces to

$$\sum_{i=1}^{p} a_{mi}^{*} a_{m'i} = \delta_{mm'}, \qquad m,m' = 1, 2, \ldots, p. \tag{4.17}$$

4.2 Symmetry Properties of the Hamiltonian

It is clear that the Hamiltonian operator for a molecule must be invariant under all the symmetry operations of the molecule. This suggests that a Hamiltonian matrix might take on an especially useful form when its basis is composed of basis functions for the irreducible representations of the symmetry group of the molecule. This inference is correct, and it is our purpose in this section to describe this situation. We first consider relationships between the symmetry of H and the properties of its eigenfunctions.

It can be shown [see Problem 4.3(c)] that the eigenfunctions of H form bases for the representations (normally irreducible[1]) of the symmetry group of H, and that partners for a given representation have the same eigenvalue. Therefore, there must be degeneracies *not lower* than the dimensions of the various irreducible representations. Thus, a molecule with $\mathbf{C}_{2v}$ symmetry might have no degeneracy, but a molecule with $\mathbf{C}_{3v}$ symmetry must have at least some double degeneracy.[2] The degeneracies may be greater than those indicated by the dimensions of the irreducible representations, however (see Example 4.2).

With its eigenfunctions as basis, H takes its diagonal form. Our task is usually to find the eigenvalues and eigenfunctions by solving a secular determinant that has as its basis some given set of functions, none of which is an eigenfunction. Often the true eigenfunctions are infinite in number, and our set of basis functions is a finite set. In such a case the diagonalization of the Hamiltonian matrix (i.e., the solving of the secular determinant) gives only approximate eigenvalues and approximate eigenfunctions of the true Hamiltonian.

In other words, we start with an approximate Hamiltonian expressed in some known basis ϕ_i, and the problem is to find the eigenvalues and the eigenfunctions ψ_m as given by approximation (4.5). This means we need to solve a pth degree polynomial equation—Eq. (4.14). We wish to use a basis set that will factor the secular determinant so that the polynomial factors can be solved separately. The principle that allows us to do this is the following.

Suppose the functions ϕ_i are a *symmetry-adapted* set, that is, are basis functions for the irreducible representations of the symmetry group of H. Then (see property (B4), Section 3.10.4) the integral

$$H_{ij} = \int \phi_i H \phi_j \, d\tau \tag{4.18}$$

is zero unless ϕ_i and ϕ_j are the same function or are from two different basis sets for the same irreducible representation and are corresponding partners in these two sets.

We can understand the significance of this principle better by considering the following hypothetical case.

Assume that the symmetry of H is $\mathbf{C}_{3v}$ and that we have a nine-dimensional basis that gives a reducible representation $\Gamma^{(9)}$ such that

$$\Gamma^{(9)} = 3E + 2A_2 + A_1. \tag{4.19}$$

Suppose by the use of projection operators (or otherwise) we obtain the

[1] They are not necessarily irreducible; see Example 4.2.

[2] We are neglecting spin and time-reversal degeneracies.

following basis sets for the irreducible representations contained in $\Gamma^{(9)}$:

$$[\phi_1^E(1), \phi_2^E(1)], \quad [\phi_1^E(2), \phi_2^E(2)], \quad [\phi_1^E(3), \phi_2^E(3)], \quad [\phi^{A_2}(1)], \quad [\phi^{A_2}(2)], \quad [\phi^{A_1}], \tag{4.20}$$

where we have placed the partners of the various representations in brackets; the numbers in parentheses designate the different basis sets for a given representation. An example of an off-diagonal matrix element that can be nonzero in this basis is

$$H_1^E(1,2) = \int \phi_1^E(1)^* H \phi_1^E(2)\, d\tau = \int \phi_1^E(2)^* H \phi_1^E(1)\, d\tau = H_1^E(2,1)^*, \tag{4.21}$$

where we have used the fact that H is Hermitian. If we now take the basis functions in the order

$$\phi_1^E(1), \quad \phi_1^E(2), \quad \phi_1^E(3), \quad \phi_2^E(1) \quad \phi_2^E(2), \quad \phi_2^E(3), \quad \phi^{A_2}(1), \quad \phi^{A_2}(2), \quad \phi^{A_1}, \tag{4.22}$$

the matrix $\mathbf{H}^\phi$ takes the *block-diagonal* form

$$\begin{array}{cc|ccccccccc|}
 & & \multicolumn{6}{c}{E} & \multicolumn{2}{c}{A_2} & A_1 \\
 & & \phi_1(1) & \phi_1(2) & \phi_1(3) & \phi_2(1) & \phi_2(2) & \phi_2(3) & \phi(1) & \phi(2) & \phi \\
 & \phi_1(1) & H_1^E(1,1) & H_1^E(1,2) & H_1^E(1,3) & 0 & 0 & 0 & 0 & 0 & 0 \\
 & \phi_1(2) & H_1^E(2,1) & H_1^E(2,2) & H_1^E(2,3) & 0 & 0 & 0 & 0 & 0 & 0 \\
E & \phi_1(3) & H_1^E(3,1) & H_1^E(3,2) & H_1^E(3,3) & 0 & 0 & 0 & 0 & 0 & 0 \\
 & \phi_2(1) & 0 & 0 & 0 & H_2^E(1,1) & H_2^E(1,2) & H_2^E(1,3) & 0 & 0 & 0 \\
 & \phi_2(2) & 0 & 0 & 0 & H_2^E(2,1) & H_2^E(2,2) & H_2^E(2,3) & 0 & 0 & 0 \\
 & \phi_2(3) & 0 & 0 & 0 & H_2^E(3,1) & H_2^E(3,2) & H_2^E(3,3) & 0 & 0 & 0 \\
A_2 & \phi(1) & 0 & 0 & 0 & 0 & 0 & 0 & H^{A_2}(1,1) & H^{A_2}(1,2) & 0 \\
 & \phi(2) & 0 & 0 & 0 & 0 & 0 & 0 & H^{A_2}(2,1) & H^{A_2}(2,2) & 0 \\
A_1 & \phi & 0 & 0 & 0 & 0 & 0 & 0 & 0 & 0 & H^{A_1}
\end{array} \tag{4.23}$$

Instead of a ninth-degree equation we now have two cubic equations, one quadratic equation, and a one-degree equation. These equations can be solved separately.

Note that the degree of an equation corresponding to a given irreducible representation is given by the coefficient in Eq. (4.19). The number of equations corresponding to that representation is equal to the dimension of the representation (two cubics for E, one quadratic for A_2, and one linear for A_1).

Let $\psi_{11}^E, \psi_{12}^E, \psi_{13}^E$ be the eigenfunctions corresponding to the top cubic in (4.23). They are linear combinations of $\phi_1^E(1), \phi_1^E(2), \phi_1^E(3)$; let their eigenvalues be E', E'', E'''. Similarly, let $\psi_{21}^E, \psi_{22}^E, \psi_{23}^E$ be the eigenfunctions from the second cubic. They are linear combinations of $\phi_2^E(1), \phi_2^E(2), \phi_2^E(3)$, and because the ψ^E functions are doubly degenerate, we know that their eigenvalues must also be E', E'', E'''. Because they have the same roots, the two cubic equations can differ only by a constant factor—there is only one cubic equation to be solved.

The eigenfunctions from the quadratic, $\psi_1^{A_2}$ and $\psi_2^{A_2}$, are linear combinations of $\phi^{A_2}(1)$ and $\phi^{A_2}(2)$, and they generally have different eigenvalues—they need not be, and usually are not, degenerate.

The eigenfunction ψ^{A_1} must be a constant times the basis function ϕ^{A_1}. In the foregoing case (4.23), the six pairs of basis functions of species E give three different eigenvalues and six eigenfunctions, which are degenerate in pairs. These eigenfunctions are each given as a linear combination of three basis functions. If we had used four pairs of basis functions of species E instead of three, we would have a fourth-degree equation to solve, leading to four different eigenvalues and eight eigenfunctions, degenerate in pairs. Each eigenfunction would be a linear combination of four basis functions. In general there should be some improvement in the approximations obtained. The idea is to choose basis functions so that the smallest possible number will give suitable approximations.

We postpone examples that indicate how basis sets are chosen until we meet them in subsequent applications (see, however, Examples 4.1 and 4.3). At the moment we make only the following observations. For a function of a basis set to affect a particular eigenvalue, the function must have a nonzero projection into the irreducible representation to which the eigenvalue belongs. The magnitude of the effect will depend upon the magnitudes of the matrix elements between this projection and the other basis functions belonging to the representation.

EXAMPLE 4.1 The one-dimensional Hamiltonian for a free particle of mass m can be written $H = -\partial^2/\partial x^2$, where the unit of energy has been taken so that $h^2/8\pi^2 m = 1$. This Hamiltonian becomes that of a particle in a one-dimensional box of length 2 with its center at the origin if we require the eigenfunctions to satisfy the following boundary conditions (these are the "auxiliary conditions" in this case)

$$\psi(-1) = \psi(1) = 0.$$

(a) Determine the eigenfunctions and eigenvalues for the particle in a box and express the Hamiltonian as a matrix with the eigenfunctions as basis.

(b) Express the Hamiltonian as a matrix with the following three functions as basis:

$$\phi_1 = (1 - x^2),$$

$$\phi_2 = x(1 - x^2),$$

$$\phi_3 = x^2(1 - x^2).$$

Calculate the eigenvalues and eigenfunctions of this matrix. Compare these eigenvalues with the true values and with the values given by individual functions ϕ_1, ϕ_2, and ϕ_3 taken separately, each as a one-dimensional problem.

SOLUTION (a) We have

$$H\psi = E\psi,$$
$$-\partial^2\psi/\partial x^2 = E\psi,$$
$$\psi(-1) = \psi(1) = 0.$$

The functions that satisfy this differential equation and the boundary conditions are

$$\psi_n = A\cos(n\pi x/2), \qquad n = 1, 3, 5 \ldots,$$
$$\psi_m = B\sin(m\pi x/2), \qquad m = 2, 4, 6 \ldots,$$

where A and B are constants. The corresponding values of E are

$$E_n = (n\pi/2)^2, \qquad E_m = (m\pi/2)^2.$$

The matrix representation of H with these eigenfunctions as basis is

$$\mathbf{H} = \frac{\pi^2}{4}\begin{bmatrix} 1 & 0 & 0 & 0 & \cdots \\ 0 & 4 & 0 & 0 & \cdots \\ 0 & 0 & 9 & 0 & \cdots \\ 0 & 0 & 0 & 16 & \cdots \\ \vdots & \vdots & \vdots & \vdots & \ddots \end{bmatrix}$$

(b) As is customary, we shall use write H_{ij} for H_{ij}^{ϕ}. From $H_{ij} = \int_{-1}^{1} \phi_i H \phi_j \, dx$ and $S_{ij} = \int_{-1}^{1} \phi_i \phi_j \, dx$ we obtain $H_{11} = \frac{8}{3}$, $H_{22} = \frac{8}{5}$, $H_{33} = \frac{88}{105}$, $H_{12} = H_{21} = 0$, $H_{13} = H_{31} = \frac{8}{15}$, $H_{23} = H_{32} = 0$, $S_{11} = \frac{16}{15}$, $S_{22} = \frac{16}{105}$, $S_{33} = \frac{16}{315}$, $S_{12} = S_{21} = 0$, $S_{13} = S_{31} = \frac{16}{105}$.

The Hamiltonian matrix is

$$[H_{ij}] = \begin{bmatrix} \frac{8}{3} & 0 & \frac{8}{15} \\ 0 & \frac{8}{5} & 0 \\ \frac{8}{15} & 0 & \frac{88}{105} \end{bmatrix}$$

The secular determinant is $|H_{ij} - \lambda S_{ij}| = 0$ or

$$\begin{vmatrix} H_{11} - \lambda S_{11} & 0 & H_{13} - \lambda S_{13} \\ 0 & H_{22} - \lambda S_{22} & 0 \\ H_{13} - \lambda S_{13} & 0 & H_{33} - \lambda S_{33} \end{vmatrix} = 0.$$

This can be factored (expanded in terms of second row or column) to give two equations:

$$H_{22} - \lambda S_{22} = 0 \qquad \text{and} \qquad \begin{vmatrix} H_{11} - \lambda S_{11} & H_{13} - \lambda S_{13} \\ H_{13} - \lambda S_{13} & H_{33} - \lambda S_{33} \end{vmatrix} = 0.$$

The first equation gives $\lambda_2 = H_{22}/S_{22} = \frac{8}{5} \times \frac{105}{16} = \frac{21}{2}$, and from the second equation

$$a\lambda^2 + b\lambda + c = 0, \qquad \lambda = \frac{-b \pm \sqrt{b^2 - 4ac}}{2a},$$

where

$$a = S_{11}S_{33} - S_{13}^2 = \frac{1024}{33{,}075},$$

$$b = -(H_{11}S_{33} + H_{33}S_{11} - 2H_{13}S_{13}) = \frac{-4096}{4725},$$

$$c = H_{11}H_{13} - H_{13}^2 = \frac{3072}{1575}.$$

These eigenvalues are listed below with the corresponding eigenvalues of the true Hamiltonian obtained in part (a).

λ	Approximate	True
λ_1	2.46744	2.46740
λ_2	10.5000	9.86960
λ_3	25.5325	22.20661

The energies corresponding to the individual functions taken separately are given by $\lambda_i = H_{ii}/S_{ii}$,

$$\lambda_1 = 2.5, \qquad \lambda_2 = 10.5, \qquad \lambda_3 = 16.5.$$

The eigenfunctions corresponding to the three-function basis are (we use the superscript ϕ to emphasize that these are approximations of the true functions ψ)

$$\psi_1^\phi = a_{11}\phi_1 + a_{12}\phi_2 + a_{13}\phi_3,$$

$$\psi_2^\phi = a_{21}\phi_1 + a_{22}\phi_2 + a_{23}\phi_3,$$

$$\psi_3^\phi = a_{31}\phi_1 + a_{32}\phi_2 + a_{33}\phi_3,$$

or in matrix notation,

$$\boldsymbol{\psi}^\phi = \mathbf{A}\boldsymbol{\phi},$$

where $\boldsymbol{\psi}^\phi$ and $\boldsymbol{\phi}$ are the vectors $(\psi_1^\phi, \psi_2^\phi, \psi_3^\phi)$ and (ϕ_1, ϕ_2, ϕ_3) and $\mathbf{A}$ is the square matrix $[a_{ij}]$. The columns of $\mathbf{A}$ are the eigenvectors of $[H_{ij} - \lambda S_{ij}]$. Thus, we have

$$\begin{bmatrix} H_{11} - \lambda_m S_{11} & 0 & H_{13} - \lambda_m S_{13} \\ 0 & H_{22} - \lambda_m S_{22} & 0 \\ H_{13} - \lambda_m S_{13} & 0 & H_{33} - \lambda_m S_{33} \end{bmatrix} \begin{bmatrix} a_{m1} \\ a_{m2} \\ a_{m3} \end{bmatrix} = 0.$$

This gives for each λ_m a set of three equations as follows [several significant figures are retained because they will be needed in Problem 4.4(b)]:

$$(H_{11} - \lambda_m S_{11})a_{m_1} + (H_{13} - \lambda_m S_{13})a_{m_3} = 0,$$
$$(H_{22} - \lambda_m S_{22})a_{m_2} = 0,$$
$$(H_{13} - \lambda_m S_{13})a_{m_1} + (H_{33} - \lambda_m)a_{m_3} = 0.$$

For $\lambda_1 = 2.46744$, these equations become

$$(0.0347364)a_{11} + (0.157344)a_{13} = 0,$$
$$a_{12} = 0,$$
$$(0.157344)a_{11} + (0.712770)a_{13} = 0.$$

These are satisfied by $a_{13} = -(0.22076)a_{11}$ and $a_{12} = 0$. Therefore, we have

$$\psi_1^\phi = a_{11}(1 - x^2) - (0.22076)a_{11}x^2(1 - x^2) = a_{11}[1 - (0.22076)x^2](1 - x^2).$$

For $\lambda_2 = 10.5$, the solution is $a_{21} = a_{23} = 0$ and $a_{22} = a_{22}$, and therefore,

$$\psi_2^\phi = a_{22}x(1 - x^2).$$

For $\lambda_3 = 25.5325$, the solution is $a_{33} = -(7.31772)a_{31}$ and $a_{32} = 0$, and these give

$$\psi_3^\phi = a_{31}[1 - (7.31772)x^2](1 - x^2).$$

These constants (e.g., a_{11}, a_{22}, and a_{31}) can be chosen for normalization.

REMARKS CONCERNING EXAMPLE 4.1 This example illustrates several principles that should be emphasized, and we do that by making the following comments:

(1) It is clear that the functions ϕ_1, ϕ_2, and ϕ_3 are approximations to the true eigenfunctions ψ_1, ψ_2, and ψ_3, respectively. They have the right parity and the right number of zeros in the range $(-1 \le x \le 1)$ for that association. It follows, provided the approximations are not too bad, that we can associate λ_1, λ_2, and λ_3 with E_1, E_2, and E_3, respectively. It also follows that ψ_i^ϕ $(i = 1, 2, 3)$, the approximation to ψ_i, should be most like ϕ_i and that a_{ii} should be the greatest (in absolute value) of the coefficients in ψ_i^ϕ.

(2) The solutions of the secular determinant satisfy the variation principle. This means that λ_1 and λ_3 cannot be less than the true value E_1 and that λ_2 cannot be less than E_2. The value of λ_2 cannot be associated with either true values E_1 or E_3, because ψ_2^ϕ and ψ_2 are odd functions, while ψ_1^ϕ and ψ_3^ϕ, as well as the true functions ψ_1 and ψ_3, are even functions. It follows that ψ_2^ϕ is orthogonal to both ψ_1 and ψ_2 and should not be considered an approximation to either of them.

(3) The point group of the Hamiltonian can be taken to be $\mathbf{C}_i$. Basis functions with different symmetry (in this case even or odd parity) do not mix. Thus ϕ_2, which belongs to species A_g, does not mix with ϕ_1 and ϕ_3, which belong to species

A_u. The matrix elements H_{12}, H_{21}, H_{23}, and H_{32} must all be zero, allowing the secular determinant to be factored.

(4) The functions ψ_1^ϕ, ψ_2^ϕ, and ψ_3^ϕ (as well as the true eigenfunctions ψ_1, ψ_2, and ψ_3) belong to the same symmetry species as ϕ_1, ϕ_2, and ϕ_3, respectively. Therefore, just as ϕ_2 must be orthogonal to ϕ_1 and ϕ_3, the function ψ_2^ϕ (ψ_2) must be orthogonal to ψ_1^ϕ and ψ_3^ϕ (ψ_1 and ψ_3). The function ψ_1^ϕ (ψ_1) is orthogonal to ψ_2^ϕ (ψ_2), because it is an eigenfunction with a different eigenvalue. [For a check that ψ_1^ϕ is orthogonal to ψ_2^ϕ, even though these functions have the same parity see Problem 4.4(b).]

(5) The addition of a new basis function with a certain symmetry will in general improve the approximation to the energy of the lowest state with that symmetry. The approximation to E_1, obtained when the function ϕ_1 was taken alone, is not as good as that obtained from ϕ_1 and ϕ_3 together. The presence of ϕ_1 and ϕ_3 does not affect the value of λ_2 (because there is no mixing of ϕ_2 with ϕ_1 and ϕ_3).

(6) If we consider only the arithmetic involved, it seems a remarkable coincidence that the nonzero matrix elements H_{13} and H_{31} are exactly equal to one another. However, this is not an accident. The differential equation and boundary conditions involved constitute a very simple Sturm–Liouville system,[3] and its matrix is Hermitian. It follows that the matrix representation in a basis that satisfies the boundary conditions will be Hermitian (symmetric when real). Recall that it is a quantum mechanical principle that all *true* Hamiltonian operators must be Hermitian. However, basis functions that do not satisfy all the auxiliary conditions (in this case the boundary conditions) can give approximate matrix representations that are not Hermitian.

EXAMPLE 4.2 The Hamiltonian for a particle in a two-dimensional square box with edges two units long and with the origin at the center of the box can be expressed (with energy units such that $h^2/8\pi^2m = 1$) as

$$H = -(\partial^2/\partial x^2 + \partial^2/\partial y^2), \qquad -1 \le x, y \le 1,$$

with the boundary conditions on $\psi(x, y)$

$$\psi(-1, y) = \psi(1, y) = \psi(x, -1) = \psi(x, 1) = 0.$$

(a) Show that the functions given in Eqs. (3.95) (Section 3.10.3) are the eigenfunctions, and calculate their eigenvalues.

(b) Correlate the degeneracies with the bases for the irreducible representations (of $\mathbf{C}_{4v}$) that were obtained in Section 3.10.3.

SOLUTION (a) The Schrödinger equation is

$$\partial^2\psi/\partial x^2 + \partial^2\psi/\partial y^2 = -E\psi,$$

with the boundary conditions given in the problem. By substituting the functions

[3] G. Arfken, *Mathematical Methods for Physicists*, 2nd ed., Academic Press, New York, 1970.

given by Eqs. (3.95) into this differential equation, it can be shown that the equation is satisfied when E takes the values

$$E^{c}_{nn'} = E^{c}_{n'n} = (\pi/2)^2[n^2 + (n')^2],$$

$$E^{s}_{mm'} = E^{s}_{m'm} = (\pi/2)^2[m^2 + (m')^2],$$

$$E^{cs}_{nm} = E^{sc}_{mn} = (\pi/2)^2[m^2 + n^2].$$

(b) The eigenfunctions, their general degeneracies g, and the symmetry species (obtained in Section 3.10.3) are tabulated as follows.

Functions	g	Symmetry species
$\psi^{c}_{nn}\ (n = n')$	1	A_1
$\psi^{s}_{mm}\ (m = m')$	1	B_2
$\psi^{c}_{nn'}, \psi^{c}_{n'n}\ (n \neq n')$	2	$A_1 + B_1$
$\psi^{s}_{mm'}, \psi^{s}_{m'm}\ (m \neq m')$	2	$A_2 + B_2$
$\psi^{cs}_{nm}, \psi^{sc}_{mn}$	2	E

REMARK CONCERNING EXAMPLE 4.2 The degeneracy of the pairs $(\psi^{c}_{nn'}, \psi^{c}_{n'n})$ and $(\psi^{s}_{mm'}, \psi^{s}_{m'm})$ is not required in terms of symmetry species. This type of degeneracy is sometimes called "accidental" degeneracy, even though it is symmetry-produced—the functions are transformed into one another by some operation of the group. Some examples of "accidental" degeneracy in a stricter sense are the following: $\psi^{c}_{5,5}$ has the same energy as the pair $\psi^{c}_{1,7}$ and $\psi^{c}_{7,1}$; $\psi^{cs}_{1,8}$, $\psi^{sc}_{8,1}$, $\psi^{cs}_{7,4}$, and $\psi^{sc}_{4,7}$ all have the same energy; $\psi^{cs}_{7,6}$, $\psi^{sc}_{6,7}$, $\psi^{cs}_{9,2}$, and $\psi^{sc}_{2,9}$ are degenerate also. Accidental degeneracy is not very common in more complicated systems.

EXAMPLE 4.3 In Example 4.2 we investigated the symmetry properties of the eigenfunctions of a particle in a square box. Because the Hamiltonian with these eigenfunctions as basis is diagonal, no problem presents itself concerning the use of symmetry-adapted basis functions to factor the secular determinant. To obtain an illustration of the factoring procedure we need some basis functions that are only approximations to the eigenfunctions.

(a) Use the approximate eigenfunctions for the one-dimensional box of Example 4.1(b) to construct nine approximate eigenfunctions for the two-dimensional box.

(b) From these functions, obtain the symmetry-adapted functions and describe the form of the factored secular determinant.

(c) The eigenfunctions of the approximate Hamiltonian (with the symmetry-adapted functions as basis) will be linear combinations of the symmetry-adapted functions. Indicate which of these symmetry-adapted functions should occur in which of the eigenfunctions.

SOLUTION (a) In the combined notation of Examples 4.1 and 4.2, the nine functions can be expressed as

$$\phi_{11}^{c} = \phi_1(x)\phi_1(y) = (1 - x^2)(1 - y^2), \qquad \phi_{12}^{cs} = (1 - x^2)y(1 - y^2),$$

$$\phi_{33}^{c} = x^2(1 - x^2)y^2(1 - y^2), \qquad \phi_{21}^{sc} = x(1 - x^2)(1 - y^2),$$

$$\phi_{13}^{c} = (1 - x^2)y^2(1 - y^2), \qquad \phi_{32}^{cs} = x^2(1 - x^2)y(1 - y^2),$$

$$\phi_{31}^{c} = x^2(1 - x^2)(1 - y^2), \qquad \phi_{23}^{sc} = x(1 - x^2)y^2(1 - y^2).$$

$$\phi_{22}^{s} = x(1 - x^2)y(1 - y^2),$$

The superscripts c and s have been used to indicate that these functions have the same symmetry as the corresponding cosine and sine eigenfunctions in Example 4.2.

(b) Using the results of Section 3.10.3, we find that the symmetry-adapted functions are as follows.

Functions	Symmetry species	Functions	Symmetry species
ϕ_{11}^{c}	A_1	ϕ_{22}^{s},	B_2
ϕ_{33}^{c}	A_1	$\phi_{12}^{cs}, \phi_{21}^{sc}$	E
$\phi_{13}^{c+} = \phi_{13}^{c} + \phi_{31}^{c}$	A_1	$\phi_{32}^{cs}, \phi_{23}^{sc}$	E
$\phi_{13}^{c-} = \phi_{13}^{c} - \phi_{31}^{c}$	B_1		

There will be a 3×3 determinant (cubic equation) obtained from the functions belonging to A_1. There will be two 1×1 determinants from ϕ_{13}^{c-} and ϕ_{22}^{s}, respectively. There will be two 2×2 determinants (quadratic equations): One will be from ϕ_{12}^{cs} and ϕ_{32}^{cs} and the other will be from ϕ_{21}^{sc} and ϕ_{23}^{sc}. These two quadratic equations must be the same equation (except for possibly a constant factor) because they must have the same roots. There are two double degeneracies. For each degenerate pair, one root comes from each of the quadratic equations.

(c) There will be three eigenfunctions belonging to A_1. They will each be linear combinations of ϕ_{11}^{c}, ϕ_{33}^{c}, and ϕ_{13}^{c+}. There will be one eigenfunction belonging to B_1. It will be ϕ_{13}^{c-}. There will be one eigenfunction belonging to B_2. It will be ϕ_{22}^{s}. There will be four eigenfunctions belonging to E. Two will be linear combinations of ϕ_{12}^{cs} and ϕ_{32}^{cs}, and the other pair will be linear combinations of ϕ_{21}^{sc} and ϕ_{23}^{sc}.

4.3 Atomic Orbitals

The group of the atom is K_h, and as we have previously stated (Section 3.13.2), the spherical harmonics $Y_{lm}(\theta, \phi)$ $(m = -l, \ldots, l)$, are a basis for the $(2l + 1)$-dimensional irreducible representation $\mathscr{D}^l$. These spherical harmonics are also the angular parts of $(2l + 1)$-fold degenerate eigenfunctions for the Hamiltonian of one-electron systems of spherical symmetry. Any unitary combination of these $(2l + 1)$ functions forms a representation equivalent to $\mathscr{D}^l$ and gives a set of eigenfunctions. The conventional atomic orbitals are combinations that give real functions. The angular part of the s orbital is Y_{00}; the angular parts of the p orbitals are combinations of $Y_{1\,-1}$, Y_{10}, and Y_{11}; etc. For example, the p orbitals are

$$p_x = -(1/\sqrt{2})(Y_{11} - Y_{1\,-1}), \qquad p_y = (i/\sqrt{2})(Y_{11} + Y_{1\,-1}), \qquad p_z = Y_{10}. \tag{4.24}$$

Polar and Cartesian forms of a conventional set of the angular parts of the s, p, d, and f orbitals are given in Table 4.1. The relationship between the polar and Cartesian form is given by the following equations relating polar and Cartesian coordinates:

$$x = r \sin\theta \cos\phi,$$

$$y = r \sin\theta \sin\phi,$$

$$z = r \cos\theta.$$

In Table 4.1, when a Cartesian form contains an r^2 in the numerator, a second, equivalent form is given in which r^2 has been replaced by $x^2 + y^2 + z^2$. This second form is more convenient for comparison with the Cartesian basis functions listed in the character tables. Note that the (sometimes shorter) subscripts on the symbols for the orbitals give the basis functions for the one-dimensional representations. However, for higher-dimensional representations the complete form is sometimes needed.

Other real forms of the atomic orbitals can be obtained by taking orthogonal combinations of those in Table 4.1. A particularly useful set is that in which the orbitals are bases for representations of the point group $\mathbf{O}_h$. In this set the p and d orbitals are the same as those given in Table 4.1, but some of the f orbitals are different. This *cubic* set of f orbitals is given in Table 4.2.

There are sets of *equivalent* orbitals in which all five d orbitals have the same shape and all seven f orbitals have the same shape.[4]

For systems of spherical symmetry, the quantum mechanical operators for the energy (i.e., the Hamiltonian), the total angular momentum, and one component of the angular momentum mutually commute. Therefore, these operators have simultaneous eigenfunctions. The $(2l + 1)$ functions Y_{lm} are the angular parts of such a simultaneous set. They have the same energy, the same

[4] L. Pauling and V. McClure, Five Equivalent d Orbitals, *J. Chem. Educ.*, 1970, **47**, 15. A. Schmelzer, Equivalent Spherical Harmonics, *Int. J. Quantum Chem.*, 1977, **11**, 561.

TABLE 4.1
Angular Factors of Conventional Atomic Orbitals

Symbol	Polar	Cartesian	Normalizing factor
s	1	1	$\frac{1}{2}\left(\frac{1}{\pi}\right)^{1/2}$
p_x	$\sin\theta\cos\phi$	x/r	$\frac{1}{2}\left(\frac{3}{\pi}\right)^{1/2}$
p_y	$\sin\theta\sin\phi$	y/r	$\frac{1}{2}\left(\frac{3}{\pi}\right)^{1/2}$
p_z	$\cos\theta$	z/r	$\frac{1}{2}\left(\frac{3}{\pi}\right)^{1/2}$
d_{z^2}	$(3\cos^2\theta - 1)$	$(3z^2 - r^2)/r^2$ $(2z^2 - x^2 - y^2)/r^2$	$\frac{1}{4}\left(\frac{5}{\pi}\right)^{1/2}$
d_{xz}	$\sin\theta\cos\theta\cos\phi$	xz/r^2	$\frac{1}{2}\left(\frac{15}{\pi}\right)^{1/2}$
d_{yz}	$\sin\theta\cos\theta\sin\phi$	yz/r^2	$\frac{1}{2}\left(\frac{15}{\pi}\right)^{1/2}$
$d_{x^2-y^2}$	$\sin^2\theta\cos 2\phi$	$(x^2 - y^2)/r^2$	$\frac{1}{4}\left(\frac{15}{\pi}\right)^{1/2}$
d_{xy}	$\sin^2\theta\sin 2\phi$	xy/r^2	$\frac{1}{4}\left(\frac{15}{\pi}\right)^{1/2}$
f_{z^3}	$(5\cos^3\theta - 3\cos\theta)$	$z(5z^2 - 3r^2)/r^3$ $[2z^3 - 3z(x^2 + y^2)]/r^3$	$\frac{1}{4}\left(\frac{7}{\pi}\right)^{1/2}$
f_{xz^2}	$(5\cos^2\theta - 1)\sin\theta\cos\phi$	$x(5z^2 - r^2)/r^3$ $[4xz^2 - x(x^2 + y^2)]/r^3$	$\frac{1}{8}\left(\frac{42}{\pi}\right)^{1/2}$
f_{yz^2}	$(5\cos^2\theta - 1)\sin\theta\sin\phi$	$y(5z^2 - r^2)/r^3$ $[4yz^2 - y(x^2 + y^2)]/r^3$	$\frac{1}{8}\left(\frac{42}{\pi}\right)^{1/2}$
f_{xyz}	$\sin^2\theta\cos\theta\sin 2\phi$	xyz/r^3	$\frac{1}{4}\left(\frac{105}{\pi}\right)^{1/2}$
$f_{z(x^2-y^2)}$	$\sin^2\theta\cos\theta\cos 2\phi$	$z(x^2 - y^2)/r^3$	$\frac{1}{4}\left(\frac{105}{\pi}\right)^{1/2}$
$f_{x(x^2-3y^2)}$	$\sin^3\theta\cos 3\phi$	$x(x^2 - 3y^2)/r^3$	$\frac{1}{8}\left(\frac{70}{\pi}\right)^{1/2}$
$f_{y(3x^2-y^2)}$	$\sin^3\theta\sin 3\phi$	$y(3x^2 - y^2)/r^3$	$\frac{1}{8}\left(\frac{70}{\pi}\right)^{1/2}$

TABLE 4.2
Angular Factors for Cubic f Orbitals

Symbol	Cartesian form	Normalizing factor	Symmetry species
f_{xyz}	(as in Table 4.1)		A_{2u}
f_{x^3}	$x(5x^2 - 3r^2)/r^3$ $[2x^3 - 3x(y^2 + z^2)]/r^3$	$\frac{1}{4}\left(\frac{7}{\pi}\right)^{1/2}$	T_{1u}
f_{y^3}	$y(5y^2 - 3r^2)/r^3$ $[2y^3 - 3y(x^2 + z^2)]/r^3$	$\frac{1}{4}\left(\frac{7}{\pi}\right)^{1/2}$	T_{1u}
f_{z^3}	(as in Table 4.1)		T_{1u}
$f_{z(x^2-y^2)}$	(as in Table 4.1)		T_{2u}
$f_{x(z^2-y^2)}$	$x(z^2 - y^2)$	$\frac{1}{4}\left(\frac{105}{\pi}\right)^{1/2}$	T_{2u}
$f_{y(z^2-x^2)}$	$y(z^2 - x^2)$	$\frac{1}{4}\left(\frac{105}{\pi}\right)^{1/2}$	T_{2u}

total angular momentum [equal to $(h/2\pi)\sqrt{l(l+1)}$], and the same z component of angular momentum (equal to $hm/2\pi$). The real atomic orbitals (except for $m = 0$) are linear combinations of harmonic functions with different values of m, so that they no longer correspond to a fixed value of the z component of angular momentum. For example, among the p orbitals, only p_z corresponds to a definite value (zero) of the z component of angular momentum.

Additional Readings

P. W. Atkins, *Molecular Quantum Mechanics*, Oxford Univ. Press, New York, 1970.
H. Eyring, J. Walter, and G. E. Kimball, *Quantum Chemistry*, Wiley, New York, 1944. Old, but still good.
R. L. Flurry, Jr., *Quantum Chemistry*, Prentice-Hall, Englewood Cliffs, New Jersey, 1983.
W. Kauzmann, *Quantum Chemistry*, Academic Press, New York, 1957.
I. N. Levine, *Quantum Chemistry*, 3rd ed., Allyn & Bacon, Boston, 1983.
J. P. Lowe, *Quantum Chemistry*, Academic Press, New York, 1978.
L. Pauling and E. B. Wilson, Jr., *Introduction to Quantum Mechanics*, McGraw-Hill, New York, 1935. A classic and good for quantum mechanical principles.
F. L. Pilar, *Elementary Quantum Chemistry*, McGraw-Hill, New York, 1968.
L. I. Schiff, *Quantum Mechanics*, 3rd ed., McGraw-Hill, New York, 1968. A physics text that chemistry students might find useful.

See also readings cited at the end of Chapter 2.

Problems

4.1 Prove that because H is Hermitian its eigenfunctions belonging to different eigenvalues must be orthogonal.

4.2 Prove that because H is Hermitian its eigenvalues must be real.

4.3 Operators that can be represented by matrices, such as symmetry transformations and Hamiltonians, are *linear* operators and as such satisfy the following type of relationship:

$$H(a\psi_1 + b\psi_2) = aH\psi_1 + bH\psi_2,$$

where a and b are constants and ψ_1 and ψ_2 are any functions on which H can operate. For the Hamiltonian operator H, prove the following:

(a) If ψ_1 and ψ_2 are two eigenfunctions of H with the same eigenvalue E, then any linear combination of ψ_1 and ψ_2 is also an eigenfunction with the eigenvalue E.

(b) If H is invariant under a transformation S (i.e., if $SHS^{-1} = H$), then H commutes with S.

(c) The eigenfunctions corresponding to the same eigenvalue E form a basis for a representation (of the symmetry group of H) with a dimensionality equal to the degeneracy of the eigenvalue E.

(d) The set of all *nonsingular* matrices that commute with H constitute a group. (A *singular* matrix is one for which the determinant vanishes. A nonsingular matrix has a nonzero determinant and so has an inverse.)

4.4 Consider the linear combinations

$$\psi_i = \sum_{j=1}^{n} a_{ij}\phi_j, \qquad i = 1, 2 \ldots, p,$$

and suppose that the set ϕ_j is not necessarily orthonormal.

(a) Prove that for ψ_i to be an orthonormal set we must have

$$\sum_{i,j} a_{ki}^* a_{lj} S_{ij} = \delta_{kl},$$

where

$$S_{ij} = \int \phi_i^* \phi_j \, d\tau.$$

(b) Use the result in part (a) to show that ψ_i and ψ_3 of Example 4.1(b) are orthogonal. (Of course, we could check the orthogonality by carrying out the integration $\int_{-1}^{1} \psi_1 \psi_3 \, d\tau_1$).

4.5 Use character tables to identify the representations for the f orbitals in $\mathbf{C}_{3h}$, $\mathbf{D}_{4h}$, and $\mathbf{O}_h$.

5

Spectral Terms and Stereoisomers

We have dealt with the symmetry properties of atomic orbitals and wave functions. Electronic energy states of isolated atoms or ions are characterized by term symbols that correspond closely in symmetry properties to atomic orbitals. We must be able to derive the term symbols for any electron configuration and obtain the symmetry species for the corresponding energy states for any environment. Before beginning major topics, we examine the usefulness of symmetry and group theory in dealing with stereoisomers.

5.1 Spectral Terms

5.1.1 RUSSELL–SAUNDERS COUPLING AND SPECTROSCOPIC TERM SYMBOLS

The orbital angular momentum l and spin angular momentum s of an electron are quantized in units of $h/2\pi$. The resultant angular momentum j is quantized also,

$$j(h/2\pi) = (l + \tfrac{1}{2})(h/2\pi) \qquad \text{or} \qquad j(h/2\pi) = (l - \tfrac{1}{2})(h/2\pi).$$

The individual orbitals of a set (characterized by a given value of l) are distinguished by the m_l values, corresponding to the components of angular momentum along one defined direction z.

$l = 0 \qquad m_l = 0$

$m_l = 1$

$l = 1 \qquad m_l = 0$

$m_l = -1$

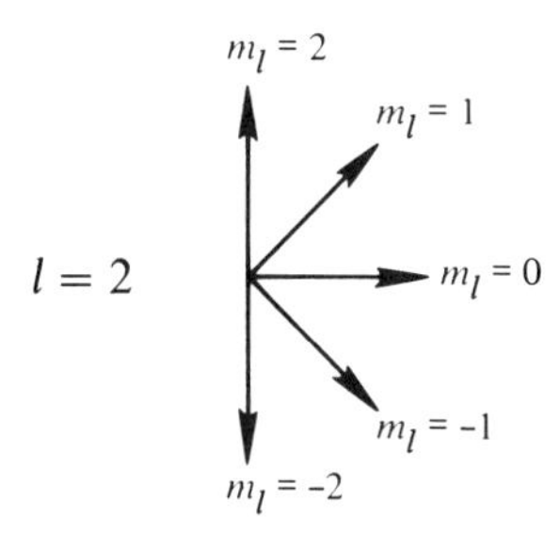

There are two ways of placing an electron in an s orbital (spin $\pm\frac{1}{2}$), but there is only one way to add a second electron since they must be paired. There are 10 ways to add an electron to the d orbitals (five orbitals and spin $\pm\frac{1}{2}$ for each) and only nine for the second electron, but since the electrons are indistinguishable, the total number of possibilities is $10 \times \frac{9}{2} = 45$. The number of *microstates* (distinguishable configurations) for d^3 and d^7 are given by

$$10 \times \frac{9}{2} \times \frac{8}{3} = 120 \quad \text{microstates} \qquad \text{for } d^3,$$

$$10 \times \frac{9}{2} \times \frac{8}{3} \times \frac{\cancel{7}}{\cancel{4}} \times \frac{\cancel{6}}{\cancel{5}} \times \frac{\cancel{5}}{\cancel{6}} \times \frac{\cancel{4}}{\cancel{7}} = 120 \quad \text{microstates} \qquad \text{for } d^7.$$

We see that d^7 is the "hole" counterpart of d^3; there are as many ways to add three holes to a complete d^{10} configuration as to add three electrons to empty d orbitals. The general expression for calculating the number of microstates for e electrons in n orbital sites (6 for p, 10 for d, etc.) is

$$\text{number of microstates} = n!/e!h!, \tag{5.1}$$

where h is the number of holes $(n - e)$.

In the Russell–Saunders coupling scheme for a polyelectron atom, the individual l values couple as vectors to give a resultant L, and s values couple to give a resultant S. For two electrons, L can have all integral values from $(l_1 + l_2)$ to $(l_1 - l_2)$. The S values differ by whole numbers from $\frac{1}{2}$ to $n/2$ (for odd n) or from 0 to $n/2$ (for even n), where n is the number of unpaired electrons. M_L is the sum of the m_l values for all electrons of the set and M_S is the sum of the m_s values. For a given L,

$$M_L = -L, -L+1, \ldots, 0, L-1, L, \qquad (2L+1) \quad \text{values},$$

$$L = \max M_L, \qquad M_L = \sum m_l.$$

For a given S there are also $(2S + 1)$ M_S values, differing by integers from $+S$ to $-S$. The total angular momentum J has all values in integral steps from $(L - S)$ to $(L + S)$, or $(2S + 1)$ values, if $L \geq S$.

Although there are 120 microstates for d^3 or d^7, there are only eight energy states. The labels used for electron energy levels were derived from atomic spectroscopy. Our labels for orbitals (s, p, d, f) were derived from the characterization of some lines in atomic emission spectra as S (sharp), P (principal), D (diffuse), and F (fundamental). The correspondence to L values is the same as for orbital labels and l values for single electrons. We follow an alphabetical sequence beyond F (omit J in the sequence),

L value	0	1	2	3	4	5	6	7	8	9	⋯
letter for term symbol	S	P	D	F	G	H	I	K	L	M	⋯

The spectroscopic term symbol $n\,{}^aT_j$ designates the L value (capital letter $S, P, D, \ldots$), the spin multiplicity a $(2S + 1)$, the value of J, and the major quantum number n (often omitted). The d^1 configuration has 10 microstates, but only one electron energy state, ${}^2D$, read as a doublet D. Here $L = 2$ and $S = \frac{1}{2}$, so J can be $\frac{5}{2}$ or $\frac{3}{2}$, giving $3^2D_{5/2}$ and $3^2D_{3/2}$ for $3d^1$. The total degeneracy, corresponding to 10 microstates of a ${}^2D$ term, is the spin multiplicity (2) times the orbital degeneracy (5 for D). The orbital degeneracy $(2L + 1)$ is the same as that of the corresponding atomic orbital set $(1, 3, 5, 7, 9, \ldots$ for $s, p, d, f, g, \ldots$, respectively).

5.1.2 DERIVATION OF SPECTRAL TERMS

Arrays of M_L and M_S for Microstates. A spectral term corresponds to a set of microstates. A microstate is described by the full electron configuration showing the population of each orbital. For a ${}^4D$ term, $L = 2$ and $S = \frac{3}{2}$, giving $M_L = 2, 1, 0, -1, -2$ $(+L$ to $-L)$ and $M_S = \frac{3}{2}, \frac{1}{2}, -\frac{1}{2}, -\frac{3}{2}$ $(+S$ to $-S)$. Each combination of M_L and M_S occurs once (enter 1 in the table), giving the array for ^4D,

M_L ↑	$\frac{3}{2}$	$\frac{1}{2}$	$-\frac{1}{2}$	$-\frac{3}{2}$
2	1	1	1	1
1	1	1	1	1
0	1	1	1	1
−1	1	1	1	1
−2	1	1	1	1

M_S

$L = 2$
$S = \frac{3}{2}$
$4 \times 5 = 20$ microstates

We can find the spectral terms for any electron configuration by writing all possible microstates (all arrangements of electrons) and tabulating the M_L and M_S values. For d^2, we have the microstates given in Table 5.1. First, we write all (10) possible ways of placing two electrons in separate orbitals, neglecting spin, and the five configurations with an electron pair in one orbital. The M_S values for each case with the electrons in separate orbitals are: $+\frac{1}{2}+\frac{1}{2}$, $+\frac{1}{2}-\frac{1}{2}$, $-\frac{1}{2}+\frac{1}{2}$, $-\frac{1}{2}-\frac{1}{2}$. M_S can be only zero for two electrons in one orbital. The number of times each combination of M_L and M_S occurs is tabulated in the array for d^2. In this array we see that the highest M_L is 4, with $M_S = 0$, so there must be an array for $L = 4$, $S = 0$, since for $L = 4$, $M_L = \pm 4, \pm 3, \pm 2, \pm 1, 0$ and for $S = 0$, $M_S = 0$. After we subtract the array for $L = 4$, $S = 0$, the highest M_L is 3 with $M_S = \pm 1$, requiring the presence of an $L = 3$, $S = 1$ (^3F) term. There must be an L value equal to the highest M_L value and an S value equal to the highest M_S value. Subtracting the array for $L = 3$, $S = 1$ leaves the maximum $M_L = 2$ with $M_S = 0$, indicating a ^1D term ($L = 2$, $S = 0$). Subtracting the array shown for ^1D leaves arrays for $L = 1$, $S = 1$ (^3P) and, finally, for $L = 0$, $S = 0$ (^1S).

At each step we have used the maximum M_L in the array to identify a value of L. From the corresponding M_S values we identify the S value for the spectral term. The spectral term ^1G is an energy state consisting of 1 (spin degeneracy) $\times$ 9 (orbital degeneracy) microstates, as tabulated above. After subtracting the array for ^1G from the full array for d^2, we identify the $L = 3$, $S = 1$ term. Continuing the process identifies the $L = 2$, $L = 1$, and $L = 0$ terms, giving

$$\begin{array}{ccccc} ^1S & ^1D & ^1G & ^3P & ^3F \\ 1 + & 5 + & 9 + & 3 \times 3 + & 3 \times 7 = 45 \text{ microstates.} \end{array}$$

Using Hund's rules we can identify the ground-state term (lowest energy term). The energy sequence of the other terms must be obtained from experiment. The ground-state term has maximum S (maximum number of unpaired electrons or multiplicity), so it is a triplet. Where there is more than one term

TABLE 5.1
Microstates[a] for d^2

$m_l =$ +2	+1	0	−1	−2	$M_L = \sum m_l$	$M_S = \sum m_s$
/	/				3	+1, 0, 0, −1
/		/			2	+1, 0, 0, −1
/			/		1	+1, 0, 0, −1
/				/	0	+1, 0, 0, −1
	/	/			1	+1, 0, 0, −1
	/		/		0	+1, 0, 0, −1
	/			/	−1	+1, 0, 0, −1
		/	/		−1	+1, 0, 0, −1
		/		/	−2	+1, 0, 0, −1
			/	/	−3	+1, 0, 0, −1
×					4	0
	×				2	0
		×			0	0
			×		−2	0
				×	−4	0

Array for d^2

M_L	M_S = 1	0	−1
4		1	
3	1	2	1
2	1	3	1
1	2	4	2
0	2	5	2
−1	2	4	2
−2	1	3	1
−3	1	2	1
−4		1	

Arrays for spectral terms contained in array for d^2

$L = 4$, $S = 0$, term ^1G

M_L	M_S = 0
4	1
3	1
2	1
1	1
0	1
−1	1
−2	1
−3	1
−4	1

$L = 3$, $S = 1$, term ^3F

M_L	M_S = 1	0	−1
3	1	1	1
2	1	1	1
1	1	1	1
0	1	1	1
−1	1	1	1
−2	1	1	1
−3	1	1	1

$L = 2$, $S = 0$, term ^1D

M_L	M_S = 0
2	1
1	1
0	1
−1	1
−2	1

$L = 1$, $S = 1$, term ^3P

M_L	M_S = 1	0	−1
1	1	1	1
0	1	1	1
−1	1	1	1

$L = 0$, $S = 0$, term ^1S

M_L	M_S = 0
0	1

[a] $10 \times 9/2 = 45$ microstrates.

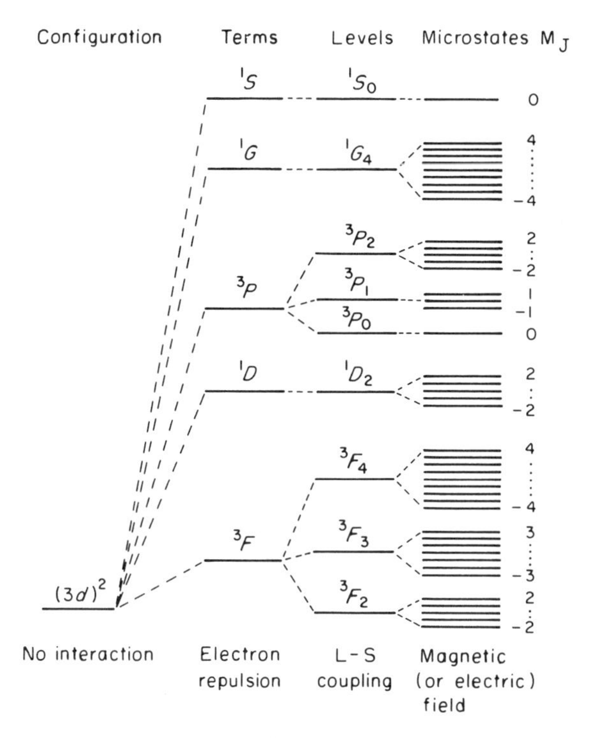

Fig. 5.1 Splitting of energy states for $3d^2$ (not to scale). Adapted from C. J. Ballhausen, *Introduction to Ligand Field Theory*, McGraw-Hill, New York, 1962, p. 31.

with maximum S, it is the one with maximum L, here ^3F. The J values for ^3F are 4 (3 + 1), 3, and 2 (3 − 1). For an orbital set less than half filled, the minimum J value is lowest in energy, so for d^2, it is $^3F_2$. The d^7 (3-hole) configuration gives the same spectral terms as for d^3; but the lowest-energy level is $^3F_4$, since the *maximum* value of J has lowest energy if the orbitals are more than half filled. The half-filled case with maximum spin-multiplicity gives an S term ($L = 0$) with only one J value.

The splitting of the energy states for the $3d^2$ *configuration* is shown in Fig. 5.1. The *terms* result from electron repulsion and L–S coupling distinguishes the J *levels*. The degeneracy of each of these levels (degeneracy is equal to $2J + 1$) is removed in a magnetic (or electric) field to distinguish the M_J values of the *microstates*. There are 45 microstates, corresponding to the total degeneracy of the d^2 configuration. We can identify the ground-state term (^3F) using Hund's rules, but the sequence of the other terms is obtained from atomic spectroscopy.

EXAMPLE 5.1 Without deriving all spectral terms, obtain the ground state levels for Cr^{2+} (d^4), Ni^{2+} (d^8), and the oxygen atom.

SOLUTION The microstate with maximum M_S and M_L must be one of the microstates of the array for the ground-state term having maximum S, and for this S, maximum L. This can be achieved by filling the orbitals individually, starting with the highest m_l, and adding one electron to each orbital before any is doubly occupied.

Configuration	Microstate with max M_S and max M_L (for max M_S)						
	m_l:	2	1	0	-1	-2	
Cr^{2+} (d^4)		↑	↑	↑	↑		$M_L = 2$, $M_S = 2$, so $L = 2$, $S = 2$
Ni^{2+} (d^8)		⇅	⇅	⇅	↑	↑	$M_L = 3$, $M_S = 1$, so $L = 3$, $S = 1$

Configuration	Ground state term	Ground state level
Cr^{2+} (d^4)	^5D	$^5D_0$ (lowest J, less than half filled)
Ni^{2+} (d^8)	^3F	$^3F_4$ (highest J, more than half filled)
O	m_l 1 0 −1 ⇅ ↑ ↑ $M_L = 1$, $M_S = 1$, so $L = 1$, $S = 1$	$^3P_2$ (highest J)

The writing of all microstates and tabulating the M_L and M_S values is tedious but instructive. It helps to convince us that a term does not correspond to one microstate, but to an array of microstates. Even ^1S, with $M_L = 0$ and $M_S = 0$, does not correspond to a unique microstate (configuration) belonging only to this term. For d^2, there are five microstates giving this combination. Procedures often given do not eliminate microstates that violate the Pauli exclusion principle (as we have done). These are usually demonstrated for the p^2 configuration only since they become terribly cumbersome for more than two electrons. Applications of the procedure just shown are available to the p^2 case,[1] d^5 case,[2] and d^3 case.[3] The terms for d^n configurations (using Russell–Saunders coupling) are given in Table 5.2.

[1] B. E. Douglas, D. H. McDaniel, and J. J. Alexander, *Concepts and Models of Inorganic Chemistry*, 2nd ed., Wiley, New York, 1983.

[2] B. E. Douglas and D. H. McDaniel, *Concepts and Models of Inorganic Chemistry*, Wiley, New York, 1965; J. Vicente, *J. Chem. Educ.* 1983, **60,** 560.

[3] K. E. Hyde, *J. Chem. Educ.* 1975, **52,** 87.

TABLE 5.2
Terms for d^n Configurations Using Russell–Saunders Coupling[a]

d^0, d^{10}	^1S
d^1, d^9	^2D
d^2, d^8	$^1(S, D, G)$ $^3(P, F)$
d^3, d^7	^2D $^2(P, D, F, G, H)$ $^4(P, F)$
d^4, d^6	$^1(S, D, G)$ $^1(S, D, F, G, I)$ $^3(P, F)$ $^3(P, D, F, G, H)$ ^5D
d^5	^2D $^2(P, D, F, G, H)$ $^2(S, D, F, G, I)$ $^4(P, D, F, S)$ ^6S

[a] In each case the ground state term is listed last.

Spectral Terms by Spin Factoring. Fortunately there is a less tedious method for deriving spectral terms. Even though the terms are tabulated in many sources, it is worthwhile to learn the method of spin factoring[1,4] because it provides a useful approach for obtaining correlation diagrams (qualitative energy-level diagrams) with the spin multiplicities assigned for any configuration in any symmetry (see Section 9.4).

In applying the spin-factoring method, we obtain the "partial terms" for each of the spin sets (the electrons with all $m_s = +\frac{1}{2}$ or $-\frac{1}{2}$ form a spin set). The partial terms of the spin sets (designated α and β) are multiplied to generate the complete terms. Since an empty, a half-filled, or a filled complete set of orbitals contributes nothing to the orbital angular momentum, corresponding to an S term, an empty spin set or a complete spin set (half-filled orbitals) also gives an S partial term. One electron in a complete orbital set gives a term with the corresponding label, $s^1 \rightarrow S$, $p^1 \rightarrow P$, $d^1 \rightarrow D$, $f^1 \rightarrow F$, etc., so a spin set (α or β) consisting of one electron gives these same partial terms. The partial terms for various numbers of electrons in a spin set are given for orbitals through g in Table 5.3. One vacancy in a spin set (one hole) gives the same partial term as for one electron, P for p_α^1 or p_α^2, D for d_α^1 or d_α^4, F for f_α^1 or f_α^6, etc. If we write the microstates for the spin set d_α^2, we get the first 10 microstates in Table 5.1, giving M_L values 3, 2, 1, 0, -1, -2, -3 and 1, 0, -1, corresponding to F and P terms. Thus the partial terms for two d electrons of one spin set are F and P. The d_α^3 configuration is the hole counterpart of d_α^2 and generates the same terms (try it). The partial terms for f^2, f^3, g^2, etc., in Table 5.3 can be generated in the same way (see ref. 4). We need only the terms to the left of the solid line since those to the right are obtained by the hole formalism.

Now let us use the partial terms to derive the spectral terms for the d^2 configuration. The possible spin configurations are $d_\alpha^1 d_\beta^1$, $d_\alpha^2 d_\beta^0$, and $d_\alpha^0 d_\beta^2$. We obtain the partial terms for each spin configuration from Table 5.3 and take the *product* of the partial terms. The product of partial terms with L values L_1 and L_2 include all integral L values from $(L_1 + L_2)$ through $(L_1 - L_2)$. For d^2

[4] D. H. McDaniel, *J. Chem. Educ.* 1977, **54**, 147.

TABLE 5.3
Partial Terms Arising from the Occupancy of a Single Spin Set[a,b]

	Orbital occupancy (electrons or holes)							
Orbital set	0	1	2	3	4	5	6	7
s	*S*	*S*						
p	*S*	*P*	*P*	*S*				
d	*S*	*D*	*PF*	*PF*	*D*	*S*		
f	*S*	*F*	*PFH*	*SDFGI*	*SDFGI*	*PFH*	*F*	*S*
g	*S*	*G*	*PFHK*	*PF*[2]*G*[c] *HIKM*	*SD*[2]*FG*[2] *HI*[2]*KLN*			

[a] Adapted from D. H. McDaniel, *J. Chem. Educ.* 1977, **54**, 147.
[b] Either α or β.
[c] [2] means the preceding partial term occurs twice.

we have

$$d_\alpha^2 d_\beta^0 \quad 5 \times \tfrac{4}{2} \times 1 = 10 \ \text{(degeneracy)} \qquad (P + F) \times S \rightarrow \begin{cases} P & L = 1 \\ F & L = 3 \end{cases} \quad M_S = +1$$

$$d_\alpha^0 d_\beta^2 \quad 10 \ \text{(degeneracy)} \qquad S \times (P + F) \rightarrow \begin{cases} P & L = 1 \\ F & L = 3 \end{cases} \quad M_S = -1$$

$$d_\alpha^1 d_\beta^1 \quad 5 \times 5 = 25 \ \text{(degeneracy)} \qquad D \times D \rightarrow \begin{cases} G & L = 4 \quad (2 + 2) \\ F & L = 3 \\ D & L = 2 \quad M_S = 0 \\ P & L = 1 \\ S & L = 0 \quad (2 - 2) \end{cases}$$

Thus we have ^3P and ^3F since $S = 1$ (from $M_S = \pm 1$). The number of microstates included in $^3P\,(3 \times 3)$ and $^3F\,(3 \times 7)$ is 30. The total degeneracy of the $d_\alpha^2 d_\beta^0$ and $d_\alpha^0 d_\beta^2$ configurations is 20, so the $M_S = 0$ components of $L = 3$ and $L = 1$ are contained in the terms generated from the $d_\alpha^1 d_\beta^1$ configuration. Thus, we eliminate the ^1F and ^1P terms generated from $d_\alpha^1 d_\beta^1$. We start with the configuration giving terms of highest spin multiplicity, realizing that the lower-M_S microstates, corresponding to terms of lower spin-multiplicity, will be replicated. The terms remaining are ^1S, ^1D, ^1G, ^3P, and ^3F. We can check the total degeneracy of these terms against the 45 microstates. McDaniel[4] applied spin factoring to the p^n configurations and to f^4 (in about 10 lines). Reference[1] deals with the p^2 and d^3 configurations.

EXAMPLE 5.2 Derive the spectral terms for the f^3 configuration by spin factoring. Identify the ground-state energy level (including J).

SOLUTION For f^3, there are $14 \times \frac{13}{2} \times \frac{12}{3} = 364$ microstates. The possible configurations are as follows.

	f^3_α	f^3_β	$f^2_\alpha f^1_\beta$	$f^1_\alpha f^2_\beta$	
Degeneracy	35	35	$21 \times 7 = 147$	$7 \times 21 = 147$	Total = 364

$f^3_\alpha f^0_\beta$ (or $f^0_\alpha f^3_\beta$) yields $^4[S(S + D + F + G + I)] = {}^4S + {}^4D + {}^4F + {}^4G + {}^4I$

$M_S = \pm\frac{3}{2}$ Degeneracy $= 140 = 4 + 20 + 28 + 36 + 52$

The degeneracy here is 140 (not 70) because the $M_S = \pm\frac{1}{2}$ terms are included. We must drop the corresponding doublets generated below.

$f^2_\alpha f^1_\beta$ (or $f^1_\alpha f^2_\beta$) yields $^2[F(P + F + H)]$ $M_S = \pm\frac{1}{2}$.

Drop ^2S, ^2D, ^2F, ^2G, ^2I,

$$F \times P = D + F + G \quad (3 - 1 \text{ to } 3 + 1),$$

$$F \times F = S + P + D + F + G + H + I,$$

$$F \times H = D + F + G + H + I + K + L.$$

The remaining terms are

$$^2P,\ ^2D[2],\ ^2F[2],\ ^2G[2],\ ^2H[2],\ ^2I,\ ^2K,\ ^2L,\ ^4S,\ ^4D,\ ^4F,\ ^4G,\ ^4I.$$

Ground-state term is ^4I ($L = 6, S = \frac{3}{2}, J = \frac{15}{2}, \frac{13}{2}, \frac{11}{2}, \frac{9}{2}$). Ground-state level is $^4I_{9/2}$.

5.1.3 *jj* COUPLING

For light atoms, the Russell–Saunders (L–S) coupling scheme is consistent with spectra. Instead of s values coupling to give S and l values coupling to give L, it is also possible for l and s to couple to give j. The j values for each electron couple to give J, the total angular momentum of the atom. This jj coupling scheme applies for heavy elements. For many transition elements, the coupling is intermediate, neither pure L–S nor pure jj. Usually the coupling will be closer to one of the model schemes, so we use that model and then consider the other effect as a perturbation. Such situations are considered under spin–orbit coupling (see Section 9.8).

5.2 Symmetry Species for Atomic Orbitals and Spectral Terms

The spherically symmetrical point group $\mathbf{K}_h$ (*Kugel Gruppe*) contains all symmetry operations. The pure rotational subgroup $\mathbf{K}$ contains all rotations (see Section 3.13.2). In either of these groups, a complete set of orbitals is a basis for an irreducible representation of the group. The s orbital is the totally

TABLE 5.4
Representations for Atomic Orbitals in the K_h Group[a]

K_h	l	E	C_4	C_2	C_3	i	σ
s	0	1	1	1	1	1	1
p	1	3	1	−1	0	−3	1
d	2	5	−1	1	−1	5	1
f	3	7	−1	−1	1	−7	1
g	4	9	1	1	0	9	1

[a] Characters are given for some common operations.

symmetric one-dimensional representation, the p orbitals give a three-dimensional representation, d orbitals are five-dimensional, etc. The character for rotation by the angle ω is given by the equation

$$\chi(\omega) = \frac{\sin(l + \frac{1}{2})\omega}{\sin(\omega/2)}, \tag{5.2}$$

where l is the quantum number for the orbital angular momentum. The characters for the identity i and σ are given by

$$\begin{aligned} \chi(E) &= 2l + 1 \quad \text{(orbital degeneracy)}, \\ \chi(\sigma) &= \pm\sin(l + \tfrac{1}{2})\pi = (-1)^l\chi(C_2), \\ \chi(i) &= \pm(2l + 1) = \pm\chi(E) = (-1)^l\chi(E), \end{aligned} \tag{5.3}$$

where the $\pm$ signs are taken as + for l even (*gerade* orbitals or states) and − for l odd (*ungerade* orbitals or states). The basis for dealing with $\chi(E)$, $\chi(i)$, and $\chi(\sigma)$ was given in Example 3.18. Derivations of expressions for $\chi(\sigma)$ and $\chi(S_n)$ can be found.[5] Characters for some l values in $\mathbf{K}_h$ are given in Table 5.4.

All other groups are subgroups of $\mathbf{K}_h$, and all rotational groups are subgroups of **K**. If we lower the symmetry to **O**, we find that the representation for the s orbital is A_1 and the representation for the p orbitals in **O** is T_1. There is no five-dimensional representation, so the representation obtained for the d orbitals is a reducible representation in the **O** group. It reduces to $E + T_2$. The representation in **K** for the f orbitals reduces to $A_2 + T_1 + T_2$ in the **O** group. Orbitals of even parity (l even) are g and those of odd parity (l odd) are u. Hence, in $\mathbf{O}_h$ symmetry the representations are s, A_{1g}; p, T_{1u}; d, $E_g + T_{2g}$; and f, $A_{2u} + T_{1u} + T_{2u}$. We can find the representations for the rotational group using Eq. (5.2) and add the g or u subscripts for the corresponding centrosymmetric group.

[5] R. L. DeKock, A. J. Kromminga, and T. S. Zwier, *J. Chem. Educ.* 1979, **56**, 510.

We can obtain the characters for a rotational group for any L value of a term using Eq. (5.2) as we did for l values of orbitals. Since we are usually concerned with C_2, C_3, or C_4 rotations and σ, the results are tabulated as

$$\chi(C_2) = \begin{cases} +1 & \text{for} \quad L = 0, 2, 4, \ldots, \\ -1 & \text{for} \quad L = 1, 3, 5, \ldots, \end{cases}$$

$$\chi(C_3) = \begin{cases} +1 & \text{for} \quad L = 0, 3, 6, 9, \ldots, \\ 0 & \text{for} \quad L = 1, 4, 7, 10, \ldots, \\ -1 & \text{for} \quad L = 2, 5, 8, 11, \ldots, \end{cases} \tag{5.4}$$

$$\chi(C_4) = \begin{cases} +1 & \text{for} \quad L = 0, 1, 4, 5, \ldots, \\ -1 & \text{for} \quad L = 2, 3, 6, 7, \ldots, \end{cases}$$

$$\chi(\sigma) = +1 \text{ for any} \quad L \quad \text{(any integer)}. \tag{5.5}$$

Because we operate on l and L in the same way, the representations for any given rotational point group are the same for a term as for the corresponding orbital set, that is, under **O** symmetry, s and S are A_1, p and P are T_1, d and D are $E + T_2$, etc. (see Table 5.5). For the $\mathbf{O}_h$ group, we add the g (even l) and u (odd l) subscripts to the representations for the orbitals. Terms derived from orbitals of even parity (s, d, and g orbitals) are g. Terms derived from orbitals of odd parity (p and f) are odd (u) for an odd number of electrons and even (g) for an even number of electrons. Since we are concerned here with the terms derived from d^n configurations, all terms are even. We add the subscript g to the representations in Table 5.5 to obtain the representations under $\mathbf{O}_h$. The spin multiplicities are the same as those of the parent terms, e.g., $^2D \rightarrow {}^2E_g + {}^2T_{2g}$ ($\mathbf{O}_h$).

EXAMPLE 5.3 Determine the representations in $\mathbf{D}_{3h}$ for (a) d orbitals, (b) f orbitals, and (c) an F term derived from a d^n configuration.

TABLE 5.5
Representations of Terms in the O Point Group

Term	L	E	$6C_4$	$3C_2\ (C_4^2)$	$8C_3$	$6C_2$	
S	0	1	1	1	1	1	A_1
P	1	3	1	−1	0	−1	T_1
D	2	5	−1	1	−1	1	$E + T_2$
F	3	7	−1	−1	1	−1	$A_2 + T_1 + T_2$
G	4	9	1	1	0	1	$A_1 + E + T_1 + T_2$
H	5	11	1	−1	−1	−1	$E + 2T_1 + T_2$
I	6	13	−1	1	1	1	$A_1 + A_2 + E + T_1 + 2T_2$

SOLUTION The representations for $l = 2$ and $l = 3$ for $\mathbf{D}_3$ are as follows.

$\mathbf{D}_3$	E	C_3	C_2	
$l = 2$	5	-1	1	$= A_1 + 2E$
$l = 3$	7	1	-1	$= A_1 + 2A_2 + 2E$

We can use the fact that reflection through a symmetry plane is an improper rotation [$\sigma_{xz} = i(C_2)_y$; see Example 3.18]. This is valid even though $\mathbf{D}_{3h}$ does not have an inversion center or a symmetry plane perpendicular to a C_2 axis. The orbitals or terms are inherently g or u, and for the $\mathbf{K}_h$ group there are σ planes and C_2 axes with all orientations. The characters for σ_h or σ_v are

$$\chi(\sigma) = (-1)^l \chi(C_2) = (+1)(+1) = +1, \qquad \text{for} \quad l = 2,$$

$$\chi(\sigma) = (-1)(-1) = +1, \qquad \text{for} \quad l = 3.$$

The representations for the orbital sets are

$$A_1' + E' + E'', \qquad \text{for} \quad l = 2$$

$$A_1' + A_2' + A_2'' + E' + E'', \qquad \text{for} \quad l = 3.$$

In each case the ′ or ″ distinctions are made using the characters for σ_h. For $L = 3$ of even parity, $\chi(i)$ is positive, giving $\chi(\sigma) = -1$. Here we need the character (-1) for σ_h *and* σ_v to obtain $A_2' + A_1'' + A_2'' + E' + E''$. Thus we see that the result is not the equivalent of changing u for f orbitals to g for an F term of even parity as it is for centrosymmetric groups (see Example 5.5).

Correlation Tables. We can obtain the representations for other symmetries by applying Eq. (5.2). However, many symmetries can be related to the commonly encountered $\mathbf{O}_h$ case. It is the most thoroughly studied case, and we can find abundant data and correlation diagrams for $\mathbf{O}_h$. It is usually convenient to treat cases of lower symmetry as the result of distortion from $\mathbf{O}_h$ symmetry. For this purpose, correlation tables are used to relate the representations in the new group to those for $\mathbf{O}_h$. The lowering of symmetry often results in removal (or partial removal) of degeneracy, so we speak of the B_{2g} and E_g representations in $\mathbf{D}_{4h}$ symmetry as "derived from" T_{2g} in $\mathbf{O}_h$ symmetry, or $B_{2g} + E_g$ ($\mathbf{D}_{4h}$) of T_{2g} ($\mathbf{O}_h$) parentage. We also speak of the splitting of T_{2g} ($\mathbf{O}_h$) to give $B_{2g} + E_g$ ($\mathbf{D}_{4h}$). The removal of degeneracy (or splitting) of representations for terms occurs in the same way as for the corresponding orbitals (allowing for any g–u difference in labels). Since we can identify the representations for p, d, and f orbitals from the character tables, this provides a way of getting representations for P, D, and F terms (except they are all g for d^n configurations).

A correlation table is derived for a group and its subgroup by treating the representations of the parent group as representations of the subgroup, tabulating the characters for these representations for the symmetry elements in common for the two groups. Some of these representations can be identified as irreducible representations of the subgroup, and some must be reduced to find the irreducible representations contained (see Example 5.4).

EXAMPLE 5.4 Find the representations in $\mathbf{D}_4$ derived from those in $\mathbf{O}$, that is, derive the correlation table for $\mathbf{O}$ and $\mathbf{D}_4$.

SOLUTION We write the character table for $\mathbf{D}_4$ and the characters for the representations of $\mathbf{O}$ for the symmetry operations corresponding to those of $\mathbf{D}_4$. C_2 and $2C_2'$ of $\mathbf{D}_4$ correspond to the $3C_2$ $(= C_4^2)$ of $\mathbf{O}$. The identical characters for these C_2 operations of $\mathbf{O}$ are written in two columns to correspond to the $\mathbf{D}_4$ classes. The totally symmetric representation is A_1 for each group. The A_2 ($\mathbf{O}$) representation corresponds to B_1 in $\mathbf{D}_4$. The E, T_1, and T_2 representations in $\mathbf{O}$ do not correspond to irreducible representations in $\mathbf{D}_4$, so they are reduced by inspection or using Eq. (3.79). The results are given below.

$\mathbf{D}_4$	E	$2C_4$	$C_2\,(z)$	$2C_2'\,(x, y)$	$2C_2''$
A_1	1	1	1	1	1
A_2	1	1	1	-1	-1
B_1	1	-1	1	1	-1
B_2	1	-1	1	-1	1
E	2	0	-2	0	0

$\mathbf{O}$	E	C_4	$C_2\,(z)$	$C_2\,(x, y)$	C_2	Representations under $\mathbf{D}_4$
A_1	1	1	1	1	1	A_1
A_2	1	-1	1	1	-1	B_1
E	2	0	2	2	0	$A_1 + B_1$
T_1	3	1	-1	-1	-1	$A_2 + E$
T_2	3	-1	-1	-1	1	$B_2 + E$

Table 5.6 is a correlation table for some of the important subgroups of $\mathbf{O}_h$. Since so many groups can be considered as subgroups of $\mathbf{O}_h$, it is convenient to tabulate the major subgroups of $\mathbf{O}_h$ and separate correlation tables for subgroups of these groups. Thus, we can go from $\mathbf{O}_h$ to $\mathbf{D}_{2h}$ in two steps, from $\mathbf{O}_h$ to $\mathbf{D}_{4h}$ and then from $\mathbf{D}_{4h}$ to $\mathbf{D}_{2h}$. We see in the correlation tables tabulated in Appendix 2 that the symmetry species differ for some subgroups depending upon which symmetry operations are retained upon lowering the symmetry. These cases involve choices between equivalent symmetry operations of different classes.

EXAMPLE 5.5 Find the representations for an F term (derived from a d^n configuration) and for the f orbitals in $\mathbf{D}_{4h}$.

TABLE 5.6
Correlation Table for $\mathbf{O}_h$

$\mathbf{O}_h$	$\mathbf{O}$	$\mathbf{T}_d$	$\mathbf{D}_{4h}$	$\mathbf{C}_{4v}$	$\mathbf{C}_{2v}$	$\mathbf{D}_3$	$\mathbf{D}_{2d}$
A_{1g}	A_1	A_1	A_{1g}	A_1	A_1	A_1	A_1
A_{2g}	A_2	A_2	B_{1g}	B_1	A_2	A_2	B_1
E_g	E	E	$A_{1g} + B_{1g}$	$A_1 + B_1$	$A_1 + A_2$	E	$A_1 + B_1$
T_{1g}	T_1	T_1	$A_{2g} + E_g$	$A_2 + E$	$A_2 + B_1 + B_2$	$A_2 + E$	$A_2 + E$
T_{2g}	T_2	T_2	$B_{2g} + E_g$	$B_2 + E$	$A_1 + B_1 + B_2$	$A_1 + E$	$B_2 + E$
A_{1u}	A_1	A_2	A_{1u}	A_2	A_2	A_1	B_1
A_{2u}	A_2	A_1	B_{1u}	B_2	A_1	A_2	A_1
E_u	E	E	$A_{1u} + B_{1u}$	$A_2 + B_2$	$A_1 + A_2$	E	$A_1 + B_1$
T_{1u}	T_1	T_2	$A_{2u} + E_u$	$A_1 + E$	$A_1 + B_1 + B_2$	$A_2 + E$	$B_2 + E$
T_{2u}	T_2	T_1	$B_{2u} + E_u$	$B_1 + E$	$A_2 + B_1 + B_2$	$A_1 + E$	$A_2 + E$

SOLUTION Table 5.5 gives the representations ($A_2 + T_1 + T_2$) for the **O** group. The representations are the same for f and F in the rotational group. We add g subscripts for the F term and u subscripts for the f orbitals under $\mathbf{O}_h$. Using the correlation table (Table 5.6) we obtain the following.

	$\mathbf{O}_h$	$\mathbf{D}_{4h}$		$\mathbf{O}_h$	$\mathbf{D}_{4h}$
F	A_{2g}	B_{1g}	f	A_{2u}	B_{1u}
	T_{1g}	$A_{2g} + E_g$		T_{1u}	$A_{2u} + E_u$
	T_{2g}	$B_{2g} + E_g$		T_{2u}	$B_{2u} + E_u$

EXAMPLE 5.6 Find the representations for the d orbitals or a D term in $\mathbf{I}_h$.

SOLUTION $\mathbf{I}_h$ is not a subgroup of $\mathbf{O}_h$, so we need to apply Eq. (5.2) to $L = 2$ to obtain the characters for the rotations of the **I** group,

$\mathbf{I}$	E	C_5	C_5^2	C_3	C_2
$L = 2$	5	0	0	-1	1

Examining the **I** character table, we see that this is the irreducible representation H. Under $\mathbf{I}_h$, a D term and the d orbitals belong to H_g. The d orbitals are fivefold degenerate in this point group; no splitting occurs.

EXAMPLE 5.7 Find the representations for the p orbitals and a P term (from a d^n configuration) in **O**, **T**, and $\mathbf{T}_d$ symmetry.

SOLUTION The appropriate rotations in Table 5.4 assign the p orbitals to the T_1 representation in the **O** group and to T in the **T** group. Going from **T** to the $\mathbf{T}_d$ group we see that T_1 is symmetric (character $+1$) with respect to S_4 and T_2 is antisymmetric (character -1) with respect to S_4. The p orbitals belong to T_2 since they are antisymmetric with respect to S_4. We see this inversion of subscripts in

Table 5.6. It occurs because a different reference is used for the subscript distinction ($T_1 - T_2$) for **O** and $\mathbf{T}_d$. A P term derived from a d^n configuration is T_1 under **O** and is T_1 under $\mathbf{T}_d$ or T_{1g} under $\mathbf{O}_h$.

5.3 Stereoisomers

5.3.1 USE OF SYMMETRY FOR DRAWING ISOMERS

The drawing of isomers can be simplified if we make use of symmetry. If we add one ligand a to an octahedron, we create a framework with $\mathbf{C}_{4v}$ symmetry (Fig. 5.2). The other position along the C_4 axis is unique, but the four "equatorial" positions are equivalent (they form a set interchanged by C_4). If we add a second a ligand, there are two possibilities (and two isomers for $[M(a)_2(b)_4]$): the second a can go into the axial position (giving a $\mathbf{D}_{4h}$ framework) or into one of the equatorial positions (giving a $\mathbf{C}_{2v}$ framework).

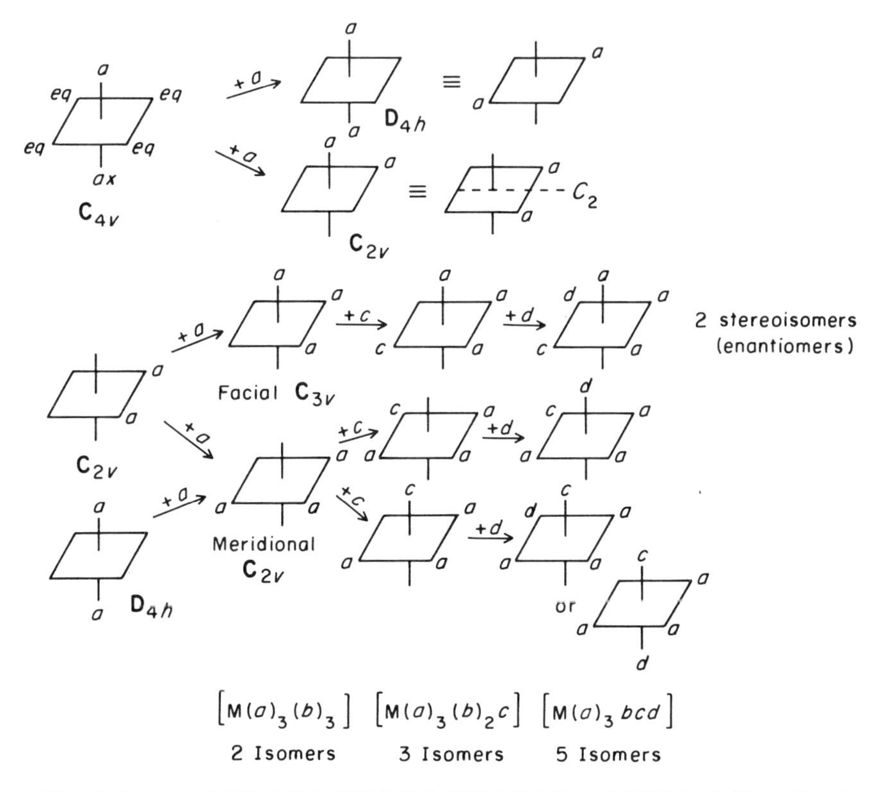

Fig. 5.2 Isomers of $[M(a)_2(b)_4]$, $[M(a)_3(b)_3]$, $[M(a)_3(b)_2(c)]$, and $[M(a)_3bcd]$. The b ligands occupy the vacant sites.

The $\mathbf{D}_{4h}$ symmetry is obvious from the way the *trans*-[M$(a)_2$] isomer is drawn. It is less apparent if we reorient the octahedron so that the *trans a* ligands are shown in equatorial positions. The $\mathbf{C}_{2v}$ symmetry of the *cis* isomer is more apparent if we orient the octahedron so the C_2 axis and symmetry planes are easier to visualize. There is just one possibility for the addition of a third ligand to the $\mathbf{D}_{4h}$ framework since the four vacant sites are equivalent. The product has $\mathbf{C}_{2v}$ symmetry whether the third ligand is *a* or *b*. This framework has a pair of equivalent sites and one unique site. There are two pairs of equivalent positions for the $\mathbf{C}_{2v}$ [(M$(a)_2$] framework. If the third ligand added is *a*, we get a *facial* (*fac*) ($\mathbf{C}_{3v}$) and a *meridional* (*mer*) ($\mathbf{C}_{2v}$) isomer. This *mer* isomer is the same as the $\mathbf{C}_{2v}$ [M$(a)_3$] isomer obtained from the $\mathbf{D}_{4h}$ framework. There are just two isomers for [M$(a)_3(b)_3$].

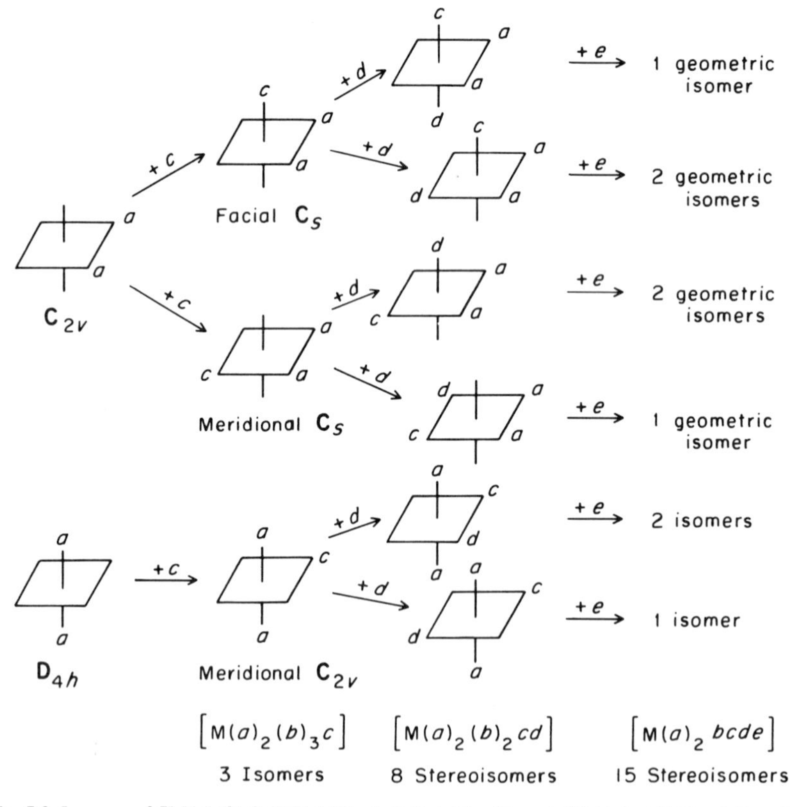

Fig. 5.3 Isomers of [M$(a)_2(b)_4$], [M$(a)_2(b)_3c$], [M$(a)_2(b)_2cd$], and [M$(a)_2bcde$]. The *b* ligands occupy the vacant sites. Each *cis*-[M$(a)_2bcde$] geometrical isomer represents two stereoisomers (enantiomers).

The number of stereoisomers of a polyhedron with n different ligands is given by

$$n!/h = \text{number of stereoisomers} \tag{5.6}$$

where h is the order of the rotational group. The order of the **O** group is 24, giving $(6 \times 5 \times 4 \times 3 \times 2 \times 1)/24 = 30$. For the addition of four different ligands to *cis*-$[M(a)_2]$ ($\mathbf{C}_{2v}$), the number of stereoisomers is $4!/2 = 12$ and for *trans*-$[M(a)_2]$ ($\mathbf{D}_{4h}$) it is $4!/8 = 3$. These isomers are generated in Fig. 5.3. For the addition of three different ligands to *fac*-$[M(a)_3]$ ($\mathbf{C}_{3v}$), the number of stereoisomers is $3!/3 = 2$, and for *mer*-$[M(a)_3]$ ($\mathbf{C}_{2v}$), $3!/2 = 3$. In cases where the geometrical isomer has no plane of symmetry, there are two optical isomers. We can verify that the isomers generated are different by identifying the point groups or checking *trans* pairs. Different geometrical isomers differ in at least one *trans* pair combination. Interchanging any single *trans* pair of different ligands generates its mirror image—the other optical isomer. If optical isomerism is not possible because of an S_n operation (including $\sigma = S_1$ and $i = S_2$), interchanging a *trans* pair of different ligands does not give a different isomer. Since optical activity is possible only in the absence of any S_n operation, only molecules belonging to rotational groups (usually just $\mathbf{C}_n$ or $\mathbf{D}_n$) are optically active.

EXAMPLE 5.8 Determine the number of stereoisomers of a $[M(a)_2bcde]$ complex with the ligands arranged around a trigonal prism. How many isomers are possible for six different ligands?

SOLUTION For a trigonal prism ($\mathbf{D}_{3h}$) with six different ligands, the number of stereoisomers is $6!/6 = 120$. Two identical ligands can give four frameworks for the addition of the other ligands. Note that the last two frameworks look similar, but they are nonsuperimposable mirror images. They are stereoisomers representing the same geometric isomer. See Problem 5.11 for a more general treatment.

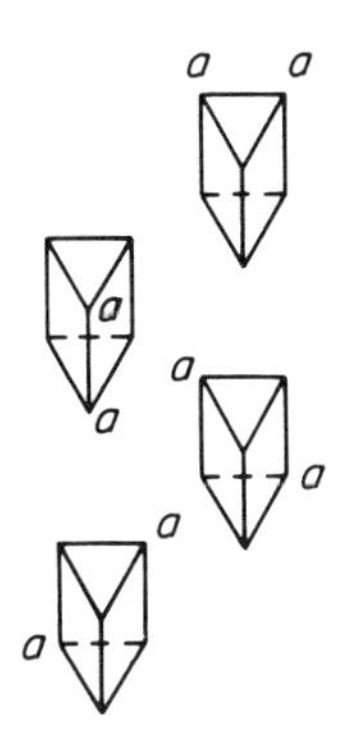

Point group	Rotational group	Number of stereoisomers
$\mathbf{C}_s$	$\mathbf{C}_1$	$4!/1 = 24$
$\mathbf{C}_{2v}$	$\mathbf{C}_2$	$4!/2 = 12$
$\mathbf{C}_2$	$\mathbf{C}_2$	$4!/2 = 12$
$\mathbf{C}_2$	$\mathbf{C}_2$	$4!/2 = 12$
		Total 60

TABLE 5.7
Maximum Number of Stereoisomers[a]

Coordination number	Geometrical figure and rotational group	Number of isomers
4	Tetrahedron, $\mathbf{T}$	$4!/12 = 2$
	Square plane, $\mathbf{D}_4$	$4!/8 = 3$
5	Trigonal bipyramid, $\mathbf{D}_3$	$5!/6 = 20$
	Square pyramid, $\mathbf{C}_4$	$5!/4 = 30$
6	Octahedron, $\mathbf{O}$	$6!/24 = 30$
	Trigonal prism, $\mathbf{D}_3$	$6!/6 = 120$
8	Square antiprism, $\mathbf{D}_4$	$8!/8 = 5040$
	Dodecahedron, $\mathbf{D}_2$	$8!/4 = 10{,}080$
12	Icosahedron, $\mathbf{I}$	$12!/60 = 7{,}983{,}360$

[a] Adapted with permission from D. H. McDaniel, *Inorg. Chem.* 1972, **11,** 2678. © 1972 American Chemical Society.

Since no planes of symmetry remain after all ligands are added, there are 30 pairs of optical isomers.

5.3.2 STEREOISOMERS FROM PÓLYA'S THEOREM

In most cases we can obtain all stereoisomers by drawing them with the help of symmetry. We see from the application of Eq. (5.6) that the number of stereoisomers becomes unmanageable for high coordination numbers (Table 5.7) without a systematic approach. McDaniel[6] provided a useful interpretation of Pólya's theorem for dealing with the enumeration of isomers. The number of stereoisomers is the number of distinguishable configurations that are invariant under each operation of the rotational group divided by the order of the rotational group. The number of geometric isomers is obtained if the full covering group (including *all* operations) is used.

All distinguishable configurations are invariant under the identity operation. The n ligands are in sets of one (no ligands are interchanged by the identity operation), and the number of permutations is given by

$$P_1^n = n!/n_a!n_b!n_c!\cdots \qquad \text{for} \quad E, \tag{5.7}$$

where P_1^n is the permutation index for "n sets of 1", n_a the number of A ligands,

[6] D. H. McDaniel, *Inorg. Chem.* 1972, **11,** 2678.

n_b the number of B ligands, etc. For a C_n operation, the equivalent ligands interchanged by the operation are in sets of n ligands. The ligands falling on the C_n axis, since they are not interchanged with others, belong to "sets of one." The permutation function for the C_n operation is the product of the permutations for these two types of sites,

$$P_1^k P_n^m \quad \text{for} \quad C_n,$$

where P_1^k indicates k sets of one along the C_n axis and P_n^m indicates m sets of n ligands interchanged by the operation. If there are no ligands along the C_n axis, the total number of ligands is $n \times m$. The number of ligands in the more general case is $n \times m + k$, where k is the number of ligands along C_n. The number of permutations for each function is given by Eqs. (5.8) and (5.9):

$$P_n^m = \frac{m!}{n_{a_n}!n_{b_n}!n_{c_n}!\cdots}, \tag{5.8}$$

$$P_1^k = \frac{k!}{n_{a_n}!n_{b_n}!n_{c_n}!\cdots} \tag{5.9}$$

where m is the number of sets of n similar ligands, n_{a_n} the number of sets of A_n, n_{b_n} the number of sets of B_n, etc., and $n_{a_n} + n_{b_n} + \cdots = m$.

For the C_3 of the trigonal bipyramid, there is a set of three ligands interchanged by C_3, and two ligands along C_3, forming "sets of one." For MA_3B_2 there is one set of three A and two sets of one along the axis,

$$P_n^m P_1^k = P_3^1 P_1^2.$$

For invariance, C_2 requires two sets of two (one pair of equatorial positions and one pair of axial positions) and one set of one (along the C_2 axis),

$$P_n^m P_1^k = P_2^2 P_1^1.$$

For MA_3B_2, the permutation functions are evaluated for $\mathbf{D}_3$ and $\mathbf{D}_{3h}$ in Table 5.8. For the identity, there are five sets of one made up of three sets of one for A and two sets of one for B. For each C_3 axis, there is one set of three (P_3^1) satisfied only by three A. For the axial positions (P_1^2), the two sets of one are satisfied by the two remaining B ligands (we have used all A). For each C_2 axis, there are two sets of two (P_2^2) satisfied by two B and two A, leaving one A for the set of one (P_1^1) along the C_2 axis. The total number of stereoisomers (from the $\mathbf{D}_3$ rotational group) is $18/6 = 3$.

We can add the operations of the full covering group $\mathbf{D}_{3h}$ to obtain the number of geometrical isomers. For σ_h, one set of two (P_2^1) can be satisfied by

TABLE 5.8
Permutation Indices for $\mathbf{D}_3$ and $\mathbf{D}_{3h}$

$\mathbf{D}_3$				$\mathbf{D}_{3h}$ add		
	E	$2C_3$	$3C_2$	σ_h	$3\sigma_v$	$2S_3$
Permutation functions for TBP	P_1^5	$2P_3^1P_1^2$	$3P_2^2P_1^1$	$P_2^1P_1^3$	$3P_2^1P_1^3$	$2P_3^1P_2^1$
Number of invariant configurations for MA_3B_2	$\frac{5!}{3!2!}$	$2 \times \left(\frac{1!}{1!}\right)\left(\frac{2!}{2!}\right)$	$3 \times \left(\frac{2!}{1!1!}\right)\left(\frac{1!}{1!}\right)$	$\left(\frac{1!}{1!}\right)\left(\frac{3!}{3!}\right)$	Same as σ_h	$2 \times \left(\frac{1!}{1!}\right)\left(\frac{1!}{1!}\right)$
				and		
		2×1	3×2	$\left(\frac{1!}{1!}\right)\left(\frac{3!}{2!1!}\right)$		
	$= 10$	$= 2$	$= 6$	$1 + 3 = 4$	$3 \times 4 = 12$	2
	$18/6 = 3$ stereoisomers			$(18 + 18)/12 = 3$ geometrical isomers		

TABLE 5.9
Permutation Indices for Example 5.9

$\mathbf{C}_4$			$\mathbf{C}_{4v}$ add	
E	$2C_4$	C_2	$2\sigma_v$	$2\sigma_d$
P_1^5	$2P_4^1P_1^1$	$P_2^2P_1^1$	$2P_2^1P_1^3$	$2P_2^2P_1^1$
$\frac{5!}{3!2!}$	0	$\left(\frac{2!}{1!1!}\right)\left(\frac{1!}{1!}\right)$	$2\times\left(\frac{1!}{1!}\right)\left(\frac{3!}{3!}\right)=2$ and $2\times\left(\frac{1!}{1!}\right)\left(\frac{3!}{2!1!}\right)=6$	$2\times\left(\frac{2!}{1!1!}\right)\left(\frac{1!}{1!}\right)=4$
$=10$	0	$=2$	$2+6=8$	
12/4 = 3 stereoisomers			24/8 = 3 geometrical isomers	

two B with the three A satisfying the three sets of one (P_1^3). This gives $(1!/1!)(3!/3!) = 1$. We can also use two A for the set of two, leaving two sets of one B and one set of one A $(1!/1!)(3!/2!\,1!) = 3$. Considering both possibilities, the number of invariant configurations is four. The same possibilities exist for each (of three) σ_v planes. For each S_3, there is only one set of three (three A), leaving the two B as a set of two ($2P_3^1P_2^1$). The total number of invariant configurations is 36 for $\mathbf{D}_{3h}$, and the order is 12, giving $36/12 = 3$ geometrical isomers. Since this is the same as the number of stereoisomers, there are no optical isomers.

EXAMPLE 5.9 Find the number of stereoisomers and geometrical isomers for a square pyramidal complex MA_3B_2 ($\mathbf{C}_{4v}$).

SOLUTION The permutation indices are given in Table 5.9. There are no configurations invariant with respect to C_4, since there is no set of four equivalent ligands. For C_2 (and for σ_d), there are two sets of two ligands (two A and two B), leaving one set of one A. For σ_v, we can have one set of two B, leaving three sets of one A $(1!/1! \times 3!/3!)$ and also one set of two A, leaving two sets of one B and one set of one A $(1!/1! \times 3!/2!\,1!)$. The number of geometrical isomers is the same as the number of stereoisomers, so there are no optical isomers.

McDaniel[6] points out that bidentate ligands generally span only adjacent (*cis*) positions so they cannot be permuted among all positions. He deals with the "octahedral" complex $[M(en)A_2BC]$ by considering the permutations of two A, B, and C on the M(en) framework having $\mathbf{C}_{2v}$ symmetry. Note that the symmetry used is that of the unsubstituted framework. It is generally lower after we add the ligands whose permutations are considered. Multidentate ligands also are best treated as part of the framework.

EXAMPLE 5.10 Determine the number of stereoisomers and geometrical isomers for [M(dien)a_2b], where dien is the tridentate ligand $H_2NC_2H_4NHC_2H_4NH_2$.

SOLUTION The ligand dien can coordinate on a face (facial) or around an edge (meridional) of the octahedron. Let us consider the two possibilities separately. First, for the *fac* isomer we have the following.

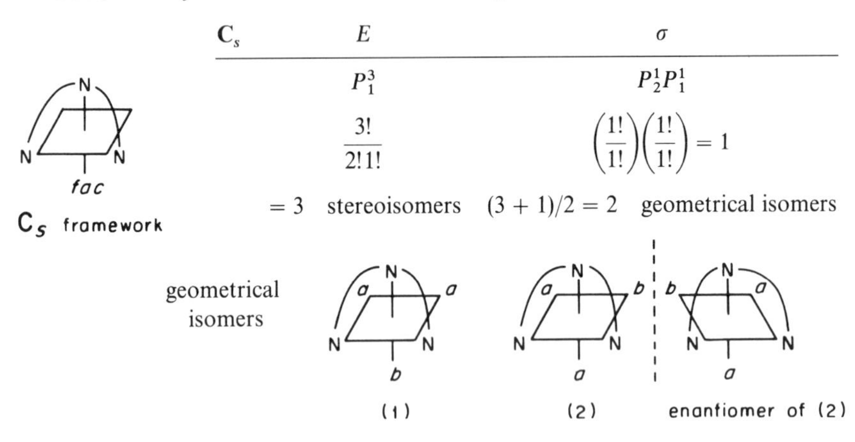

$\mathbf{C}_s$	E	σ
	P_1^3	$P_2^1P_1^1$
	$\frac{3!}{2!1!}$	$\left(\frac{1!}{1!}\right)\left(\frac{1!}{1!}\right) = 1$
	$= 3$ stereoisomers	$(3 + 1)/2 = 2$ geometrical isomers

The number of stereoisomers is obtained from the rotational subgroup, here $\mathbf{C}_1$ ($C_1 = E$). Since there are three invariant configurations for C_1 (E), there are $3/1 = 3$ stereoisomers. One of the geometrical isomers [the one without a plane of symmetry (2)] is optically active. These nonsuperimposable mirror images (enantiomers) are shown above.

For the *mer* isomer we have the following.

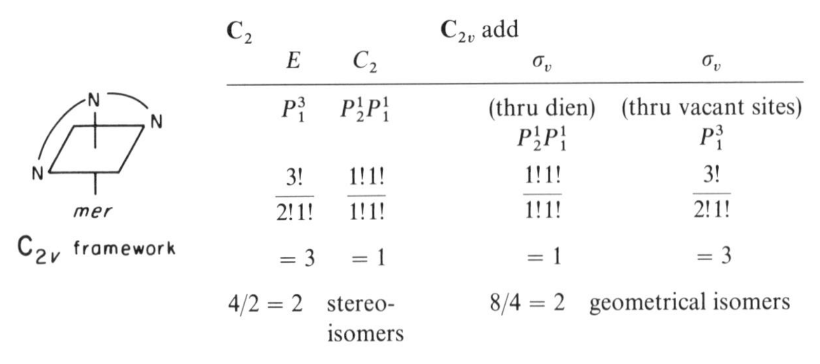

	$\mathbf{C}_2$		$\mathbf{C}_{2v}$ add	
	E	C_2	σ_v	σ_v
	P_1^3	$P_2^1P_1^1$	(thru dien) $P_2^1P_1^1$	(thru vacant sites) P_1^3
	$\frac{3!}{2!1!}$	$\frac{1!1!}{1!1!}$	$\frac{1!1!}{1!1!}$	$\frac{3!}{2!1!}$
	$= 3$	$= 1$	$= 1$	$= 3$
	$4/2 = 2$	stereo-isomers	$8/4 = 2$	geometrical isomers

Here there are just two isomers (b *trans* to NH or b *trans* to NH_2).

Additional Reading

E. U. Condon and G. H. Shortley, *The Theory of Atomic Spectra*, Cambridge Univ. Press, New York, 1935.

B. Douglas, D. H. McDaniel, and J. J. Alexander, *Concepts and Models of Inorganic Chemistry*, Wiley, New York, 1983. Background for derivation of term symbols and isomerism.

G. Herzberg, *Atomic Spectra and Atomic Structure*, Dover, New York, 1944.

Problems

5.1 Obtain the ground-state levels for N, Fe^{2+}, and Cr^{3+}.

5.2 Derive the spectral terms and identify the ground-state level for d^4 and f^2 configurations.

5.3 Show that only *D*, *F*, and *S* terms are possible for the ground states for d^n configurations.

5.4 Derive the irreducible representations for the *f* orbitals in $\mathbf{I}_h$. Verify the results from the character table.

5.5 (a) Identify the symmetry species for *p* orbitals in $\mathbf{D}_{3h}$ and $\mathbf{C}_{4v}$ from the character tables.

(b) Obtain the symmetry species for a *P* term derived from a d^n configuration (see Example 5.3) by calculating the characters in $\mathbf{D}_{3h}$ and $\mathbf{C}_{4v}$.

5.6 Determine the symmetry species by calculating the characters in $\mathbf{C}_{4v}$ for (a) *d* orbitals, (b) *f* orbitals, and (c) an *F* term derived from a d^n configuration (see Example 5.3).

5.7 (a) Verify the results in 5.5(a) and (b) from the character table.

(b) Using the symmetry species of an *F* term derived from a d^n configuration in $\mathbf{D}_{4h}$ (Example 5.5) and the correlation tables in Appendix 3, verify the results in 5.6(c).

5.8 From Table 5.5 obtain the symmetry species for the *g* orbitals in $\mathbf{O}_h$. Use the correlation table (Table 5.6) to find the species in $\mathbf{D}_{4h}$.

5.9 Determine the number of stereoisomers for a distorted tetrahedral complex M(*abcd*) with $\mathbf{C}_{3v}$ symmetry and with $\mathbf{D}_{2d}$ symmetry (with respect to the framework).

5.10 How many isomers are possible for the "octahedral" complex $[M(\widehat{aa})bcde]$? How many pairs of optical isomers are there? Consider the framework with the bidentate *aa* ligand coordinated along an octahedral edge.

5.11 Apply Pólya's theorem to obtain the number of geometrical isomers and stereoisomers for a trigonal prismatic complex ($\mathbf{D}_{3h}$) with the formula [$\mathrm{M}a_2bcde$].

5.12 Apply Pólya's theorem to obtain the number of geometrical isomers and stereoisomers for a trigonal prismatic complex ($\mathbf{D}_{3h}$) with the formula [$\mathrm{M}a_3b_2c$]. Sketch the optically inactive isomers.

5.13 Apply Pólya's theorem to determine the number of stereoisomers and the number of geometrical isomers for the "octahedral" complex [$\mathrm{M(en)}a_2b_2$]. Sketch the optical isomers. (Consider the framework with the symmetrical bidentate en ligand coordinated along an octahedral edge.)

5.14 Determine the number of stereoisomers for a binuclear complex

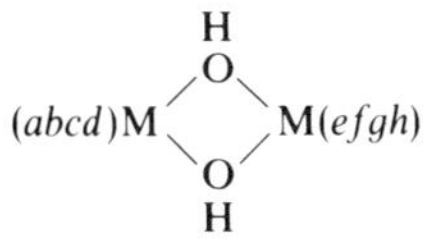

```
          H
          O
        /   \
(abcd)M       M(efgh)
        \   /
          O
          H
```

involving two octahedra sharing an edge (bridging OH^-). How many pairs of optical isomers are there?

6

Bonding in Simple AX_n Molecules

The molecular orbital treatments of diatomic molecules (Section 6.1) and molecular geometry (Section 6.2) are brief reviews of essentials assumed as background. These review sections are not intended as introductions to the topics. For that purpose, consult a basic inorganic text. Molecular orbital theory is applied, after the review, to simple molecules for which adequate valence bond or electron dot structures can be drawn.

6.1 Diatomic Molecules

6.1.1 SYMMETRY OF MOLECULAR ORBITALS FOR X_2

The molecular orbital description of a molecule treats the orbitals as the property of the whole molecule, considering the influence of the various nuclei and the interactions among all of the electrons. The task is too formidable to start from scratch (the *ab initio* approach) each time. The usual simplification is to recognize that while electrons are near a particular nucleus, they should be influenced primarily by that nucleus. This permits us to use the familiar descriptions of atomic orbitals (AOs). The molecular orbitals are obtained as *linear combinations of atomic orbitals* (LCAOs). The allowed combinations of AOs are limited by the symmetry of the molecule and the symmetry properties of the atomic orbitals. The molecular orbitals must transform as symmetry species (irreducible representations) of the point group.

In the familiar case of the H_2 molecule, the two 1*s* orbitals combine (where the signs match, $1s_A + 1s_B$) to give a bonding molecular orbital (MO) and an antibonding MO (where the signs are opposite, $1s_A - 1s_B$). The H_2 molecule is stabilized by having two electrons in the bonding MO, which is lower in energy than the atomic orbitals from which it is obtained. The LCAOs for He_2 are the same, but the molecule is unstable because although two electrons go into the lower energy bonding MO, the other two electrons must enter the higher energy antibonding MO. Simple MO energy-level diagrams (Fig. 6.1) usually show the antibonding orbital being raised in energy by the same amount as the bonding orbital is lowered in energy, giving no net bonding if both orbitals are filled. Actually, the antibonding orbital is raised more than the bonding orbital is lowered, giving a net antibonding interaction for the unstable He_2. The normalized wave functions for the two MOs are

$$\begin{aligned} \Psi(\sigma) &= N_b(s_A + s_B) \qquad \text{(bonding)}, \\ \Psi(\sigma^*) &= N^*(s_A - s_B) \qquad \text{(antibonding)}. \end{aligned} \tag{6.1}$$

Since the square of a normalized wave function is unity,

$$\begin{aligned} \int[\Psi(\sigma)]^2\,d\tau &= 1 = N_b^2\left[\int(s_A)^2\,d\tau + 2\int s_A s_B\,d\tau + \int(s_B)^2\,d\tau\right], \\ \int[\Psi(\sigma^*)]^2\,d\tau &= 1 = N^{*2}\left[\int(s_A)^2\,d\tau - 2\int s_A s_B\,d\tau + \int(s_B)^2\,d\tau\right], \end{aligned} \tag{6.2}$$

and using normalized AO's, for which $\int(s_A)^2\,d\tau = \int(s_B)^2\,d\tau = 1$, and substituting S for the *overlap integral* $\int(s_A)(s_B)\,d\tau$,

$$N_b^2[2 + 2S] = 1 = N^{*2}[2 - 2S], \tag{6.3}$$

giving

$$N_b = \sqrt{1/2(1 + S)} \qquad \text{and} \qquad N^* = \sqrt{1/2(1 - S)}. \tag{6.4}$$

Neglecting the overlap integral, the normalization constants are both $\sqrt{1/2}$, but including S, N_b is less than N^*. The energies of the MOs are given by

$$E = \int \Psi H \Psi\,d\tau, \tag{6.5}$$

and since N_b and N^* are constants entering Eq. (6.5), the energy of σ^* will be

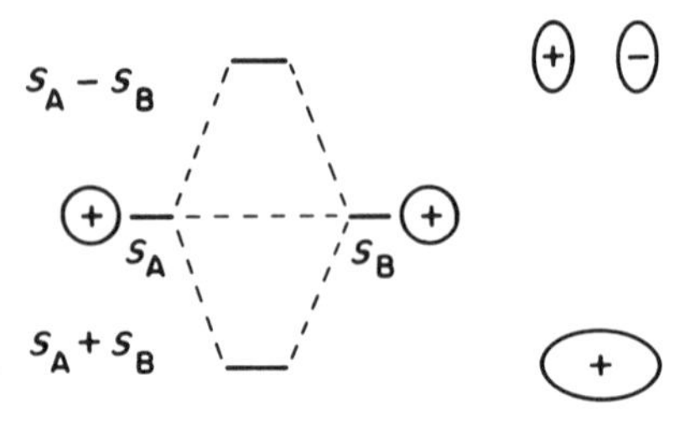

Fig. 6.1 Molecular orbital energy-level diagram for H_2.

raised more than that of σ will be lowered. In the applications to follow, the overlap integral will be neglected so that for an LCAO, the normalization constant is the square root of the sum of the squares of the orbital coefficients.

The MO description of bonds for diatomic molecules is based on symmetry. *Sigma* bonds have maximum electron density along the line joining the two nuclei, with axial symmetry about the bond and no nodes through the bond. The antibonding σ^* orbital has a nodal plane perpendicular to the bond direction. Sigma bonds can be formed from combinations of any AOs that have components along the bond direction (defined here as z): $s + s$, $s + p_z$, $s + d_{z^2}$, $p_z + p_z$, $p_z + d_{z^2}$, etc. (see Fig. 6.2). *Pi*-bonding orbitals have a nodal

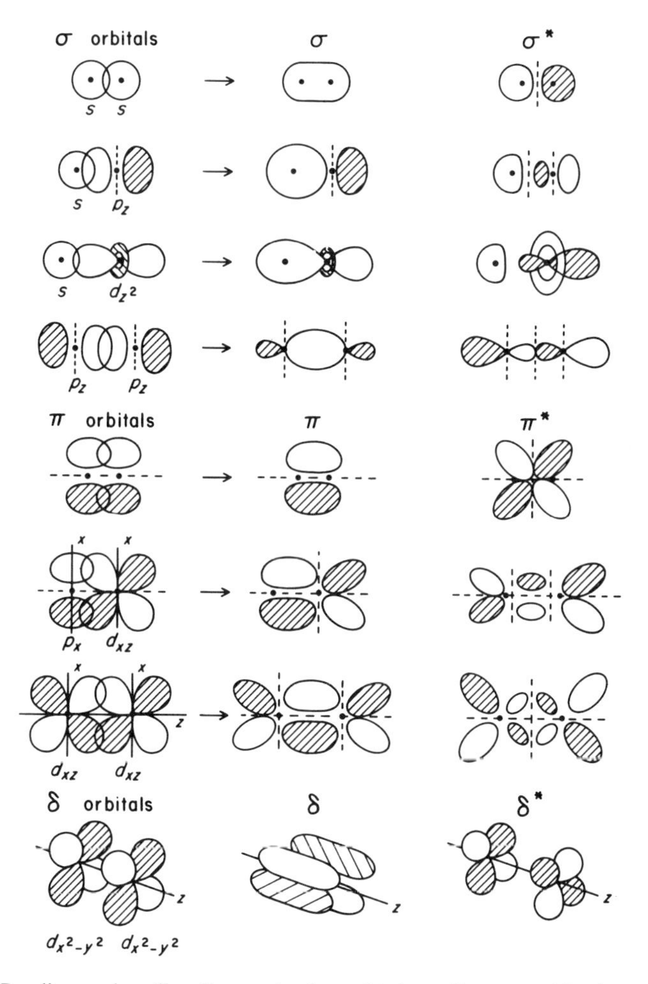

Fig. 6.2 Bonding and antibonding molecular orbitals as linear combinations of atomic orbitals.

plane through the line joining the two nuclei. Antibonding π^* orbitals have an additional nodal plane perpendicular to the line joining the nuclei. Pi bonds can be formed from combinations of AOs with at least one nodal plane: $p_x + p_x$ (or $p_y + p_y$), $p_x + d_{xz}$ (or $p_y + d_{yz}$), and $d_{xz} + d_{xz}$ (or $d_{yz} + d_{yz}$) (see Fig. 6.2). *Delta*-bonding orbitals have two perpendicular nodal planes through the line joining the nuclei (two eclipsed four-leafed clovers), as shown in Fig. 6.2 for $d_{x^2-y^2} + d_{x^2-y^2}$ (or $d_{xy} + d_{xy}$).

The σ, π, and δ designations for bonding MOs are derived from the *s*, *p*, and *d* AOs with the corresponding symmetry properties about the bond. These descriptions apply specifically to diatomic or linear ($\mathbf{D}_{\infty h}$ or $\mathbf{C}_{\infty v}$) molecules, although they also are used to describe the local bond symmetry in other polyatomic molecules. We shall use them in this way.

In the H_2 molecule the two AOs combine to form two MOs, σ and σ^*. The HHe^+ ion has two electrons that could be accommodated in the σ-bonding orbital. Although the symmetry requirements are satisfied, there is no significant bond in this case because of the very great difference in the energies of the 1*s* orbitals for H and He (the first ionization energy is a reasonable indication of the orbital energy). That is, the He 1*s* orbital energy is so much lower (electrons are held so much more strongly) that there is no effective electron sharing. The electron density remains unchanged about the He. Thus, one very important limitation in addition to symmetry requirements is that *only orbitals of comparable energy* combine effectively.

6.1.2 HYDROGEN FLUORIDE

In the HF molecule, the F 1*s* orbital is very low in energy and is screened effectively by the outer shell of electrons. Thus, inner shells are generally nonbonding and are treated as part of the "core." This offers the great advantage of allowing us to treat the bonding interactions of all members of a family (e.g., the halogens) in the same way; only the valence shell, which they have in common, is important. Of course, an important distinction arises where for one member of a family (e.g., P or S) the 3*d* orbitals might participate as part of the valence shell, while there are no *d* orbitals for the second shell (e.g., N or O).

For HF, the 2*s* and $2p_z$ orbitals have the proper symmetry for forming σ bonds with the H 1*s* orbital. However, once again the F 2*s* orbital is so much lower in energy than the H 1*s* orbital that the F 2*s* orbital is effectively nonbonding. The F orbital of the proper symmetry and energy for σ bonding is p_z. Since even the F p_z orbital is considerably lower in energy (higher ionization energy) than the H 1*s* orbital, an unsymmetrical energy-level diagram results (Fig. 6.3). The bonding σ MO is closer in energy to the F $2p_z$ orbital (it is said to have more "F character" or be more "F-like") and the σ^*

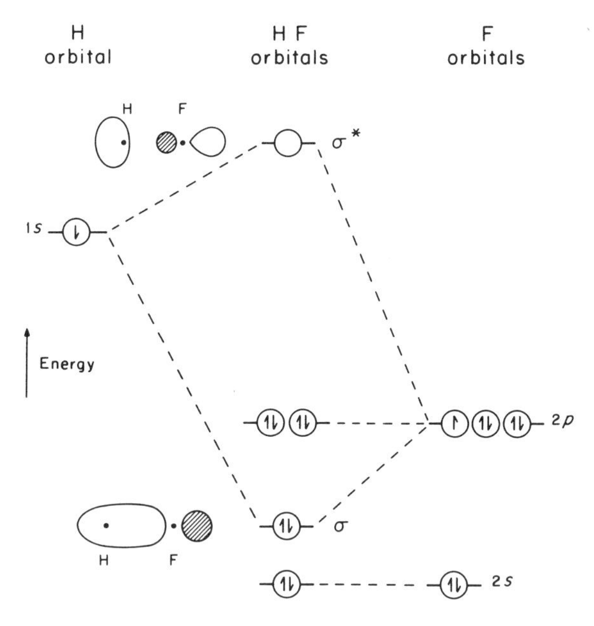

Fig. 6.3 Bonding in HF.

MO is closer in energy to the H 1*s* orbital. The effect is that in the bonding σ orbital, the electron density is greater near the more electronegative F atom, and if there is an electron in the σ^* orbital (such as for HF^-), then the electron density for this orbital would be greater near H. The F p_x and p_y orbitals are nonbonding.

For an HX molecule, the p_x and p_y orbitals are completely excluded from σ bonding because of symmetry. The participation of the *ns* orbital (on X) depends on the relative energies of the X *ns* orbital and the H 1*s* orbital. The closer the orbitals are in energy, the greater the bonding interaction.

Considering only the valence shells for HF, there is one σ MO, one σ^* MO, two nonbonding *p* orbitals, and 2s (which is nonbonding), for a total of five MOs. The H has one orbital, and F has four valence shell orbitals. This is a trivial case, but it illustrates the *conservation of orbitals*—the number of MOs must be the same as the number of AOs being combined.

6.1.3 X_2 MOLECULES OF THE SECOND PERIOD

For a second-period X_2 molecule, the *ns* and np_z orbitals have the proper symmetry for σ bonding. For periodic group IA (ns^1), the σ bonding results from $s + s$ interaction. The dissociation energies decrease from 100.9 kJ/mole for Li_2 to 38.0 kJ/mole for Cs_2. The M_2 molecules for group IIA (Be–Ra) are unstable, as might be expected, since the σ and σ^* orbitals are both filled.

The group III elements have the outer-electron configuration ns^2np^1. If the interaction of the s orbitals were excluded as nonbonding because of filled σ_s and σ_s^* orbitals, then the bonding MO would be σ_{p_z}, giving a diamagnetic B_2 molecule ($\sigma_{p_z}^2$). In fact, B_2 is paramagnetic with two unpaired electrons, consistent with the single occupancy of each of the two π orbitals, $\pi_x^1\pi_y^1$. These experimental facts indicate that we cannot simply exclude the s orbitals as nonbonding. Here we have a case of the mixing of σ orbitals belonging to the same representation (Σ^+ or A_1 for $\mathbf{D}_{\infty h}$). The net effect of the mixing is that the lower-energy σ orbital ($s + s$) is lowered still further and the higher-energy σ orbital ($p_z + p_z$) is raised *above* the level of the π orbitals (Fig. 6.4). In other words, there is s–p mixing, or hybridization. In this case, on each atom the s and p orbitals combine to give nonequivalent sp hybrids ($s + \lambda p$ and $\lambda s - p$, where $\lambda \neq 0$). The two hybrids on each atom combine to form four σ MOs. The lower-energy sp hybrid orbital (the one with more s character) on each atom contributes more to the two lower-energy σ orbitals, and both of these are lower in energy than $\pi_x\pi_y$. The third σ MO is higher in energy than $\pi_x\pi_y$, and the fourth σ MO is more strongly antibonding than $\pi_x^*\pi_y^*$.

The magnetic properties and bond order of B_2 (B.O. = 1) and C_2 (B.O. = 2) are consistent with the population of energy levels as in the MO diagram in Fig. 6.4. The N_2 molecule is diamagnetic with BO = 3, but this would be expected irrespective of whether the $\pi_x^2\pi_y^2$ orbitals are higher (no s–p mixing) or lower (s–p mixing) in energy than $3\sigma_g^2$. The significant experimental result is that N_2^+ has a single σ electron, so for N_2^+, and presumably for N_2, the energy-level sequence involving s–p mixing in Fig. 6.5 applies.

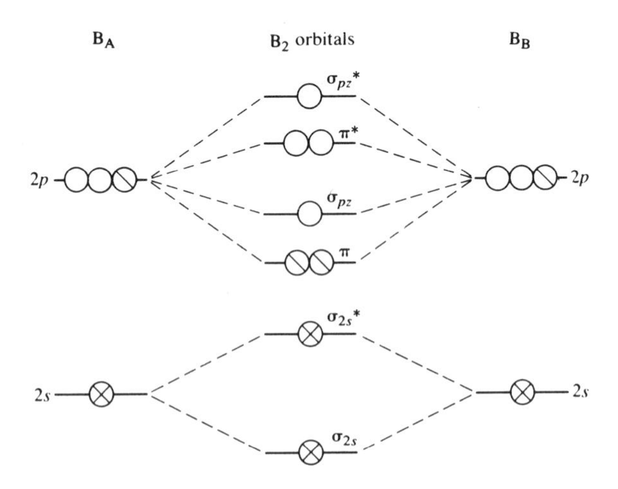

Fig. 6.4 Molecular orbital energy-level diagram for B_2.

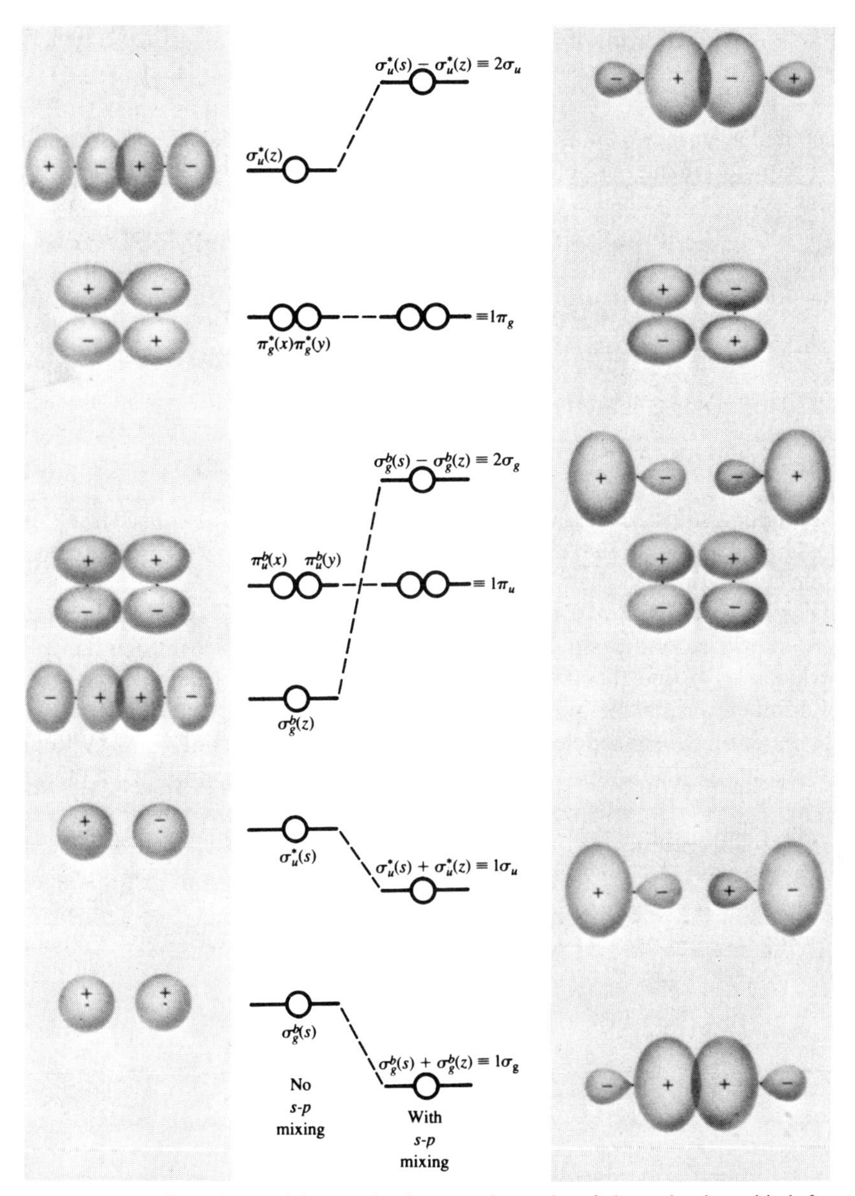

Fig. 6.5 The effect of s–p mixing on the shapes and energies of the molecular orbitals for a homonuclear diatomic molecule. Notice that the $1\sigma_u$ and $2\sigma_u$ orbitals become relatively nonbonding, and that the energy order of $1\pi_u$ and $2\sigma_g$ interchanges. The extent of s–p mixing becomes smaller as the energy separation of the valence s and p atomic levels increases. From R. L DeKock and H. B. Gray, *Chemical Structure and Bonding*, Benjamin/Cummings, Menlo Park, California 1980, p. 226. © 1980 Benjamin/Cummings Publishing Company.

The magnetic properties and bond order of O_2 (two unpaired *e*, B.O. 2), O_2^+ (one unpaired *e*, B.O. = $2\frac{1}{2}$), O_2^- (one unpaired *e*, B.O. = $1\frac{1}{2}$), and O_2^{2-} (or F_2) (zero unpaired *e*, B.O. = 1) are those expected from occupancy of the next higher-energy orbitals, $\pi_x^*\pi_y^*$. These results do not tell us anything about the sequence in energies of the next lower π and σ orbitals. Nevertheless, it is believed that there is less *s*–*p* mixing for O_2 and F_2 than for N_2, C_2, and B_2 and that the sequence is $\sigma_{p_z} < \pi_x\pi_y < \pi_x^*\pi_y^*$. Note that the reversal of the sequence for the bonding orbitals is determined only by the relative energies of the atomic and hybrid orbitals involved. Symmetry arguments are useful only in determining which orbitals *can* mix.

6.2 Molecular Geometry

6.2.1 HYBRIDIZATION AND MOLECULAR SHAPE

Although some MO treatments optimize atomic positions considering any possible geometry, the more usual approach begins with an assumed molecular geometry. We want to review some of the simple approaches used to deduce the structure of a molecule. Lewis structures are useful for elements through the second period and even beyond. Although elements in the later periods can exceed the octet rule, it is often obeyed. Molecular geometry is determined primarily by directed orbitals forming sigma bonds or accommodating unshared electron pairs on the central atom of an AX_n molecule (generally A is the more electropositive element). In most cases the directional characteristics of these localized orbitals are described conveniently by familiar hybrids. For this purpose, we can neglect π bonding. Thus, all of the following are described as sp^3 hybrids. The molecular *shape* (angular, pyramidal, tetrahedral) describes the positions of *atoms*, ignoring unshared electrons.

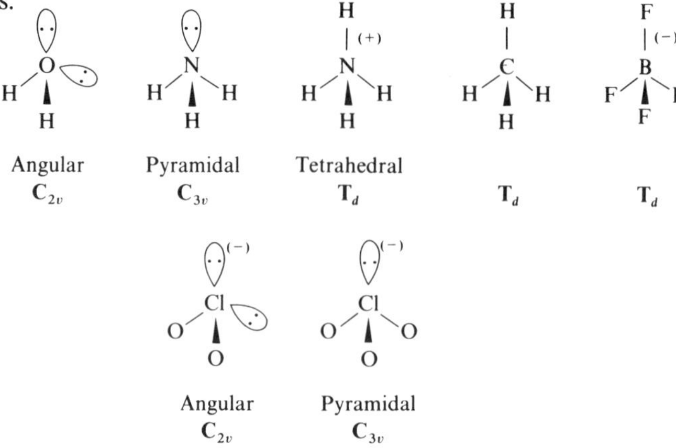

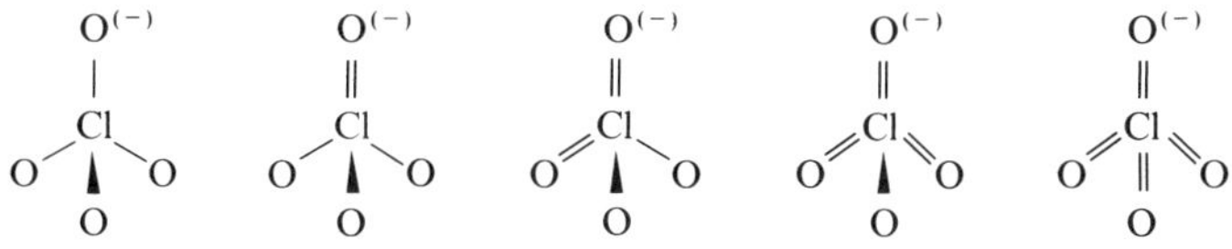

All $\mathbf{T}_d$

The following are described as sp^2 hybrids.

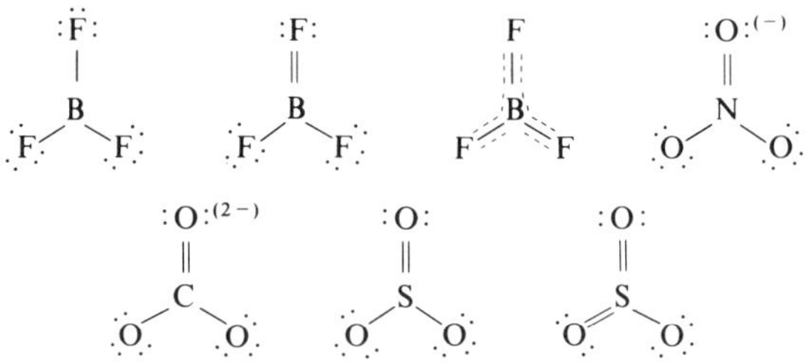

All trigonal planar ($\mathbf{D}_{3h}$)

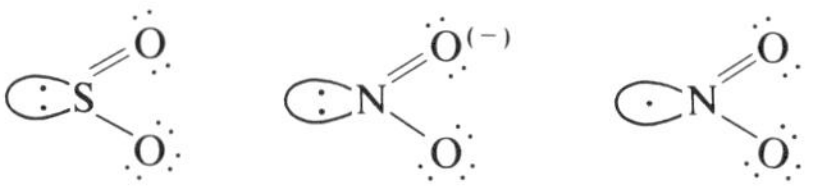

Angular ($\mathbf{C}_{2v}$)

Since π bonding involves the sidewise overlap of orbitals perpendicular to the σ bond between two atoms (also perpendicular to the molecular plane for planar AX_3 molecules), the presence or the extent of π bonding has little effect on the σ framework. Thus the BF_3 molecule is described as trigonal planar with sp^2 hybridization whether we consider σ bonding only (with an incomplete octet on B) or include one π bond (delocalized in the MO description or by resonance in a valence bond description), corresponding to the familiar descriptions of the isoelectronic NO_3^- and CO_3^{2-}. Two or even three π bonds can be written for SO_3 by using empty d orbitals on S.

The extent of π bonding in molecules such as SO_3 and oxoanions (i.e., ClO_4^-, SO_4^{2-}, PO_4^{3-}, ClO_3^-) is determined by the relative energies of the orbitals involved and by the charge distribution within the molecule (or ion). A useful guide is Pauling's *electroneutrality principle*, the tendency for atoms to become neutral or for excess negative charge to be shared among electronegative atoms and excess positive charge to be shared among electropositive atoms. We can assess charge distribution using the *formal charge*, the excess or deficiency of charge assuming the equal sharing of electrons in all bonds. Ordinary covalent bonds, formed by each atom furnishing one electron, do not contribute to the formal charge. Dative bonding (σ or π), in which one atom furnishes an electron pair to be shared, gives the donor a +1 charge and the acceptor a −1 charge:

(0) H:Cl: (0)   (0) :C::O (0)   (−1) :C:::O: (+1)

BF_3 all (0)   :NH_3 all (0) $F_3B^{(-1)}$—$N^{(+1)}H_3$

Single-bonded ClO_4^- involves three Cl—O dative bonds, giving formal charges of +3 on Cl and −1 on each oxygen. One π bond, with O as the electron pair donor, gives that oxygen zero formal charge and reduces the formal charge on Cl to +2. The Cl has +1 formal charge with two double bonds, no formal charge with three double bonds, and −1 formal charge with four double bonds. Of course, formal charges are assigned assuming nonpolar bonds (equal sharing). Polarity of the bonds (resulting from the greater electronegativity of oxygen) makes Cl more positive and tends to offset the decrease in positive charge accompanying π donation from O. Bond lengths indicate that the bond order for ClO_4^- is about 1.5, corresponding to two (delocalized) double bonds with the negative charge delocalized among the four oxygens.

EXAMPLE 6.1 Neglecting bond polarity, how many double bonds are required for zero charge on the central atom for SO_4^{2-} and PO_4^{3-}?

SOLUTION Because of the lower oxidation states (S^{VI} and P^{V}), less double bonding is required compared to ClO_4^-. The following structures give zero formal charge on the central atom.

SO_4^{2-}: S with two S=O and two S—O(−1); PO_4^{3-}: P with one P=O and three P—O(−1); ClO_4^-: Cl with three Cl=O and one Cl—O(−1).

In cases of *sp* hybridization on the central atom, the possibilities are X—A—X (linear) and ⊙A—X, a diatomic molecule. The following are examples of linear hybridized AX_2 (or XAY) molecules.

H—Be—H   O=C=O   H—C≡N: O=N=O (+ on N)   :O—C≡N: (− on O)   N=N=O (− on N, + on central N)

Other resonance contributing structures (e.g., N≡N—O:) involve different charge distribution, but still correspond to linear *sp* hybridization.

Table 6.1 summarizes the molecular shapes for simple AX_n molecules, including molecules for which the number of σ pairs plus lone pairs is greater than four. Although bonding can be explained (see Section 7.2) without using *d* orbitals, molecular shapes can be predicted by simply assuming electron promotion to *d* orbitals in cases such as the following.

		s	*p*	*d*	Hybrid
	Cl	··	·· ·· ·	_ _ _ _ _	
Cl—F	Cl	··	·· ·· ··	_ _ _ _ _	
Promote one *e*	Cl	··	·· · ·	· _ _ _ _	
ClF_3	Cl	··	·· ·· ··	·· _ _ _ _	sp^3d
Promote two *e*	Cl	··	· · ·	· · _ _ _	
ClF_5	Cl	··	·· ·· ··	·· ·· _ _ _	sp^3d^2
Promote three *e*	I	·	· · ·	· · · _ _	
IF_7	I	··	·· ·· ··	·· ·· ·· _ _	sp^3d^3

(IF_7 is included because ClF_7 is unstable.) Here the hybridized orbitals accommodate all shared and unshared electrons. The molecular shapes are described in Table 6.1 (the more stable BrF_5 is used instead of ClF_5). The addition of a halide ion involves donation of another electron pair occupying a *d* orbital.

$$Cl{-}F + F^- \longrightarrow [F{-}Cl{-}F]^{(-)}$$ (XeF$_2$ is isoelectronic)

$$ClF_3 + F^- \longrightarrow [ClF_4]^{(-)}$$ (XeF$_4$ is isoelectronic)

$$IF_5 + F^- \longrightarrow IF_6^-$$ (Irregular structure, not octahedral)

6.2.2 VALENCE SHELL ELECTRON PAIR REPULSION (VSEPR)

The approximate molecular shape is determined by the number of electron pairs on the central atom and hybridization. This is often adequate for a molecular orbital treatment. In the trigonal bipyramid the axial and equatorial positions are geometrically nonequivalent. Unshared electron pairs, acting as very large substituents, occupy the less crowded equatorial positions preferentially. This leads to the structures for XeF_2, ClF_3, and SF_4 shown in Table 6.1.

Distortions from the regular geometry of the tetrahedron, trigonal bipyramid, or octahedron occur because the repulsion among electron pairs on the central atom decreases in the order, lone pair–lone pair > lone pair–bonding pair > bonding pair–bonding pair. In NH_3 the bond angle (107.3°) is

TABLE 6.1
Molecular Geometries for AX_n Molecules

Number of σ pairs plus lone pairs	Arrangement of directed orbitals	Hybridization	Unshared electron pairs	Molecule	Molecular shape
2	Linear	*sp*	None	X — A — X X — A ≡ X X = A = X	Linear
3	Trigonal	sp^2	One		Angular
			None		Trigonal planar
4	Tetrahedral	sp^3	Two		Angular
			One		Pyramidal
			None		Tetrahedral
5	Trigonal bipyramidal	dsp^3	Three		Linear (XeF_2)

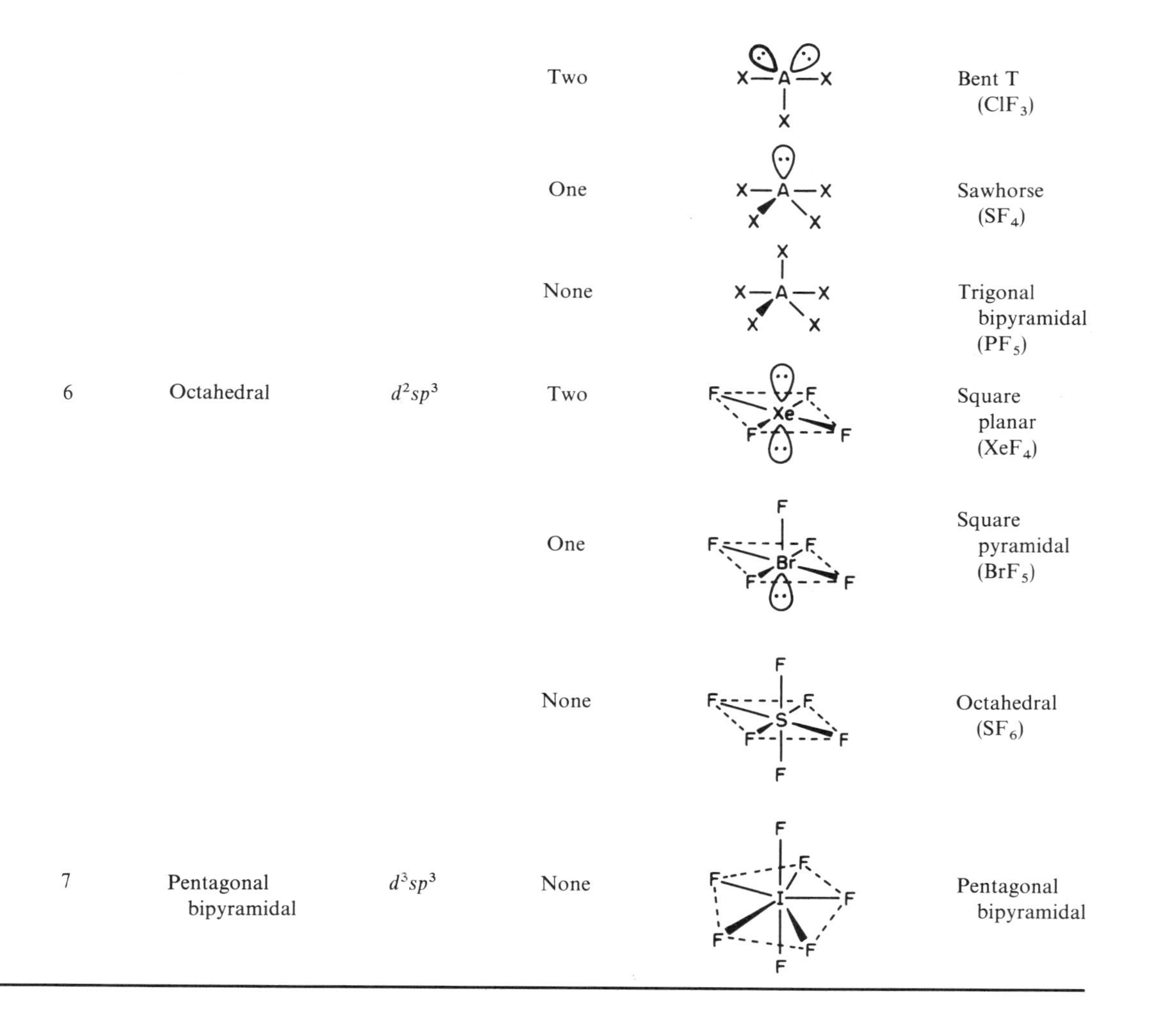

			Two		Bent T (ClF_3)
			One		Sawhorse (SF_4)
			None		Trigonal bipyramidal (PF_5)
6	Octahedral	d^2sp^3	Two		Square planar (XeF_4)
			One		Square pyramidal (BrF_5)
			None		Octahedral (SF_6)
7	Pentagonal bipyramidal	d^3sp^3	None		Pentagonal bipyramidal

less than the tetrahedral angle (109.5°) because of repulsion between the lone pair and the bonding pairs. The bond angle for H_2O (104.5°C) is decreased further because of the repulsion between the two lone pairs. Fluorine substitution for H normally *decreases* the bond angles (NF_3, 102.4° and OF_2, 101.5°) because of the decrease in repulsion between the bonding electron pairs in the longer A—F bonds and because of displacement of electron density away from A toward the highly electronegative F. Bond angles are also much smaller for the hydrides of third-period elements (PH_3, 93.3° and H_2S, 92.2°) because of the longer bonds and lower electronegativity of A (P compared to N and S compared to O).

The cases considered above involved only σ bonding. Double bonding increases the electron density in the bond, with the electron density of the π orbitals being concentrated in lobes on each side of the σ bond. Thus the bond angle *increases* from PH_3 (93.3°C) to PF_3 (97.8°) because of some $d \leftarrow p$ π bonding in PF_3 (not possible for PH_3 or NF_3). The molecule $FClO_2$ is expected to have higher bond order for the Cl—O bonds than for the Cl—F bond because π donation lowers the formal charge on oxygen. The most important repulsion is between the lone pair and the (partial) double bonds. The smallest bond angle is $Cl\angle^{O}_{F}$ (102°).

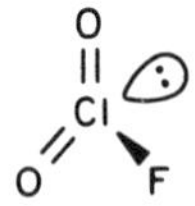

6.3 Sigma Bonding in AX_n Molecules

6.3.1 SYMMETRY SPECIES AND HYBRIDS FOR XeF_4

Next let us consider the case for σ bonding in AX_n molecules, with XeF_4 as an example. We choose a coordinate system with the central atom A at the origin and the unique direction (if there is one) vertical along the z axis. XeF_4 is square planar ($\mathbf{D}_{4h}$), so we orient the XeF_4 in the xy plane with the σ bonds along the x and y axes. We must find the representations for the combinations of the vectors along the bonds. For one operation of each class, we write the number of vectors left unchanged. The numbers obtained are the characters describing the transformation properties of this set of vectors (Fig. 6.6). Γ_σ is a reducible representation of the group. We can reduce Γ_σ by inspection using the characters for σ_h, σ_v, and i as useful clues or using Eq. (3.79) to obtain $\Gamma_\sigma = A_{1g} + B_{1g} + E_u$.

The σ bonds for XeF_4 transform as A_{1g}, B_{1g}, and E_u. If we want to interpret this in terms of familiar atomic orbital descriptions, we can identify the AOs of

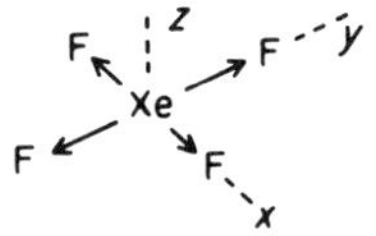

$\mathbf{D}_{4h}$	E	C_4	C_2	C_2'	C_2''	i	S_4	σ_h	σ_v	σ_d	
Γ_σ	4	0	0	2	0	0	0	4	2	0	$= A_{1g} + B_{1g} + E_u$

Fig. 6.6 Orientation of the XeF_4 molecule and the transformation properties of the vectors along the σ bonds.

Xe belonging to these representations from the character table.

$$A_{1g}: \quad s, d_{z^2}$$

$$B_{1g}: \quad d_{x^2-y^2}$$

$$E_u: \quad p_x, p_y$$

Only these AOs can participate in σ bonding; but we have five orbitals and we need only four. Since there is only one orbital with B_{1g} symmetry, $d_{x^2-y^2}$ must participate. The E_u representation is two-dimensional so it consists of two equivalent (degenerate) orbitals—here p_x and p_y; we have orbitals directed right where we need them. In addition, we need *one* A_{1g} orbital and there are two that meet the symmetry requirements. Two additional criteria to be considered are overlap and relative energies. The most favorable (lowest energy) bonding orbitals are obtained from the lowest energy AOs (considering only those in the range of energies of the orbitals on the other atom). Here the energy of d_{z^2} is much higher than that of s, making s more favorable. This also precludes (or it makes rather unimportant) another possibility—the mixing (hybridization) of s and d_{z^2}. Since the s orbital is spherically symmetric, it can interact equally with a ligand along any direction. The d_{z^2} orbital has the major lobes along z, orthogonal to the Xe—F bonds, and only the small donut in the xy plane. On the grounds of overlap and energy, s is more favorable than d_{z^2}. Thus, the favored combination is s, $d_{x^2-y^2}$, p_x, p_y. This is the familiar dsp^2 hybrid. The designation dsp^2 is specific—only p_x and p_y will do (p_z is orthogonal to the bond directions), and although there are two d orbitals in the xy plane, only $d_{x^2-y^2}$ has lobes along the axes.

6.3.2 LIGAND GROUP ORBITALS USING THE PROJECTION OPERATOR METHOD

We know which orbitals on the central atom are suitable for σ bonding; now let us focus on the ligand orbitals. Each F has an orbital directed toward Xe for σ bonding. From the description of bonding in HF we can conclude that it is a

p orbital, but it does not matter. It could be an sp, sp^2, or sp^3 hybrid, or even an s orbital. It is only necessary that there be a component of σ symmetry along the bonding direction. In effect we can deal with the orbitals as a set of vectors directed toward Xe. Since the transformation properties with respect to the origin in $\mathbf{D}_{4h}$ of the four equivalent vectors does not depend on whether they all point in or all point out, this is an *equivalent* set belonging to the *same* representations. Examining the set of vectors along the bonds give us the appropriate orbitals on the central atom *and* the representations of the combinations of ligand orbitals. These combinations of ligand orbitals are called the *ligand group orbitals* (LGOs). They are the *symmetry adapted LCAOs* for the group of ligands since they belong to the group representations.

We find the descriptions of the LGOs by using the *projection operator method* (see Section 3.10.5). For a one-dimensional representation, each character is the matrix for a symmetry operation. We apply Eq. (3.103) to the B_{1g} representation by operating on an orbital using every operation of the group (see Table 6.2) with the appropriate sign from the character table to give

$$\begin{aligned}\psi_{b_{1g}} = O_{B_{1g}}(\phi_1) = \tfrac{1}{16}[&(1)E\phi_1 - 1C_4\phi_1 - 1C_4^{-1}\phi_1 + 1C_2\phi_1 + 1(C_2')_x\phi_1 \\ &+ 1(C_2')_y\phi_1 - 1(C_2'')_1\phi_1 - 1(C_2'')_2\phi_1 + 1(i)\phi_1 - 1S_4\phi_1 - 1S_4^{-1}\phi_1 \\ &+ 1\sigma_h\phi_1 + 1(\sigma_v)_x\phi_1 + 1(\sigma_v)_y\phi_1 - 1(\sigma_d)_1\phi_1 - 1(\sigma_d)_2\phi_1].\end{aligned}$$

The signs of the characters for B_{1g} are taken from the character table ($\mathbf{D}_{4h}$). The tedious part is carrying out each of the symmetry operations on ϕ_1 to determine which orbital is produced by the operation. Since the result will be the same for all representations, except for the characters, it is simpler to carry out all of these operations just once and tabulate the results, as shown in the first line of Table 6.2. Then we merely have to multiply the ϕ_i, obtained by an operation, by the corresponding character for that operation. The results for A_{1g}, B_{1g}, E_u, and B_{2g} are shown in the table. The factor l/h ($\frac{1}{16}$ for one-dimensional representations with $h = 16$) can be dropped and the coefficients simplified to the smallest integers since we must normalize.

We can summarize the steps involved in applying the projection operator method as follows:

(1) Identify the orbital obtained by operating on one ligand orbital using *every* symmetry operation of the group.

(2) To obtain the LCAO for each representation, multiply each orbital obtained in (1) by the character for the operation used to obtain the orbital, using the characters for the representation being examined. This determines the projection of the orbital set in the direction of each representation.

(3) Take the sum of the orbitals for each representation and divide by a number to give the smallest integral coefficients. These are the LCAOs. They

TABLE 6.2
Ligand Group Orbitals for σ Bonding in XeF_4

Representations for σ bonds
A_{1g}
B_{1g}
E_u

$\mathbf{D}_{4h}$	E	C_4	C_4^{-1}	C_2	$(C_2')_x$	$(C_2')_y$	$(C_2'')_1$	$(C_2'')_2$	i	S_4	S_4^{-1}	σ_h	$(\sigma_v)_x$	$(\sigma_v)_y$	$(\sigma_d)_1$	$(\sigma_d)_2$
ϕ_1	ϕ_1	ϕ_2	ϕ_4	ϕ_3	ϕ_1	ϕ_3	ϕ_4	ϕ_2	ϕ_3	ϕ_2	ϕ_4	ϕ_1	ϕ_1	ϕ_3	ϕ_4	ϕ_2
A_{1g}	ϕ_1	$+\phi_2$	$+\phi_4$	$+\phi_3$	$+\phi_1$	$+\phi_3$	$+\phi_4$	$+\phi_2$	$+\phi_3$	$+\phi_2$	$+\phi_4$	$+\phi_1$	$+\phi_1$	$+\phi_3$	$+\phi_4$	$+\phi_2$
B_{1g}	ϕ_1	$-\phi_2$	$-\phi_4$	$+\phi_3$	$+\phi_1$	$+\phi_3$	$-\phi_4$	$-\phi_2$	$+\phi_3$	$-\phi_2$	$-\phi_4$	$+\phi_1$	$+\phi_1$	$+\phi_3$	$-\phi_4$	$-\phi_2$
E_u	$2\phi_1$			$-2\phi_3$					$-2\phi_3$			$+2\phi_1$				
B_{2g}	ϕ_1	$-\phi_2$	$-\phi_4$	$+\phi_3$	$-\phi_1$	$-\phi_3$	$+\phi_4$	$+\phi_2$	$+\phi_3$	$-\phi_2$	$-\phi_4$	$+\phi_1$	$-\phi_1$	$-\phi_3$	$+\phi_4$	$+\phi_2$

$a_{1g} = 4\phi_1 + 4\phi_2 + 4\phi_3 + 4\phi_4$ or $\phi_1 + \phi_2 + \phi_3 + \phi_4$ $\qquad \psi_{a_{1g}} = (1/2)(\phi_1 + \phi_2 + \phi_3 + \phi_4)$

$b_{1g} = 4\phi_1 - 4\phi_2 + 4\phi_3 - 4\phi_4$ or $\phi_1 - \phi_2 + \phi_3 - \phi_4$ $\qquad \psi_{b_{1g}} = (1/2)(\phi_1 - \phi_2 + \phi_3 - \phi_4)$

$e_u = 4\phi_1 - 4\phi_3$ or $\phi_1 - \phi_3$ $\qquad \psi_{e_u} = (1/\sqrt{2})(\phi_1 - \phi_3)$

b_{2g} vanishes using $\mathbf{D}_{4h}$ or $\mathbf{D}_4$

can be normalized by dividing by the square root of the sum of the squares of the coefficients (see Section 6.1.1).

We note that if we had identified one of the representations incorrectly and used it in the procedure, here B_{2g} as an example, the sum vanishes. The notation used follows the convention of using lowercase letters for the symmetry species of orbitals. Uppercase letters are used for representations in the general sense and for energy states.

It is apparent from examination of Table 6.2 that we could have obtained these LCAOs using only the proper rotations, the rotational subgroup $\mathbf{D}_4$. This reduces the effort considerably. In general, we can use the rotational subgroup. For this purpose, the direct products (or semidirect products) of subgroups are identified with the character tables in Appendix I. Often we can obtain the LCAOs using the cyclic subgroup ($\mathbf{C}_4$ here). In this case, the $\mathbf{C}_4$ subgroup would suffice for A_{1g}, B_{1g}, and E_u, but it would not have rejected B_{2g}. The reason, of course, is that there is only one B representation in the $\mathbf{C}_4$ group. The subscript distinction requires the full rotational (dihedral) group. We shall see later that in some cases we can use the cyclic ($\mathbf{C}_n$) group, but more generally, it is safer to use the rotational subgroup, provided that all of the orbitals involved are generated for the one-dimensional representations. That is, each ligand orbital must be included in the sum. Of course, those orbitals belonging to another symmetry set will have to be dealt with separately.

Getting back to the LCAOs obtained, those for the one-dimensional representations are complete, but not those for e_u. The LCAO for the e_u LGO involves only orbitals along the x axis. In this group, x and y are equivalent and belong to the E_u representation. Hence, there must be another equivalent e_u LGO along the y axis. If we perform the C_4 operation on this LGO, we get $(\phi_2 - \phi_4)$:

$$(\phi_1 - \phi_3) \xrightarrow{C_4} (\phi_2 - \phi_4),$$

$$(\Psi_{e_u})_a = \frac{1}{\sqrt{2}}(\phi_1 - \phi_3), \qquad (\Psi_{e_u})_b = \frac{1}{\sqrt{2}}(\phi_2 - \phi_4).$$

This is the other e_u LGO—it is equivalent to $(\Psi_{e_u})_a$, but involves the other pair of ligand orbitals. If we had started out by applying all symmetry operations to ϕ_2, we would have gotten the same result for a_{1g}, b_{1g} would be $\phi_2 - \phi_1 + \phi_4 - \phi_3$, and e_u would be $\phi_2 - \phi_4$. We thus could have gotten the second e_u LGO in this way. The two sign combinations for b_{1g} are equivalent because we can multiply by -1 without changing the wave function since it is Ψ^2 that is physically significant.

This is a case where performing a symmetry operation generates another LGO that we can recognize as the equivalent partner needed for a two-dimensional representation (see Section 8.2.1 for a case involving a three-dimensional representation). We can get both partners using the projection operator also, and this works when the symmetry transformation does not give the partner directly. Using the projection operator, we get one LCAO for e_u by using the characters for E_u. This involves the *trace (character) projection operator* [Eq. (3.106)]. There is no distinction for one-dimensional representations where the character is the full (1×1) matrix. We can apply the projection operator for the E_u representation directly using the matrix elements as follows:

	E	C_4	C_4^{-1}	C_2	$(C_2')_x$	$(C_2')_y$	$(C_2'')_1$	$(C_2'')_2$
$E_u \begin{bmatrix} x \\ y \end{bmatrix}$	$\begin{bmatrix} 1 & 0 \\ 0 & 1 \end{bmatrix}$	$\begin{bmatrix} 0 & -1 \\ 1 & 0 \end{bmatrix}$	$\begin{bmatrix} 0 & 1 \\ -1 & 0 \end{bmatrix}$	$\begin{bmatrix} -1 & 0 \\ 0 & -1 \end{bmatrix}$	$\begin{bmatrix} 1 & 0 \\ 0 & -1 \end{bmatrix}$	$\begin{bmatrix} -1 & 0 \\ 0 & 1 \end{bmatrix}$	$\begin{bmatrix} 0 & 1 \\ 1 & 0 \end{bmatrix}$	$\begin{bmatrix} 0 & -1 \\ -1 & 0 \end{bmatrix}$
$P(\phi_1) \rightarrow$	ϕ_1	ϕ_2	ϕ_4	ϕ_3	ϕ_1	ϕ_3	ϕ_4	ϕ_2
a_{11}	ϕ_1			$-\phi_3$	$+\phi_1$	$-\phi_3$		$= 2\phi_1 - 2\phi_3$
a_{22}	ϕ_1			$-\phi_3$	$-\phi_1$	$+\phi_3$		$= 0$
a_{12}		$-\phi_2$	$+\phi_4$				$+\phi_4$	$-\phi_2 = 2\phi_4 - 2\phi_2$

Application of a_{11} gives $2\phi_1 - 2\phi_3$ or $\phi_1 - \phi_3$. This is one of the e_u LGOs. Applying a_{22} gives zero. The second diagonal element can give the partner LGO, the same LGO, or zero. Using an off-diagonal element involves the *transfer operator* [Eq. (3.105)], and this gives the partner. Or at least one of these gives the partner; a_{21} causes the LGO to vanish. Application of a_{12} gives $-\phi_2 + \phi_4$, which we can write as $\phi_2 - \phi_4$, the second partner.

We have the representations for the Xe orbitals suitable for σ bonding and the LCAOs for the LGOs. Let us proceed to combine them (Fig. 6.7). The combinations of perfect matches of the signs of the amplitudes of the wave functions are the bonding σ MOs. Reversing the signs of one set (here the LGO) gives the antibonding σ^* MOs.

A qualitative energy-level MO diagram (Fig. 6.8) can be obtained by considering the relative energies of the orbitals and the extent of orbital overlap based on the angular distribution functions. The extension of the orbitals (radial part) is difficult to assess. We recognize that the F valence shell orbital is lower in energy than the Xe orbitals because we know that F is *much* more electronegative than Xe. The energies of the Xe orbitals increase in the order $5s < 5p \ll 5d$. The b_{1g} and e_u MOs involve ideal overlap for σ bonding, making full use of the lobes directed along the bonds. The Xe d orbitals other than $d_{x^2-y^2}$ are nonbonding, but in a $\mathbf{D}_{4h}$ field they split into the groups a_{1g} (d_{z^2}), e_g (d_{xz}, d_{yz}), and b_{2g} (d_{xy}). Of these, the d_{z^2} should have lowest energy because the major lobe is along the z axis and, hence, farthest from the xy plane. The d_{xy} orbital should be highest in energy because it is contained in the crowded xy plane.

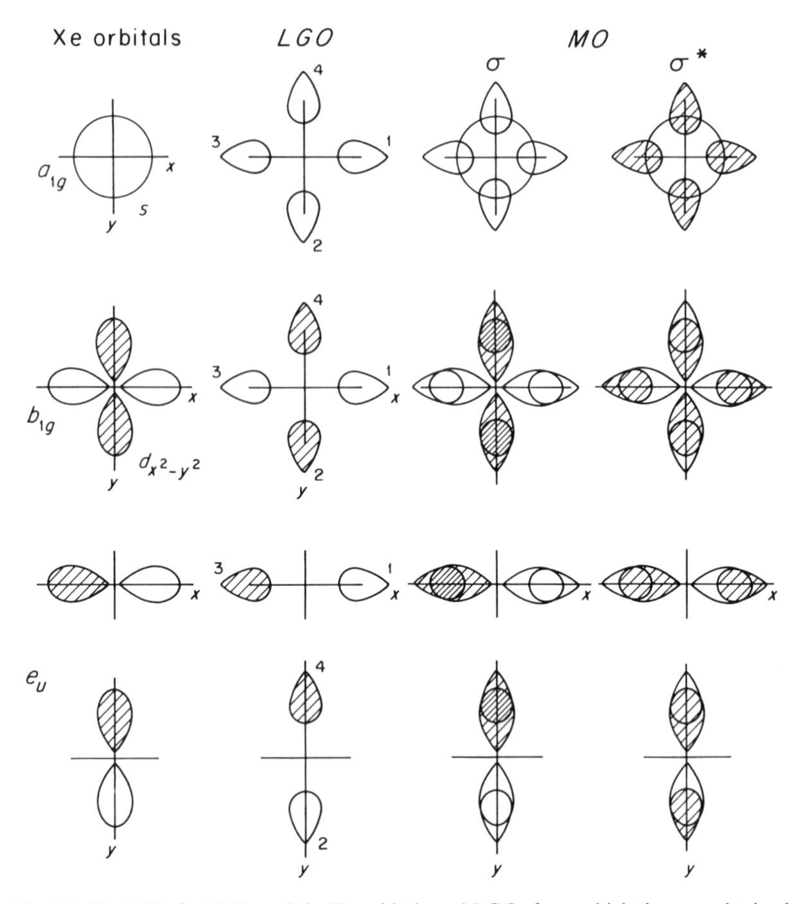

Fig. 6.7 The MOs for XeF_4 and the Xe orbitals and LGOs from which they are obtained.

The square planar XeF_4 molecule has four bonding electron pairs and two nonbonding pairs on Xe. In this description, assuming that the Xe d orbitals are low enough in energy for participation in bonding, the two Xe nonbonding pairs would be in the $a_{2u}(p_z)$ and $a_{1g}(d_{z^2})$ orbitals. It is reasonable to expect the lone pairs to occupy orbitals perpendicular to the xy plane; of course, d_{z^2} does have a minor lobe (the donut) in the xy plane.

This treatment of XeF_4 applies also to the σ bonding in square planar transition metal complexes. The symmetry considerations are the same. The treatment would differ primarily in the energy-level diagram because of differences in the energies of the metal orbitals (particularly the expected lower energy of the d orbitals) and the energies of the ligand orbitals. Square planar complexes are not usually encountered with the most highly electronegative ligands such as F^-.

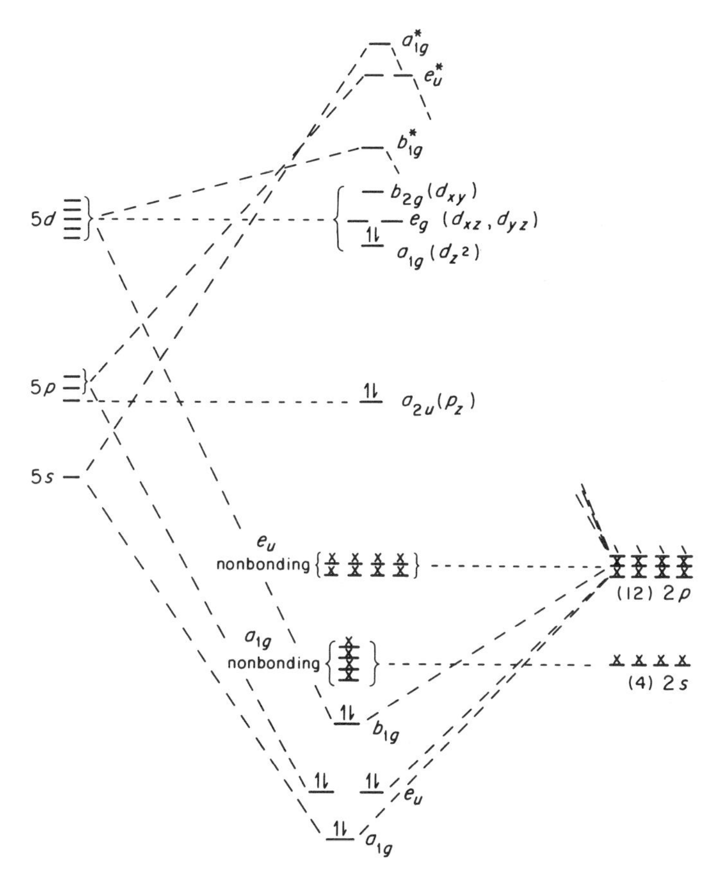

Fig. 6.8 Qualitative MO energy-level diagram for XeF_4.

6.3.3 TRIGONAL BIPYRAMIDAL AX_5 ($\mathbf{D}_{3h}$) MOLECULES

Let us consider σ bonding in a $\mathbf{D}_{3h}$ molecule such as PF_5. If we examine the σ vectors, we see that they are in two independent sets. That is, the σ_h and C_2 operations interchange ϕ_1 and ϕ_5, and the C_3, S_3, and σ_v operations interchange ϕ_2, ϕ_3, and ϕ_4, but no operations interchange ϕ_1 or ϕ_5 (the axial

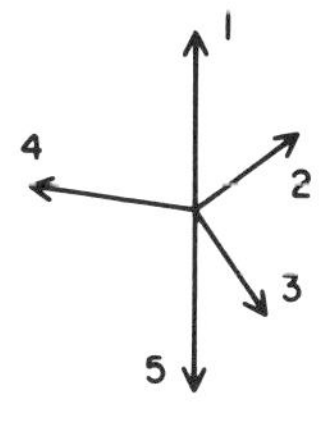

σ bonds in PF_5

set) with the others (the equatorial set). We should treat the two sets separately.[1] The number of vectors of each set left unchanged by each operation (using one operation of each class here) is tabulated in Table 6.3. The results are

$$\Gamma_{ax} = A_1' + A_2'' \qquad \text{and} \qquad \Gamma_{eq} = A_1' + E'.$$

The representations of the AOs can be identified from the character table. We see that the $(d_{xz}, d_{yz}; e'')$ orbitals are nonbonding because of symmetry. There are two A_1' orbitals, so we need *both* s and d_{z^2}. In addition to s, d_{z^2}, and p_z, we need one pair of orbitals with E' symmetry. On symmetry grounds, either (p_x, p_y) or $(d_{xy}, d_{x^2-y^2})$ will do. The *possible* hybrids are dsp^3 or d^3sp. Here the dsp^3 is more realistic because the p orbitals are more favorable for bonding since they are lower in energy than the d orbitals.

As we have seen, the σ vectors for the LGOs belong to the same representations as those on the central atom. To get the linear combination, we apply the projection operator method separately to one orbital of each set and multiply the results by the corresponding characters of the appropriate representations. Once again we see from the results in Table 6.3 that the rotational group ($\mathbf{D}_3$) would suffice. The LGOs are unique for both a_1' and for a_2'', but the e' LGO is one of a pair. If we try the same approach used for the $\mathbf{D}_{4h}$ case and operate on this e' LGO using C_3, then we get $2\phi_3 - \phi_4 - \phi_2$. This result is not an independent LGO. The three equatorial σ orbitals are equivalent, and we have merely permuted the arbitrary subscripts. Performing the C_3^2 operation gives $2\phi_4 - \phi_2 - \phi_3$, still another permutation of subscripts. However, if we take a linear combination of these two permutations, the sum gives $-2\phi_2 + \phi_3 + \phi_4$ (-1 times our original e' LGO), but the difference is $3\phi_3 - 3\phi_4$:

$$2\phi_3 - \phi_4 - \phi_2 - (2\phi_4 - \phi_2 - \phi_3) = 3\phi_3 - 3\phi_4 \qquad \text{or} \qquad \phi_3 - \phi_4.$$

This is an independent LGO that is orthogonal to e_a'.

$$\Psi_{e_a'} = \frac{1}{\sqrt{6}}(2\phi_2 - \phi_3 - \phi_4), \qquad \Psi_{e_b'} = \frac{1}{\sqrt{2}}(\phi_3 - \phi_4)$$

We obtained the other e' LGO by a trial-and-error search for another LGO. One systematic way to do this is to apply the Gram–Schmidt method for orthogonalization.[2] A simple, direct approach is to use the projection operator method and the real characters derived from the $\mathbf{C}_3$ subgroup, for

[1] It is possible to combine the sets, but there can be complications. It is much easier to see how to handle the two A_1' combinations if we deal with the sets separately.

[2] See M. Orchin and H. H. Jaffe, *Symmetry, Orbitals, and Spectra*, Wiley (Interscience), New York, 1971, p. 153.

TABLE 6.3
Ligand Group Orbitals for σ Bonding in PF_5

D_{3h}	E	C_3	C_2	σ_h	S_3	σ_v	
Γ_{ax}	2	2	0	0	0	2	$= A_1' + A_2''$
Γ_{eq}	3	0	1	3	0	1	$= A_1' \quad E'$

Representations of Orbitals on P	
s, d_{z^2}	A_1'
p_z	A_2''
$(p_x, p_y)(d_{xy}, d_{x^2-y^2})$	E'
(d_{xz}, d_{yz})	E''

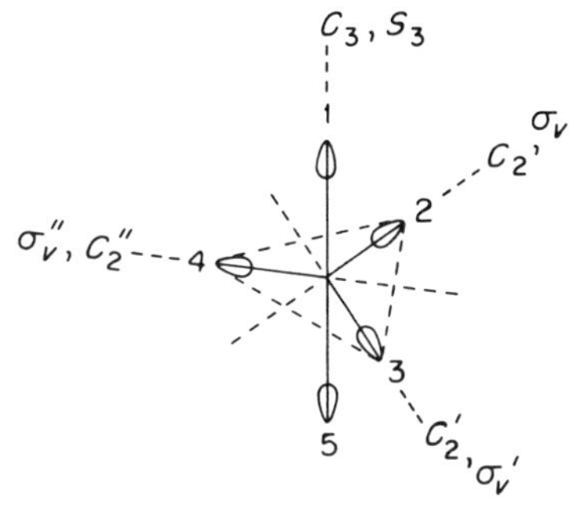

Apply the projection operator

D_{3h}	E	C_3	C_3'	C_2	C_2'	C_2''	σ_h	S_3	S_3'	σ_v	σ_v'	σ_v''
ϕ_1	ϕ_1	ϕ_1	ϕ_1	ϕ_5	ϕ_5	ϕ_5	ϕ_5	ϕ_5	ϕ_5	ϕ_1	ϕ_1	ϕ_1
ϕ_2	ϕ_2	ϕ_3	ϕ_4	ϕ_2	ϕ_4	ϕ_3	ϕ_2	ϕ_3	ϕ_4	ϕ_2	ϕ_4	ϕ_3

Axial σ LGO

a_1' $\quad 6\phi_1 + 6\phi_5$ or $\phi_1 + \phi_5$ $\quad \psi_{a_1'} = (1/\sqrt{2})(\phi_1 + \phi_5)$

a_2'' $\quad 6\phi_1 - 6\phi_5$ or $\phi_1 - \phi_5$ $\quad \psi_{a_2''} = (1/\sqrt{2})(\phi_1 - \phi_5)$

Equatorial LGO

a_1' $\quad 4\phi_2 + 4\phi_3 + 4\phi_4$ or $\phi_2 + \phi_3 + \phi_4$ $\quad \psi_{a'_1} = (1/\sqrt{3})(\phi_2 + \phi_3 + \phi_4)$

e' $\quad 4\phi_2 - 2\phi_3 - 2\phi_4$ or $2\phi_2 - \phi_3 - \phi_4$ $\quad (\psi_{e'})_a = (1/\sqrt{6})(2\phi_2 - \phi_3 - \phi_4)$

$(\psi_{e'})_b = (1/\sqrt{2})(\phi_3 - \phi_4)$

which the E representation consists of two one-dimensional representations involving imaginary characters. The real characters for the E representation can be obtained from the sum and the difference of the one-dimensional representations:

$$\varepsilon = \cos 120^\circ + i \sin 120^\circ = -0.5 + 0.866i,$$

$$\varepsilon^* = \cos 120^\circ - i \sin 120^\circ = -0.5 - 0.866i,$$

$$\varepsilon + \varepsilon^* = -1.0, \qquad \varepsilon - \varepsilon^* = 2(0.866i), \qquad \varepsilon^* - \varepsilon = -2(0.866i).$$

Simplifying, since we normalize later, we get

$$\frac{\varepsilon - \varepsilon^*}{2(0.866)i} = 1, \qquad \frac{\varepsilon^* - \varepsilon}{2(0.866)i} = -1.$$

The real characters are as follows.

	E	C_3	C_3^2
$E_a = E_1 + E_2$	2	-1	-1
$E_b = E_1 - E_2$	0	1	-1

Multiplying the orbitals generated in Table 6.3 by operating on ϕ_2 by these characters, we obtain

$$e'_a = 2\phi_2 - \phi_3 - \phi_4 \qquad \text{and} \qquad e'_b = \phi_3 - \phi_4, \qquad \text{not normalized.}$$

We could also use the matrices for the E representation in $\mathbf{C}_3$. The characters for the two one-dimensional E representations are the diagonal elements of the matrices. We would still have to take linear combinations to eliminate complex numbers. It is simpler to obtain the sets of real characters first, as we did above.

Fortunately there are helpful symmetry guidelines to aid our search for the second independent LCAO by an intuitive approach and to recognize it when we have it. First, we note that with AOs or MOs, the orbitals belonging to a given representation have the same number of nodal planes (zero for s and σ, 1 for p and π, 2 for d and δ, etc.). The independent orbitals are orthogonal and, in the simple case of orbitals with a single nodal plane, the planes are orthogonal for the independent orbitals, e.g., p. In the case of a pair of e orbitals (a and b), the nodal planes are as follows.

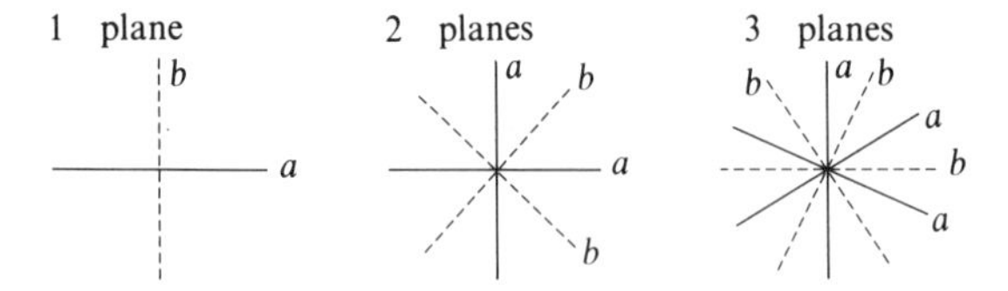

The nodal planes of the second set in each case bisect the angles formed by those of the first set. The case of one plane is special, but the nodal plane for the second orbital is perpendicular to that for the first (it bisects the 180° angle). There is an important energetic consequence of the pattern of nodal planes; *the greater the number of nodal planes, the more localized the electrons and the higher the energy of the orbital.* This is true for radial nodes as well. For atomic orbitals, there are other important considerations; but the energies increase as the number of radial nodes increases (see Fig. 1.1).

Getting back to our e' equatorial orbitals, we note that e'_a has a single nodal plane separating ϕ_3 and ϕ_4 from ϕ_2. The other e' orbital must have a single nodal plane perpendicular to this one and necessarily slicing through ϕ_2. This means that ϕ_3 and ϕ_4 must have opposite signs in the LCAO, and the coefficient of ϕ_2 must be zero (it is in the nodal plane). Thus, we are seeking a linear combination that eliminates ϕ_2. Here the C_3 operation rotates the nodal plane by 120° so the resulting linear combination does not have a nodal plane perpendicular to the first one. In the case of XeF_4, the C_4 operation rotates $(e_u)_a$ by 90°, giving $(e_u)_b$ directly.

The P orbitals, LGOs, and MOs for σ bonding in PF_5 are depicted in Fig. 6.9. The antibonding orbitals are obtained by reversing the signs of the P

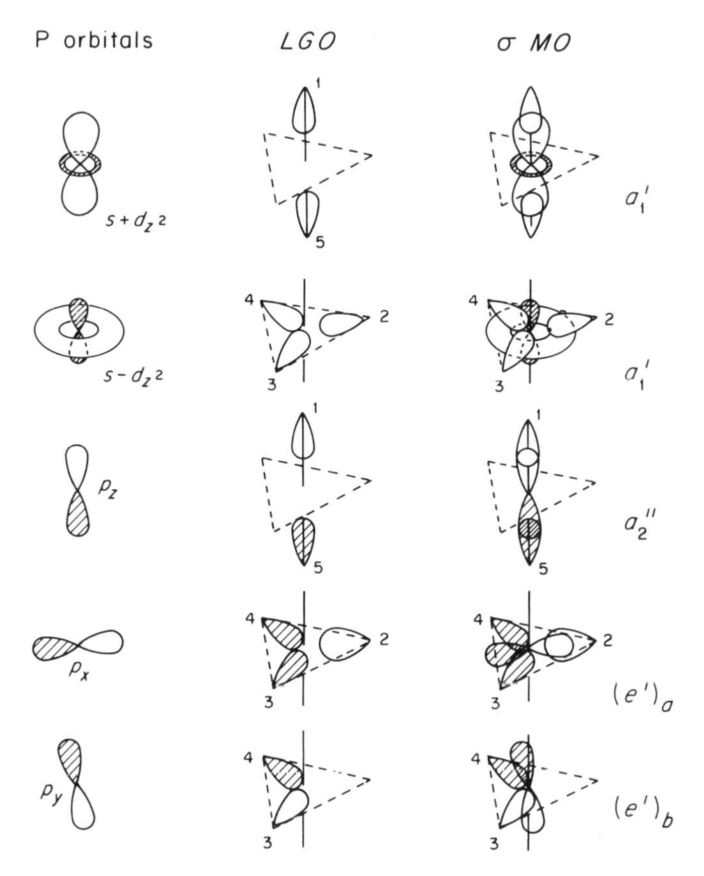

Fig. 6.9 Molecular orbitals for σ bonding in PF_5.

orbitals or those of the LGOs. Since s and d_{z^2} belong to a_1', they can both participate in bonding to the axial and equatorial ligands. We can see that s–d_{z^2} hybrids can maximize the overlap with each set. The extent of hybridization will be limited by the rather large difference in the energies of the orbitals.

A qualitative energy-level diagram for PF_5 is shown in Fig. 6.10. The a_1' orbital involving d_{z^2} to the greater extent is expected to have high energy because of the high energy of the d orbitals themselves. The energy of a_2'' is expected to be lower than that of e' because of the more favorable overlap for the case where the AOs are aligned ideally. The remaining nonbonding d orbitals belong to the representations e' and e''. The e' pair in the more crowded xy plane should be higher in energy than the e'' pair (d_{xz}, d_{yz}).

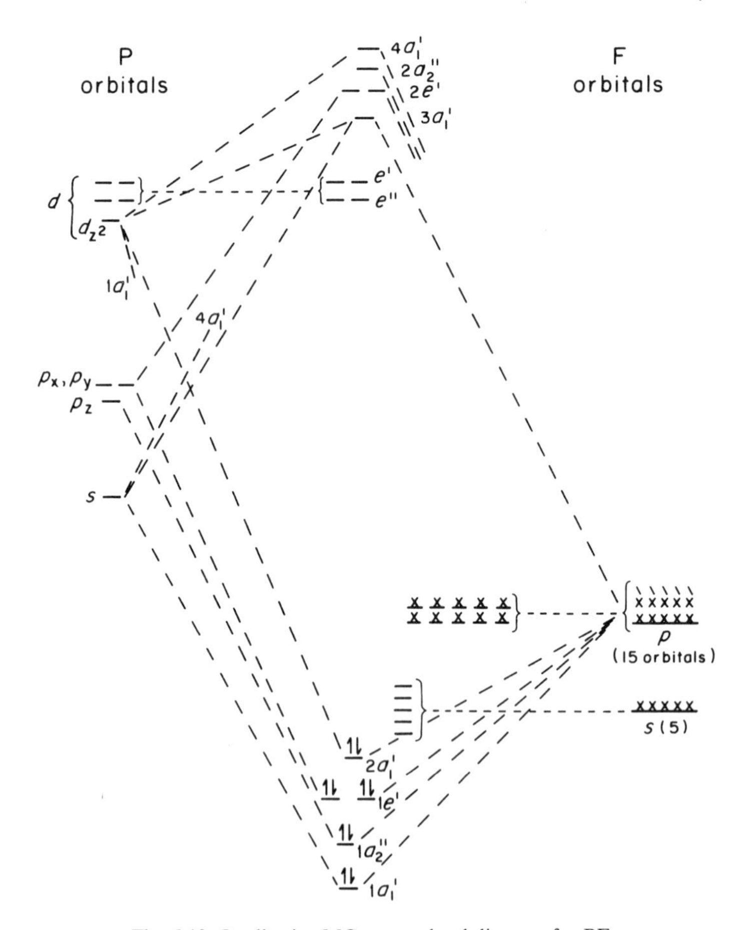

Fig. 6.10 Qualitative MO energy-level diagram for PF_5.

EXAMPLE 6.2 Find the possible and likely σ hybrid combinations for CO_3^{2-} ($\mathbf{D}_{3h}$) and BrF_5 ($\mathbf{C}_{4v}$).

SOLUTION The CO_3^{2-} ion is trigonal planar. The vectors for σ bonding are the same as those of the equatorial set for PF_5 (Table 6.3), so the representations are $A_1' + E'$.

Representation	Orbitals
A_1'	s, d_{z^2}
E'	$(p_x, p_y), (d_{x^2-y^2}, d_{xy})$

The *possible* hybrid combinations involve one A_1' and one E': sp^2, sd^2, dp^2, or d^3. Since C is in the second period, the valence shell contains only 2*s* and 2*p* orbitals. The *d* hybrids would require participation of 3*d*, which are unrealistically high in energy. The only reasonable hybrid is sp^2.

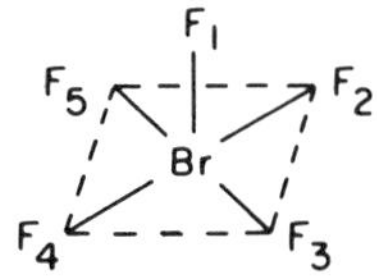

The BrF_5 molecule with $\mathbf{C}_{4v}$ symmetry is square pyramidal. The transformation of the σ vectors for the axial and basal sets gives the representations $2A_1 + B_1 + E$.

$\mathbf{C}_{4v}$	E	C_4	C_2	σ_v	σ_d	
(basal) Γ_σ	4	0	0	2	0	$= A_1 + B_1 + E$
(axial) Γ^σ	1	1	1	1	1	$= A_1$

The orbitals belonging to these representations are A_1 s, p_z, d_{z^2}, B_1 $d_{x^2-y^2}$, and E $(p_x, p_y), (d_{xz}, d_{yz})$. Possible hybrids involve any combinations of two A_1, B_1, and either E pair, including dsp^3, d^2sp^2, d^3sp, d^2p^3, sd^4, etc. Here dsp^3 must mean $d_{x^2-y^2}sp^3$, while for the $\mathbf{D}_{3h}$ case (PF_5) it is $d_{z^2}sp^3$. Since the *s* and *p* orbitals are lower in energy than the *d* orbitals, dsp^3 is most favorable because it utilizes the lower-energy orbitals to the greatest extent.

6.4 Hybrid Orbitals as Linear Combinations of Atomic Orbitals

For the CH_4 molecule ($\mathbf{T}_d$), the hybridization is sp^3. This permits us to use familiar atomic orbitals as a basis set, but the hybrids refer to equivalent carbon orbitals directed along the bonds. Each of these orbitals can be described as a linear combination of *s* and *p* orbitals.

For the square planar XeF_4 molecule, we obtained the LGOs belonging to the A_{1g}, B_{1g}, and E_u representations. We also found the hybrid oritals on Xe to be dsp^2, specifically involving the $d_{x^2-y^2}$, p_x, and p_y orbitals.

The four equivalent hybrid orbitals (ϕ_i) on Xe must be linear combinations of $d_{x^2-y^2}$, s, p_x, and p_y with the same symmetry as the σ LGOs. We want to find the coefficients necessary for the following LCAOs:

$$\begin{aligned} \phi_1 &= c_{11}s + c_{12}d + c_{13}p_x + c_{14}p_y, \\ \phi_2 &= c_{21}s + c_{22}d + c_{23}p_x + c_{24}p_y, \\ \phi_3 &= c_{31}s + c_{32}d + c_{33}p_x + c_{34}p_y, \\ \phi_4 &= c_{41}s + c_{42}d + c_{43}p_x + c_{44}p_y. \end{aligned} \tag{6.6}$$

These equations can be written in matrix form as

$$\begin{bmatrix} \phi_1 \\ \phi_2 \\ \phi_3 \\ \phi_4 \end{bmatrix} = \begin{bmatrix} c_{11} & c_{12} & c_{13} & c_{14} \\ c_{21} & c_{22} & c_{23} & c_{24} \\ c_{31} & c_{32} & c_{33} & c_{34} \\ c_{41} & c_{42} & c_{43} & c_{44} \end{bmatrix} \begin{bmatrix} s \\ d \\ p_x \\ p_y \end{bmatrix}. \tag{6.7}$$

Our σ orbitals are the same as ϕ, and we have found the linear combinations of these LGOs to get them to match the symmetry of the atomic s, d, and p orbitals, or

$$\psi_{a_{1g}} = \frac{1}{2}(\phi_1 + \phi_2 + \phi_3 + \phi_4), \qquad \psi_{b_{1g}} = \frac{1}{2}(\phi_1 - \phi_2 + \phi_3 - \phi_4),$$

$$\psi_{e_{u_a}} = \frac{1}{\sqrt{2}}(\phi_1 - \phi_3), \qquad \psi_{e_{u_b}} = \frac{1}{\sqrt{2}}(\phi_2 - \phi_4),$$

or

$$\begin{bmatrix} \psi_{a_{1g}} \\ \psi_{b_{1g}} \\ \psi_{e_{u_a}} \\ \psi_{e_{u_b}} \end{bmatrix} = \begin{bmatrix} s \\ d \\ p_x \\ p_y \end{bmatrix} = \begin{bmatrix} 1/2 & 1/2 & 1/2 & 1/2 \\ 1/2 & -1/2 & 1/2 & -1/2 \\ 1/\sqrt{2} & 0 & -1/\sqrt{2} & 0 \\ 0 & 1/\sqrt{2} & 0 & -1/\sqrt{2} \end{bmatrix} \begin{bmatrix} \phi_1 \\ \phi_2 \\ \phi_3 \\ \phi_4 \end{bmatrix}. \tag{6.8}$$

Since $\psi_{a_{1g}}$ on the central atom is the s orbital, $\psi_{b_{1g}}$ is $d_{x^2-y^2}$, etc., this equation gives the matrix **M** to convert the $\phi_1, \phi_2, \phi_3, \phi_4$ vector into the s, d, p_x, p_y vector. The matrix **C** [Eq. (6.7)] is the inverse of matrix **M**. Since these are orthogonal matrices, the inverse is the transpose, or **C** is the transpose of **M**.

$$\begin{bmatrix} 1/2 & 1/2 & 1/2 & 1/2 \\ 1/2 & -1/2 & 1/2 & -1/2 \\ 1/\sqrt{2} & 0 & -1/\sqrt{2} & 0 \\ 0 & 1/\sqrt{2} & 0 & -1/\sqrt{2} \end{bmatrix} \xrightarrow[\text{(transpose)}]{\text{take inverse}} \begin{bmatrix} 1/2 & 1/2 & 1/\sqrt{2} & 0 \\ 1/2 & -1/2 & 0 & 1/\sqrt{2} \\ 1/2 & 1/2 & -1/\sqrt{2} & 0 \\ 1/2 & -1/2 & 0 & -1/\sqrt{2} \end{bmatrix}. \tag{6.9}$$

Substituting in Eq. (6.7) gives

$$\begin{bmatrix} \phi_1 \\ \phi_2 \\ \phi_3 \\ \phi_4 \end{bmatrix} = \begin{bmatrix} 1/2 & 1/2 & 1/\sqrt{2} & 0 \\ 1/2 & -1/2 & 0 & 1/\sqrt{2} \\ 1/2 & 1/2 & -1/\sqrt{2} & 0 \\ 1/2 & -1/2 & 0 & -1/\sqrt{2} \end{bmatrix} \begin{bmatrix} s \\ d \\ p_x \\ p_y \end{bmatrix} \tag{6.10}$$

or

$$\begin{aligned} \phi_1 &= \frac{1}{2}s + \frac{1}{2}d_{x^2-y^2} + \frac{1}{\sqrt{2}}p_x, \\ \phi_2 &= \frac{1}{2}s - \frac{1}{2}d_{x^2-y^2} + \frac{1}{\sqrt{2}}p_y, \\ \phi_3 &= \frac{1}{2}s + \frac{1}{2}d_{x^2-y^2} - \frac{1}{\sqrt{2}}p_x, \\ \phi_4 &= \frac{1}{2}s - \frac{1}{2}d_{x^2-y^2} - \frac{1}{\sqrt{2}}p_y. \end{aligned} \tag{6.11}$$

These are the LCAOs required to give four equivalent dsp^2 orbitals for the square planar case.

EXAMPLE 6.3 Obtain the LCAOs for sp^2 ($\mathbf{D}_{3h}$) bonding as in BF_3, NO_3^-, and CO_3^{2-}.

SOLUTION The p orbitals involved in sp^2 ($\mathbf{D}_{3h}$) hybridization are p_x and p_y. We want the LCAOs to give three equivalent hybrid orbitals with 120° angles as shown in the sketch for BF_3:

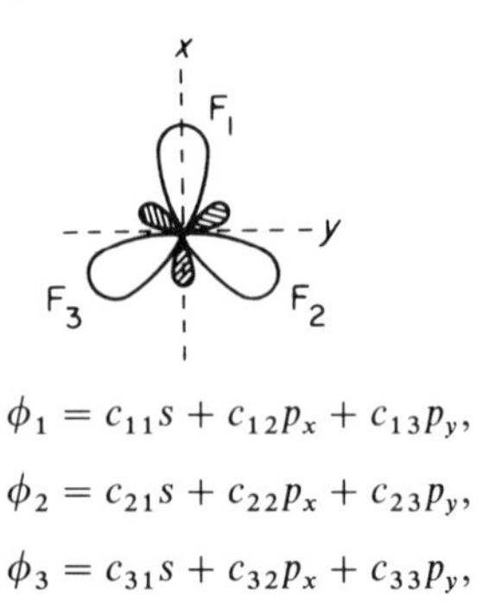

$$\phi_1 = c_{11}s + c_{12}p_x + c_{13}p_y,$$
$$\phi_2 = c_{21}s + c_{22}p_x + c_{23}p_y,$$
$$\phi_3 = c_{31}s + c_{32}p_x + c_{33}p_y,$$

or

$$\begin{bmatrix} \phi_1 \\ \phi_2 \\ \phi_3 \end{bmatrix} = \begin{bmatrix} c_{11} & c_{12} & c_{13} \\ c_{21} & c_{22} & c_{23} \\ c_{31} & c_{32} & c_{33} \end{bmatrix} \begin{bmatrix} s \\ p_x \\ p_y \end{bmatrix}. \tag{6.12}$$

We know the LGOs for the σ bonds for BF_3 (these are the same as the LGOs for the equatorial set for PF_5):

$$\psi_1(a_1') = \frac{1}{\sqrt{3}}(\phi_1 + \phi_2 + \phi_3),$$

$$\psi_2(e_a') = \frac{1}{\sqrt{6}}(2\phi_1 - \phi_2 - \phi_3),$$

$$\psi_3(e_b') = \frac{1}{\sqrt{2}}(\phi_2 - \phi_3),$$

or

$$\begin{bmatrix} \psi_1(a_1') \\ \psi_2(e_a') \\ \psi_3(e_b') \end{bmatrix} = \begin{bmatrix} 1/\sqrt{3} & 1/\sqrt{3} & 1/\sqrt{3} \\ 2/\sqrt{6} & -1/\sqrt{6} & -1/\sqrt{6} \\ 0 & 1/\sqrt{2} & -1/\sqrt{2} \end{bmatrix} \begin{bmatrix} \phi_1 \\ \phi_2 \\ \phi_3 \end{bmatrix}. \tag{6.13}$$

Taking the inverse of Eq. (6.13) and applying it to the s, p_x, p_y vector,

$$\begin{bmatrix} 1/\sqrt{3} & 2/\sqrt{6} & 0 \\ 1/\sqrt{3} & -1/\sqrt{6} & 1/\sqrt{2} \\ 1/\sqrt{3} & -1\sqrt{6} & -1/\sqrt{2} \end{bmatrix} \begin{bmatrix} s \\ p_x \\ p_y \end{bmatrix} \quad \text{gives}$$

$$\begin{aligned} (1/\sqrt{3})s + (2/\sqrt{6})/p_x \qquad &= \phi_1, \\ (1/\sqrt{3})s - (1\sqrt{6})p_x + (1/\sqrt{2})p_y &= \phi_2, \\ (1/\sqrt{3})s - (1/\sqrt{6})p_x - (1/\sqrt{2})p_y &= \phi_3. \end{aligned} \tag{6.14}$$

In summary, we obtain the LGOs, write them in matrix form, take the inverse, and apply the inverse matrix to the atomic orbitals written in the order corresponding to the representations of the LGOs.

6.5 Pi Bonding in Square Planar MX_4 (D_{4h}) Molecules

So far we have considered only σ bonding. Let us now consider π bonding for a square planar complex MX_4. Pi bonding is not important for XeF_4, because F has only filled orbitals (it can only serve as a donor) and Xe has no empty low-energy orbitals for accepting electrons. We proceed in the same way as for σ bonding, sketching the orbitals as vectors, choosing the orientation of axes, and labeling the vectors and symmetry elements. The vectors at the origin (central atom) are oriented with z (the unique direction) along C_4 and with C_2' and σ_v along x and y. We can check the character table to see that this is the choice made for the $d_{x^2-y^2}$ orbital (B_{1g}). For the ligands, z is chosen as the σ bond direction. The x and y axes on the ligands are oriented so that some operation, here C_4, takes one set into another.

The p orbitals are represented as the two sets of vectors perpendicular to the σ bond direction. One set is parallel to C_4, $\pi(||)$, the "out-of-plane" set, and one set is perpendicular to C_4, $\pi(\perp)$, the "in-plane" set. In the case of the σ vectors (pointing away or toward the origin), any operation either left a vector unchanged (character 1) or transformed it into something else (character 0). We see that the $\pi(\perp)$ vectors are left unchanged (character 1 for each) by E or σ_h or changed into another vector (character 0) by C_4 or C_2, but the $\pi(||)$ vectors are reversed in direction or sign (character -1 for each) by σ_h. In Table 6.4, we record the sum of the number of vectors left unchanged ($+1$), changed into another vector (0), or reversed in sign (-1) for each operation. Here we need the full $\mathbf{D}_{4h}$ group to distinguish the subscripts. These sums are the characters for the representations for the two sets of ligand p orbitals. Reducing these and checking the representations of the orbitals on M, we see that of the parallel set, A_{2u} and E_g are of the proper symmetry for π bonding, and B_{2u} is nonbonding. For the perpendicular set, E_u and B_{2g} have proper symmetry for π bonding, and A_{2q} is nonbonding. The LCAO for the bonding LGOs are obtained by the projection operator method in Table 6.4. As for the case of two-dimensional σ LGOs, we can get the second one of the pair by performing a C_4 operation on the first. The bonding and nonbonding orbitals are sketched from the LCAOs in Fig. 6.11. The nonbonding orbitals are sketched from their symmetry properties as given in the character table (e.g., use C_4 on π_1 or π_5 and then check the resulting combination using other operations). The p_z, d_{xy}, and (d_{xz}, d_{yz}) orbitals on the central atom were found to be σ nonbonding, so these

TABLE 6.4
Treatment of the π Vectors for $MX_4(D_{4h})$

D_{4h}	M orbitals
A_{1g} s, d_z^2	E_u p_x, p_y
B_{1g} $d_{x^2-y^2}$	B_{2g} d_{xy}
A_{2u} p_z	E_g (d_{xz}, d_{yz})

D_{4h}	E	C_4	C_2	C_2'	C_2''	i	S_4	σ_h	σ_v	σ_d	
$\Gamma_\pi(\|\|)$	4	0	0	−2	0	0	0	−4	2	0	$\Gamma_\pi(\|\|) = A_{2u} + E_g + B_{2u}$
$\Gamma_\pi(\perp)$	4	0	0	−2	0	0	0	4	−2	0	$\Gamma_\pi(\perp) = E_u + B_{2g} + A_{2g}$

D_{4h}	E	C_4	C_4'	C_2	$(C_2')_a$	$(C_2')_b$	$(C_2'')_a$	$(C_2'')_b$	i	S_4	S_4'	σ_h	σ_v	σ_v'	σ_d	σ_d'
ϕ_1	ϕ_1	ϕ_2	ϕ_4	ϕ_3	$-\phi_1$	$-\phi_3$	$-\phi_4$	$-\phi_2$	$-\phi_3$	$-\phi_2$	$-\phi_4$	$-\phi_1$	ϕ_1	ϕ_3	ϕ_4	ϕ_2
ϕ_5	ϕ_5	ϕ_6	ϕ_8	ϕ_7	$-\phi_5$	$-\phi_7$	$-\phi_8$	$-\phi_6$	ϕ_7	ϕ_6	ϕ_8	ϕ_5	$-\phi_5$	$-\phi_7$	$-\phi_8$	$-\phi_6$

$\Gamma_\pi(\|\|)$:

a_{2u}: $4\phi_1 + 4\phi_2 + 4\phi_3 + 4\phi_4$ or $\phi_1 + \phi_2 + \phi_3 + \phi_4$

$\psi_{a_{2u}} = (1/2)(\phi_1 + \phi_2 + \phi_3 + \phi_4)$

$2\phi_1 - 2\phi_3$ or $\phi_1 - \phi_3$

e_g: $(\psi_{e_g})_a = (1/\sqrt{2})(\phi_1 - \phi_3)$ $\quad (\psi_{e_g})_b = (1/\sqrt{2})(\phi_2 - \phi_4)$

$\Gamma_\pi(\perp)$:

b_{2g}: $4\phi_5 - 4\phi_6 + 4\phi_7 - 4\phi_8$ or $\phi_5 - \phi_6 + \phi_7 - \phi_8$

$\psi_{b_{2g}} = (1/2)(\phi_5 - \phi_6 + \phi_7 - \phi_8)$

$4\phi_5 - 4\phi_7$ or $\phi_5 - \phi_7$

$(e_u)_a$: $(\psi_{e_u})_a = (1/\sqrt{2})(\phi_5 - \phi_7)$ $\quad (\psi_{e_u})_b = (1/\sqrt{2})(\phi_6 - \phi_8)$

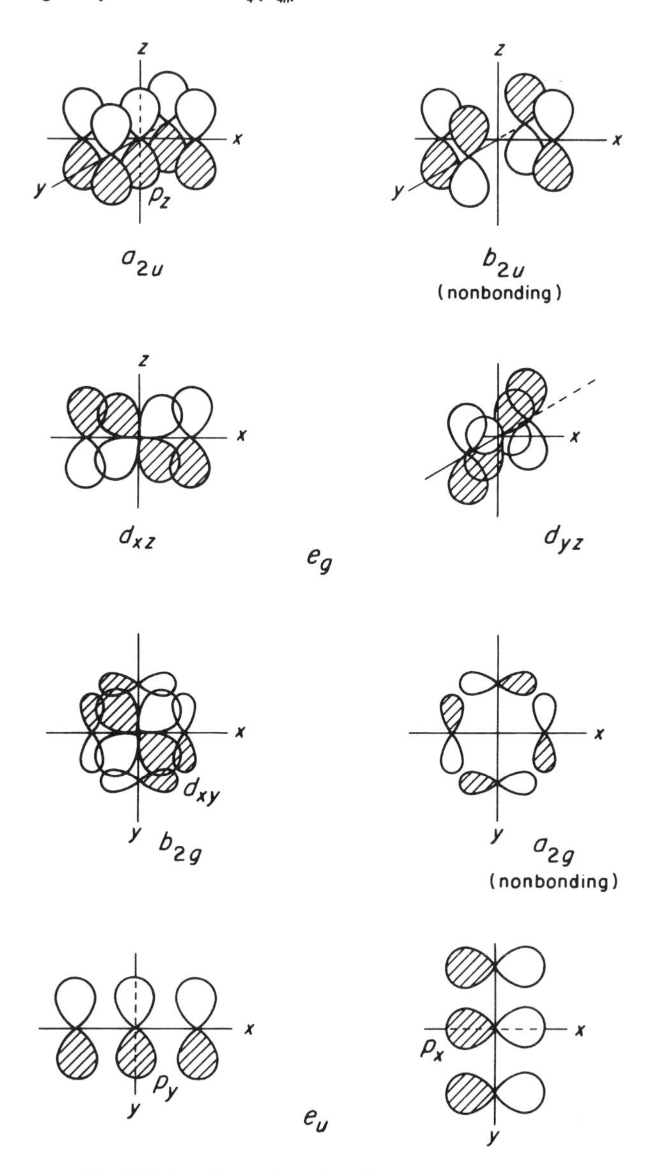

Fig. 6.11 Bonding and nonbonding MOs for MX_4 ($\mathbf{D}_{4h}$).

can participate only in π bonding. The d orbitals give very good overlap. The (p_x, p_y) orbitals have proper symmetry for both σ and π bonding. The p orbitals provide much better overlap for σ bonding than for the sidewise π interaction, so they are expected to participate primarily in σ bonding.

Here we have dealt with the π-type orbitals on the ligands as vectors. These could be p orbitals or, in the case of ligands such as CN^-, π^* orbitals. Pi bonding is important in many square planar $[M(CN)_4]^{n-}$ complexes, involving donation from the filled M orbitals into the empty π^* orbitals of CN^-.

EXAMPLE 6.4 Give the MO description of π bonding in a planar trigonal ion such as CO_3^{2-} ($\mathbf{D}_{3h}$). Sketch the π MOs and give a qualitative energy-level diagram for σ and π bonding.

SOLUTION Carbon forms three σ bonds for which we can use sp^2 hybridization. This leaves only the p_z orbital (perpendicular to the trigonal plane) for π bonding.

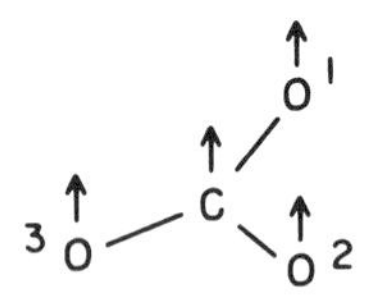

As a consequence, only the O p orbitals parallel to this direction are available for π interaction. Examining the effects of one symmetry operation of each class on the O π vectors, we find the following.

$\mathbf{D}_{3h}$	E	C_3	C_2	σ_h	S_3	σ_v	
Γ_π	3	0	-1	-3	0	1	$= A_2'' + E''$

Applying the projector operator, we get the following.

$\mathbf{D}_{3h}$	E	C_3	C_3'	C_2	C_2'	C_2''	σ_h	S_3	S_3'	σ_v	σ_v'	σ_v''
ϕ_1	ϕ_1	ϕ_2	ϕ_3	$-\phi_1$	$-\phi_3$	$-\phi_2$	$-\phi_1$	$-\phi_2$	$-\phi_3$	ϕ_1	ϕ_3	ϕ_2

$$\psi_{a_2''} = \frac{1}{\sqrt{3}}(\phi_1 + \phi_2 + \phi_3),$$

$$(\psi_{e''})_a = \frac{1}{\sqrt{6}}(2\phi_1 - \phi_2 - \phi_3), \qquad (\psi_{e''})_b = \frac{1}{\sqrt{2}}(\phi_2 - \phi_3)$$

The second e'' LGO is obtained as for the case of the σ LGOs. The e'' LGO is nonbonding since there is no e'' orbital on C. The π MOs are sketched and a qualitative energy-level diagram, including σ bonding (Example 6.2), are given in Fig. 6.12.

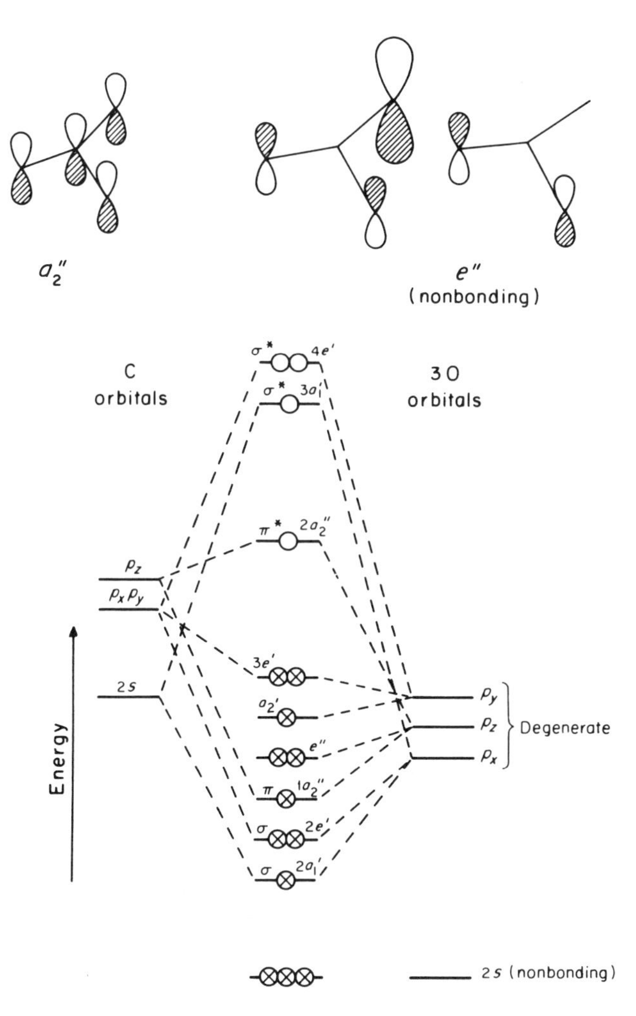

Fig. 6.12 Sketches of the π MOs and a qualitative MO energy-level diagram for CO_3^{2-} ($\mathbf{D}_{3h}$).

EXAMPLE 6.5 Give an MO description of CO_2, sketch the MOs and give a qualitative energy-level diagram.

SOLUTION CO_2 is linear ($\mathbf{D}_{\infty h}$). The vectors on C for σ bonding transform as $\cdots O \leftarrow C \rightarrow O \cdots z$, $A_{1g} + A_{1u}$. The C orbitals belonging to thcsc rcprcscntations are s and p_z. The O s orbitals are much lower in energy than the C orbitals and are expected to be nonbonding. The O σ bonding orbitals are p_z. The C a_{1g} (s) and a_{1u} (p_z) combine with two O LGOs of the same symmetry to give two σ bonding and two σ antibonding orbitals.

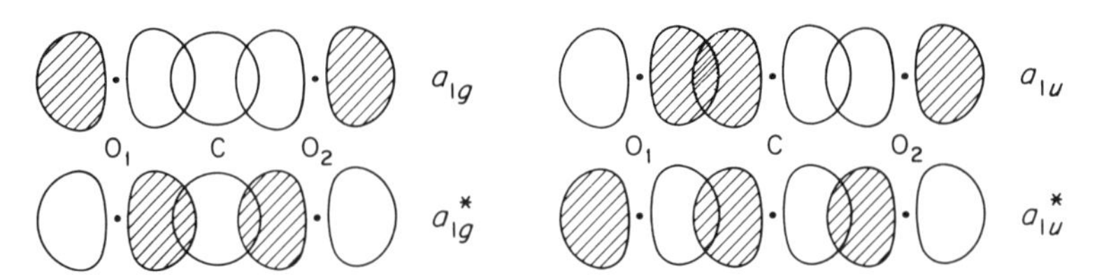

The remaining orbitals available for π bonding are p_x and p_y on each atom. The C p_π orbitals belong to the two-dimensional representation $E_{1u}(\pi_u)$. The O LGOs are obtained as follows.

$\mathbf{D}_{\infty h}$	E	C_∞	σ_v	i	C_2	
Γ_π	4	$4 \cos \phi$	0	0	0	$= E_{1g}(\pi_g) + E_{1u}(\pi_u)$ (by inspection)

Where there are two AOs combined (or two pairs, as here) the combinations are the sum and the difference. The e_{1u} LGOs ($p_{x,y} + p_{x,y}$) can combine with the C e_{1u} orbitals to give a bonding pair and an antibonding pair.

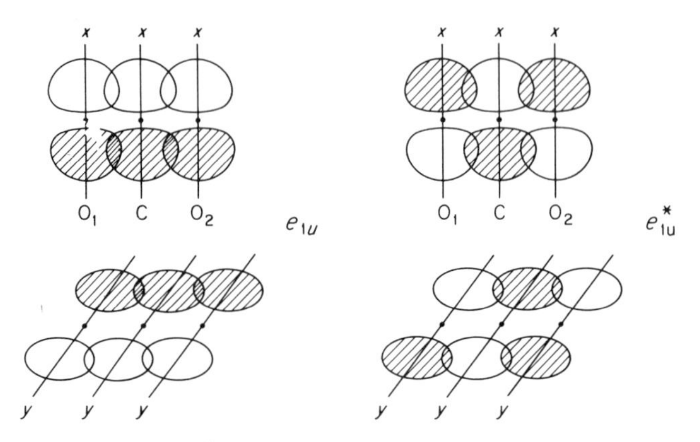

The nonbonding e_{1g} (centrosymmetric) O orbitals ($p_{x,y} - p_{x,y}$) are as follows.

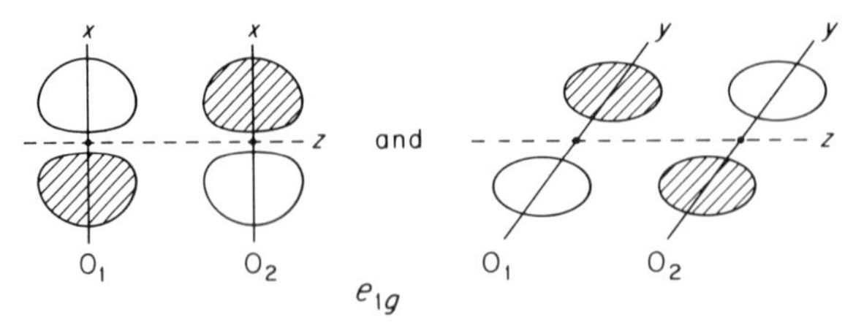

These are the real descriptions of the combinations of the p_x and p_y orbitals. Actually, since the x and y directions are not defined [there is only one direction (z)

defined], the e_{1g} and e_{1u} orbitals have cylindrical symmetry. The e_{1u} sausage-shaped lobes coalesce to form a cylindrical sheath.

There are $3 \times 4 = 12$ valence orbitals. The two O 2*s* orbitals are primarily nonbonding—they also can give a_{1g} and a_{1u} combinations, and as such, could mix with the a_{1g} and a_{1u} combinations of the O p_z orbitals. There are two σ bonding orbitals (a_{1g} and a_{1u}) and two corresponding σ^* orbitals. There is a pair of e_{1u} π bonding orbitals and a pair of π^* orbitals. The remaining e_{1g} orbitals are nonbonding. The 16 electrons are accommodated in the bonding and nonbonding orbitals.

Linear molecules are described usually in terms of *sp* hybridization for the central atom. For a delocalized bonding description, we must use the symmetry-adapted combinations, $++$ for a_{1g} and $+-$ for a_{1u} in $\mathbf{D}_{\infty h}$. For a noncentrosymmetric molecule, OCS ($\mathbf{C}_{\infty v}$) the g and u subscripts do not apply.

6.6 Basis Sets of Orbitals

We have used the atomic orbitals on the central atom as our basis sets. Symmetry determines which of these can be used. After we find the AOs belonging to the appropriate symmetry species, we can obtain suitable linear combinations of the AOs to give a set of equivalent orbitals—the hybrids.

In cases where we recognize that the geometry of the molecule matches that of a hybrid and the proper orbitals are available, we might reasonably start with the hybrid orbitals as the basis set. Thus, we recognize that the tetrahedral CH_4 molecule has four equivalent bonds. Each bond can be described as the overlap of a C sp^3 hybrid orbital and the s orbital of H. Each bond is a localized two-center bond. We would have a set of four equivalent σ bonds accommodating the four bonding electron pairs. There would be a corresponding set of four antibonding (σ^*) orbitals. These are the eight MOs resulting from the four C and four H orbitals. Such a *localized MO description* does not differ significantly from a valence bond description. It is suitable in cases where there are only localized two-center bonds and localized nonbonding electrons.

The water molecule has $\mathbf{C}_{2v}$ symmetry. If we orient the molecule in the xz plane, the representations are A_1 (s and p_z), B_1 (p_x), and B_2 (p_y). The H (1*s*) combination orbitals arc a_1 and b_1. The σ MOs are b_1 (p_x) and a_1 (involving a combination of s and p_z). The a_1 MO involving more s character is lowest in energy (because of the low energy of the s orbital itself), the a_1 MO involving

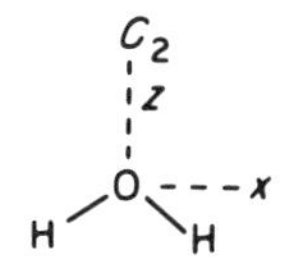

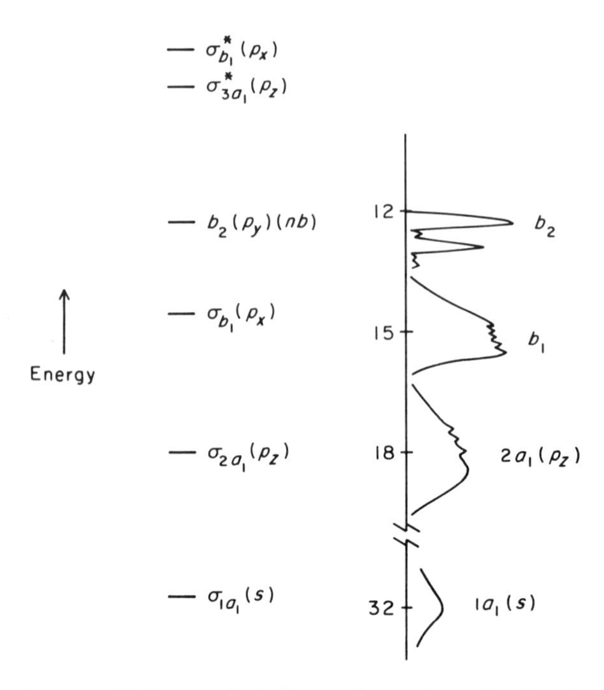

Fig. 6.13 Molecular orbital energy-level diagram for H_2O and its photoelectron spectrum (PES). Adapted from A. W. Potts and W. C. Price. *Proc. Roy. Soc. London* 1972, **A326**, 181.

more p_z character is higher in energy, and the b_1 orbital is still higher in energy because of poor overlap and no participation by s. The b_2 (p_y) orbital is nonbonding because of symmetry. The relative order of these levels is obtained from the photoelectron spectrum shown in Fig. 6.13 beside the energy-level diagram.

The localized MO approach recognizes that since the bond angle (104.5°) more closely matches the tetrahedral angle than the 90° angle for the p orbitals, the sp^3 hybrid provides a reasonable basis set. We can form two localized two-center bonds and have two localized lone pairs. The essentially tetrahedral distribution of hydrogen bonds about H_2O in ice provides strong evidence that the lone pairs are highly localized and might be treated as occupying a tetrahedral set of orbitals. The deviation of the bond angles from the tetrahedral angle can be explained as the result of repulsions involving the lone pairs or using greater s character for the orbitals occupied by the lone pairs.[3] Using the sp^3 hybrid orbitals as a basis set, we still use the sign combinations for the symmetry-adapted σ MOs.

[3] M. B. Hall, *J. Am. Chem. Soc.* 1978, **100,** 6333; *Inorg. Chem.* 1978, **17,** 2261.

The molecular geometry of H_2O is accounted for satisfactorily by the localized bond description. However, the photoelectron spectrum matches the energy-level diagram derived from the delocalized MO treatment. Generally, the localized bonding approach is adequate for dealing with geometry in such simple molecules, but the delocalized MO treatment, using symmetry-adapted MOs, is better for dealing with energy states and spectra. Often the localized bonding approach is useful in predicting the molecular geometry for use in the delocalized MO treatment.

Additional Reading

A. D. Baker and D. Betteridge, *Photoelectron Spectroscopy*, Pergamon, Oxford, 1972.

R. E. Ballard, *Photoelectron Spectroscopy and Molecular Orbital Theory*, Wiley, New York, 1978. Excellent treatment, good figures.

C. J. Ballhausen and H. B. Gray, *Molecular Orbital Theory*, Benjamin, New York, 1965.

R. L. DeKock and H. B. Gray, *Chemical Structure and Bonding*, Benjamin-Cummings, Menlo Park, California, 1980. Good nonmathematical treatment.

B. Douglas, D. H. McDaniel, and J. J. Alexander, *Concepts and Models of Inorganic Chemistry*, 2nd ed., Wiley, New York, 1983. Assumed background.

J. E. Huheey, *Inorganic Chemistry*, 3rd ed., Harper and Row, New York, 1984. Assumed background.

W. L. Jorgensen and L. Salem, *The Organic Chemist's Book of Orbitals*, Academic Press, New York, 1973. Good pictures.

J. P. Lowe, *Quantum Chemistry*, Academic Press, New York, 1978. Good level to accompany this text.

J. A. Pople and D. L. Beveridge, *Approximate Molecular Orbital Theory*, McGraw-Hill, New York, 1970.

A. Streitwieser, Jr., *Molecular Orbital Theory for Organic Chemists*, Wiley, New York, 1961.

D. W. Turner, C. Baker, A. D. Baker, and C. R. Brundle, *Molecular Photoelectron Spectroscopy*, Wiley, New York, 1970.

Problems

6.1 Give the population of σ, π, and π^* orbitals for CO and NO. Predict the effect on the bond energy of each for the removal of an electron and for the addition of an electron. Explain. How many unpaired electrons are there in each case?

6.2 The bond energy of Cl_2^+ (415 kJ/mole) is considerably greater than that for Cl_2 (239 kJ/mole). Account for the relative bond energies, and give the bond orders.

6.3 What is the expected bond order for diamagnetic Nb_2? Explain in terms of the appropriate MOs.

6.4 What is the expected bond order for diamagnetic Mo_2? Explain in terms of the appropriate MOs. [See W. Klotzbücher and G. A. Ozin, *Inorg. Chem.* 1977, **16,** 984; J. G. Norman, Jr., J. J. Kolari, H. B. Gray, and W. C. Trogler, *Inorg. Chem.* 1977, **16,** 987. Consider two s–d_{z^2} hybrids on each Mo to combine to give four MOs.

6.5 Not surprisingly, NF_5 exists only as an ionic species. Describe the bonding and structure expected.

6.6 What effects on the bond angle are expected for the oxidation of NO_2 to NO_2^+ and for the reduction to NO_2^-?

6.7 Give the expected structures for (a) AX_3, (b) $:AX_3$, (c) $\dot{:}AX_3$.

6.8 Predict the gross geometry (from the σ orbital hybridization) and the fine geometry (from electron-pair repulsion, etc.) of the following species: (a) $ClOF_2^+$, (b) $ClOF_3$, (c) $ClOF_4^-$, (d) $ClO_2F_2^-$.

6.9 Give an MO description of σ and $\bar{\pi}$ bonding in NO_2. Sketch the MOs involved, and give a qualitative energy-level diagram showing the occupancy of the orbitals. Neglect nonbonding electrons on oxygens.

6.10 Give an MO description of σ bonding in ClO_3^-. Sketch the MOs involved and give a qualitative energy-level diagram. Neglect nonbonding electrons on oxygens.

6.11 Give an MO description of π bonding in ClO_3^-. Sketch the MOs involved.

6.12 Give an MO description of σ bonding in ClF_3 (see Table 6.1). Neglect nonbonding electrons on fluorines. Sketch the bonding and nonbonding MOs.

6.13 Obtain the LCAOs for the sp^3 hybrid orbitals of CH_4. (*Hint*: Use $\mathbf{D}_2$ and a correlation table to get four nondegenerate LGOs.)

6.14 Using the LGOs derived in Section 8.2.1, obtain the LCAOs for the d^2sp^3 hybrid orbitals for an octahedral metal complex.

7

Multicenter Bonding

In this chapter we deal with bonding in molecules involving delocalized or multicenter bonding. These include molecules with too few low-energy orbitals or too few electrons to form two-center bonds between all bonded atoms. Bonding descriptions of planar cyclic molecules involving delocalized π bonding generate the molecular orbitals required for bonding in metal sandwich compounds. Since σ-bond hybridization and electron-pair repulsion considerations are inadequate for dealing with molecular shapes of electron-deficient compounds, other rules for structural classification are presented.

7.1 Simple Examples of Multicenter Bonding

The compounds discussed in Chapter 6 are ones that can be described adequately by simple valence bond structures. The framework is held together by σ bonds, each of which can be described as a localized bond involving an electron pair shared between two atoms: a two-center, two-electron bond. The delocalized MO approach treats equivalent bonds in symmetry-adapted sets, but for all of the compounds, the net result is that there is an electron pair shared between each pair of bonded atoms, excluding π bonding.

The π-bonding orbital in CO_3^{2-} (Example 6.3) is delocalized over all four atoms, so this is a four-center bond. There are two π bonds in CO_2, and they are represented as two-center bonds in the VB description, O=C=O. However, in the MO description (Example 6.4) we see that each of the π bonds is a three-center bond. Delocalization is much more commonly encountered for π bonding. If the primary σ bonding holds the molecule together, then only the proper orientation of π-type orbitals is required for π bonding. The greater the delocalization, the more stable the molecule. We can still use valence bond or electron dot structures to account for all of the bonds and unshared electrons. Any π delocalization requires resonance contributing structures or other notation, such as the circle drawn in the center of a benzene ring to represent delocalization around the ring. There are other molecules for which the primary bonding holding the framework together is delocalized.

In the gas phase a proton bonds to any molecule; the energy released is the proton affinity. Even H_2 adds H^+ to form H_3^+ with the release of about 418 kJ/mole. The H atoms form an equilateral triangle. The bonding MO is the result of a combination of three *s* orbitals.

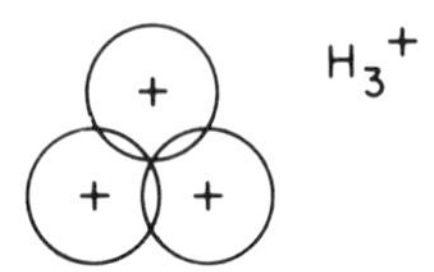

This is the simplest three-center, two-electron bond—a multicenter bond. Since there are three atomic orbitals, there will be three LCAOs. The one shown has A_1' symmetry ($\mathbf{D}_{3h}$). Because of the C_3 symmetry axis, the other two LCAOs will be degenerate, belonging to E'. These correspond to the E' MOs for CO_3^{2-} (Example 6.1) and the in-plane σ bonds for PF_5 (Section 6.2.3).

Three-center, two-electron bonds are involved in bridge bonding in boron hydrides (see Section 7.4). Some of the boron hydrides involve larger numbers of atoms in the multicenter bonds. For example, one of the bonding orbitals of B_5H_9 involves five borons. The boron hydrides are described as electron-deficient compounds because there are too few electrons for the usual valence bond structures. Some molecules might be described as "orbital deficient." Many compounds that exceed the octet rule, such as the xenon halides, interhalogens, SF_4, and SF_6, have more electron pairs than atomic valence orbitals unless we use higher-energy *d* orbitals.

7.2 Bonding in XeF_4 without Using *d* Orbitals

In Sections 6.2.1 and 6.2.2 we examined the σ bonding in XeF_4 assuming that we could use all orbitals of proper symmetry, including the 5*d* orbitals. The 5*d* orbitals are so much higher in energy than the 5*s* and 5*p* orbitals that the extent of participation of the 5*d* orbitals is very uncertain. If we exclude them as too high in energy, then the Xe orbitals available are 5*s* and 5*p*. However, $5p_z$ is perpendicular to the plane of the molecule and, as such, is unavailable for σ bonding. This leaves an *s* and two *p* orbitals available for bonding to four F atoms.

We have examined the symmetry species of the LGOs (Section 6.2.1) and the representations of the Xe orbitals. The low-energy *s* orbital and the nonbonding p_z orbital are suitable for the lone pairs on Xe. If we prefer to visualize two localized lone pairs, then we could consider the orbitals to be *sp* hybrids. The two bonding MOs are a result of the combination of the Xe p_x and p_y orbitals with the e_u LGOs. Each of these is a three-center bond involving one Xe *p* orbital and a *trans* pair of F atoms. Each is a "half-bond" since one pair of bonding electrons is delocalized over what we draw as two Xe—F σ bonds. The energy-level diagram using *d* orbitals (see Fig. 6.8) would be modified as follows:

(1) omit the *d* orbitals,
(2) show 5*s* and 5p_z (Xe) as nonbonding (lone pairs),
(3) show the b_{1g} LGO as nonbonding.

The nonbonding orbitals for the lone pairs can be considered as occupying *s* and p_z (or *sp*) orbitals or symmetry-adapted a_{1g} and a_{2u} orbitals. If the energy of the Xe *s* orbital is suitable for some degree of mixing with the a_{1g} LGO, the LGO becomes somewhat bonding and the Xe *s* (a_{1g}) becomes somewhat antibonding. There is no net gain in bonding. The energy sequence is

$$e_u^4 \quad \underset{\text{LGOs}}{(a_{1g}^2 b_{1g}^2)} \quad \underset{\text{Xe}}{(a_{1g}^2 a_{2u}^2)} \quad e_u^*.$$

The Xe—F bonds are weak; the average Xe—F bond energy is about 150 kJ/mole, compared to 432 kJ/mole for H_2, but only 154.6 kJ/mole for F_2. Although the participation of Xe *d* orbitals is controversial, the extreme view requiring no participation is probably as far from reality as the simple VB description that implies equal participation of the dsp^2 hybrid orbitals. The *d* orbitals are very high in energy for the free atom, but in combination with highly electronegative atoms, the energy of the *d* orbitals should be lowered. Here the primary choice is between occupancy of a strictly antibonding a_{1g}^* orbital (excluding the *d* orbitals) or, with some participation of the *d* orbitals, of the primarily nonbonding d_{z^2} (a_{1g}) orbital. The greater the participation of the *d* orbitals, the lower the energy of the bonding a_{1g} orbital (some mixing with d_{z^2}) and the lower the energy of the b_{1g} orbital, because it becomes bonding with participation of $d_{x^2-y^2}$.

7.3 Planar Cyclic π Molecules

7.3.1 PI BONDS IN CARBON CHAINS

In the ethylene molecule (C_2H_4), the C—C σ bond is a localized two-center bond. Likewise, for each carbon, there is just one *p* orbital perpendicular to the MO. These combine to give a π MO (a two-center bond) and a π^* MO. In the allyl ion ($H_2CCHCH_2^-$) with $\mathbf{C}_{2v}$ symmetry, the p orbitals for π bonding are in two sets.

ϕ_2 C ϕ_1 C ϕ_3 C C_2

3 C in xz plane

$\mathbf{C}_{2v}$	E	C_2	σ_{xz}	σ_{yz}	
$\Gamma_\pi(1-3)$	2	0	-2	0	$= A_2 + B_2$
$\Gamma_\pi(2)$	1	-1	-1	1	$= B_2$

The three combinations are

$$\psi_{a_2} = \frac{1}{\sqrt{2}}(\phi_1 - \phi_3) \qquad \text{(nonbonding);}$$

the bonding combination $b_2(\phi_{1-3}) + b_2(\phi_2)$,

$$\psi_{b_2} = \frac{1}{\sqrt{2}}\left(\frac{1}{\sqrt{2}}\phi_1 + \frac{1}{\sqrt{2}}\phi_3 + \phi_2\right) = \frac{1}{2}\left(\phi_1 + \sqrt{2}\phi_2 + \phi_3\right);$$

and the antibonding combination $b_2(\phi_{1-3}) - b_2(\phi_2)$,

$$\psi_{b_2^*} = \frac{1}{2}(\phi_1 - \sqrt{2}\phi_2 + \phi_3).$$

The two electron pairs can be accommodated in the bonding and nonbonding orbitals. The bonding electrons are delocalized over all three carbon atoms, and for the nonbonding electrons, the electron density is distributed between the two terminal carbon atoms.

For butadiene ($CH_2{=}CH{-}CH{=}CH_2$), there are four p orbitals for π bonding and four combinations. Following are the sign combinations of the p

ϕ_4 (+)(−)(+)(−) —

ϕ_3 (−)(+)(+)(−) —

ϕ_2 (+)(+)(−)(−) ↿⇂

ϕ_1 (+)(+)(+)(+) ↿⇂

orbitals for butadiene, showing only the upper lobes of the orbitals. Here the terms bonding, nonbonding, and antibonding are inadequate, except for the most favorable combination (bonding) and least favorable combination (antibonding). Of the other two combinations, one is somewhat more bonding than antibonding (two favorable combinations and one unfavorable) and the other is somewhat more antibonding than bonding (one favorable and two unfavorable combinations). The two electron pairs occupy the two lower-energy orbitals (ϕ_1 and ϕ_2).

7.3.2 CYCLOBUTADIENE

Let us join the end carbon atoms to form the cyclic compound C_4H_4 and assume a planar ring. The symmetry is $\mathbf{D}_{4h}$ and each carbon atom has one p orbital perpendicular to the molecular plane. The representations are as follows, giving $\Gamma_\pi = A_{2u} + B_{2u} + E_g$.

$\mathbf{D}_{4h}$	E	$2C_4$	C_2	$2C_2'$	$2C_2''$	i	$2S_4$	σ_h	$2\sigma_v$	$2\sigma_d$
Γ_π	4	0	0	-2	0	0	0	-4	2	0

Applying the projection operator, we get the following.

$\mathbf{D}_{4h}$	E	C_4	C_4'	C_2	$(C_2')_a$	$(C_2')_b$	$(C_2'')_a$	$(C_2'')_b$	i	S_4	S_4'	σ_h	σ_v	σ_v'	σ_d	σ_d'
ϕ_1	ϕ_1	ϕ_2	ϕ_4	ϕ_3	$-\phi_3$	$-\phi_1$	$-\phi_2$	$-\phi_4$	$-\phi_3$	$-\phi_2$	$-\phi_4$	$-\phi_1$	ϕ_3	ϕ_1	ϕ_2	ϕ_4

$$a_{2u}: \quad 4\phi_1 + 4\phi_2 + 4\phi_3 + 4\phi_4 \quad \text{or} \quad \phi_1 + \phi_2 + \phi_3 + \phi_4$$

$$b_{2u}: \quad 4\phi_1 - 4\phi_2 + 4\phi_3 - 4\phi_4 \quad \text{or} \quad \phi_1 - \phi_2 + \phi_3 - \phi_4$$

$$(e_g)_a: \quad 4\phi_1 - 4\phi_3 \quad \text{or} \quad \phi_1 - \phi_3$$

We see that we can get the same combinations considering only the rotational group $\mathbf{D}_4$, or even the cyclic group $\mathbf{C}_4$. For planar cyclic π molecules we can use the cyclic group. The C_n rotations generate all of the orbitals and there is one for each representation of the $\mathbf{C}_n$ group (remembering that the E representation consists of a pair of one-dimensional representations).

$\mathbf{C}_4$	E	C_4	C_4^2	C_4^3
ϕ_1	ϕ_1	ϕ_2	ϕ_3	ϕ_4

$$a_2: \quad \phi_1 + \phi_2 + \phi_3 + \phi_4$$

$$b: \quad \phi_1 - \phi_2 + \phi_3 - \phi_4$$

$$(e)_1: \quad \phi_1 + i\phi_2 - \phi_3 - i\phi_4$$

$$(e)_2: \quad \phi_1 - i\phi_2 - \phi_3 + i\phi_4$$

Thus, we do not have to go through the application of the projection operator stepwise. We can write the representations (all of those in the table) and the LCAOs by applying the characters to the orbitals directly.

These LCAOs are good descriptions, but for applications to physical problems, we can eliminate imaginary numbers by adding and subtracting,

$$e_a = (e)_1 + (e)_2 = 2\phi_1 - 2\phi_3 \quad \text{or} \quad \phi_1 - \phi_3,$$

$$e_b = (e)_1 - (e)_2 = 2i\phi_2 - 2i\phi_4,$$

or dividing by $2i$, we get $e_b = \phi_2 - \phi_4$. Here we obtain the two LCAOs for e

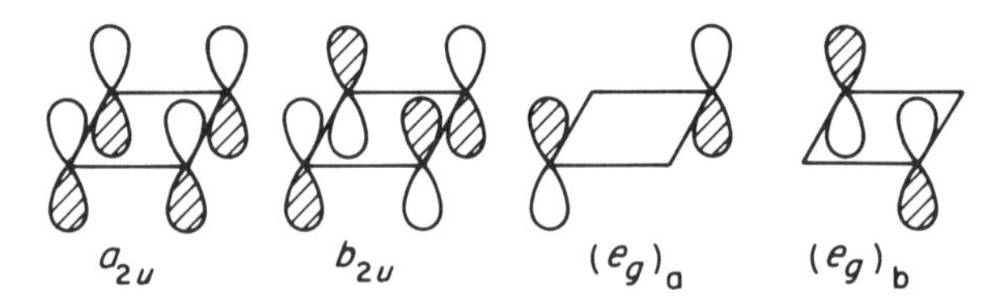

Fig. 7.1 Sketches of the π molecular orbitals for planar C_4H_4.

directly, without performing some operation on the first LCAO and then perhaps taking a linear combination.

The result does not give us the subscripts for the $\mathbf{D}_{4h}$ group, but we can add these by examining the sketches shown in Fig. 7.1, as drawn from the LCAOs. A pair of electrons would be in the bonding a_{2u} orbital, and the two electrons in the doubly degenerate e_g orbitals should be unpaired.

Energy ↑		
	b_{2u}	—
	e_g	↑ ↑
	a_{2u}	↑↓

Diradicals are usually unstable. Distortion to remove the degeneracy of the e_g orbitals could result in stabilization of C_4H_4. The unsubstituted free molecule is not known, but there are stable metal complexes containing planar C_4H_4 as a π donor. In these cases the donor can be considered as $C_4H_4^{2-}$ with filled e_g orbitals (see Section 7.3.2).

7.3.3 PATTERNS FOR PLANAR CYCLIC π MOLECULES

Other planar cyclic π molecules can be treated similarly. In the $\mathbf{C}_n$ group for C_nH_n, the energy levels for odd values of n are a and $(n - 1)/2$ e levels and for even values of n, a, $(n - 2)/2$ e levels, and b (see Fig. 7.2). The reasons for the stability of six π electrons (and 10 π electrons for higher values of n) is apparent from the energy-level diagrams. As noted, planar C_4H_4 is stabilized in metal complexes with the e orbitals filled. Benzene is stable as the neutral molecule, but $C_5H_5^-$ and $C_7H_7^+$ are formed readily. The sequence of energy levels, and their approximate relative energies, correspond to the spacing of the vertices of the regular polygons standing on end.

The sketches of the π MOs involve complete delocalization of the π electron cloud around the ring for a, alternating signs around the ring for b (odd n only), one nodal plane perpendicular to the plane of the molecule for e_1, two such nodal planes for e_2, three for e_3, etc. These cyclic patterns are superimposed on the rings in the most symmetrical way. The second e orbital of a pair has the same number of nodes as the first and is positioned so that the nodal planes of the second set bisect the angles between the nodal planes of the first set

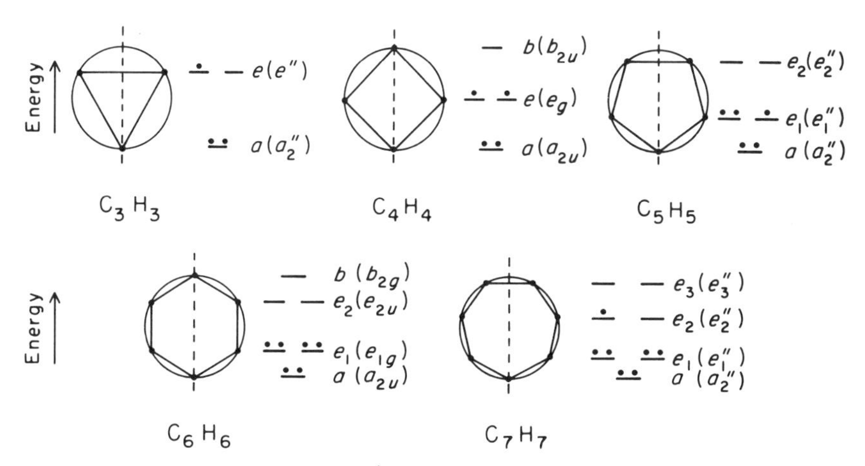

Fig. 7.2 Energy levels for planar cyclic C_nH_n molecules. The representations for $\mathbf{D}_{nh}$ are given in parentheses.

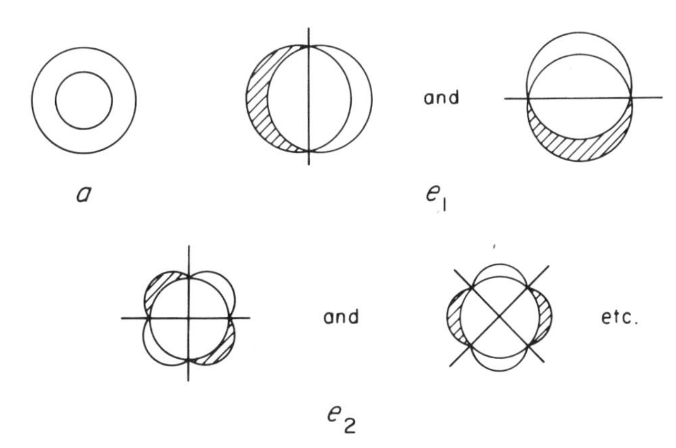

Fig. 7.3 Wave patterns of the π molecular orbitals for planar cyclic π systems, showing the lobes on one side of the molecular plane. These are superimposed on the rings oriented as shown in Fig. 7.2.

(Fig. 7.3). These can be seen to correspond to the sketches for C_4H_4 in Fig. 7.1. For C_4H_4, there is only one π MO with two nodal planes (*b*) since the other possible orientation would have the nodal planes slice through each carbon *p* orbital, corresponding to coefficients of zero in the LCAO.

7.3.4 CYCLOPENTADIENE

For C_5H_5 the orbitals corresponding to those in Fig. 7.3 are sketched in Fig. 7.4. The LCAOs can be written directly from the character table as

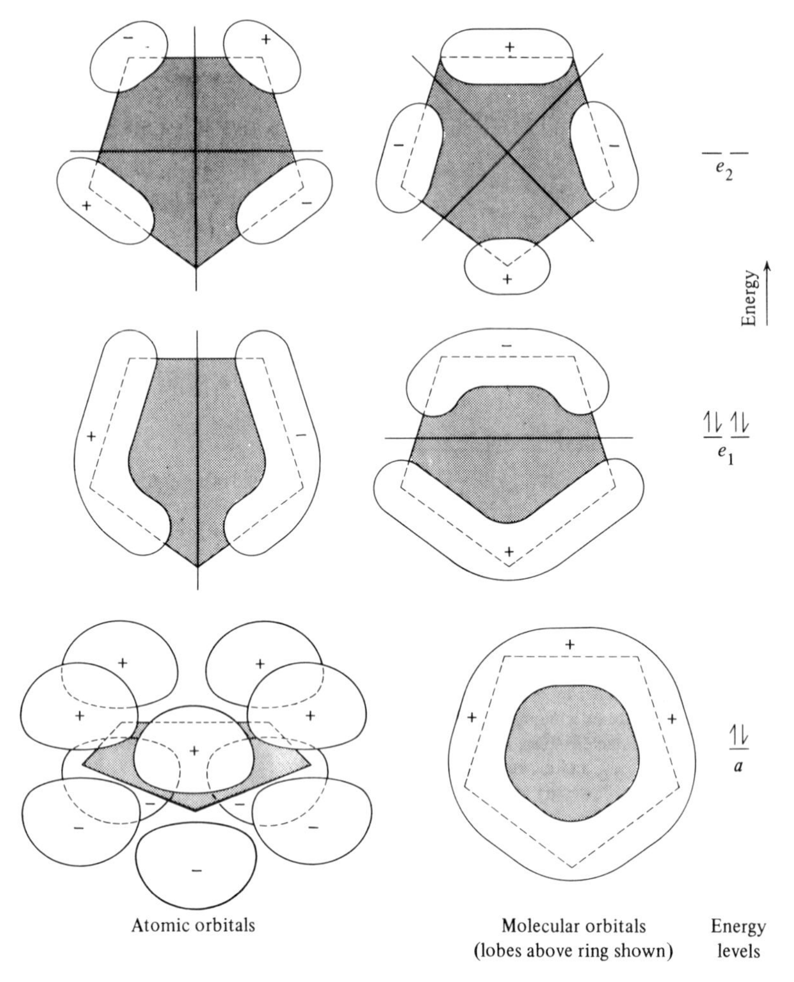

Fig. 7.4 Pi molecular orbitals for C_5H_5. The orbitals are designated for the $\mathbf{C}_5$ point group. From B. Douglas, D. H. McDaniel, and J. J. Alexander. *Concepts and Models of Inorganic Chemistry*, 2nd ed., Wiley, New York, 1983, p. 168. © 1983 John Wiley & Sons, Inc. Reproduced by permission.

follows.

$$a: \quad \phi_1 + \phi_2 + \phi_3 + \phi_4 + \phi_5$$

$$e_1 \begin{cases} e_1(1): & \phi_1 + \varepsilon\phi_2 + \varepsilon^2\phi_3 + \varepsilon^{2*}\phi_4 + \varepsilon^*\phi_5 \\ e_1(2): & \phi_1 + \varepsilon^*\phi_2 + \varepsilon^{2*}\phi_3 + \varepsilon^2\phi_4 + \varepsilon\phi_5 \end{cases}$$

$$e_2 \begin{cases} e_2(1): & \phi_1 + \varepsilon^2\phi_2 + \varepsilon^*\phi_3 + \varepsilon\phi_4 + \varepsilon^{2*}\phi_5 \\ e_2(2): & \phi_1 + \varepsilon^{2*}\phi_2 + \varepsilon\phi_3 + \varepsilon^*\phi_4 + \varepsilon^2\phi_5 \end{cases}$$

We can add and subtract the e pairs to eliminate complex numbers since

$$\varepsilon = \cos(2\pi/5) + i\sin(2\pi/5), \qquad \varepsilon^* = \cos(2\pi/5) - i\sin(2\pi/5),$$

$$\varepsilon + \varepsilon^* = 2\cos(2\pi/5) = 0.618, \qquad \varepsilon - \varepsilon^* = 2i\sin(2\pi/5),$$

$$\varepsilon^2 = \cos(4\pi/5) + i\sin(4\pi/5), \qquad \varepsilon^{2*} = \cos(4\pi/5) - i\sin(4\pi/5),$$

$$\varepsilon^2 + \varepsilon^{2*} = 2\cos(4\pi/5) = -1.618, \qquad \varepsilon^2 - \varepsilon^{2*} = 2i\sin(4\pi/5)$$

$$\begin{aligned} e_{1a} &= e_1(1) + e_1(2) \\ &= 2\phi_1 + (\varepsilon + \varepsilon^*)\phi_2 + (\varepsilon^2 + \varepsilon^{2*})\phi_3 + (\varepsilon^{2*} + \varepsilon^2)\phi_4 + (\varepsilon^* + \varepsilon)\phi_5 \\ &= 2\phi_1 + 0.618\phi_2 - 1.618\phi_3 - 1.618\phi_4 + 0.618\phi_5. \end{aligned}$$

Dividing by 2 and normalizing gives

$$\psi(e_{1a}) = \frac{1}{\sqrt{2.5}}(\phi_1 + 0.314\phi_2 - 0.809\phi_3 - 0.809\phi_4 + 0.314\phi_5),$$

$$\begin{aligned} e_{1b} &= e_1(1) - e_1(2) \\ &= 2i\sin(2\pi/5)\phi_2 + 2i\sin(4\pi/5)\phi_3 - 2i\sin(4\pi/5)\phi_4 - 2i\sin(2\pi/5)\phi_5. \end{aligned}$$

Dividing by $2i\sin(4\pi/5)$ and normalizing gives

$$\psi(e_{1b}) = \frac{1}{\sqrt{7.24}}(1.62\phi_2 + \phi_3 - \phi_4 - 1.62\phi_5),$$

Similarly for e_2, we get

$$\psi(e_{2a}) = \frac{1}{\sqrt{2.5}}(\phi_1 - 0.809\phi_2 + 0.314\phi_3 + 0.314\phi_4 - 0.809\phi_5),$$

$$\psi(e_{2b}) = \frac{1}{\sqrt{7.24}}(\phi_2 - 1.62\phi_3 + 1.62\phi_4 - \phi_5).$$

7.3.5 THE HÜCKEL MOLECULAR ORBITAL APPROACH: THE ALLYL RADICAL

Pi bonding in the allyl radical (C_3H_5, $\mathbf{C}_{2v}$) is a result of the overlap of the p orbitals perpendicular to the plane of the carbon atoms.

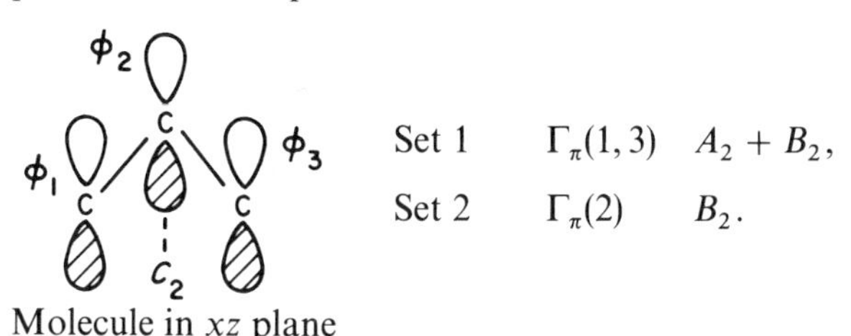

Molecule in xz plane

The three π MOs are

$$\psi_{b_2} = \frac{1}{\sqrt{2}}\left[\phi_2 + \frac{1}{\sqrt{2}}(\phi_1 + \phi_3)\right] \qquad \text{(bonding)},$$

$$\psi_{a_2} = \frac{1}{\sqrt{2}}[\phi_1 - \phi_3] \qquad \text{(nonbonding)},$$

$$\psi_{b_2^*} = \frac{1}{\sqrt{2}}\left[\phi_2 - \frac{1}{\sqrt{2}}(\phi_1 + \phi_3)\right]. \qquad \text{(antibonding)}.$$

Each π MO is a linear combination of atomic orbitals for which we have just evaluated the coefficients c_{ij} in

$$\begin{aligned} \psi_1 &= c_{11}\phi_1 + c_{12}\phi_2 + c_{13}\phi_3, \\ \psi_2 &= c_{21}\phi_1 + c_{22}\phi_2 + c_{23}\phi_3, \\ \psi_3 &= c_{31}\phi_1 + c_{32}\phi_2 + c_{33}\phi_3. \end{aligned} \tag{7.1}$$

Now we are interested in finding the relative energies of these MOs. We use the wave equation in the form

$$H\psi - E\psi = (H - E)\psi = 0, \tag{7.2}$$

and the energy is given by

$$E = \int \psi H \psi \, d\tau \bigg/ \int \psi^2 \, d\tau. \tag{7.3}$$

Interactions of atomic orbitals are expressed as

$$\begin{aligned} H_{ii} &= \int \phi_i H \phi_i \, d\tau \qquad \text{(Coulomb integral)}, \\ H_{ij} &= \int \phi_i H \phi_j \, d\tau \qquad \text{(resonance integral)}, \\ S_{ij} &= \int \phi_i \phi_j \, d\tau \qquad \text{(overlap integral)} \end{aligned} \tag{7.4}$$

The Coulomb integral H_{ii} in Eq. (7.4) is taken as the energy of the atomic orbital ϕ_i (carbon p_y). The energy of the kth LCAO is given by

$$E_k = \sum_i \sum_j c_{ki}c_{kj}H_{ij} \bigg/ \sum_i \sum_j c_{ki}c_{kj}S_{ij}. \tag{7.5}$$

Cross multiplying and rearranging, we get

$$\sum_i \sum_j c_{ki}c_{kj}(H_{ij} - E_k S_{ij}) = 0. \tag{7.6}$$

The LCAO for ψ_1 [Eq. (7.1)] gives

$$\begin{aligned} c_{11}(H_{11} - ES_{11}) + c_{12}(H_{12} - ES_{12}) + c_{13}(H_{13} - ES_{13}) = 0, \\ c_{11}(H_{21} - ES_{21}) + c_{12}(H_{22} - ES_{22}) + c_{13}(H_{23} - ES_{23}) = 0, \\ c_{11}(H_{31} - ES_{31}) + c_{12}(H_{32} - ES_{32}) + c_{13}(H_{33} - ES_{33}) = 0. \end{aligned} \tag{7.7}$$

These *secular equations* can be simplified since $S_{ii} = \int \phi_i \phi_i \, d\tau = 1$ for normalized orbitals. For homogeneous linear equations, nontrivial solutions exist if the determinant (the *secular determinant* here) of the coefficient c is equal to zero,

$$\begin{vmatrix} H_{11} - E & H_{12} - ES_{12} & H_{13} - ES_{13} \\ H_{21} - ES_{21} & H_{22} - E & H_{23} - ES_{23} \\ H_{31} - ES_{31} & H_{32} - ES_{32} & H_{33} - E \end{vmatrix} = 0. \tag{7.8}$$

The determinant is made less cumbersome by replacing the Coulomb integral (H_{ii}) by α and the resonance integral (H_{ij}) by β for adjacent (bonded) atoms. Only adjacent atoms are bonded, so $H_{ij} = 0$ for more remote atoms (C_1 and C_3 in C_3H_5). The Hückel approximation that simplifies the treatment greatly is that $S_{ij} = 0$, or that we can neglect overlap integrals. The determinant becomes

$$\begin{vmatrix} \alpha - E & \beta & 0 \\ \beta & \alpha - E & \beta \\ 0 & \beta & \alpha - E \end{vmatrix} = 0. \tag{7.9}$$

Further simplification for purposes of manipulation is achieved by dividing by β and letting $x = \alpha - E/\beta$ to give

$$\begin{vmatrix} x & 1 & 0 \\ 1 & x & 1 \\ 0 & 1 & x \end{vmatrix} = 0, \tag{7.10}$$

$$x\begin{vmatrix} x & 1 \\ 1 & x \end{vmatrix} - 1\begin{vmatrix} 1 & 1 \\ 0 & x \end{vmatrix} + 0\begin{vmatrix} 1 & x \\ 0 & 1 \end{vmatrix} = x(x^2 - 1) - 1x = x^3 - 2x = 0,$$

for which $x = 0$ or

$$x^2 \quad 2 = 0, \qquad x - \pm\sqrt{2}.$$

The energies are:

$$E = \alpha, \qquad E = \alpha - \sqrt{2}\beta, \qquad E = \alpha + \sqrt{2}\beta.$$

These are the energies of the three MOs for the allyl radical. Since β, the interaction energy between adjacent orbitals, is negative, $\alpha + \sqrt{2}\beta$ is the energy of the bonding (lowest-energy) orbital and $\alpha - \sqrt{2}\beta$ the energy of the antibonding MO. The energy of the other MO is α, the same as that of the AO, so this is a nonbonding MO. We obtain the energies of all MOs from one of the LCAOs. We could have used ψ_2 or ψ_3 to obtain them. Once the energies are known, we can use the secular equations to determine the coefficients for the LCAOs, but we already obtained them based upon the combinations of the symmetry group orbitals.

Application of the Hückel Approach to Cyclic C_nH_n π Systems. For cylobutadiene (C_4H_4), the lowest energy LCAO is

$$\psi_a = c_1\phi_1 + c_2\phi_2 + c_3\phi_3 + c_4\phi_4.$$

Since only adjacent pairs of atoms are bonded, the secular determinant is

$$\begin{array}{c} \\ \phi_1 \\ \phi_2 \\ \phi_3 \\ \phi_4 \end{array} \begin{array}{c} \begin{array}{cccc} \phi_1 & \phi_2 & \phi_3 & \phi_4 \end{array} \\ \begin{vmatrix} \alpha - E & \beta & 0 & \beta \\ \beta & \alpha - E & \beta & 0 \\ 0 & \beta & \alpha - E & \beta \\ \beta & 0 & \beta & \alpha - E \end{vmatrix} \end{array} = 0. \tag{7.11}$$

Dividing by β and letting $x = \alpha - E/\beta$ gives

$$\begin{vmatrix} x & 1 & 0 & 1 \\ 1 & x & 1 & 0 \\ 0 & 1 & x & 1 \\ 1 & 0 & 1 & x \end{vmatrix} = 0. \tag{7.12}$$

Solving by cofactors gives

$$x\begin{vmatrix} x & 1 & 0 \\ 1 & x & 1 \\ 0 & 1 & x \end{vmatrix} - 1\begin{vmatrix} 1 & 0 & 1 \\ 1 & x & 1 \\ 0 & 1 & x \end{vmatrix} + 0 - 1\begin{vmatrix} 1 & 0 & 1 \\ x & 1 & 0 \\ 1 & x & 1 \end{vmatrix} = 0$$

or

$$x^4 - 4x^2 = 0, \qquad x^2(x^2 - 4) = 0.$$

The solutions are $x = 0$ and $x = \pm 2$ or $E = \alpha$, $E = \alpha + 2\beta$, and $E = \alpha - 2\beta$. The energy of the bonding orbital ($\psi_{a_{1u}}$) is $\alpha + 2\beta$, the pair of e_g orbitals is nonbonding ($E = \alpha$), and the energy of the antibonding b_{1u} orbital is $\alpha - 2\beta$.

For some C_nH_n molecules, higher-order equations (x^5 for C_5H_5) are obtained. Often the equations can be factored, simplifying their solutions. However, just as we can evaluate the c_i coefficients from the secular equations once the energies are known, we can calculate the energies from the LCAOs having evaluated the coefficients for the LCAOs using group theory.

Let us proceed to obtain the energies for the π MOs of C_5H_5 (Section 7.24), using just one orbital of each e pair:

$$\begin{aligned}\psi_{a_1} &= \frac{1}{\sqrt{5}}(\phi_1 + \phi_2 + \phi_3 + \phi_4 + \phi_5),\\ \psi_{e_{1b}} &= \frac{1}{\sqrt{7.25}}(1.62\phi_2 + \phi_3 - \phi_4 - 1.62\phi_5),\\ \psi_{e_{2b}} &= \frac{1}{\sqrt{7.25}}(\phi_2 - 1.62\phi_3 + 1.62\phi_4 - \phi_5).\end{aligned} \tag{7.13}$$

The energy of each orbital is given by $E = \int \psi_i H \psi_i \, d\tau$. The square of a wave function gives α, and the product of wave functions for adjacent orbitals gives β, and for a_1 (in abridged form),

$$[(1/\sqrt{5})(\phi_1 + \phi_2 + \phi_3 + \phi_4 + \phi_5)]^2 \rightarrow E_{a_1}, \qquad \phi_i^2 = \alpha, \quad \phi_i\phi_{i\pm 1} = \beta,$$

$$E_{a_1} = \tfrac{1}{5}(5\alpha + 10\beta) = \alpha + 2\beta.$$

For the other orbitals we get

$$\begin{aligned}E_{e_1} &= (1/7.25)\,[2(1.62)^2\alpha + 2\alpha + 4(1.62)\beta - 2\beta]\\ &= \alpha + 0.618\beta \quad \text{or} \quad [\alpha + 2(\cos 2\pi/5)\beta],\\ E_{e_2} &= (1/7.25)\,[7.25\alpha - 4(1.62)\beta - 2(1.62)^2\beta]\\ &= \alpha - 1.62\beta \quad \text{or} \quad [\alpha + 2(\cos 4\pi/5)\beta].\end{aligned}$$

The a_1 orbital is lowest in energy (β is negative), and e_2 is highest in energy.

In Problem 7.4 you are asked to calculate the energies of the π orbitals for benzene. The energies of these planar cyclic π systems, C_nH_n, also are obtained without calculation simply by inscribing the regular polygon in a circle of radius 2β, such that one apex is pointed down (see Fig. 7.2). The energies of the π orbitals, in units of β, correspond to the vertical distance above the lowest apex. Thus, all cyclic C_nH_n molecules have the lowest orbital at $+2\beta$ and those with even numbers of carbon atoms have the highest orbital at -2β.

For carbon compounds containing delocalized π systems, the π orbitals are of greatest interest. The filled σ bonding orbitals are very low in energy, and the σ antibonding orbitals are extremely high in energy. The π orbitals are the *frontier orbitals*, containing the *highest occupied* MOs (HOMOs) and the *lowest unoccupied* MOs (LUMOs). These are the orbitals involved in electronic absorption spectra and for chemical reactions. The HMO approach is used widely for such compounds because of its simplicity and great success. For more complex cases, the secular determinants can be factored using symmetry to simplify their evaluation. Inorganic compounds are often treated using the *extended Hückel molecular orbital method.*

Organic compounds containing delocalized π systems and heteroatoms (N, O, etc.), such as pyridine, pyrrole, furan, and phenol, can be treated using the HMO method. The Coulomb integral for the heteroatom is expressed as $\alpha + h\beta$ and the C—X bond integral is expressed as $k\beta$. The values of h and k are those found suitable for X in a particular type of compound. Since N contributes two electrons to the π system in pyrrole but only one electron in pyridine, the parameters differ.[1] The complication of added parameters is offset by the fact that the carbon atoms occur in sets or symmetry groups that can be dealt with separately. This reduces the sizes of the determinants to be solved.

7.4 Sandwich Compounds of Metals

7.4.1 FERROCENE

Ferrocene $[Fe(C_5H_5)_2]$ is a sandwich compound with two C_5H_5 rings serving as π donors. Considering the molecule to consist of Fe(II) and two $C_5H_5^-$, each $C_5H_5^-$ donates six π electrons. The LGOs are combinations of the π MOs (Fig. 7.4) for the two rings. The orientation of the π orbitals of the two rings is shown in Fig. 7.5, with the positive lobes for each ring directed toward the metal atom for the most symmetrical bonding.

Ruthenocene $[Ru(C_5H_5)_2]$ has the eclipsed $\mathbf{D}_{5h}$ configuration in the solid, and ferrocene has the staggered $\mathbf{D}_{5d}$ configuration. These results indicate little difference in energy between the two configurations, as might be expected. For ferrocene, if we perform the symmetry operations of the $\mathbf{D}_{5d}$ group (using one C_5 axis, one σ_d plane, etc.) on each orbital separately, we obtain the following characters: 10 for the identity operation (each orbital is left unchanged), 2 for σ_d (only the two orbitals cut by the plane are left unchanged), and all others are zero. Looking at the π molecular orbitals in Figs. 7.4 and 7.5, it is apparent that the a orbitals will combine to give a orbitals (one g and one u) and the e orbitals will combine to give e orbitals (one g and one u in each case).

$\mathbf{D}_{5d}$	E	C_5	C_5^2	C_2	i	σ_d
Γ_π	10	0	0	0	0	2

With this information, it is easy to find, by inspection of the $\mathbf{D}_5$ character table, that the irreducible representations contained are

$$\Gamma_\pi = A_{1g} + A_{2u} + E_{1g} + E_{1u} + E_{2g} + E_{2u}.$$

[1] Values of suitable parameters are tabulated in many sources. For example, A. Streitweiser, *Molecular Orbital Theory for Organic Chemistry*, Wiley, New York, 1961.

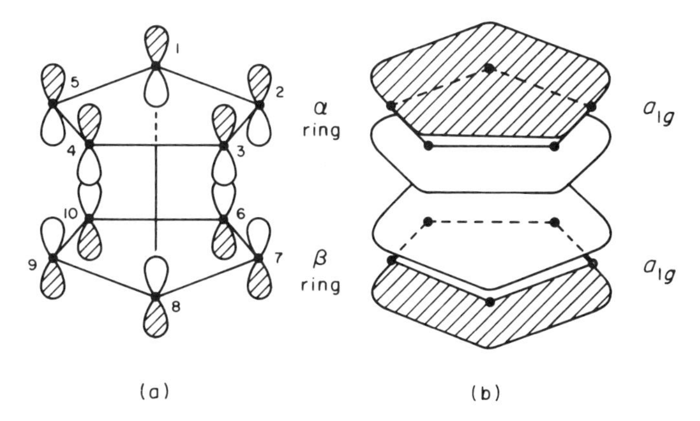

Fig. 7.5 (a) Orbitals of π orbitals for consideration of bonding in ferrocene, and (b) a_{1g} orbitals for the two ring.

If the two rings are designated α and β, the 10 LGOs are

$$a_{1g} = a_\alpha + a_\beta, \qquad a_{2u} = a_\alpha - a_\beta,$$

$$(e_{1g})_a = (e_{1a})_\alpha + (e_{1a})_\beta, \qquad (e_{1g})_b = (e_{1b})_\alpha + (e_{1b})_\beta,$$

$$(e_{1u})_a = (e_{1a})_\alpha - (e_{1a})_\beta, \qquad (e_{1u})_b = (e_{1b})_\alpha - (e_{1b})_\beta,$$

$$(e_{2g})_a = (e_{2a})_\alpha + (e_{2a})_\beta, \qquad (e_{2g})_b = (e_{2b})_\alpha + (e_{2b})_\beta,$$

$$(e_{2u})_a = (e_{2a})_\alpha - (e_{2a})_\beta, \qquad (e_{2u})_b = (e_{2b})_\alpha - (e_{2b})_\beta.$$

The metal orbitals are

a_{1g}: $4s$ and $3d_{z^2}$, e_{1g}: $3d_{xz}$ and $3d_{yz}$,

a_{2u}: $4p_z$, e_{1u}: $4p_x$ and $4p_y$,

e_{2g}: $3d_{xy}$ and $3d_{x^2-y^2}$.

The ligand e_{2u} orbitals are nonbonding since there are no metal e_{2u} orbitals.

The combinations of metal and LGOs and an approximate energy-level diagram are shown in Fig. 7.6. The rings are presumed to be far enough apart so that any interaction between them (+ + or + −) can be neglected, and the g and u LGOs of each set are treated as having the same energies. The a_{1g} molecular orbital for the metallocene is shown as lowest in energy because it is from combination of the lowest-energy metal and LGO. Because of poor overlap, the energy of the a_{2u} MO is only slightly lower than that of the a_{2u} LGO, and the energy of the a_{2u}^* MO is only slightly higher than that of the metal $4p_z$ orbital. The overlap is most favorable for the e_{1g} orbitals—the combination shown for d_{xz} and d_{yz} taken together in Fig. 7.6. There are three

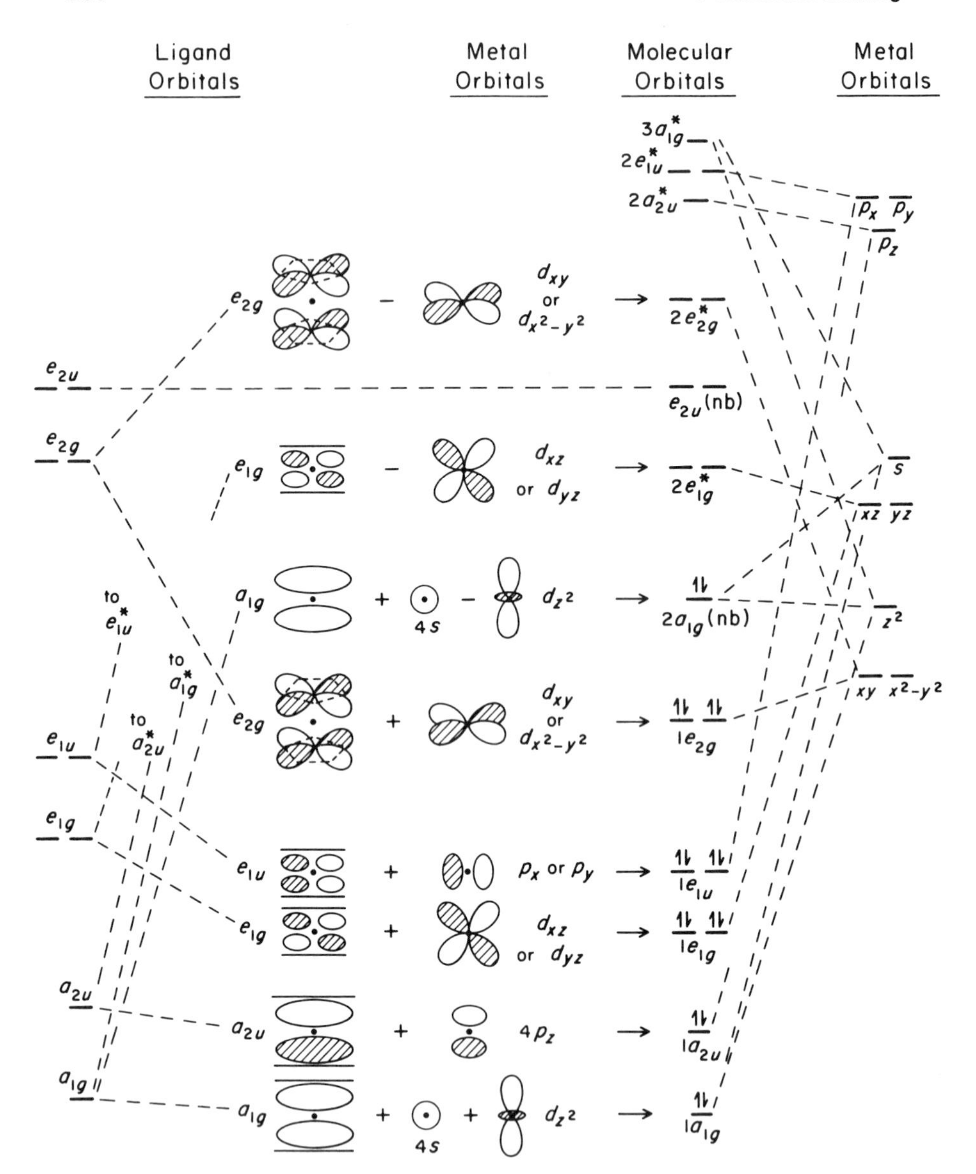

Fig. 7.6 Molecular orbitals for ferrocene.

a_{1g} molecular orbitals since one LGO and two metal orbitals are combined. The bonding orbital has the best combination of signs, and a_{1g}^* has negative lobes of metal orbitals touching positive lobes of LGOs. The other combination is nonbonding and is primarily a metal orbital with electron density concentrated in a large "donut" in the xy plane. The nonbonding e_{2u} and antibonding e_{1u}^*, e_{2g}^*, and e_{1g}^* orbitals are not sketched. Their relative energies might differ from those shown. An 18-electron metallocene, such as ferrocene, has all of the bonding and low-energy nonbonding orbitals filled. The e_{2u} orbitals have very high energy. As a check, we note that there are 19 MOs corresponding to 10 ligand orbitals and nine metal orbitals.

7.4.2. OTHER SANDWICH COMPOUNDS

Ferrocene was the first compound of this type to be discovered, and it is the most stable one. The effective atomic number (EAN) principle is seen to hold, because the sum of the number of electrons for Fe(II) (24) plus those donated by two $C_5H_5^-$ (12) is 36 [the atomic number of the next higher noble gas (Kr)]. This is also known as the *rule of eighteen* since the valence shell (4s, 4p, and 3d) is filled (18 electrons) with six electrons from Fe^{2+} and 12 from two $C_5H_5^-$. The bookkeeping works also, considering Fe^0 and two neutral C_5H_5 rings.

Fe^0,	8 valence e		26 e	for Fe^0
2 C_5H_5	10 e		10 e	
	18 e	total	36 e	total

For electron counting, it is perhaps more common to consider the cyclic π donors as six-electron donors when possible: $C_4H_4^{2-}$, $C_5H_5^-$, C_6H_6, $C_7H_7^+$.

According to the EAN, we expect $Co(C_5H_5)_2$ to be oxidized readily to $Co(C_5H_5)_2^+$, as observed. There are many "piano stool" compounds of the type $(OC)_3M{-}L$, where L is a cyclic π donor. Since three CO ligands donate six electrons, the most stable neutral $(CO)_3M(Cp)$ (Cp is cyclopentadiene) is $(CO)_3Mn(Cp)$. Planar cyclobutadiene is stabilized in the compound $(CO)_3Fe(C_4H_4)$, as expected from the EAN rule.

Since C_6H_6 is a neutral six-electron planar cyclic π system, we expect the most stable sandwich compound to be $Cr(C_6H_6)_2$, as observed. The C_6H_6 π orbitals for the two rings combine to give the LGOs for the eclipsed $\mathbf{D}_{6h}$ configuration,

$$a_{1g}, a_{2u}, e_{1g}, e_{1u}, e_{2g}, e_{2u}, b_{2g}, b_{1u}.$$

There are metal orbitals belonging to all these representations except e_{2u}, b_{2g}, and b_{1u}; so these are nonbonding. In Problem 7.5 you are asked to give descriptions for the MOs and to obtain a qualitative energy-level diagram for $Cr(C_6H_6)_2$.

7.5 Boron Hydrides

7.5.1 DIBORANE

Diborane (B_2H_6) seemed to challenge simple bonding descriptions. The formula looks like that of C_2H_6 where there are seven bonds, yet B_2H_6 has only six electron pairs (3e/B and 1e/H). Such molecules are commonly referred to as electron deficient. Actually B_2H_6 has a planar ethylene-like framework (B_2H_4), with two bridging H atoms in a plane perpendicular to this plane. The symmetry is $\mathbf{D}_{2h}$. The usual sketch of the molecule (Fig. 7.7) shows lines joining all atoms bonded to one another, but we recognize that bridge bonds cannot represent two-electron bonds. H atoms cannot accommodate more than two electrons. A bridge bond is sometimes represented as

H
B B

The plane containing the bridge bonds is chosen arbitrarily as the xz plane. The bond angles about each B do not match the 90° for atomic p orbitals nor the angles for a regular tetrahedron. However, the latter description is closer, so we will consider the sp^3 hydrids of B as the basis sets. The terminal B—H bonds are localized two-electron, two-center bonds, so we neglect them to focus on the delocalized bridge bonding. There are four B orbitals and two H

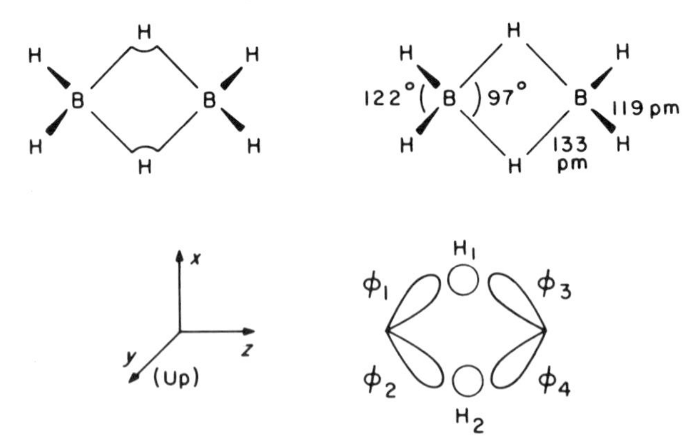

Fig. 7.7 Structure and orientation of the orbitals involved in the bridges for B_2H_6.

orbitals involved, designated as in Fig. 7.7. Eight electrons are involved in the four terminal B—H bonds, so we have four electrons for bridge bonding.

We examine the symmetry properties of the B and H orbitals in the $\mathbf{D}_{2h}$ group, recording the number of orbitals left unchanged to give [$\Gamma_{\text{boron}} = A_g + B_{2g} + B_{1u} + B_{3u}$, $\Gamma_H = A_g + B_{3u}$].

$\mathbf{D}_{2h}$	E	$C_2(z)$	$C_2(y)$	$C_2(x)$	i	σ_{xy}	σ_{xz}	σ_{yz}
Γ_{boron}	4	0	0	0	0	0	4	0
Γ_H	2	0	0	2	0	2	2	6

We can use the rotational group $\mathbf{D}_2$ to obtain the LCAOs using the projection operator.

$\mathbf{D}_2$	E	$C_2(z)$	$C_2(y)$	$C_2(x)$
ϕ_1	ϕ_1	ϕ_2	ϕ_4	ϕ_3
H_1	H_1	H_2	H_2	H_1

$$\psi_{a_g} = \frac{1}{2}(\phi_1 + \phi_2 + \phi_3 + \phi_4) \qquad \psi_{a_g} = \frac{1}{\sqrt{2}}(H_1 + H_2)$$

$$\psi_{b_{3u}} = \frac{1}{\sqrt{2}}(\phi_1 - \phi_2 + \phi_3 - \phi_4) \qquad \psi_{b_{3u}} = \frac{1}{\sqrt{2}}(H_1 - H_2)$$

$$\psi_{b_{2g}} = \frac{1}{2}(\phi_1 - \phi_2 - \phi_3 + \phi_4) \qquad \psi_{b_{1u}} = \frac{1}{2}(\phi_1 + \phi_2 - \phi_3 - \phi_4)$$

The b_{2g} and b_{1u} orbitals are nonbonding. The bonding MOs result from the sums of the B and H orbitals of a_g and b_{3u} symmetry. The LCAOs give the sign combinations necessary for the sketches in Fig. 7.8. A qualitative energy-level diagram can be drawn (Fig. 7.9) based on the extent of delocalization involved in the orbitals. Thus, we might expect the entirely delocalized bonding orbital a_{1g} (no nodes) to be lower in energy than b_{3u} (one node), and the nonbonding b_{1u} (one node) to be lower in energy than b_{2g} (two nodes).

The original description of B_2H_6 as "electron deficient" referred to the fact that there seemed to be too few electrons for the expected number of bonds. BH_3 might be expected to be stable from simple valence rules, and it was not apparent how two, electron-precise, BH_3 molecules could be bridged without another pair of electrons. In the delocalized MO description, there are two bridging MOs involving both B and both H. This description would probably be useful for dealing with spectra. In describing the structural aspects of the bonding, we usually speak of two bridge bonds, each a two-electron, three-center bond. In the energy-level diagram (Fig. 7.10), we see the important characteristic of electron-deficient compounds is that only the bonding

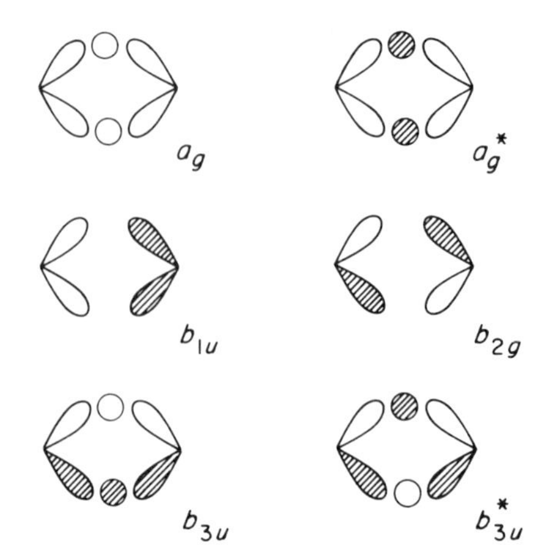

Fig. 7.8 Molecular orbitals for bridge bonding in B_2H_6.

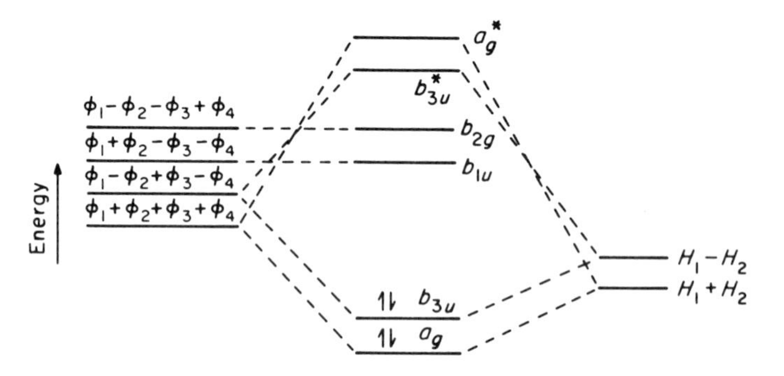

Fig. 7.9 Qualitative energy-level diagram for the bridge bonding in B_2H_6.

orbitals are filled, not the nonbonding orbitals. Accordingly, we might expect boron hydrides to accept electrons readily. Many boron hydrides can form anions but only with substantial changes in structure and bonding. Electron-precise molecules usually have all low-energy orbitals—bonding and nonbonding—filled.

7.5.2 TETRABORANE-10

Tetraborane-10 (B_4H_{10}) has $\mathbf{C}_{2v}$ symmetry as shown in Fig. 7.10. Two borons have two terminal H each, and the other two borons, each having one terminal H, are joined by a single bond. We can use sp^3 hybrid orbitals as a basis set for each boron. Each boron uses two of the orbitals for bridge bonds and two for two-center bonds (to two H or one H and one B). The two-center

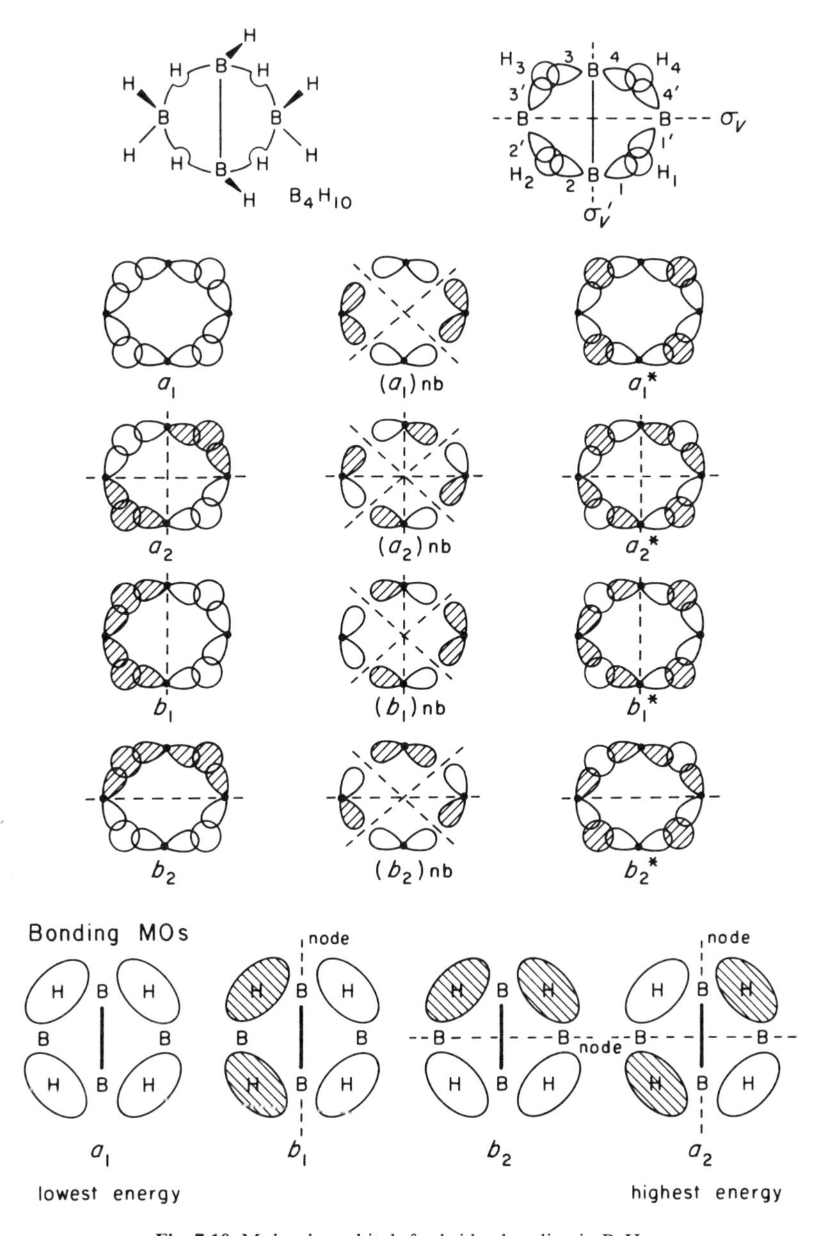

Fig. 7.10 Molecular orbitals for bridge bonding in B_4H_{10}.

bonds can be treated as localized bonds. The bonding orbitals are very low in energy, and the antibonding orbitals are very high in energy. This is not "where the action is." The orbitals of importance for spectral transitions and chemical reactions are the *frontier orbitals*, the HOMOs and LUMOs. We will focus our attention on the bridge bonding to generate the orbitals in the middle of the energy-level diagram.

The boron orbitals (ϕ) and H orbitals (H) are numbered as in Fig. 7.10 for convenience. The orbitals are grouped by sets, a set consisting of a group of orbitals interchanged (mixed) by some symmetry operation of the group. Thus the four H orbitals are in one set. The orbitals of the borons joined by a single bond are in one set since they are interchanged by C_2 and σ. Those of the borons with two terminal H each are in a separate set since no symmetry operation of the group interchanges them with those of the other set.

For the boron orbitals of set 1, we have the following.

$\mathbf{C}_{2v}$	E	C_2	σ_v	σ_v'	
$\Gamma_{\text{Set 1}}$	4	0	0	0	$= A_1 + A_2 + B_1 + B_2$
ϕ_1	ϕ_1	ϕ_3	ϕ_4	ϕ_2	

Applying the projection operator, for set 1,

$$A_1 = \phi_1 + \phi_2 + \phi_3 + \phi_4, \qquad B_1 = \phi_1 - \phi_2 - \phi_3 + \phi_4,$$

$$A_2 = \phi_1 - \phi_2 + \phi_3 - \phi_4, \qquad B_2 = \phi_1 + \phi_2 - \phi_3 - \phi_4.$$

Treating the boron orbitals of set 2 and the orbitals of H similarly, we find the same combinations for set 2,

$$(A_1)' = \phi_{1'} + \phi_{2'} + \phi_{3'} + \phi_{4'}, \qquad (B_1)' = \phi_{1'} - \phi_{2'} - \phi_{3'} + \phi_{4'},$$

$$(A_2)' = \phi_{1'} - \phi_{2'} + \phi_{3'} - \phi_{4'}, \qquad (B_2)' = \phi_{1'} + \phi_{2'} - \phi_{3'} - \phi_{4'}.$$

For H,

$$A_1 = \mathrm{H}_1 + \mathrm{H}_2 + \mathrm{H}_3 + \mathrm{H}_4, \qquad B_1 = \mathrm{H}_1 - \mathrm{H}_2 - \mathrm{H}_3 + \mathrm{H}_4,$$

$$A_2 = \mathrm{H}_1 - \mathrm{H}_2 + \mathrm{H}_3 - \mathrm{H}_4, \qquad B_2 = \mathrm{H}_1 + \mathrm{H}_2 - \mathrm{H}_3 - \mathrm{H}_4.$$

We have three group orbitals for each symmetry species $[A_1, (A_1)', (A_1)_{\mathrm{H}}, \text{etc.}]$. For each of these, as for other three-center bonds, there are three linear combinations to be considered:

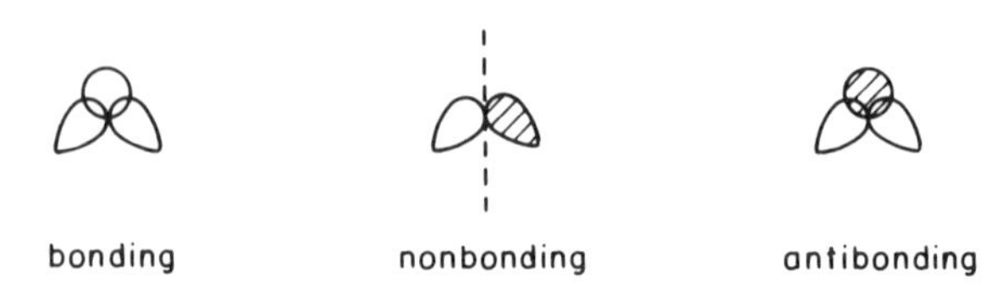

$$a_1 = A_1 + (A_1)_H + (A_1)', \quad (a_1)_{nb} = A_1 - (A_1)', \quad a_1^* = A_1 - (A_1)_H + (A_1)',$$

$$a_2 = A_2 + (A_2)_H + (A_2)', \quad (a_2)_{nb} = A_2 - (A_2)', \quad a_2^* = A_2 - (A_2)_H + (A_2)',$$

$$b_1 = B_1 + (B_1)_H + (B_1)', \quad (b_1)_{nb} = B_1 - (B_1)', \quad b_1^* = B_1 - (B_1)_H + (B_1)',$$

$$b_2 = B_2 + (B_2)_H + (B_2)', \quad (b_2)_{nb} = B_2 - (B_2)', \quad b_2^* = B_2 - (B_2)_H + (B_2)'.$$

Applying the sign combinations from the LCAOs, the MO sketches in Fig. 7.10 are obtained.

The MOs for the B—B single bond (a_1 and b_2^*) are depicted in Fig. 7.11. We check the number of AOs used to be sure that we conserve orbitals, that is, that we obtain the same number of MOs. Also, we determine the number of electrons available for the framework bonding. The bookkeeping is shown in Fig. 7.10. The qualitative energy-level diagram is drawn by separating the orbitals into three groups: bonding, nonbonding, and antibonding. Because of favorable overlap of the orbitals directed toward one another, the a_1 orbital for the B—B single bond is assumed to be lowest in energy, and, correspondingly, b_2^* is highest. For the bridging orbitals of each group (bonding,

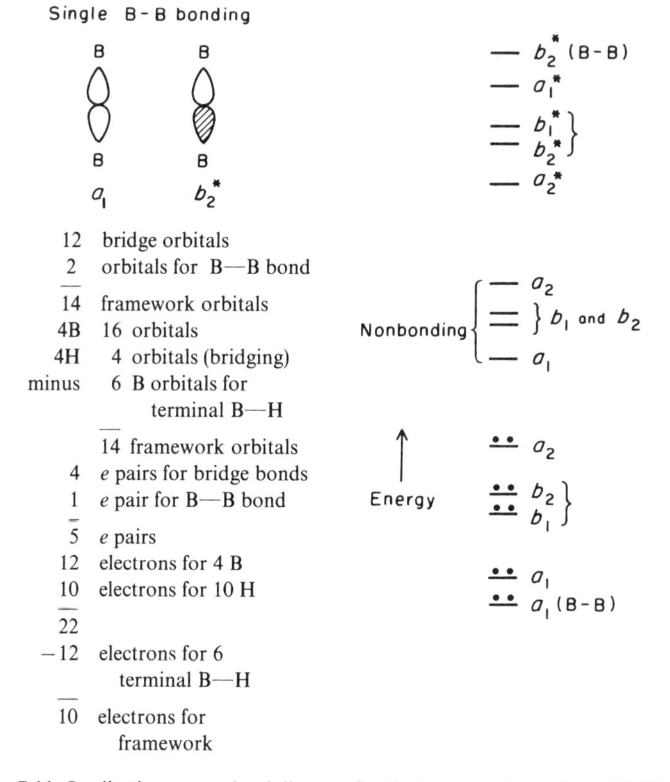

Fig. 7.11 Qualitative energy-level diagram for the framework bonding of B_4H_{10}.

nonbonding, or antibonding), a_1, with no nodes (bonding case), is assumed lowest in energy and a_2, with two nodes (bonding case), is highest. The b_1 and b_2 orbitals each have one node (bonding case) and are assumed to be intermediate in energy. They are not degenerate, since the borons are paired differently in the two cases because of the B—B bond. We cannot determine which is lower in energy from symmetry considerations. The bonding orbitals are filled (10 electrons), leaving the nonbonding orbitals empty. This is usually the case for electron-deficient compounds.

7.5.3 BORON HYDRIDE ANIONS

The simplest boron hydride anion is BH_4^-, a tetrahedral, electron-precise anion involving only two-center bonds. There are many singly charged anions that can be classified on the basis of electron count (see Section 7.6). There is also a series of $B_pH_p^{2-}$ anions having structures of closed polyhedra (*closo* structures). The very symmetrical ones include $B_5H_5^{2-}$ (trigonal bipyramid), $B_6H_6^{2-}$ (octahedron), $B_7H_7^{2-}$ (pentagonal bipyramid), $B_8H_8^{2-}$ (dodecahedron with triangular faces), $B_{10}H_{10}^{2-}$ (bicapped square antiprism), and $B_{12}H_{12}^{2-}$ (icosahedron). Let us examine the bonding in $B_6H_6^{2-}$ ($\mathbf{O}_h$).

In $B_6H_6^{2-}$, each boron is bonded to four neighboring borons and a terminal H. Since boron has only four orbitals, multicenter delocalized bonding must be involved, except for the localized two-center B—H bonds. A reasonable basis set of orbitals for boron consists of the *sp* hybrid orbitals plus two *p* orbitals. For each boron, one of the *sp* hybrids would be used for the B—H bond. This leaves an *sp* hybrid orbital directed toward the center of the octahedron and two *p* orbitals perpendicular to this direction, we can consider them to be directed along the octahedral edges.

If we view these orbitals as vectors, then the vectors pointing inward (radial) are the same as those considered for the σ-bonding LGOs for an octahedral MX_6 complex, and the vectors along the edges (tangential) are the same as those considered for the π-bonding LGOs. These are obtained as in Sections 6.3.3 and 6.5 and directly in Section 8.2. The big difference is that these can only combine with one another since there is no atom at the center of the octahedron.

The orbital combinations corresponding to those in Figs. 8.4 and 8.6 are sketched in Fig. 7.12. The radial a_{1g} is strongly bonding. The e_g orbitals are antibonding, and the t_{1u} orbitals are clearly antibonding. There are two types of tangential combinations, those contained in a σ_h plane (t_{1g} and t_{2g}) and those perpendicular to a σ_h plane (t_{1u} and t_{2u}). One of each of these combinations is shown in Fig. 7.12.

The most favorable (lowest-energy) bonding orbital is a_{1g}, with t_{2g} (directed overlap) next, and tangential t_{1u} (sidewise overlap) highest of the bonding

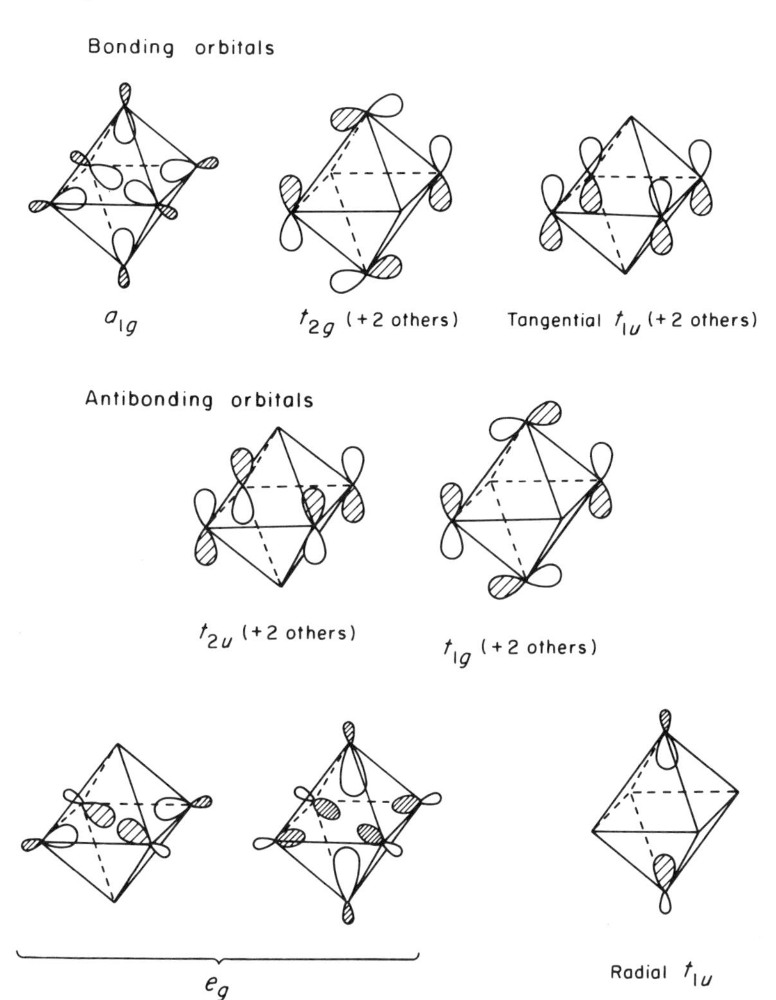

Fig. 7.12 Molecular orbitals for bonding in $B_6H_6^{2-}$.

combinations. The antibonding orbitals increase in energy in the order $t_{2u} < t_{2g} < e_g < t_{1u}$ (radial). The electron count is as follows.

$$6B + 6H - 2 \quad \text{charge.}$$

$$18e + 6e + 2e = 26e$$

$$26e - 12e = 14 \quad \text{electrons for framework bonding}$$

$$6\ B\text{—}H$$

The seven pairs fill the bonding a_{1g}, t_{2g}, and t_{1u} orbitals. The remaining orbitals are antibonding to varying extents.

The "emtpy cage" structure of $B_6H_6^{2-}$ is not limited to boron hydride anions. The tetrahedral P_4 molecule involves bonding similar to that of MnO_4^- without the central atom. Metal clusters can be approached in the same way.

7.6 Cluster Compounds

7.6.1 CLASSIFICATION OF BORON HYDRIDES

Boron hydrides can be classified structurally as *closo* (closed), *nido* (nest), *arachno* (web), or *hypho* (net) based on the formulas shown in Table 7.1. For a general formula of the type $[(BH)_pH_{q+c}]^c$ representing a polyhedron (or fragment) of p vertices with the following electron count

$2e$ from each BH unit,

$1e$ from each second terminal H,

$-1e$ from each + charge,

TABLE 7.1
Classification of Born Hydrides

Type	General formula	Examples
Closo	B_pH_{p+2}	$B_6H_6^{2-}$, $B_8H_8^{2-}$, $B_{10}H_{10}^{2-}$, $B_{11}H_{11}^{2-}$, $B_{12}H_{12}^{2-}$
Nido	B_pH_{p+4}	B_2H_6, B_5H_9, B_6H_{10}, $B_{10}H_{14}$, $B_4H_7^-$, $B_5H_8^-$
Arachno	B_pH_{p+6}	B_4H_{10}, B_5H_{11}, $B_3H_8^-$, $B_4H_9^-$, $B_{10}H_{15}^-$, $B_{10}H_{14}^{2-}$
Hypho	B_pH_{p+8}	$B_5H_{12}^-$

TABLE 7.2
Electron Count for Boron Hydrides

Structural type	Number of vertices of the parent deltahedron left vacant	Skeletal electrons $2p + q$
Closo	0	$2p + 2$
Nido	1	$2p + 4$
Arachno	2	$2p + 6$
Hypho	3	$2p + 8$

the electron balance is given by

$$2p + (q + c) - c = 2p + q,$$

and the structural type is determined by the skeletal electron count given in Table 7.2. Thus, for B_2H_6, $p = 2$ and $q = 4$, so $2p + q = 8 = 2p + 4$, corresponding to a *nido* structure. For $B_6H_6^{2-}$, $p = 6$, $q + c = 0$, and since $c = -2$, $q = +2$, giving $2p + q = 2p + 2$ for a *closo* structure. The $B_3H_8^-$ ion has $p = 3$, $(q + c) = 5$, and $q = 6$, to give an *arachno* structure. The relationships among the structural classification are shown in Fig. 7.13.

7.6.2 WADE'S RULES FOR CLUSTER COMPOUNDS

The correlations between structure and electron count have progressed from Lipscomb's *styx* rules[2] to approaches covered in reviews appearing at about the same time.[3,4,5] The generalized skeletal electron-count correlations are known as *Wade's Rules.* For carboranes with the general formula $[(CH)_a(BH)_pH_{q+c}]^c$, the CH is assumed to contribute three electrons to the framework, giving the number of skeletal electrons to be $3a + 2p + (q + c) - c = 2(a + p) + (a + q)$, where $(a + p)$ is the number of polyhedral vertices. The $2p + q$ in Table 7.2 is replaced by $2(a + p) + (a + q)$. The very stable carborane $C_2B_{10}H_{12}$ is rewritten as $(CH)_2(BH)_{10}$ to give $a = 2$, $p = 10$, $(q + c) = 0$, and 26 skeletal electrons, corresponding to $2(a + p) + 2 = 26$ or the *closo* structure. The stable anion $B_9C_2H_{11}^{2-}$ $[(CH)_2(BH)_9]^{2-}$ gives $a = 2$, $p = 9$, $(q + c) = 0$, and $q = 2$, for $2(a + p) + 4$ skeletal electrons and a *nido* structure.

This approach can be applied to metal clusters also. A transition metal with nine valence orbitals is assumed to use three of these for framework bonding, the others being nonbonding or used for bonding to ligands. The metal contributes the number of valence electrons (v) in excess of the 12 required to fill the six orbitals not involved in the framework. Since we have used metal electrons to fill these orbitals, ligands are assumed to supply x electrons to skeletal bonding (2 for each CO, 5 for η^5–C_5H_5, etc). The number of skeletal bonding electrons is $v + x - 12$, as tabulated in Table 7.3.

The cluster $(CpCo)_2B_4H_6$ with six vertices has

$$2\,CpCo + 4BH + 2q,$$

$$4e \quad + \quad 8e \quad + 2e \quad = 14e.$$

[2] W. N. Lipscomb, *Boron Hydrides*, Benjamin, New York, 1963.
[3] K. Wade, *Adv. Inorg. Chem. Radiochem.* 1976, **18,** 1.
[4] R. E. Williams, *Adv. Inorg. Chem. Radiochem.* 1976, **18,** 67.
[5] R. W. Rudolph, *Acc. Chem. Res.* 1976, **9,** 446.

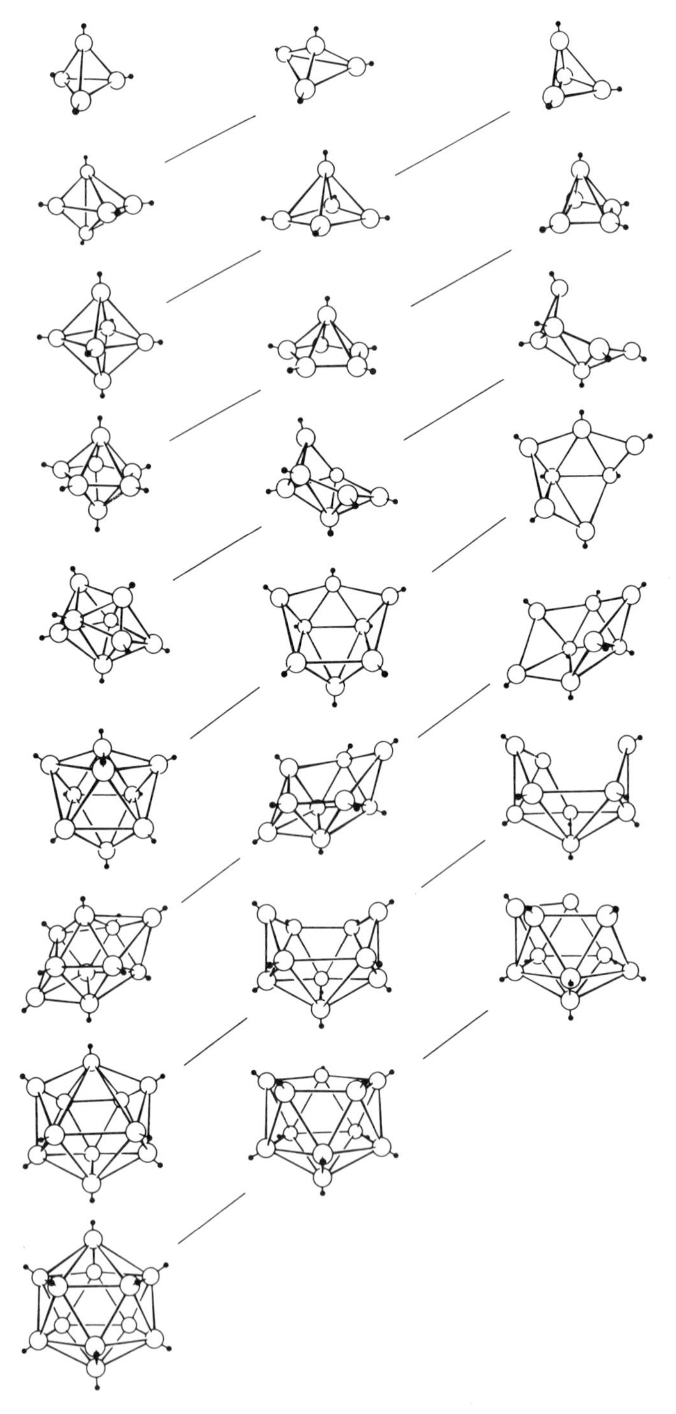

Fig. 7.13 Idealized deltahedra and deltahedral fragments for boranes. The diagonal progressions represent excision of successive BH groups generating *nido* from *closo* and *arachno* from *nido* species. Reproduced with permission from R. W. Rudolph, *Acc. Chem. Res.* 1976, **9**, 446. Copyright © 1976, American Chemical Society.

TABLE 7.3
Number of Skeletal Bonding Electrons ($v + x - 12$) Contributed by Some Transition Metal Cluster Units[a]

		Cluster unit			
v	M	$M(CO)_2$	$M(\eta^5\text{-}C_5H_5)$	$M(CO)_3$	$M(CO)_4$
6	Cr, Mo, W	−2	−1	0	2
7	Mn, Tc, Re	−1	0	1	3
8	Fe, Ru, Os	0	1	2	4
9	Co, Rh, Ir	1	2	3	5
10	Ni, Pd, Pt	2	3	4	6

[a] Adapted from K. Wade. *Adv. Inorg. Chem. Radiochem* 1976, **18,** 1.

For six vertices, this is 2(6) + 2, corresponding to a *closo* structure. For $Fe_5(CO)_{15}C$, there are five $Fe(CO)_3$ groups. The C (not CH) contributes four electrons:

$$5\,Fe(CO_3) + C,$$

$$10e + 4e = 14e.$$

If we assume six vertices (octahedron), this would be 2(6) + 2, a *closo* structure. However, the C is in the base of a square pyramid, with five vertices, giving 2(5) + 4, a *nido* structure. Carbides generally occupy interstitial positions, and CH groups occupy vertices.

EXAMPLE 7.1 What is the expected structural type for $Re_4(CO)_{16}^{2-}$?

SOLUTION Assuming four $Re(CO)_4$ units and four vertices, an *arachno* structure is predicted:

$$4\,Re(CO)_4 + -2 \quad \text{charge},$$

$$12e + 2e = 14e = 2(4) + 6.$$

EXAMPLE 7.2 What is the expected structural type for $(CO)_3FeC_2B_3H_7$?

SOLUTION Writing the formula as $(OC)_3Fe(CH)_2(BH)_3H_2$, we get

$$Fe(CO)_3 + 2CH + 3BH + 2H,$$

$$2e + 6e \quad + 6e \quad + 2e = 16e.$$

For six vertices, 16 = 2(6) + 4 or a *nido* structure is expected, with one vertex of the polyhedron left vacant.

Additional Reading

M. J. S. Dewar, *The Molecular Orbital Theory of Organic Chemistry*, McGraw-Hill, New York, 1969.

B. F. G. Johnson (ed.), *Transition Metal Clusters*, Wiley (Interscience), New York, 1980. The chapter by K. Wade on bonding is particularly useful.

W. L. Jorgensen and L. Salem, *The Organic Chemist's Book of Orbitals*, Academic Press, New York, 1973. Good pictures.

A. Liberles, *Introduction to Molecular Orbital Theory*, Holt, New York, 1966. Good treatment of Hückel theory.

W. N. Lipscomb, *Boron Hydrides*, Benjamin, New York, 1963.

J. P. Lowe, *Quantum Chemistry*, Academic Press, New York, 1978. Good level to accompany this text.

E. L. Muetterties (ed.), *Boron Hydride Chemistry*, Academic Press, New York, 1975.

J. A. Pople and D. L. Beveredge, *Approximate Molecular Orbital Theory*, McGraw-Hill, New York, 1961.

A. Streitwieser, Jr., *Molecular Orbital Theory for Organic Chemists*, Wiley, New York, 1961.

Problems

7.1 (a) The Jahn–Teller theorem (see Section 8.1.3) predicts that the unequal population of degenerate orbitals, such as the e' orbitals of H_3^+, should lead to distortion, causing removal of the degeneracy. In the case of one electron in an e' orbital, the energies of the orbitals would split, with the electron going into the lower-energy orbital. On this basis, what would you expect to be the effect on molecular shape caused by adding an electron to H_3^+ to form H_3?
(b) Considering the great bond energies of H_2 and H_3^+ and the tendency for maximum charge delocalization, what would you expect for the shape of H_3^-? Sketch the MOs and give a qualitative MO energy-level diagram.

7.2 Give MO descriptions for bonding in SF_4 with, and without, the use of d orbitals

7.3 The cyclic $C_4O_4^{2-}$ ion is planar. What is the σ hybridization expected for the C atoms in the four-membered ring? Give an MO treatment for the π bonding involving all eight atoms. Obtain the representations and LCAOs for all of the π orbitals and give simplified sketches.

7.4 Obtain the LCAOs for the π MOs for benzene. Calculate the energies of the orbitals.

7.5 Obtain the MOs for the metal–ligand interactions in $Cr(C_6H_6)_2$. Sketch the MOs and give a qualitative energy-level diagram.

7.6 Obtain the MOs for the metal–ligand interactions in $M(C_4H_4)_2$ using the π MOs for $C_4H_4^{2-}$ described in Section 7.3.2. Sketch the MOs for the sandwich compound. What metal should be most favorable for $M(C_4H_4)_2$?

7.7 The symmetry of B_5H_9 is $\mathbf{C}_{4v}$. The borons form a square pyramid with one terminal H for each B and four bridging H in the base of the pyramid. Give an MO description of the bonding for all except the two-center, two-electron terminal B—H bonds. A convenient basis set consists of sp^3 hybrids for the four B in the base of the pyramid and *sp* for the one at the apex. Obtain the representations and the LCAOs; give sketches and a qualitative energy-level diagram.

7.8 What is the expected structural type for $B_3H_8^-$? Give a bonding description assuming three BH_2 groups and two bridging H's.

7.9 Give the structural types expected for the following clusters:
(a) $B_{10}H_{10}^{2-}$, (b) B_5H_9, (c) B_5H_{11}, (d) $B_5H_{12}^-$, (e) $B_{10}H_{14}$.

7.10 Give the structural types expected for the following clusters:
(a) $Rh_6(CO)_{16}$, (b) $Os_5(CO)_{16}$, (c) $Rh_2(FeCp)_2(CO)_8$, (d) $(CoCp)_2C_2B_6H_8$, (e) $C_2B_9H_{11}(CoCp)$.

7.11 Give an MO description of bonding in square planar cyclic S_4^{2+}. Indicate the basis set of orbitals chosen and describe the σ and π bonding.

8

Bonding in Transition Metal Complexes

Ligand field theory is reviewed as background for later treatment of spectra of transition metal complexes. The molecular orbital description of metal complexes is presented as an extension of the approach used for simple binary compounds in Chapter 6. The angular overlap model is presented as an alternative bonding description.

8.1 Ligand Field Theory

8.1.1 OCTAHEDRAL AND TETRAHEDRAL FIELDS

The d orbitals are fivefold degenerate in an isolated ion. In a chemical environment the energy levels split as dictated by the symmetry of the local field surrounding the ion. The representations (symmetry species) for the d orbitals can be identified from the character table for the point group. In the case of an octahedral complex the representations are E_g and T_{2g}. Symmetry does not give information about the relative energies of the orbitals. Since metal complexes are formed by donation of electrons from ligands, the e_g orbitals directed toward the ligands are raised in energy. Since ligand electrons occupy the bonding e_g MOs, the e_g orbitals to accommodate metal electrons are, in fact, antibonding e_g^* MOs. The sum of the energies of the d orbitals, the *baricenter*, remains the same regardless of the symmetry of the field. In the $\mathbf{O}_h$ case the ratio of the splitting e_g/t_{2g} is $\frac{3}{2}$, or as a matter of convenience for 10 possible electrons, $\frac{6}{4}$. That is, e_g is raised by $6Dq$ and t_{2g} is lowered by $4Dq$. The total splitting is $10Dq$ (or Δ), where Dq (or Δ) is treated as an empirical parameter. In a tetrahedral field the d orbitals are grouped in the same way, but the splitting is reversed: the t_2 orbitals are *raised* in energy by $4Dq$ and the e orbitals are lowered by $6Dq$ (see Fig. 8.1).

The parameter Dq is defined by

$$Dq = \frac{1}{6}\frac{Ze^2}{a^5}\langle r^4 \rangle,$$

where Z is the charge on the central ion, e is the charge on the electron, $\langle r^4 \rangle$ is an integral dependent on the fourth power of the mean radius of the d electron considered, and a is the metal–ligand distance. Obviously Dq is strongly dependent on the charge of the metal ion since Z increases and a decreases with increasing charge on the metal ion. The magnitude of the splitting is determined by the field strength of the ligands. The splitting increases (Dq increases) from left to right in the *spectrochemical series*

$$I^- < Br^- < Cl^- < S^{2-} < F^- < OH^- < C_2O_4{}^{2-}$$

$$< H_2O < NH_3 < en < NO_2{}^- \sim phen < CN^-.$$

Each electron entering the lower-energy orbitals is lowered in energy by $4Dq$ for $\mathbf{O}_h$ and by $6Dq$ for $\mathbf{T}_d$, and those entering the higher-energy orbitals are raised in energy by $6Dq$ for $\mathbf{O}_h$ and $4Dq$ for $\mathbf{T}_d$. The *ligand field stabilization energy* (LFSE) is the sum of the stabilization and destabilization energies for all electrons. For the d^3 case shown in Fig. 8.1, the LFSE is $(3)(-4Dq) = -12Dq$ for $\mathbf{O}_h$ and $(2)(-6) + (1)(4)Dq = -8Dq$ for $\mathbf{T}_d$. For the high-spin d^5 case (or for d^{10}), the LFSE is 0 for $\mathbf{O}_h$, $\mathbf{T}_d$, or for *any* symmetry. For the d^4 case, the LFSE is $-6Dq$ ($\mathbf{O}_h$) unless the splitting is large enough so that the gain in energy ($-10Dq$) by placing the fourth electron in a t_{2g} orbital is greater than the pairing energy. This is true for ligands high in the spectrochemical series. The result for the spin-paired complex is known as the *strong-field* case or, since there are two unpaired electrons instead of four, the low-spin case.

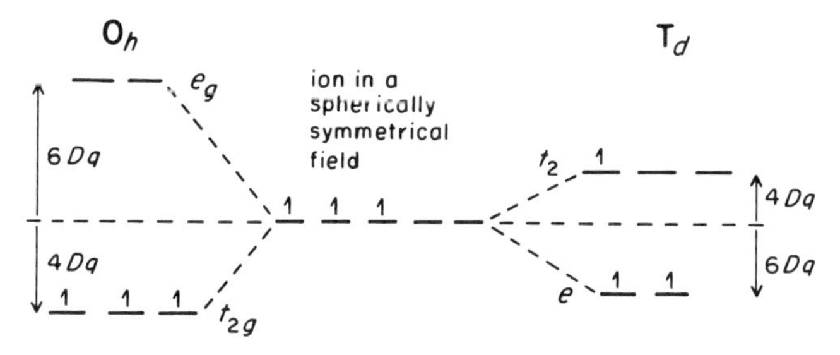

Fig. 8.1 Splitting of the energies of the d orbitals in $\mathbf{O}_h$ and $\mathbf{T}_d$ fields.

TABLE 8.1
Ligand Field Stabilization Energies for Octahedral and Tetrahedral Complexes

		Octahedral		Tetrahedral	
Configuration	Examples	Weak field $(-Dq)$	Strong field $(-Dq)$	Weak field $(-Dq)$	Strong field $(-Dq)$
d^0	Ca^{2+}, Sc^{3+}	0	0	0	0
d^1	Ti^{3+}	4	4	6	6
d^2	V^{3+}	8	8	12	12
d^3	Cr^{3+}, V^{2+}	12	12	8	(18)[a]
d^4	Cr^{2+}, Mn^{3+}	6	16	4	(24)[a]
d^5	Mn^{2+}, Fe^{3+}	0	20	0	(20)[a]
d^6	Fe^{2+}, Co^{3+}	4	24	6	(16)[a]
d^7	Co^{2+}	8	18	12	12
d^8	Ni^{2+}	12	12	8	8
d^9	Cu^{2+}	6	6	4	4
d^{10}	Cu^+, Zn^{2+}	0	0	0	0

[a] These cases are not realized for regular tetrahedral complexes.

Table 8.1 gives the LFSE for various d^n configurations for the weak-field (high-spin) and strong-field (low-spin) cases for $\mathbf{O}_h$ and $\mathbf{T}_d$ fields.

Since the ligands do not approach along the direction of the d orbitals for a tetrahedron, and since there are only four ligands, the splitting for the tetrahedral case is much smaller than for the octahedral case, $(10Dq)_{\text{tet}} = \frac{4}{9}(10Dq)_{\text{oct}}$. Hence the LFSE for a d^2 ion in a tetrahedral field $(-12Dq_{\text{tet}})$ is smaller than for an octahedral field $(-8Dq_{\text{oct}})$. The important consequence is that the splitting is never great enough for regular tetrahedral complexes to cause spin pairing. Tetrahedral $MX_4{}^{n-}$ complexes are high-spin complexes, and the strong-field case is not encountered; these LFSE values are given in parentheses in Table 8.1.

The pairing energy is greater for the d^5 case (~ 30 kK for Fe^{3+}) than for d^6 (~ 23 kK for Co^{3+}), so the field strength required to bring about spin pairing is smaller for d^6. Most Co^{3+} complexes have low spin, but only very strong field ligands form low-spin complexes for Fe^{3+} and Mn^{2+}. Values of Dq increase with increasing charge on the metal ion, as might be expected, because of stronger interaction with the ligands. The values of Dq increase and pairing energies decrease as we proceed to the second and then to the third transition series. Spin pairing is much more commonly encountered for complexes of the second and third transition series.

EXAMPLE 8.1 From the following pairing energies and $10Dq$ values for $[M(H_2O)_6]^{n+}$, predict which metal ions should give low-spin aqua complexes.

Considering that the Dq values for NH_3 and CN^- are about 1.25 times and 1.7 times the values for H_2O, respectively, predict which metal ions might give low-spin complexes with these ligands.

Ion	Pairing energy for free ions[a] (kK)	$10Dq$ for $M(H_2O)_6^{n+}$ (kK)
Mn^{3+} (d^4)	26.0	21.0
Mn^{2+} (d^5)	24.0	8.5
Fe^{3+}	30.0	14.3
Fe^{2+} (d^6)	18.0	10.4
Co^{3+}	23.0	20.7

[a] Pairing energies for complexes are likely to be 10–30% lower because of decreased interelectron repulsion. 1 kK = 1,000/cm.

SOLUTION The value of $10Dq$ is smaller than the pairing energy for all of these ions. Actually $[Co(H_2O)_6]^{3+}$ has low spin, indicating that the pairing energy is lowered enough for spin pairing in this case. An increase in $10Dq$ by about 25% would make $10Dq$ greater than the pairing energy only for Mn^{3+} (as well as for Co^{3+}). The complex $[Mn(NH_3)_6]^{3+}$ would be expected to have low spin, but the complex is unstable because of competing reactions. A ligand such as CN^- with a Dq value 1.7 times that of H_2O would be expected to give low-spin complexes with the d^4 and d^6 ions. Actually $[Fe(CN)_6]^{3-}$ and even $[Mn(CN)_6]^{4-}$ are low-spin complexes. The strong σ bonding and π-acceptor role of the CN^- ligand lowers the pairing energy sufficiently to accomplish spin pairing.

8.1.2 DISTORTED OCTAHEDRAL AND SQUARE PLANAR COMPLEXES

If the effective field symmetry is lowered from $\mathbf{O}_h$ to $\mathbf{D}_{4h}$, the new representations for the d orbitals can be found from the $\mathbf{D}_{4h}$ character table or from a correlation table. The changes are $e_g \rightarrow a_{1g} + b_{1g}$ and $t_{2g} \rightarrow b_{2g} + e_g$. If the octahedron is distorted by elongating the bonds to the ligands along z, or if two weaker field ligands are substituted along z, the a_{1g} (d_{z^2}) orbital will be lowered in energy and b_{1g} ($d_{x^2-y^2}$) will be raised in energy since it interacts with the ligands exerting a stronger field. Also the e_g (d_{xz}, d_{yz}) orbitals will be lower in energy than b_{2g} (d_{xy}) (see Fig. 8.2). If the field is made stronger along z, these splittings are reversed. The splitting is much greater for the orbitals (d_{z^2} and $d_{x^2-y^2}$) directed along the axes than for the t_{2g} set.

This situation with one unique axis is known as the tetragonal case. The greater the difference in field strength along z and in the xy plane, the greater

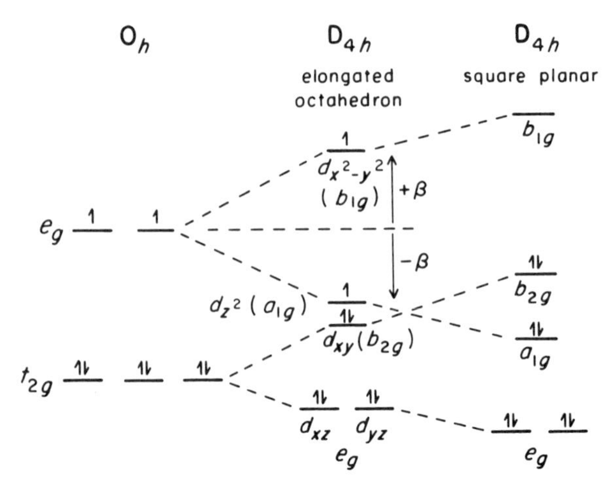

Fig. 8.2 Effect of $\mathbf{D}_{4h}$ distortion on the energies of the d orbitals.

the splitting. The limiting case involves the complete removal of the ligands along z to give a square planar complex. Since the symmetry is still $\mathbf{D}_{4h}$, the splitting pattern remains the same, but the a_{1g} orbital is lowered and e_g is raised sufficiently so they cross. The $a_{1g}\,(d_{z^2})$ orbital never becomes the lowest-energy orbital because of the interaction of the donut in the xy plane with the four remaining ligands. The important feature of the splitting for the square planar case is that one orbital $(d_{x^2-y^2})$ goes to very high energy. The most favorable configuration for forming square planar complexes is low-spin d^8 (Ni^{2+}, Pd^{2+}, Pt^{2+}, Ir^{+}, Au^{3+}, etc.). These square planar complexes are diamagnetic, indicating the complete filling of all of the lower-energy orbitals, with $d_{x^2-y^2}$ unoccupied. The d^9 and d^7 cases are not quite as favorable.

8.1.3 JAHN–TELLER DISTORTION

The Jahn–Teller theorem predicts that unsymmetrical occupancy of degenerate orbitals should lead to distortion. This is the situation for a d^9 (Cu^{2+}) octahedral complex, where one of the e_g orbitals is filled and one is half-filled. Distortion of the octahedron along z, such as is shown in Fig. 8.2, removes the degeneracy of the e_g orbitals so we would have $d_{z^2}^2\, d_{x^2-y^2}^1$. The distortion results in further stabilization by an amount given by the splitting parameter β since two electrons are lowered in energy by β and one is raised by β. Six-coordinate Cu^{2+} complexes commonly have two bonds longer than the other four. We see from Fig. 8.2 that there is no net stabilization upon distortion for the d^8 configuration and $[NiL_6]^{2+}$ complexes have $\mathbf{O}_h$ symmetry. In the case of Ni^{2+}, only very strong field ligands (usually π-acceptor ligands such as CN^-) form low-spin, square planar complexes. Jahn–Teller distortion can occur for unequal occupancy of the e_g or t_{2g} orbitals, as for high-spin d^4 ($t_{2g}^3e_g^1$), d^1, d^2, or high-spin d^6 and d^7 configurations. The distortion is expected to be greater for unequal occupancy of the e_g orbitals that are directed toward the ligands.

EXAMPLE 8.2 Would Jahn–Teller splitting be expected for a "tetrahedral" d^8 ML_4 complex? If so, how would the tetrahedron distort?

SOLUTION Jahn–Teller distortion is expected for an unequally populated degenerate orbital set. For a regular tetrahedron, we expect the configuration $e^4t_2^4$, with two vacancies in t_2. If the tetrahedron is compressed toward the xy plane, then we expect d_{xy} to be higher in energy than d_{xz} and d_{yz}. The reverse is expected for elongation along z.

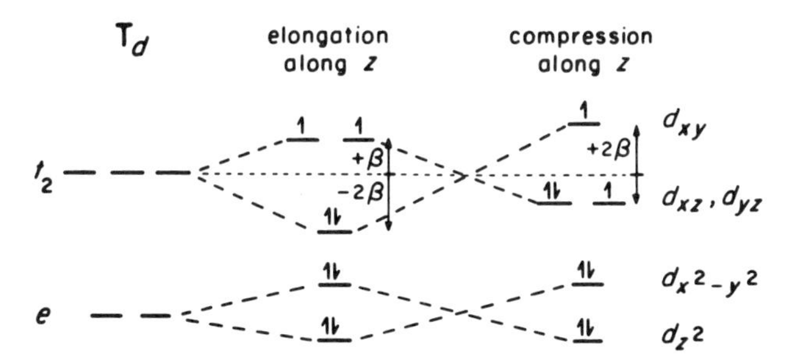

For d^8 elongation along z is expected (net stabilization $-2 \times 2\beta + 2\beta = -2\beta$) since compression gives smaller stabilization ($-3\beta + 2\beta = -\beta$).

8.1.4 ENERGIES OF THE *d* ORBITALS IN FIELDS OF LOWER SYMMETRY

If an ion is placed in the center of a spherical shell and a negative charge is applied to the sphere, then the energies of the d orbitals are all raised without removing the degeneracy. If equal magnitudes of the charge are localized at $\pm x$, $\pm y$, and $\pm z$, then the relative energies of the d orbitals are those for an $\mathbf{O}_h$ field, but the total potential energy is unchanged. Moving the charges around on the surface of the sphere changes the relative energies of the d orbitals as required by the field symmetry, but the net energy, the "baricenter," remains unchanged. The effects are additive, and since a cube can be regarded as two tetrahedra superimposed (both are "cubic" groups), the d orbital splitting for a cubic complex ML_8 is twice that of a tetrahedral complex ML_4, assuming

TABLE 8.2
Relative Energies of the *d* Orbitals for Three Primary Geometric Configurations

Coordination number (C.N.) and configuration of primary group		Relative energy of *d* orbitals[a]				
		d_{z^2}	$d_{x^2-y^2}$	d_{xy}	d_{xz}	d_{yz}
(I) C.N. = 1 One ligand or charge located on the z axis		5.14	−3.14	−3.14	0.57	0.57
(II) C.N. = 2 Two ligands at right angles in the xy plane (on the x and y axes)		−2.14	6.14	1.14	−2.57	−2.57
(III) C.N. = 4	θ					
Four ligands in a staggered, tetrahedral-like arrangement, two above and two below the xy plane; each at a variable angle θ with respect to the $\pm z$ axes	45°	0.64	−5.14	−2.64	3.57	3.57
	54.74°	−2.67	−2.67	1.78	1.78	1.78
	60°	−3.70	−1.43	4.30	0.465	0.465
	62.47°	−4.11	−0.68	5.71	−0.46	−0.46

[a] In Dq units.

the same ligands at the same distance—$(10Dq)_{\text{cube}} = 2 \times (10Dq)_{\text{tet}} = 2 \times \frac{4}{9}(10Dq)_{\text{oct}}$.

We can get the relative energies of the d orbitals for a linear ML_2 complex ($\mathbf{D}_{\infty h}$) as twice the values for one ligand along z. The relative energies of the d orbitals for an $\mathbf{O}_h$ field can be obtained by adding the relative energies of each orbital for the square planar ($\mathbf{D}_{4h}$) case and for the linear ($\mathbf{D}_{\infty h}$) case.

	d_{z^2}	$d_{x^2-y^2}$	d_{xy}	d_{xy}	d_{yz}
ML_4, square planar (xy) $\mathbf{D}_{4h}$	−4.28	12.28	2.28	−5.14	−5.14Dq
ML_2, linear (z) $\mathbf{D}_{\infty h}$	10.28	−6.28	−6.28	1.14	1.14Dq
ML_6, Octahedral $\mathbf{O}_h$	6.00	6.00	−4.00	−4.00	−4.00Dq

Krishnamurthy and Schaap[1] showed that we can calculate the relative energies of the d orbitals for almost any symmetry from the relative energies calculated for three primary ligand groups (Table 8.2). The groups shown in

[1] R. Krishnamurthy and W. B. Schaap, *J. Chem. Educ.* 1969, **46,** 799.

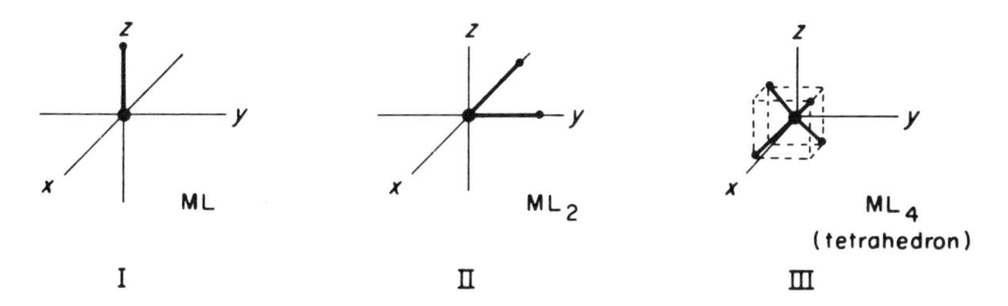

Fig. 8.3 The three primary ligand groups.

Fig. 8.3 are for a single ligand along z, ML; two ligands at 90° in the xy plane, ML_2; and four ligands disposed tetrahedrally, ML_4. In the ML_4 case relative energies are given for the regular tetrahedron (the angle with respect to the z axis is $\theta = 54.74°$) and for $\mathbf{D}_{2h}$ distortion corresponding to elongation or flattening of the tetrahedron along the z direction.

We examined the additivity of the d-orbital energies for the linear and square planar fields to give the energies for the $\mathbf{O}_h$ field. Since the effects of two ligands at $+x$ and $+y$ are the same as for two ligands at $-x$ and $-y$, the energies of the d orbitals for a square planar complex are given by $2V_{\mathrm{II}}$. We can obtain the d-orbital energies for an octahedral field from primary groups I and II: $V_{\mathrm{oct}} = 2V_{\mathrm{I}} + 2V_{\mathrm{II}}$.

We can obtain the relative energies of the d orbitals for any symmetrical arrangement of charges in the xy plane from primary group II and for equal numbers of ligands in planes above and below the xy plane from primary group III. It is sometimes necessary to average the resulting energies of d orbitals that symmetry requires to belong to a degenerate set. Thus, with three units of charge in the xy plane, the total potential energy is $\frac{3}{2}$ times primary group II. This applies to the case with three units of charge distributed symmetrically along the x and y axes or at other points in the plane. The result

	d_{z^2}	$d_{x^2-y^2}$	d_{xy}	d_{xz}	d_{yz}
ML_3 or $\frac{3}{2} \times ML_2$(case II)	-3.21	9.21	1.71	-3.85	$-3.85Dq$

applies to the case where two charges of $-\frac{3}{2}$ are at $+x$ and $+y$, or four charges of $-\frac{3}{4}$ are at $\pm x$ and $\pm y$. If there are unit charges arranged at 120° to form a plane triangle, the same energies apply except that the $\mathbf{D}_{3h}$ symmetry requires that d_{xy} and $d_{x^2-y^2}$ belong to an E_g representation and hence are degenerate. We allow for the degeneracy by averaging the energies for these orbitals,

$(9.21 + 1.71)/2 = 5.46$. Adding two ligands along the z axis (add 2 times case I) gives the energies for a regular trigonal bipyramid ML_5. Since the symmetry is the same, $\mathbf{D}_{3h}$, the degeneracies are correct.

	d_{z^2}	$d_{x^2-y^2}$	d_{xy}	d_{xz}	d_{yz}
Planar ML_3, $\mathbf{D}_{3h}$	−3.21	5.46	5.46	−3.85	−3.85Dq
Linear ML_2	10.28	−6.28	−6.28	1.14	1.14Dq
Trigonal bipyramid ML_5, $\mathbf{D}_{3h}$	7.07	−0.82	−0.82	−2.71	−2.71Dq

For a pentagonal antiprism, $\mathbf{D}_{5d}$, such as ferrocene, we can calculate the energies of the d orbitals as $\frac{5}{2}$ times primary group III with $\theta = 62.47°$ and then average the energies for $\mathbf{D}_{5d}$ symmetry.

	d_{z^2}	$d_{x^2-y^2}$	d_{xy}	d_{xz}	d_{yz}
ML_{10}, $\frac{5}{2}$ times case III	−10.28	−1.70	14.28	−1.14	−1.14Dq
ML_{10}, $\mathbf{D}_{5d}$	−10.28	+6.28	6.28	−1.14	−1.14Dq
Add 2 L along z	10.28	−6.28	−6.28	+1.14	+1.14Dq
Icosahedron, $\mathbf{I}_h$	0	0	0	0	0 Dq

If we add to the pentagonal antiprism two equivalent ligands, we obtain an icosahedron for which d orbitals are fivefold degenerate! In the paper by Krishnamurthy and Schaap, the zero splitting for $\mathbf{I}_h$ is used (check the character table), and the splitting for two ligands along z is subtracted to obtain the orbital energies for the pentagonal antiprism.

A useful application of this approach is to determine the relative energies of the d orbitals for complexes such as *cis*- and *trans*-$[MX_4Y_2]$. This permits us to evaluate the splittings relative to the familiar octahedral MX_6 case. We can obtain the orbital energies by building up the complex using the X and Y groups and their Dq values, or by subtracting the appropriate groups and adding the corresponding Y groups. Thus, *trans*-$[MX_4Y_2]$ is

$$[MX_4]_{\text{sq. planar}} + [MY_2]_{\text{linear}} \text{ or } [MX_6]_{\text{oct}} - [MX_2]_{\text{linear}} + [MY_2]_{\text{linear}}.$$

The Krishnamurthy–Schaap approach is useful for generating energy-level diagrams as we shall see (Section 9.4.5). The treatment applies specifically to the d^1 case, or using the hole formalism, to the d^9 case. Interelectron repulsion is not included so the results are not exact. Nevertheless, this is a useful approximation that permits us to deal with cases which might be too formidable otherwise. Approximate methods can be extremely helpful as long as we remember the limitations.

EXAMPLE 8.3 Calculate the relative energies of the d orbitals for a trigonal prismatic complex ML_6 with the ligands at an angle $\theta = 45°$ relative to the z axis.

SOLUTION We start with an elongated tetrahedral arrangement of four ligands with $\theta = 45°$ and multiply by $\frac{3}{2}$ to correspond to three ligands above and three below the xy plane. Then we check for $\mathbf{D}_{3h}$ symmetry to find that the d orbitals belong to a_1' (z^2), e' ($x^2 - y^2, xy$), and e'' (xz, yz). We see that xz and yz are degenerate, and we can make $x^2 - y^2$ and xy degenerate by averaging their energies.

	d_{z^2}	$d_{x^2-y^2}$	d_{xy}	d_{xz}	d_{yz}
$ML_4, \theta = 45°$	0.64	−5.14	−2.64	3.57	$3.57Dq$
Multiply by $\frac{3}{2}$	0.96	−7.71	−3.96	5.36	$5.36Dq$
$\mathbf{D}_{3h}$					
$ML_6, \theta = 45°$	0.96	−5.84	−5.84	5.36	$5.36Dq$

8.2 Molecular Orbital Theory

8.2.1 SIGMA BONDING IN OCTAHEDRAL COMPLEXES

The MO treatment for metal complexes is the same as for AX_n compounds dealt with in Chapter 6. We examined the σ and π bonding for a square planar compound (Section 6.5). The results of the transformation of the σ vectors for an octahedral complex (see Fig. 8.4) are shown in Table 8.3. The representation for the σ vectors reduces to $A_1 + E + T_1$ in the $\mathbf{O}$ group. Since there will be an a_{1g} LGO and the character for the center of symmetry for Γ_σ is zero, the representations must be $A_{1g} + E_g + T_{1u}$. The hybrid combination required is d^2sp^3. The t_{2g} orbitals are nonbonding for σ interaction. Using the projection operators and the characters we obtain a_{1g}, $(e_g)_a$, and $(t_{1u})_a$ directly. Operating on $(\psi_{t_{1u}})_a$ by C_3 gives another of the LCAOs and operating by C_3^2 gives the third LCAO. We recognize these as the equivalent partners since two ligands interact with one of the p orbitals along x, y, and z, respectively. We could use the matrix elements from the T_{1u} matrices (for all 24 operations) for the transformation of x, y, and z. A sketch of the $(\psi_{e_g})_a$ LCAO (Fig. 8.4) matches the d_{z^2} orbital perfectly. Obviously the equivalent LCAO partner should

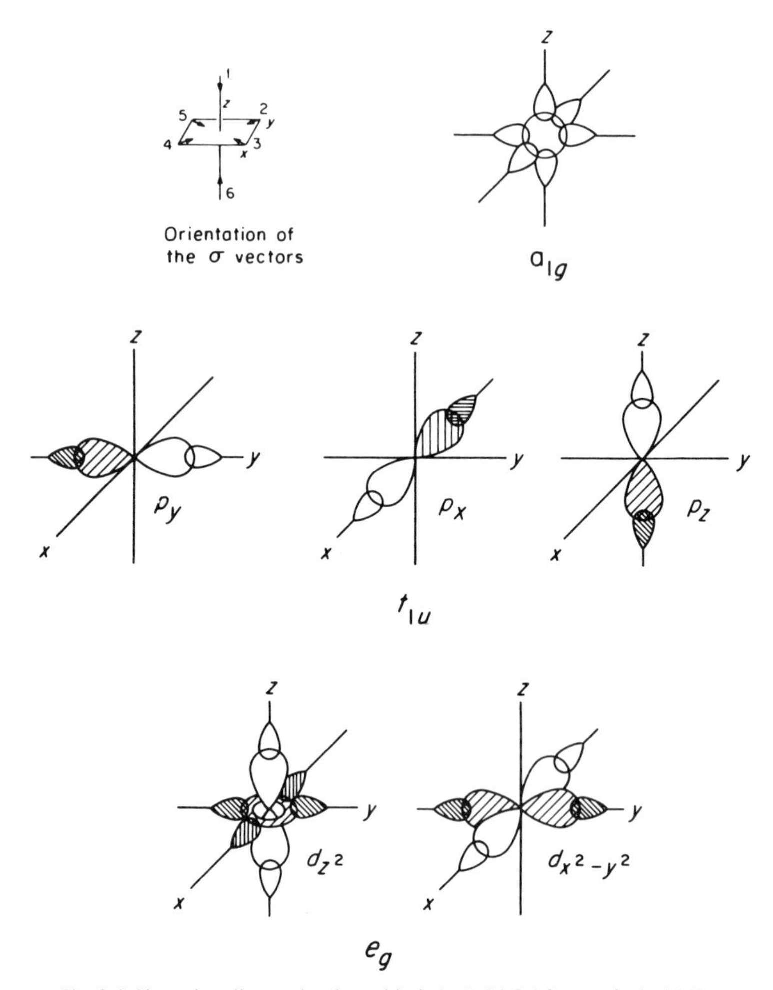

Fig. 8.4 Sigma bonding molecular orbitals (as LCAOs) for octahedral ML_6.

match the other metal e_g orbital, $d_{x^2-y^2}$. If we perform a C_2 operation to interchange ϕ_1 and ϕ_2, we get $2\phi_2 + 2\phi_4 - \phi_1 - \phi_3 - \phi_5 - \phi_6$. This is another LCAO involving merely a permutation of subscripts, not the required partner. Since we know that the partner we seek does not involve ligands 1 and 6, we can see that adding two times the new LCAO to the LCAO for $(e_g)_a$ gives $\phi_2 - \phi_3 + \phi_4 - \phi_5$, matching the $d_{x^2-y^2}$ metal orbital. The normalized LCAO $\frac{1}{2}(\phi_2 - \phi_3 + \phi_4 - \phi_5)$ is $(\psi_{e_g})_b$. The e_g LCAO pair can be obtained also using the projection operator method, but this requires the matrices for all 24 operations of the **O** group. Where we have AOs to serve as templates to aid

TABLE 8.3
Derivation of the LCAOs for the σ Bonds in an Octahedral Complex

Effects of the symmetry operations of the **O** point group on the σ orbitals of an octahedral complex

O	E	$(C_4)_z$	$(C_2)_z$	C_3	C_2'	i
Γ_σ	6	2	2	0	0	0

$\Gamma_\sigma = A_1 + E + T_1(\mathbf{O})$

$\Gamma_\sigma = A_{1g} + E_g + T_{1u}\,(\mathbf{O}_h)$

M orbitals	
s	A_{1g}
p	T_{1u}
$d_{z^2}, d_{x^2-y^2}$	E_g
d_{xz}, d_{yz}, d_{xy}	T_{2g}

Results of applying all symmetry operations of the **O** point group to ϕ_1

E	$(C_4)_1$	$(C_4)_2$	$(C_4)_3$	$(C_4)_4$	$(C_4)_5$	$(C_4)_6$	$(C_2)_1$	$(C_2)_2$	$(C_2)_3$
ϕ_1	ϕ_1	ϕ_5	ϕ_2	ϕ_3	ϕ_4	ϕ_1	ϕ_1	ϕ_6	ϕ_6

$(C_3)_1$	$(C_3)_2$	$(C_3)_3$	$(C_3)_4$	$(C_3)_5$	$(C_3)_6$	$(C_3)_7$	$(C_3)_8$
ϕ_2	ϕ_3	ϕ_4	ϕ_5	ϕ_5	ϕ_2	ϕ_3	ϕ_4

$(C_2')_1$	$(C_2')_2$	$(C_2')_3$	$(C_2')_4$	$(C_2')_5$	$(C_2')_6$
ϕ_6	ϕ_6	ϕ_2	ϕ_3	ϕ_5	ϕ_4

$$\psi_{a_{1g}} = \frac{1}{\sqrt{6}}(\phi_1 + \phi_2 + \phi_3 + \phi_4 + \phi_5 + \phi_6)$$

$$(\psi_{e_g})_a = \frac{1}{\sqrt{12}}(2\phi_1 + 2\phi_6 - \phi_2 - \phi_3 - \phi_4 - \phi_5)$$

$$(\psi_{e_g})_b = \frac{1}{2}(\phi_2 - \phi_3 + \phi_4 - \phi_5)$$

$$(\psi_{t_{1u}})_a = \frac{1}{\sqrt{2}}(\phi_1 - \phi_6)$$

$$(\psi_{t_{1u}})_b = \frac{1}{\sqrt{2}}(\phi_3 - \phi_5)$$

$$(\psi_{t_{1u}})_c = \frac{1}{\sqrt{2}}(\phi_2 - \phi_4)$$

in obtaining the partner LCAO, it is easier to take advantage of symmetry rather than rely on a more tedious mathematical approach.

The bonding MOs as combinations of the LCAOs and appropriate metal orbitals are shown in Fig. 8.4. A qualitative energy-level diagram for σ bonding only is shown in Fig. 8.5 for an octahedral complex such as $[Co(NH_3)_6]^{3+}$.

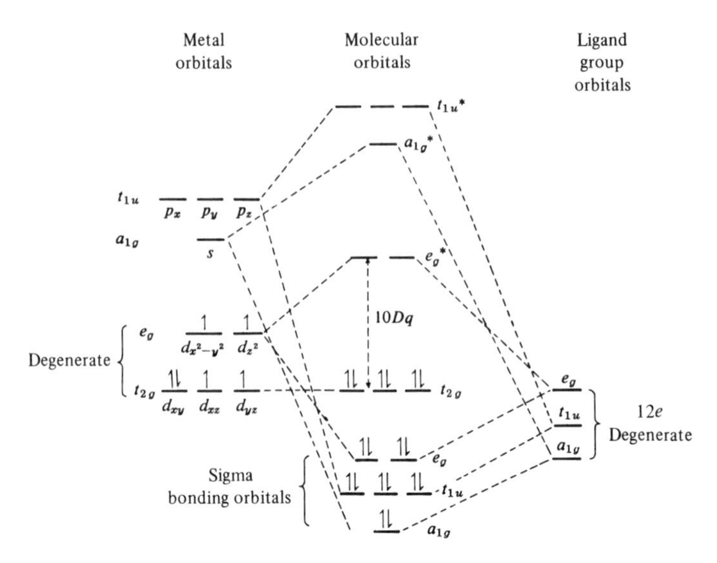

Fig. 8.5 Qualitative energy-level diagram for an octahedral low-spin complex such as $[Co(NH_3)_6]^{3+}$ considering σ bonding only.

8.2.2 PI BONDING IN AN OCTAHEDRAL COMPLEX

The π orbitals of the six ligands in an octahedral complex can be represented by a set of vectors oriented as shown in Fig. 8.6. In this case, all 12 vectors belong to the same set since they can be interchanged by some operation of the group. In Table 8.4, we record the result of operating on the 12π vectors with one operation of each group for **O**. The irreducible representations are $2T_1 + 2T_2$. The character is zero for i ($\mathbf{O}_h$), so the representations are $T_{1g} + T_{2g} + T_{1u} + T_{2u}$. Since there are no metal t_{1g} or t_{2u} orbitals, these LGOs must be nonbonding. The $t_{1u}\,(p)$ metal orbitals are used for σ bonding and, since they give better σ overlap, they are expected to participate primarily in σ bonding. The t_{2g} orbitals are more important for π bonding. The t_{1u} and t_{2u} MOs are illustrated in Fig. 8.6 as combinations of the metal and LGOs. Application of the projection operator is straightforward to obtain the LCAOs, but it is tedious because there are many orbitals and many symmetry operations. Sketches describing the LCAOs are obtained easily using the metal orbitals as templates. Ligand π orbitals are drawn in with the signs matching those of the metal orbital. The ligand orbitals are shown as p orbitals; they can be the π^* orbitals of a ligand such as CN^-.

Effects of π Bonding. Metal–ligand π bonding can involve either partner as the π donor. If the metal is the donor, the metal t_{2g} electrons go into the

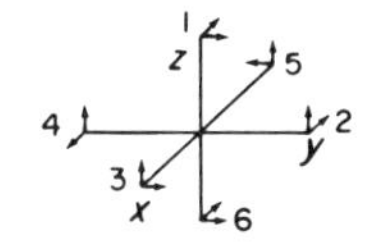

orientation of the vectors for π bonding

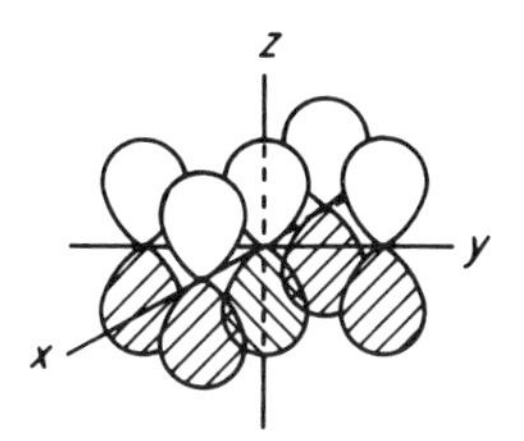

one of the t_{1u} π orbitals, there are identical orbitals along the x and y directions

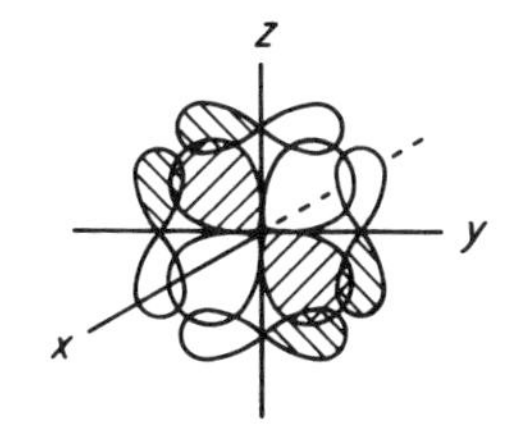

one of the t_{2g} orbitals, there are identical orbitals in the xy and xz planes

Fig. 8.6 The two symmetry types of π bonds in octahedral complexes.

TABLE 8.4
Pi Ligand Orbitals in an Octahedral Complex

Transformation properties of the π vectors for **O**

O	E	C_3	C_2	C_4	C_2^z	$i(\mathbf{O}_h)$
Γ_π	12	0	0	0	−4	0

$$\Gamma_\pi = 2T_1 + 2T_2\ (\mathbf{O})$$
$$\Gamma_\pi = T_{1g} + T_{2g} + T_{1u} + T_{2u}\ (\mathbf{O}_h)$$

bonding t_{2g} MOs. If the metal is the acceptor, the ligand electrons fill the bonding t_{2g} MOs, and any metal t_{2g} electrons must go into the antibonding t_{2g}^* orbitals.

Metals early in each transition series and those in high oxidation states have few d electrons and can serve as π acceptors. Ligands such as F^-, Cl^-, OH^-, and O^{2-} are good π donors. Since $10Dq$ is the separation between the t_{2g} orbitals occupied by metal electrons (this is now t_{2g}^*) and the metal e_g orbitals (really e_g^* with respect to the σ bonding), ligand-to-metal π donation decreases $10Dq$ (Fig. 8.7).

If the metal has filled, or nearly filled, t_{2g} orbitals and the ligands have empty low-energy orbitals, then metal-to-ligand donation is favored. Metals

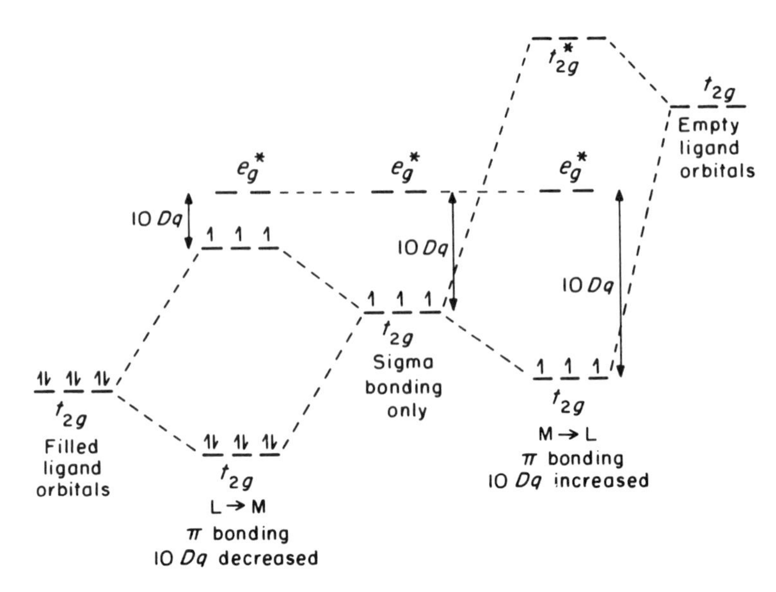

Fig. 8.7 Comparison of the effects of π bonding using (a) filled low-energy π ligand-orbitals for L → M donation and (b) empty ligand-orbitals of π symmetry for M → L donation.

late in each transition series and those in low oxidation states can serve as π donors with ligands having empty low-energy orbitals, such as the *d* orbitals of P or S or the π^* orbitals of CN^-, CO, NO^+, bipyridine, or 1,10-phenanthroline. Metal-to-ligand π bonding increases the M–L bond order and bond strength, it *increases* $10Dq$ (Fig. 8.7), and in the case of π^* acceptors such as CN^-, it lowers the bond order for the C—N bond. Since CO is a weak Lewis base, the π acceptor role is important in placing CO high in the spectrochemical series. The lowering of the CO stretching frequency has been used as an indication of the extent of M–L π bonding.

8.2.3 COMPARISON OF THE MOLECULAR ORBITAL AND LIGAND FIELD THEORY DESCRIPTIONS

The ligand field theory (LFT) is not really a *bonding* theory. It focuses attention on the splitting of the energies of the *d* orbitals. The splitting parameter $10Dq$ is evaluated empirically. The magnitude of the splitting is determined partly by the electrostatic field imposed by the ligands (the *crystal field splitting*) and partly by the covalent bonding. The greater the covalent interaction, the greater the lowering of the bonding e_g orbitals and the greater the elevation of the antibonding e_g^* orbitals. The LF splitting diagram is the central portion of the MO diagram. The e_g orbitals of the LF diagram are the e_g^* MOs.

For a full treatment, including bond energies, the MO treatment is required. Of course, π bonding (see Section 8.2.2) must be considered also where it is important. However, the LF approach using $10Dq$ as an empirical parameter is quite successful in treating electronic absorption spectra and magnetic properties.

8.3 Angular Overlap Model

8.3.1 APPLICATIONS INVOLVING σ INTERACTION OF p ORBITALS

The interaction of orbitals (ϕ_i and ϕ_j, with ϕ_i of lower energy) on two atoms gives a bonding orbital somewhat lower in energy than ϕ_i and an antibonding orbital somewhat higher in energy than ϕ_j. The interaction increases, and the energies of the AOs (H_{ii} and H_{jj}) are altered to a greater extent, the closer the energies of the two AOs are or the smaller $|H_{ii} - H_{jj}|$. Including the overlap integral and assuming that the energy changes are small so $E \approx H_{ii}$, we can write the secular determinant as[2].

$$\begin{vmatrix} H_{ii} - E & H_{ij} - ES_{ij} \\ H_{ij} - ES_{ij} & H_{jj} - E \end{vmatrix} = 0. \tag{8.1}$$

The roots in the form of a power series are

$$\begin{aligned} E_b &= H_{ii} + \frac{(H_{ij} - S_{ij}H_{ii})^2}{H_{ii} - H_{jj}} - \frac{(H_{ij} - S_{ij}H_{ii})^4}{(H_{ii} - H_{jj})^3} + \cdots, \\ E_{ab} &= H_{jj} - \frac{(H_{ij} - S_{ij}H_{jj})^2}{H_{ii} - H_{jj}} + \frac{(H_{ij} - S_{ij}H_{jj})^4}{(H_{ii} - H_{jj})^3} + \cdots. \end{aligned} \tag{8.2}$$

Using the Wolfsberg–Helmholz approximation

$$H_{ij} \simeq S_{ij}(H_{ii} + H_{jj}) \tag{8.3}$$

and substituting for H_{ij}, we obtain

$$\begin{aligned} E_b &= H_{ii} + \frac{H_{jj}^2 S_{ij}^2}{H_{ii} - H_{jj}} - \frac{H_{jj}^4 S_{ij}^4}{(H_{ii} - H_{jj})^3} + \cdots, \\ E_{ab} &= H_{jj} - \frac{H_{ii}^2 S_{ij}^2}{H_{ii} - H_{jj}} + \frac{H_{ii}^4 S_{ij}^4}{(H_{ii} - H_{jj})^3} + \cdots. \end{aligned} \tag{8.4}$$

[2] J. K. Burdett, *Molecular Shapes*, Wiley (Interscience), New York, 1980, p. 30.

Since H^n_{ii} is greater than H^n_{jj} (H_{ii} and H_{jj} are taken as the valence-shell ionization energies, $|H_{ii}| > |H_{jj}|$), the antibonding orbital is destabilized more than the bonding orbital is stabilized, as we have seen (see Section 6.1.1). The stabilization energy $\varepsilon_{\text{stab}}$ of the bonding orbital is given by

$$\varepsilon_{\text{stab}} = \frac{H^2_{jj}S^2_{ij}}{H_{ii} - H_{jj}} - \frac{H^4_{jj}S^4_{ij}}{(H_{ii} - H_{jj})^3} + \cdots, \qquad (8.5)$$

and the destabilization energy can be obtained from Eq. (8.4).

The *angular overlap model* deals with the interaction energy (ε) between orbitals on two atoms as the square of the overlap integral S times a constant β. Here we are considering the quadratic term only of Eq. (8.5) and we see that the constant β $(H^2_{jj}/H_{ii} - H_{jj})$ varies inversely with the energy separation of the two orbitals. For our present considerations, we assume $\varepsilon_{\text{stab}} \simeq \varepsilon_{\text{destab}}$, as represented in Fig. 8.8. The overlap integral can be separated into a radial part S_{R} and an angular part S_{A}:

$$S^2 = S^2_{\text{R}}S^2_{\text{A}} \quad \text{and} \quad \varepsilon = \beta S^2 = \beta S^2_{\text{R}}S^2_{\text{A}}. \qquad (8.6)$$

For a given molecule, with fixed internuclear distance, the radial part can be taken as constant, though we might want to vary the angular part for the consideration of various geometries. Combining the constants β and S^2_{R} as e, we get

$$\varepsilon = (\beta S^2_{\text{R}})S^2_{\text{A}} = \text{e}S^2_{\text{A}}. \qquad (8.7)$$

For many purposes, the first term (S^2) of Eq. (8.5) is sufficient, so our expression reduces to Eq. (8.7). If the S^4 term is needed, Eq. (8.5) can be simplified using γ for the coefficient of S^4_{ij} to give

$$\varepsilon_{\text{stab}} = \beta S^2 - \gamma S^4 = \beta S^2_{\text{R}}S^2_{\text{A}} - \gamma S^4_{\text{R}}S^4_{\text{A}}. \qquad (8.8)$$

Burdett simplified the expression by using $\text{f} = \gamma S^4_{\text{R}}$ to give

$$\varepsilon_{\text{stab}} = \text{e}S^2_{\text{A}} - \text{f}S^4_{\text{A}}. \qquad (8.9)$$

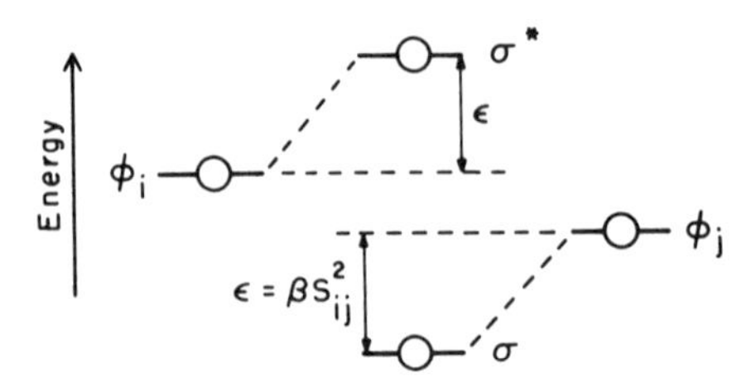

Fig. 8.8 Energy of interaction between two orbitals involved in σ bonding.

Now, if we consider the interaction between a p orbital on one atom and a directed orbital of the other atom, we need to examine the angular dependence of the p orbitals in polar coordinates:

$$\begin{aligned} &r\cos\theta && \text{for } p_z, \\ &r\sin\theta\cos\phi && \text{for } p_x, \\ &r\sin\theta\sin\phi && \text{for } p_y, \end{aligned} \tag{8.10}$$

where θ is the angle relative to the z axis and ϕ is the angle in the xy plane relative to the x axis. For a molecule A–L using p_z on A and a directed orbital on the ligand L (Fig. 8.9), we can use the trigonometric functions above as S_A, giving the stabilization energy as

$$\begin{aligned} &\varepsilon = \mathrm{e}\cos^2\theta - \mathrm{f}\cos^4\theta, \\ &\varepsilon = \mathrm{e} - \mathrm{f} \qquad \text{for} \quad \theta = 0°, \end{aligned} \tag{8.11}$$

or

$$\varepsilon = 2(\mathrm{e} - \mathrm{f}) \qquad \text{for two electrons.}$$

The interaction is greatest along z ($\theta = 0°$) and decreases to zero for $\theta = 90°$. With one ligand bonded along z, a second ligand would bond preferentially with one of the other p orbitals, if it is available (i.e., not filled). The interaction with one of these (let us use p_x) would be the same as for p_z for a bond angle of 90°. For two pairs of electrons the stabilization energy from Eq. (8.9) is $4(\mathrm{e} - \mathrm{f})$ (see Fig. 8.10).

If a second p orbital is not available, the second ligand can interact with p_z also. The maximum stabilization energy is for approach from the opposite direction along z ($\theta = 180°$), giving a linear AL_2 molecule. Because of the opposite signs of the lobes of the p orbital, bonding occurs for the ψ_u LCAO, with ψ_g nonbonding:

$$\psi_u = \frac{1}{\sqrt{2}}(\phi_1 - \phi_2), \qquad \psi_g = \frac{1}{\sqrt{2}}(\phi_1 + \phi_2). \tag{8.12}$$

Fig. 8.9 Interaction between a p_z orbital on one atom and an orbital on another atom.

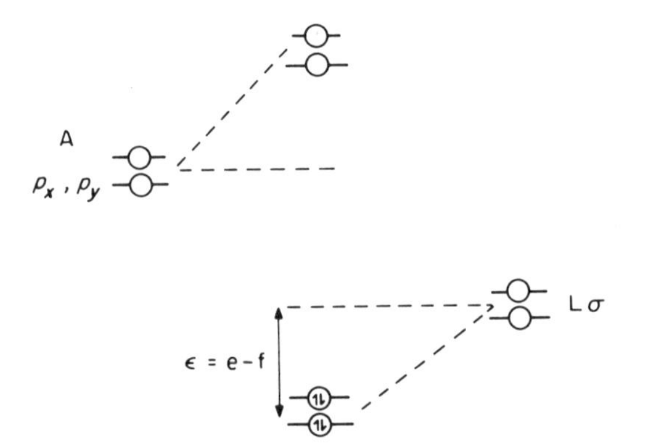

Fig. 8.10 Bonding interaction for AL_2 considering only σ bonding and a bond angle of 90°.

The overlap integral for interaction of the two orbitals of the ψ_u LGO with p_z is $2(1/\sqrt{2})S$ (Fig. 8.11), giving the stabilization energy

$$\begin{aligned} \varepsilon &= (2/\sqrt{2})^2 \beta S^2 - (2/\sqrt{2})^4 \gamma S^4 \\ &= 2eS_A^2 - 4fS_A^4 \qquad (8.13) \\ &= 2e - 4f \qquad \text{for} \quad \theta = 0^\circ \text{ and } 180^\circ. \end{aligned}$$

Here, for the quadratic term, the stabilization energy (2e) is exactly twice that for one ligand [e, see Eq. (8.11)]. An important *sum rule* applies to the quadratic term only: the contributions to the stabilization energy are additive for all ligands. We see that this is not true for the S^4 term. The larger negative coefficient for f means that the stabilization energy for two ligands sharing the same orbital is less than twice that for one ligand (A–L). Thus, the three-center, four-electron bond, with one nonbonding pair, is weaker than the usual two-

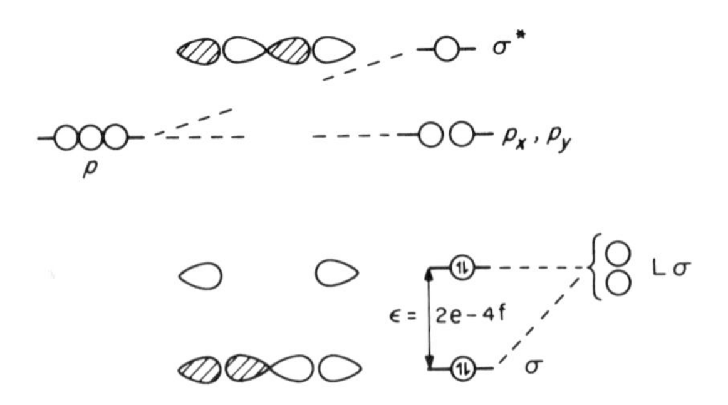

Fig. 8.11 Bonding interaction for linear AL_2 using only one p orbital (p_z).

center, two-electron bond. The three-center, four-electron bond is encountered for XeF_2 (without using *d* orbitals) and for XeF_4 (see Section 7.2).

The angular overlap model has been shown to provide a useful approach to molecular geometry for molecules formed by main group elements.[3]

EXAMPLE 8.4 Account for the **T** shape of ClF_3 using the angular overlap model and *p* orbitals only.

SOLUTION The Cl—F molecule corresponds to the A–L molecule using one *p* orbital. Since the two remaining *p* orbitals on Cl are filled, the two additional F atoms must share one of the *p* orbitals involving a three-center, four-electron bond as for linear AL_2. The second *p* orbital used is perpendicular to the A–L bond, giving the **T** shape.

The electron pair in the nonbonding MO is localized on the two F atoms involved in the 3c, 4e (three-center, four-electron) bond. Chlorine has two lone pairs that would be expected to be in the same plane as the F involved in the 2c, 2e bond. Minimum repulsion leads to the trigonal bipyramidal (TBP) arrangement for the five (bonding plus unshared) pairs, with the lone pairs in equatorial positions. The atoms form a **T**, but with the top of the **T**, bent down because of repulsion involving the lone pairs. The weaker 3c, 4e bond leads to longer bonds to the two axial (with respect to the TBP) F atoms.

8.3.2 APPLICATIONS OF THE ANGULAR OVERLAP MODEL INVOLVING *d* ORBITALS

For many purposes we can use LFT rather than the full MO treatment because useful properties of transition metal compounds and complexes depend on the population, arrangement, and relative energies of the *d* orbitals. The LFT diagram focuses on the HOMO and LUMO of the full MO treatment, and this is where the action is. The energy separations among the *d* orbitals are treated as empirical parameters combining the effects of the electrostatic field and covalent bonding of any type. The angular overlap model permits the separation of the effects caused by σ- and π- bonding interactions. Since the contribution of each ligand is considered in the angular overlap model (AOM), complexes of low symmetry are treated more readily than by LFT.

In the case of an M–L compound, the d_{z^2} orbital is most favorable for bonding because of the larger lobes along the *z* axis. The overlap integral of d_{z^2} for σ bonding is $\frac{1}{2}(3\cos^2\theta - 1)(S_{\mathrm{R}})_\sigma$, giving $(S_{\mathrm{R}})_\sigma$ for $\theta = 0°$, decreasing to zero for $\theta = 109.48°/2$ (corresponding to the nodal plane), and increasing in

[3] J. K. Burdett, *Str. Bonding*, 1976, **31**, 67; *Molecular Shapes*, Wiley (Interscience), New York, 1980.

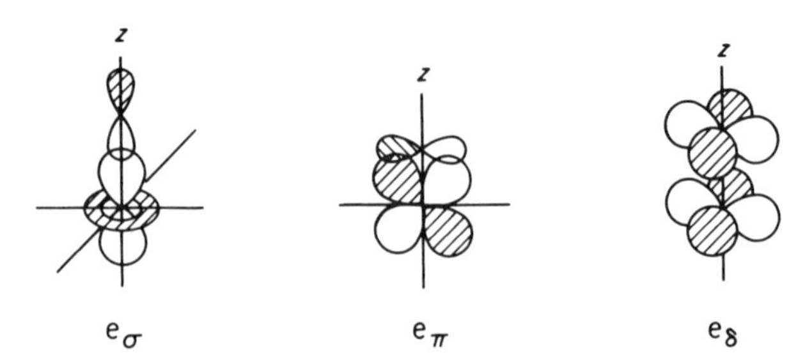

Fig. 8.12 Orientation of a ligand along z for maximum interaction for σ, π, and δ bonding.

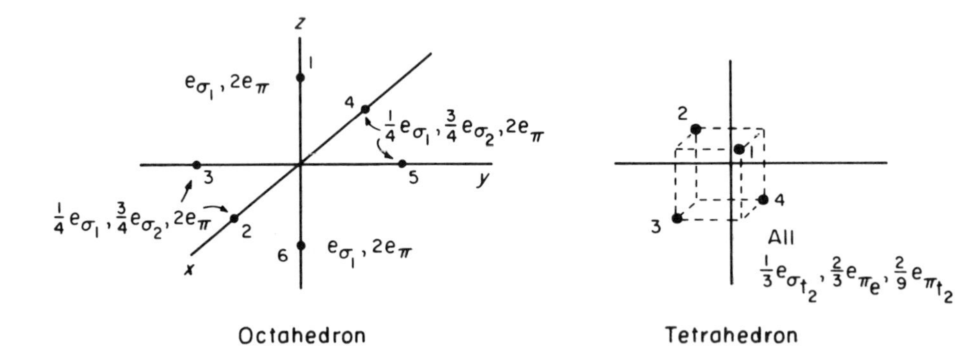

Fig. 8.13 Angular scaling factors for each ligand interacting with one metal orbital. For the octahedron, σ_1 refers to the d_{z^2} orbital and σ_2 refers to the $d_{x^2-y^2}$ orbital; $2e_\pi$ refers to the π interaction with two of the t_{2g} orbitals for each ligand. For the tetrahedron, the scaling factors are the same for each ligand and refer to the t_2 and e orbitals.

magnitude to $-\frac{1}{2}(S_R)_\sigma$, giving $(-\frac{1}{2})^2(S_R)^2 = \frac{1}{4}e_\sigma$ for $\theta = 90°$. A ligand approaching along z with an orbital of π symmetry (p_x, p_y, d_{xz}, or d_{yz}) can form a π bond using the metal d_{xz} or d_{yz} orbital. The angular dependence for the d_{yz} orbital is $(\cos\theta)(S_R)_\pi$, giving maximum interaction at $\theta = 0°$ (Fig. 8.12). Similarly, a ligand with a d orbital available (perhaps another metal) can form a δ bond using $d_{x^2-y^2}$ or d_{xy} on M. We shall consider only σ and π interaction here.

For an octahedral complex (axes defined in Fig. 8.13), we can evaluate the angular part of the overlap integrals for each ligand interacting (σ and π) with each d orbital using the trigonometric relationships given in Table 8.5. The interaction energy of an orbital with each ligand is the square of the overlap integral times a constant β [Eq. (8.6)], as given earlier:

$$\varepsilon = \beta S_R^2 S_A^2 = eS_A^2. \tag{8.7}$$

The interaction energies are the same for d_{z^2} and for $d_{x^2-y^2}$.

TABLE 8.5
Some Angular Overlap Integrals between Central Atom *s*, *p*, and *d* Orbitals and Ligand σ and π Bonding Orbitals[a]

	σ	π_y	π_x
s	1		
p_z	$\cos\theta$	0	$-\sin\theta$
p_x	$\sin\theta\cos\phi$	$-\sin\phi$	$\cos\theta\cos\phi$
p_y	$\sin\theta\sin\phi$	$\cos\phi$	$\cos\theta\sin\phi$
d_{z2}	$\frac{1}{2}(3\cos^2\theta - 1)$	0	$-\frac{\sqrt{3}}{2}\sin 2\theta$
$d_{x^2-y^2}$	$\frac{\sqrt{3}}{4}(\cos 2\phi)(1 - \cos 2\theta)$	$-\sin\theta\sin 2\phi$	$\frac{1}{2}\sin 2\theta\cos 2\phi$
d_{xy}	$\frac{\sqrt{3}}{4}(\sin 2\phi)(1 - \cos 2\theta)$	$\sin\theta\cos 2\phi$	$\frac{1}{2}\sin 2\theta\sin 2\phi$
d_{xz}	$\frac{\sqrt{3}}{2}\cos\phi\sin 2\theta$	$-\cos\theta\sin\phi$	$\cos 2\theta\cos\phi$
d_{yz}	$\frac{\sqrt{3}}{2}\sin\phi\sin 2\theta$	$\cos\theta\cos\phi$	$\cos 2\theta\sin\phi$

[a] Adapted from C. E. Schäffer, *Pure Appl. Chem.* 1970, **24,** 361.

Ligand	1	2	3	4	5	6
d_{z^2}	e_σ	$+\frac{1}{4}e_\sigma$	$+\frac{1}{4}e_\sigma$	$+\frac{1}{4}e_\sigma$	$+\frac{1}{4}e_\sigma$	$+e_\sigma = 3e_\sigma$
$d_{x^2-y^2}$	0	$+\frac{3}{4}e_\sigma$	$+\frac{3}{4}e_\sigma$	$+\frac{3}{4}e_\sigma$	$+\frac{3}{4}e_\sigma$	$+0 = 3e_\sigma$

We expect the doubly degenerate (e_g) orbitals to have the same energy. In like manner, we evaluate the σ- and π-interaction energies of all ligands with all orbitals in Table 8.6. The results are summarized in Fig. 8.13. We see that the e_g orbitals are σ bonding and π nonbonding, while the (equivalent) t_{2g} orbitals participate in π bonding only. The difference in the energies of the orbitals ($e_g - t_{2g}$) is $10Dq$ in LFT and $3e_\sigma - 4e_\pi$ using the AOM:

$$3e_\sigma - 4e_\pi = 10Dq. \tag{8.14}$$

As in the case of LFT diagrams, the AOM diagrams describe the relative energies of the nonbonding and antibonding *d* orbitals—those accommodating the metal electrons. Thus, e_σ is positive, while e_π is positive for π donors

TABLE 8.6
Overlap Integrals and Interaction Energies of the Orbitals for σ and π Bonding in an Octahedral Complex

Ligand		d_{z^2}		$d_{x^2-y^2}$		d_{xy}		d_{xz}		d_{yz}	
		Overlap integral	xe_λ	Overlap integral	xe_λ	Overlap integral	xe_λ	Overlap integral	xe_λ	Overlap integral	xe_λ
1	σ	S_R	e_σ	0	0	0	0	0	0	0	0
$\theta = 0°$	π	0	0	0	0	0	0	0	0	S_R	e_π
$\phi = 0°$	π	0	0	0	0	0	0	S_R	e_π	0	0
			e_σ		0		0		e_π		e_π
2	σ	$-\frac{1}{2}S_R$	$\frac{1}{4}e_\sigma$	$\frac{\sqrt{3}}{2}S_R$	$\frac{3}{4}e_\sigma$	0	0	0	0	0	0
$\theta = 90°$	π	0	0	0	0	S_R	e_π	0	0	0	0
$\phi = 0°$	π	0	0	0	0	0	0	$-S_R$	e_π	0	0
			$\frac{1}{4}e_\sigma$		$\frac{3}{4}e_\sigma$		e_π		e_π		0
3	σ	$-\frac{1}{2}S_R$	$\frac{1}{4}e_\sigma$	$-\frac{\sqrt{3}}{2}S_R$	$\frac{3}{4}e_\sigma$	0	0	0	0	0	0
$\theta = 90°$	π	0	0	0	0	$-S_R$	e_π	0	0	0	0
$\phi = 90°$	π	0	0	0	0	0	0	0	0	$-S_R$	e_π
			$\frac{1}{4}e_\sigma$		$\frac{3}{4}e_\sigma$		e_π		0		e_π

4	σ	$-\frac{1}{2}S_R$	$\frac{1}{4}e_\sigma$	$\frac{\sqrt{3}}{2}S_R$	$\frac{3}{4}e_\sigma$	0	0	0	0	0	0
$\theta = 90°$	π	0	0	0	0	S_R	e_π	0	0	0	0
$\phi = 180°$	π	0	0	0	0	0	0	S_R	e_π	0	0
			$\frac{1}{4}e_\sigma$		$\frac{3}{4}e_\sigma$		e_π		e_π		0
5	σ	$-\frac{1}{2}S_R$	$\frac{1}{4}e_\sigma$	$-\frac{\sqrt{3}}{2}S_R$	$\frac{3}{4}e_\sigma$	0	0	0	0	0	0
$\theta = 90°$	π	0	0	0	0	$-S_R$	e_π	0	0	0	0
$\phi = 270°$	π	0	0	0	0	0	0	0	0	$+S_R$	e_π
			$\frac{1}{4}e_\sigma$		$\frac{3}{4}e_\sigma$		e_π		0		e_π
6	σ	S_R	e_σ	0	0	0	0	0	0	0	0
$\theta = 180°$	π	0	0	0	0	0	0	0	0	$-S_R$	e_π
$\phi = 0°$	π	0	0	0	0	0	0	S_R	e_π	0	0
			e_σ		0		0		e_π		e_π
Total			$3e_\sigma$		$3e_\sigma$		$4e_\pi$		$4e_\pi$		$4e_\pi$

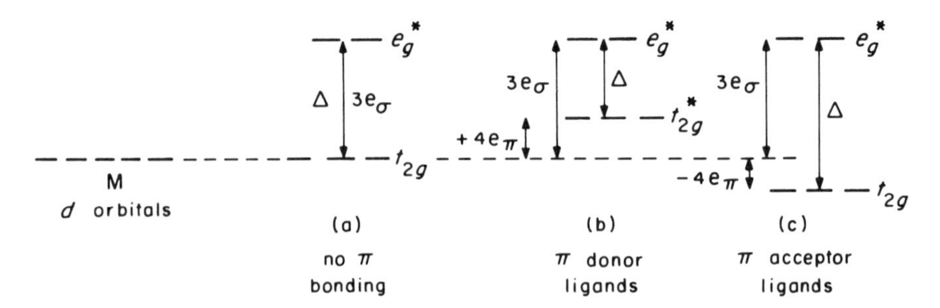

Fig. 8.14 AOM energy-level diagrams for octahedral complexes (a) without π bonding, (b) with π-donor ligands, and (c) with π-acceptor ligands.

but negative for π acceptors (see Figs. 8.7 and 8.14). One of the great advantages of the AOM over LFT is that we can separate σ and π contributions and even π donor or π acceptor effects. Hence, we can understand why CO and C_2H_4 (π acceptors) are much higher in the spectrochemical series than we would expect for their roles as σ donors. Similarly, the anomaly of OH^- being lower in the spectrochemical series (smaller Δ) than H_2O is the result of e_π being positive for OH^- as a π donor.

We can construct an AOM energy-level diagram for a square planar complex ($\mathbf{D}_{4h}$) by combining interactions of ligands 2, 3, 4, and 5 (see Fig. 8.15) from the $\mathbf{O}_h$ case. The σ interaction for $d_{x^2-y^2}$ (b_{1g}) is the same as for $\mathbf{O}_h$, $3e_\sigma$. For d_{z^2} (a_{1g}), the four ligands interact only with the donut, $(4 \times \frac{1}{4})e_\sigma = e_\sigma$. The other d orbitals (e_g and b_{2g}) are σ nonbonding, but d_{xy} (b_{2g}) can form π bonds with all four ligands as in the $\mathbf{O}_h$ case, giving $4e_\pi$. Each of the e_g orbitals (d_{xz} and d_{yz}) can interact with only two ligands, giving $2e_\pi$. A square planar

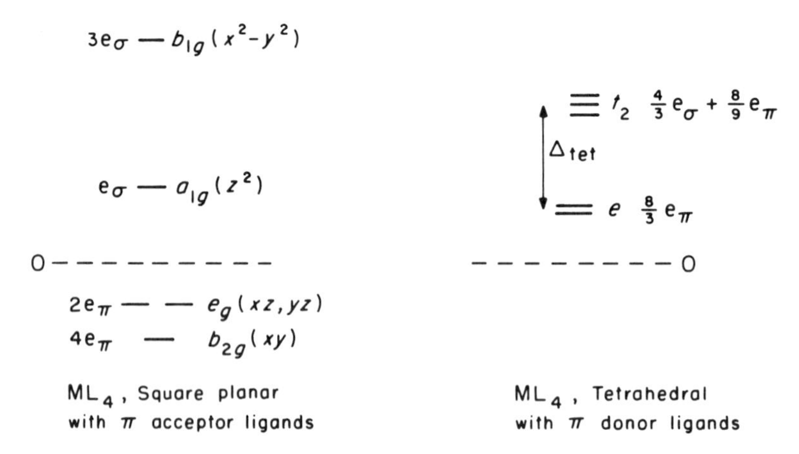

Fig. 8.15 AOM energy-level diagrams for square planar and tetrahedral complexes.

complex such as $[Ni(CN)_4]^{2-}$ with π acceptor ligands involves negative values of e_π (Fig. 8.15).

In the LFT treatment the tetrahedral case is the inverse of the octahedral case (with smaller splitting), and this is the case for the MO or AOM treatment. The metal t_2 orbitals are raised in energy by $\frac{4}{3}e_\sigma$ (see Table 8.7), and considering σ interaction only, the e orbitals are σ nonbonding. However, since we can deal with π_g interaction easily using the AOM, we see an important difference since all d orbitals can form π bonds for the tetrahedral case. In Table 8.7 the angular overlap integrals are worked out for the tetrahedral case. Although one orientation of π bonds is possible for d_{xy} and two orientations are possible for d_{xz} and d_{yz}, the total interaction energy for each orbital of the t_2 set for four ligands is $\frac{8}{9}e_\pi$. Thus, for π donors, the metal t_2 orbitals are raised in energy by $\frac{8}{9}e_\pi$ while the e orbitals are raised by $\frac{8}{3}e_\pi$. The total energy separation Δ_{tet} is $\frac{4}{9}\Delta_{oct}$ [Eq. (8.14)]:

$$\begin{aligned}\Delta_{tet} = t_2 - e &= \tfrac{4}{3}e_\sigma + \tfrac{8}{9}e_\pi - \tfrac{8}{3}e_\pi \\ &= \tfrac{12}{9}e_\sigma - \tfrac{16}{9}e_\pi = \tfrac{4}{9}(3e_\sigma - 4e_\pi).\end{aligned} \tag{8.15}$$

We can use the AOM diagrams for any symmetry as we have used the LFT diagrams—for calculating AOM stabilization energies and for predicting (or rationalizing) geometries.

The angular overlap d orbital σ and π interaction energies are given in Table 8.8 for various geometries. The σ AOM stabilization energies (AOMSEs) (Table 8.9) are given by the sum of the stabilization energies of all of the bonding electrons, from ligands in the e_g orbitals ($3e_\sigma$ for each of four electrons) for octahedral complexes and t_2 orbitals ($\frac{4}{3}e_\sigma$ for each of six electrons) for tetrahedral complexes, minus the sum of the destabilization energies of the antibonding d electrons. The number of antibonding electrons is n.

$$\text{AOMSE (oct)} = 12e_\sigma - 3(n_{e_g^*})e_\sigma \tag{8.16}$$

$$\text{AOMSE (tet)} = 8e_\sigma - \tfrac{4}{3}n_{t_2^*}e_\sigma \tag{8.17}$$

We are considering that the destabilization of the antibonding orbitals is the same in magnitude as the stabilization of the bonding orbitals, so the net stabilization energy is zero if the antibonding orbitals are filled. Hence, we can get the net stabilization energy from the number of electron holes h_i in the antibonding orbitals:

$$\sum_i(\sigma) = \sum h_i\beta_\sigma S^2 = \sum h_i e_\sigma S_A^2. \tag{8.18}$$

Thus, the high-spin octahedral AOMSE is $12e_\sigma$ ($h_i = 4$) for d^0 to d^3, $9e_\sigma$ ($h_i = 3$) for d^4, $6e_\sigma$ ($h_i = 2$) for d^5 to d^8, $3e_\sigma$ ($h_i = 1$) for d^9, and zero ($h_i = 0$) for

TABLE 8.7
Angular Overlap Integral and Interaction Energies of the d Orbitals for σ and π Bonding by Four Ligands in
Tetrahedral Complex

	Ligand 1	Ligand 2	Ligand 3	Ligand 4
θ	54.74°	54.74°	125.26°	125.26°
$\sin\theta$	0.817	0.817	0.817	$0.817 = \sqrt{\frac{2}{3}}$
$\cos\theta$	0.577	0.577	-0.577	$-0.577 = -\frac{1}{\sqrt{3}}$
$\sin 2\theta$	0.943	0.943	-0.943	-0.943
$\cos 2\theta$	$-\frac{1}{3}$	$-\frac{1}{3}$	$-\frac{1}{3}$	$-\frac{1}{3}$
ϕ	45°	225°	135°	315°
$\sin\phi$	0.707	-0.707	0.707	-0.707
$\cos\phi$	0.707	-0.707	-0.707	$0.707 = \frac{1}{\sqrt{2}}$
$\sin 2\phi$	1	1	-1	-1
$\cos 2\phi$	0	0	0	0

		Overlap	xe_λ	Overlap	xe_λ	Overlap	xe_λ	Overlap	xe_λ	Total
d_{z^2}	σ	0	0	0	0	0	0	0	0	
	π	0	0	0	0	0	0	0	0	
	π	$\sqrt{\frac{2}{3}}$	$\frac{2}{3}e_\pi$	$\sqrt{\frac{2}{3}}$	$\frac{2}{3}e_\pi$	$\sqrt{\frac{2}{3}}$	$\frac{2}{3}e_\pi$	$\sqrt{\frac{2}{3}}$	$\frac{2}{3}e_\pi$	$\frac{8}{3}e_\pi$
$d_{x^2-y^2}$	σ	0	0	0	0	0	0	0	0	
	π	$-\sqrt{\frac{2}{3}}$	$\frac{2}{3}e_\pi$	$-\sqrt{\frac{2}{3}}$	$\frac{2}{3}e_\pi$	$\sqrt{\frac{2}{3}}$	$\frac{2}{3}e_\pi$	$\sqrt{\frac{2}{3}}$	$\frac{2}{3}e_\pi$	$\frac{8}{3}e_\pi$
	π	0	0	0	0	0	0	0	0	
d_{xy}	σ	$\frac{1}{\sqrt{3}}$	$\frac{1}{3}e_\sigma$	$\frac{1}{\sqrt{3}}$	$\frac{1}{3}e_\sigma$	$-\frac{1}{\sqrt{3}}$	$\frac{1}{3}e_\sigma$	$-\frac{1}{\sqrt{3}}$	$\frac{1}{3}e_\sigma$	$\frac{4}{3}e$
	π	0	0	0	0	0	0	0	0	
	π	$\frac{\sqrt{2}}{3}$	$\frac{2}{9}e_\pi$	$\frac{\sqrt{2}}{3}$	$\frac{2}{9}e_\pi$	$\frac{\sqrt{2}}{3}$	$\frac{2}{9}e_\pi$	$\frac{\sqrt{2}}{3}$	$\frac{2}{9}e_\pi$	$\frac{8}{9}e$
d_{xz}	σ	$\frac{1}{\sqrt{3}}$	$\frac{1}{3}e_\sigma$	$-\frac{1}{\sqrt{3}}$	$\frac{1}{3}e_\sigma$	$\frac{1}{\sqrt{3}}$	$\frac{1}{3}e_\sigma$	$-\frac{1}{\sqrt{3}}$	$\frac{1}{3}e_\sigma$	$\frac{4}{3}e_\sigma$
	π	-0.408	$0.166e_\pi$	0.408	$0.166e_\pi$	0.408	$0.166e_\pi$	-0.408	$0.166e_\pi$	$\frac{8}{9}e$
	π	-0.236	$0.056e_\pi$	0.236	$0.056e_\pi$	0.236	$0.056e_\pi$	-0.236	$0.056e_\pi$	
d_{yz}	σ	$\frac{1}{\sqrt{3}}$	$\frac{1}{3}e_\sigma$	$-\frac{1}{\sqrt{3}}$	$\frac{1}{3}e_\sigma$	$\frac{1}{\sqrt{3}}$	$\frac{1}{3}e_\sigma$	$-\frac{1}{\sqrt{3}}$	$\frac{1}{3}e_\sigma$	$\frac{4}{3}e$
	π	0.408	$0.166e_\pi$	-0.408	$0.166e_\pi$	0.408	$0.166e_\pi$	-0.408	$0.166e_\pi$	$\frac{8}{9}e$
	π	-0.236	$0.056e_\pi$	0.236	$0.056e_\pi$	-0.236	$0.056e_\pi$	0.236	$0.056e_\pi$	

TABLE 8.8
Angular Overlap d Orbital σ and π Interaction Energies for Various Geometries[a,b,c]

		z^2		x^2-y^2		xy		xz		yz	
	Geometry	e_σ	e_π	e_σ	e_π	e_σ	e_π	e_σ	e_π	e_σ	e_π
MY	linear, $\mathbf{C}_{\infty v}$	1	0	0	0	0	0	0	1	0	1
MY_2	linear, $\mathbf{D}_{\infty h}$	2	0	0	0	0	0	0	2	0	2
	bent (90° angle), $\mathbf{C}_{2v}$	$\frac{1}{2}$	0	$\frac{3}{2}$	0	0	2	0	1	0	1
MY_4	tetrahedron, $\mathbf{T}_d$	0	$\frac{8}{3}$	0	$\frac{8}{3}$	$\frac{4}{3}$	$\frac{8}{9}$	$\frac{4}{3}$	$\frac{8}{9}$	$\frac{4}{3}$	$\frac{8}{9}$
	square plane, $\mathbf{D}_{4h}$	1	0	3	0	0	4	0	2	0	2
	cis-divacant, $\mathbf{C}_{2v}$	$\frac{5}{2}$	0	$\frac{3}{2}$	0	0	2	0	3	0	3
MY_5	trigonal bipyramid, $\mathbf{D}_{3h}$	$\frac{11}{4}$	0	$\frac{9}{8}$	$\frac{3}{2}$	$\frac{9}{8}$	$\frac{3}{2}$	0	$\frac{7}{2}$	0	$\frac{7}{2}$
	square pyramid, $\mathbf{C}_{4v}$	2	0	3	0	0	4	0	3	0	3
MY_6	octahedron, $\mathbf{O}_h$	3	0	3	0	0	4	0	4	0	4
MY_8	cube, $\mathbf{O}_h$	0	$\frac{16}{3}$	0	$\frac{16}{3}$	$\frac{8}{3}$	$\frac{16}{9}$	$\frac{8}{3}$	$\frac{16}{9}$	$\frac{8}{3}$	$\frac{16}{9}$
	square antiprism[d], $\mathbf{D}_{4d}$	0	$\frac{16}{3}$	$\frac{4}{3}$	$\frac{32}{9}$	$\frac{4}{3}$	$\frac{32}{9}$	$\frac{8}{3}$	$\frac{37}{9}$	$\frac{8}{3}$	$\frac{37}{9}$
	hexagonal bipyramid, $\mathbf{D}_{6h}$	$\frac{7}{2}$	0	$\frac{9}{4}$	3	$\frac{9}{4}$	3	0	5	0	5
MY_{10}	pentagonal antiprism[e], $\mathbf{D}_{5d}$	$\frac{2}{5}$	$\frac{24}{5}$	$\frac{12}{5}$	$\frac{24}{5}$	$\frac{12}{5}$	$\frac{24}{5}$	$\frac{12}{5}$	$\frac{14}{5}$	$\frac{12}{5}$	$\frac{14}{5}$
MY_{12}	icosahedron, $\mathbf{I}_h$	$\frac{12}{5}$	$\frac{24}{5}$	$\frac{12}{5}$	$\frac{24}{5}$	$\frac{12}{5}$	$\frac{24}{5}$	$\frac{12}{5}$	$\frac{24}{5}$	$\frac{12}{5}$	$\frac{24}{5}$

[a] Adapted from J. K. Burdett, *Molecular Shapes*, Wiley (Interscience), New York, 1980, pp. 148–149.
[b] The coefficients p_σ and p_π are given for the quadratic interaction energy $\varepsilon = p_\sigma e_\sigma + p_\pi e_\pi$.
[c] z is along figure axis except where noted.
[d] MY bond makes angle of 54.7° with z axis.
[e] Ligands in same position as for icosahedron ($\theta = 63.43°$).

d^{10}. The AOMSE for square-planar low-spin d^8 using the equivalent of Eqs. (8.16) and (8.17) is $8e_\sigma - 2(1)e_\sigma = 6e_\sigma$, while it would be only $8e_\sigma - \frac{4}{3}(4)e_\sigma = 2\frac{2}{3}$ for a tetrahedral arrangement [Eq. (8.17)].

The $Cr(CO)_4$ molecule (low-spin d^6, characterized by matrix isolation) has the "sawhorse" structure that can be described as *cis*-divacant octahedral (an octahedron with two *cis* ligands missing). Here the σ stabilization energies are

TABLE 8.9
Angular Overlap Model Stabilization Energies (σ Only) for Octahedral and Tetrahedral Complexes

		Octahedral		Tetrahedral	
Configuration	Examples	Weak field (xe_σ)	Strong field (xe_σ)	Weak field (xe_σ)	Strong field (xe_σ)
d^0	Ca^{2+}, Sc^{3+}	12	12	8	8
d^1	Ti^{3+}	12	12	8	8
d^2	V^{3+}	12	12	8	8
d^3	Cr^{3+}, V^{2+}	12	12	6.67	(8.0)[a]
d^4	Cr^{2+}, Mn^{3+}	9	12	5.33	(8.0)[a]
d^5	Mn^{2+}, Fe^{3+}	6	12	4.0	(6.67)[a]
d^6	Fe^{2+}, Co^{3+}	6	12	4.0	(5.67)[a]
d^7	Co^{2+}	6	9	4.0	4.0
d^8	Ni^{2+}	6	6	2.67	2.67
d^9	Cu^{2+}	3	3	1.33	1.33
d^{10}	Cu^+, Zn^{2+}	0	0	0	0

[a] Cases not realized.

the same ($8e_\sigma$) for the square planar and *cis*-divacant structures (Table 8.10). The metal electrons are in nonbonding orbitals, and the stabilization energies for the four ligand electrons in d_{z^2} and $d_{x^2-y^2}$ are $2e_\sigma + 2(3e_\sigma)$ for the square planar case and $2(\frac{3}{2}e_\sigma) + 2(\frac{5}{2}e_\sigma)$ for the *cis*-divacant case. The e_π interactions are slightly different but generally much smaller than e_σ. Thus far we have considered only the quadratic (S^2) terms since the contributions to the S^2 term are additive and usually dominant. If there are large differences in the S^2 terms, the smaller S^4 terms can be neglected. Although the contributions to the S^4 terms are not additive, once we evaluate the coefficient for e (S_A^2), we can square it to obtain the coefficient for f (S_A^4) [Eq. (8.5)]. The energies of the vacant high-energy orbitals and the stabilization energies including S^4 are given in Table 8.10.

TABLE 8.10
Stabilization energies[a]

Square planar	*cis*-Divacant octahedral
$3e_\sigma - 9f_\sigma$ — $x^2 - y^2$	$\frac{5}{2}e_\sigma - \frac{25}{4}f_\sigma$ — z^2
$e_\sigma - f_\sigma$ — z^2	$\frac{3}{2}e_\sigma - \frac{9}{4}f_\sigma$ — $x^2 - y^2$
$8e_\sigma - 20f_\sigma$	Stabilization for low-spin d^6 $8e_\sigma - 17f_\sigma$

[a] These orbitals accommodate four ligand electrons. The six d electrons are in nonbonding orbitals.

Since e_σ and f_σ are both negative (corresponding to the stabilization of the bonding orbitals), the square planar case is less favorable because of the larger negative f_σ term.

EXAMPLE 8.5 Determine the angular overlap d orbital σ- and π-interaction energies for ML_3 ($\mathbf{D}_{3h}$, trigonal planar with one ligand along x).

SOLUTION In $\mathbf{D}_{3h}$ symmetry, the d orbitals belong to a_1' (d_{z^2}), e' (d_{xy}, $d_{x^2-y^2}$), and e'' (d_{xz}, d_{yz}) representations. For d_{z^2}, there are three with $\theta = 90°$ (using equations from Table 8.5 to calculate overlap integrals that are squared to give pe_σ or pe_π), giving $3\frac{1}{4}e_\sigma$. For $d_{x^2-y^2}$, the contributions are $\frac{3}{4}e_\sigma$ (along x) $+\frac{3}{16}e_\sigma$ ($\phi = 120°$) $+\frac{3}{16}e_\sigma$ ($\phi = 240°$) $= \frac{9}{8}e_\sigma$. For d_{xy}, the contributions are 0 (along x) $+\frac{9}{16}e_\sigma + \frac{9}{16}e_\sigma = \frac{9}{8}e_\sigma$. Actually, we need to calculate $S_A^2\, e_\sigma$ for only one orbital of a degenerate set. For d_{xz} and d_{yz}, there is no σ interation (the xy plane is a nodal plane). There is no π interaction of ligands in the xy plane with d_{z^2}. The π interaction for $d_{x^2-y^2}$ is zero along x and $\frac{3}{4}e_\pi$ for the ligands at 120 and 240°, giving $1.5e_\pi$. For d_{xy} (and d_{xz}), the contributions are [1 (along x) $+\frac{1}{4}$ (at 120°) $+\frac{1}{4}$ (at 240°)]$e_\pi = 1.5e_\pi$. For d_{yz}, we get [0 along x + $\frac{3}{4}$ (at 120°) $+\frac{3}{4}$ (at 240°)]$e_\pi = 1.5e_\pi$.

EXAMPLE 8.6 Determine the angular overlap d orbital σ- and π-interaction energies of ML_5 ($\mathbf{D}_{3h}$, trigonal bipyramidal) from the results for trigonal planar ML_3 and the octahedral ML_6.

SOLUTION The contributions of each ligand for each orbital for the trigonal planar complex and two ligands along z (from the octahedral case) are additive.

	a_1'		e'				e''			
	z^2		x^2-y^2		xy		xz		yz	
	e_σ	e_π	e_σ	e_π	e_σ	e_π	e_σ	e_π	e_σ	e_π
ML_3, $\mathbf{D}_{3h}$	$\frac{3}{4}$	0	$\frac{9}{8}$	$\frac{3}{2}$	$\frac{9}{8}$	$\frac{3}{2}$	0	$\frac{3}{2}$	0	$\frac{3}{2}$
2L along z	2	0	0	0	0	0	0	2	0	2
Sum										
ML_5, $\mathbf{D}_{3h}$	2.75	0	$\frac{9}{8}$	1.5	$\frac{9}{8}$	1.5	0	3.5	0	3.5

Additional Reading

C. J. Ballhausen, *Introduction to Ligand Field Theory*, McGraw-Hill, New York, 1962. An excellent early source.

J. K. Burdett, *Molecular Shapes: Theoretical Models of Inorganic Stereochemistry*, Wiley (Interscience), New York, 1980. Good treatment of AOM, MO, and LFT.

B. Douglas, D. H. McDaniel, and J. J. Alexander, *Concepts and Models of Inorganic Chemistry*, 2nd ed., Wiley, New York, 1983. The text covers the background assumed in this book.
B. N. Figgis, *Introduction to Ligand Fields*, Wiley, New York, 1966.
M. Gerloch and R. G. Woolley, The Empirical and Theoretical Status of the Angular Overlap Model, *Prog. Inorg. Chem.* 1984, **31**, 371. This is a good review of the AOM as a ligand field model. The authors would like the molecular orbital version of AOM, used structurally in this chapter, to be kept separate.
H. L. Schlafer aand G. Glieman, *Basic Principles of Ligand Field Theory*, Wiley (Interscience), New York, 1969.

Problems

8.1 For what d^n configuration could maximum values of ligand field stabilization energy be achieved for octahedral and for tetrahedral complexes? Are these situations actually encountered?

8.2 If the tetragonal distortion of octahedral complexes is not sufficiently great so as to bring about spin pairing, as for square planar d^8 complexes, what d^n configuration is most favorable?

8.3 What kind of Jahn–Teller distortion of a tetrahedral ML_4 complex would be expected for a d^9 ion?

8.4 Calculate the energies of the d orbitals from the Krishnamurthy–Schaap treatment for
(a) a cube, ML_8 ($\mathbf{O}_h$);
(b) a square pyramid, ML_5 ($\mathbf{C}_{4v}$);
(c) a pentagonal pyramid, ML_6 ($\mathbf{C}_{5v}$).

8.5 Calculate the relative energies of the d orbitals for the following complexes, assuming $Dq(\text{Y}) = 0.70\ Dq(\text{X})$. Let z be the unique axis.
(a) MX_6; (b) MY_6; (c) MX_5Y; (d) *trans*-$[MX_4Y_2]$; (e) *cis*-$[MX_4Y_2]$; (f) *fac*-$[MX_3Y_3]$.

8.6 Give an MO description for σ and π bonding in a trigonal bipyramidal complex, ML_5. Obtain the LCAOs, sketch the bonding and nonbonding MOs, and give a qualitative energy-level diagram.

8.7 Give an MO description for σ and π bonding in a trigonal prismatic complex, ML_6. Obtain the LCAOs, sketch the bonding and antibonding MOs, and give a qualitative energy-level diagram.

8.8 Give an AOM description of bonding using p orbitals for XeF_4. Account for the geometry and bond strength and give an energy-level diagram.

8.9 Give an AOM description of bonding using p orbitals for PF_5. Account for the geometry and bond strength and give an energy-level diagram.

8.10 Give an AOM description of SF_4 without using d orbitals.

8.11 From the overlap integrals and interaction energies for the d orbitals in a trigonal bipyramidal complex, ML_5 (Example 8.6), determine which high-spin d^n configuration would give maximum AOMSE.

8.12 Calculate the total σ and π d-orbital AOMSE for square planar $[Ni(CN)_4]^{2-}$.

8.13 Consider a pyramidal ML_3 complex with ligands along x, y, and z (a facial trivacant octahedron). What d^n configuration would give the maximum AOMSE?

8.14 Calculate the d-orbital σ and π interaction energies for a compressed tetrahedron ($\theta = 45°$). For what high-spin d^1 to d^5 configuration would this distortion be favored relative to a regular tetrahedron? What are the major changes in σ and π interactions for the orbitals?

9

Electronic Spectroscopy

Spectra are treated in the framework of ligand field theory. Selection rules are considered for electric-dipole and magnetic-dipole transitions and for optical activity. Electronic energy-level diagrams are generated for octahedral and lower symmetries using descending symmetry and spin factoring. Spectra of octahedral and tetrahedral complexes of various d^n configurations are treated briefly. The spectra of complexes of lower symmetry are discussed in terms of ligand field theory and angular overlap parameters. Double groups are applied to cases of spin–orbit coupling.

9.1 Electric-Dipole Transitions

9.1.1 ABSORPTION INTENSITY FROM EXPERIMENT AND THEORY

The absorption spectra of gaseous substances often consist of sharp lines. In such cases, the peak height ε_{max}, the *molar absorptivity*, is a direct indication of the intensity of the transition and reflects the probability of the corresponding electronic transition. The lines are frequently broadened to give band spectra because the electronic transitions are coupled to vibrational transitions (see Section 9.1.2). In solution or in the liquid or solid phase, broad bands are usually observed. Each broad band consists of an envelope of transitions from the various occupied vibrational levels of the electronic ground state to various vibrational levels of the electronic excited state. The significant measure of the intensity of the electronic transitions, and hence the probability of the electronic transition, is not ε_{max} but rather the area under the band—the integrated intensity I.

$$I = \int \varepsilon \, d\nu. \tag{9.1}$$

A useful theoretical measure of intensity is the *oscillator strength*,

$$f = 0.102 \frac{mc^2}{N\pi e^2} \int \varepsilon \, d\nu = 4.315 \times 10^{-9} \int \varepsilon \, d\nu, \tag{9.2}$$

where e and m are the charge and mass of an electron, respectively, c is the velocity of light, and N is Avogadro's number. The oscillator strength can be calculated from $\varepsilon_{\max}$ as an approximation by

$$f \cong 4.6 \times 10^{-9} \, \varepsilon_{\max} \Delta\nu_{1/2} \tag{9.3}$$

where $\Delta\nu_{1/2}$ is the "half-width," the width at $\varepsilon_{\max}/2$.

The oscillator strength is proportional to the *dipole strength* D, which is the square of the *transition-moment integral*

$$f \propto D = \left| \int_{-\infty}^{+\infty} \psi_{\mathrm{e}}^{\mathrm{gr}} \mu_{\mathrm{e}} \Psi_{\mathrm{e}}^{\mathrm{ex}} \, d\tau \right|^2 \tag{9.4}$$

where $\Psi_{\mathrm{e}}^{\mathrm{gr}}$ and $\Psi_{\mathrm{e}}^{\mathrm{ex}}$ are the electronic wave functions for the ground and excited states, respectively, and μ_{e} is the *electric-dipole moment operator*. This applies for *electric-dipole transitions* (with which we will be concerned primarily). The integral can be regarded as corresponding roughly to charge displacement during the transition, resulting in a change in electric-dipole moment. For a single atom, the dipole moment is the sum of the products of the charge on the electron times the distance of the electron from the nucleus. The summation for all electrons and all nuclei in a molecule is the electric-dipole moment vector $\boldsymbol{\mu}_{\mathrm{e}}$. The dipole moment is defined by $\boldsymbol{\mu}_{\mathrm{e}} = \sum_i q_i \mathbf{r}_1$, where q_i is the ith charge and $\mathbf{r}_i$ is the vector from the origin to the ith charge. If the total charge is zero (i.e., $\sum_i q_i = 0$), then $\boldsymbol{\mu}_{\mathrm{e}}$ is independent of the choice of the origin. If the system is not electrically neutral, $\boldsymbol{\mu}_{\mathrm{e}}$ does depend upon position of the origin. In symmetry considerations, the origin should be the center of mass.

9.1.2 SELECTION RULES FOR ELECTRIC-DIPOLE TRANSITIONS

In the introductory section we examined just the part of the transition-moment integral dealing with orbital wave functions [Eq. (9.4)]. The transition-moment integral also includes a Franck–Condon factor dependent on the nuclear coordinates and an integral dealing with the spin wave functions:

$$M = \underbrace{\int \Psi_v^{gr}\Psi_v^{ex}\,d\tau_n}_{\text{Frank–Condon factor}} \underbrace{\int \Psi_e^{gr}\,\mu_e\,\Psi_e^{ex}\,d\tau_e}_{\text{orbital wave functions}} \underbrace{\int \Psi_s^{gr}\,\Psi_s^{ex}\,d\tau_s}_{\text{spin wave functions}}. \tag{9.5}$$

The operator μ_e does not operate on the spin wave functions or the Franck–Condon factor. M is zero if any one of the integrals is zero. If M is zero, then so are f and ε_{max}. The transition is *forbidden*. As we shall see later, transitions that are forbidden, in terms of the approximations we use, sometimes occur, but with low probability and hence low intensity.

Symmetry Selection Rule. Let us focus attention again on the integral of the orbital wave functions as the basis of the orbital or symmetry selection rule. Even without good wave functions we can use symmetry to determine whether the integral is nonzero and, therefore, corresponds to an *orbitally* (or *symmetry*) *allowed transition.* Since the integral of an odd function over all space is zero, the orbital wave function product in Eq. (9.5) for an allowed transition must be an even function. For electric-dipole transitions, the operator μ_e has components along x, y, and z, and these components (μ_x, μ_y, and μ_z) belong to the same corresponding representations as x, y, and z. In effect, we examine three integrals,

$$\begin{aligned} &\int \Psi_e^{gr}\,\mu_x\,\Psi_e^{ex}\,d\tau_e, \\ &\int \Psi_e^{gr}\,\mu_y\,\Psi_e^{ex}\,d\tau_e, \\ &\int \Psi_e^{gr}\,\mu_z\,\Psi_e^{ex}\,d\tau_e. \end{aligned} \tag{9.6}$$

Of course, in some point groups, such as $\mathbf{D}_{4h}$, the operators belong to two representations (E_u and A_{2u}), and in some they belong to only one representation (T_{1u} for $\mathbf{O}_h$). A transition is allowed if any one, or more, of the integrals is nonzero. The transition-moment operators for a centrosymmetric molecule belong to u representations (the same as vectors along x, y, and z). Since $g \times g = g$ or $u \times u = g$ and μ_e is u, the product of the representations for the electronic ground and excited state wave functions (this product is known as the *symmetry of the transition*) must be u also. This is the basis for the *Laporte*, *symmetry*, *orbital*, or *parity* selection rule for electric-dipole transitions, $\Gamma_{gr} \times \Gamma_{ex} = u$. In terms of atomic orbitals this means $\Delta l \neq 0$ or 2, but $\Delta l = 1$. Thus, $s \rightarrow s$, $p \rightarrow p$, $s \rightarrow d$, and $d \rightarrow d$, etc., transitions are forbidden, but $s \rightarrow p$, $p \rightarrow d$, $d \rightarrow f$, etc., transitions are allowed.

In more general terms we see that the symmetry selection rule is more restrictive than $\Delta l = 1$, and it applies to noncentrosymmetric groups as well. The integral will be nonzero only if the integrand is *totally symmetric*, that is, it must belong to the symmetry species A_{1g}, A_1, etc. Thus, the direct product of the symmetry of the transition $(\Gamma_{\Psi gr} \times \Gamma_{\Psi ex})$ times μ_x, μ_y, or μ_z must be A_{1g}, A_1, etc., or, if the product is a reducible representation, it must contain A_{1g}, A_1, etc. In terms of symmetry, the integrand must be unchanged by any of the symmetry operations of the group.

EXAMPLE 9.1 The electronic ground states are $^1A_1$ for $[Co(en)_3]^{3+}$ (d^6, $\mathbf{D}_3$) and $^4A_2$ for $[Cr(en)_3]^{3+}$ (d^3, $\mathbf{D}_3$). Are the transitions to the A_1 excited states (of the same spin multiplicity as the ground state in each case) allowed for these cases, and if so, for which μ components?

SOLUTION $[Co(en)_3]^{3+}$ $A_1 \times A_1 = A_1$, (transition symmetry), $[Cr(en)_3]^{3+}$ $A_2 \times A_1 = A_2$, (μ_x, μ_y) belong to E and μ_z belongs to A_2.

$A_1 \times A_2 = A_2$
$A_1 \times E = E$
Therefore for $[Co(en)_3]^{3+}$, $A_1 \rightarrow A_1$ is forbidden.

$A_2 \times A_2 = A_1$
$A_2 \times E = E$
Therefore for $[Cr(en)_3]^{3+}$, $A_2 \rightarrow A_1$ is allowed for μ_z.

EXAMPLE 9.2 The electronic ground states are $^1A_{1g}$ for $[Co(NH_3)_6]^{3+}$ and $^4A_{2g}$ for $[Cr(NH_3)_6]^{3+}$ (both $\mathbf{O}_h$). Are the transitions to T_{1g} (of the same multiplicity as the ground state in each case) electric-dipole allowed?

SOLUTION Even without taking the direct products of the representations, we see that for the centrosymmetric $\mathbf{O}_h$ group, the ground and excited states are g (these are the $d \rightarrow d$ transitions) and the operators are u, so $g \times u \times g = u$, and the transitions are symmetry- or parity-forbidden. The transitions are observed with relatively low intensity, as we shall see.

Spin Selection Rule. The integral of the spin wave functions determines the spin selection rule. Wave functions of different spin multiplicity are orthogonal (See Section 9.9, which deals with the spin quantum number), so the integral is zero unless the ground and excited states have the same spin. Transitions that involve a change in the number of unpaired electrons are *spin-forbidden.* A spin-allowed electronic transition involves promotion of an electron without change in its spin. For a singlet to triplet transition, we go from two electrons with opposite spins to two with the same spin occupying different orbitals in the excited state.

Forbidden Bands. The selection rules apply strictly to the idealized models which they describe. As we saw for the case of $[Cr(en)_3]^{3+}$ (Example 9.1), the

$A_2 \rightarrow A_1$ transition is symmetry-allowed for the noncentrosymmetric point group, even though this is a $d \rightarrow d$ transition. Here the (p_x, p_y) orbitals are E, as are (d_{xz}, d_{yz}) and $(d_{x^2-y^2}, d_{z^2})$, and there can be d–p mixing. The d orbitals can acquire some p character in a noncentrosymmetric point group, and the selection rule prohibiting $d \rightarrow d$ $(g \rightarrow g)$ transitions is not strictly applicable. The probability of the transition, and the intensity, depend upon the extent of orbital mixing. Actually, for a noncentrosymmetric system, the true ψ_e cannot have a definite parity and must be some mixture of orbitals of different parity.

For centrosymmetric point groups, d–p (i.e., g–u) orbital mixing is prohibited for the orbital wave functions in Eq. (9.5). However, the symmetry selection rule can be "relaxed" through coupling between vibrational and orbital wave functions. This can be represented by rearranging Eq. (9.5) to give

$$M = \int \Psi_e^{gr} \Psi_v^{gr} \mu_e \Psi_e^{ex} \Psi_v^{ex} \, d\tau_{en} \int \Psi_s^{gr} \Psi_s^{ex} \, d\tau_s. \tag{9.7}$$

The first integral corresponds to a transition from an electronic ground state in a particular vibrational level (perhaps 0 or 1) to an electronic excited state in a different vibrational level (perhaps 1 or 0). Because of the coupling of a *vib*rational transition (here $0 \rightarrow 1$ or $1 \rightarrow 0$) with an elect*ronic* transition, this is known as a *vibronic* transition. For this integral to have a nonzero value, the product must be totally symmetric. The product is more simple than it appears. For the $0 \rightarrow 1$ vibrational transition, v_0 is A_{1g} (or A_1, etc., for the particular point group), so we need examine only the vibrationally excited state. A transition is vibronically allowed (from v_0) if the product of the representation of the transition symmetry and μ_e (or any of its components) is the same as the representation for one of the vibrational modes.[1]

$$(\Gamma_{\Psi_e^{gr}} \times \Gamma_{\Psi_e^{ex}}) \times \Gamma_{\mu_e} = \Gamma_i, \tag{9.8}$$

transition symmetry — any μ_e component

$$\Gamma_i \times \Gamma_i = A_{1g} \text{ (or the product contains } A_{1g}) \tag{9.9}$$

vibrational mode

Thus, we take the product of the representation for the transition symmetry with Γ_{μ_e} (or its components). If there is a vibration belonging to any representation obtained from this product (or these products), then the

[1] We examined the symmetry of molecular vibrations briefly in Chapter 1. These modes are derived in Chapter 10. For the present applications of symmetry rules, the representations (symmetry species) will be identified for the vibrations of a complex in a particular point group.

transition is vibronically allowed. For $d \to d$ transitions, the transition symmetry is g, and μ_e (or any of its components) is u. Hence, only vibrations of odd parity are effective. The beautiful varied colors of transition metal complexes are usually the result of vibronic $d \to d$ transitions. Often this coupling is referred to as the *vibronic mechanism* or the mechanism for relaxation of the selection rule. Actually, the orbital selection rule is not relaxed, as it applies specifically to the $0 \to 0$ pure electronic transition.

Using a simplified physical picture, we can view the coupling with an odd (u) vibration as removing the center of symmetry. That is, an odd vibration causes distortion of the molecule in such a way as to destroy the center of symmetry, producing a complex to which the parity selection rule does not apply strictly.

EXAMPLE 9.3 For $[Co(NH_3)_6]^{3+}$ ($\mathbf{O}_h$) the spin-allowed transitions ${}^1A_{1g} \to {}^1T_{1g}$ and ${}^1A_{1g} \to {}^1T_{2g}$ are symmetry-forbidden. The M–N odd vibrational modes for $\mathbf{O}_h$ are T_{1u} and T_{2u}. Are these transitions vibronically allowed?

SOLUTION The transition symmetries are T_{1g} ($A_{1g} \times T_{1g}$) and T_{2g} ($A_{1g} \times T_{2g}$) and the μ_e components are T_{1u}. The resulting products are

$$T_{1g} \times T_{1u} = A_{1u} + E_u + T_{1u} + T_{2u},$$

$$T_{2g} \times T_{1u} = A_{2u} + E_u + T_{1u} + T_{2u}.$$

Since both products contain the representations of the odd vibrations (T_{1u} and T_{2u}), both are vibronically allowed since $T_{1u} \times T_{1u}$ contains A_{1g}, and $T_{2u} \times T_{2u}$ contain A_{1g}.

The spin-selection rule prohibits a transition involving a change in spin multiplicity. This applies to the Russell–Saunders or L–S coupling scheme, where S is a valid quantum number. The other extreme is j–j coupling which applies to heavier elements. Intermediate coupling schemes, involving some degree of spin–orbit coupling, are common. This kind of coupling involves the mixing of spin states (so spin is not a good quantum number), and the spin selection rule is not strictly applicable. In such cases, spin-forbidden transitions can occur, but usually with very low intensities.

Intensities. The intensities of allowed electric-dipole transitions are generally in the range $\varepsilon \simeq 10^4$–10^5 L cm^{-1} mole^{-1}. Vibronically allowed but symmetry-forbidden transitions usually have $\varepsilon \simeq 10$–10^2. Spin-forbidden transitions, if they appear at all, are usually very weak ($\varepsilon \simeq 10^{-2}$–1). Thus, for large differences in intensities, we can often distinguish among these classes. Small differences in intensities must be interpreted with caution because they often depend on the extent of orbital mixing or spin-state mixing, and these are effects that are not readily predictable.

One feature of the vibronic mechanism is the involvement of various vibrational levels. As noted earlier (see Section 9.9.1), this results in broadening of the bands. For octahedral complexes of transition metals, the absorption bands are quite broad with $\varepsilon_{max} = 50\text{–}100$. If we are considering bands of about the same width, ε_{max} is a reasonable basis for comparing intensities. The absorption bands for tetrahedral complexes, with no center of symmetry, are usually more intense, with $\varepsilon = 100\text{–}200$. Because the vibronically allowed bands are broad, they usually cover the much weaker spin-forbidden bands unless the spin-forbidden bands are well removed from these more intense bands.

For metal complexes, the usually observed electronically allowed transitions are ligand transitions, that is, transitions within the ligand molecules, or charge-transfer transitions (metal-to-ligand or ligand-to-metal). For the first transition series metals, the ligand and charge-transfer bands are usually at higher energy than the $d \rightarrow d$ transitions (also called *ligand field* transitions). If the intense allowed bands extend into the visible region, they are likely to cover one or more of the $d \rightarrow d$ bands.

Table 9.1 gives data for the spin-allowed $d \rightarrow d$ transitions for $[Co(NH_3)_6]^{3+}$, $[Co(en)_3]^{3+}$, $[Cr(NH_3)_6]^{3+}$, and $[Cr(en)_3]^{3+}$ for comparison. We see that the intensities are about twice as great for the $\mathbf{D}_3$ complexes of each metal. The small difference (any difference much less than $10\times$ is small) indicates that the bands are vibronic in all cases with little additional intensity gained from orbital mixing for $[M(en)_3]^{3+}$. Also, we find two symmetrical bands for each complex, with no apparent splitting into the components of T_{1g} and T_{2g} that might be expected for the lower ($\mathbf{D}_3$) symmetry. The lack of apparent splitting and the comparable intensities seem to indicate that $\mathbf{O}_h$ is the *effective symmetry* for these cases. Hence, the bands for $[M(en)_3]^{3+}$ are identified using the $\mathbf{O}_h$ symmetry species. In effect, the intensities and apparent splitting patterns from the absorption spectra reflect the essentially octahedral $[M(N)_6]^{3+}$ chromophore. The addition of the ethylene groups to complete the chelate rings has little effect here. In general, we use the apparent degree of splitting as the indication of the effective symmetry. We cannot deal with individual components about which we have no information.

TABLE 9.1
Absorption Data for the Spin-Allowed Transitions of Co(III) and Cr(III) Complexes[a]

Complex	${}^1A_{1g} \rightarrow {}^1T_{1g}$	${}^1A_{1g} \rightarrow {}^1T_{2g}$	Complex	${}^4A_{2g} \rightarrow {}^4T_{2g}$	${}^4A_{2g} \rightarrow {}^4T_{1g}$
$[Co(NH_3)_6]^{3+}$	21,200 (56)	29,550 (46)	$[Cr(NH_3)_6]^{3+}$	21,550 (30)	28,500 (27)
$[Co(en)_3]^{3+}$	21,550 (88)	29,600 (78)	$[Cr(en)_3]^{3+}$	21,850 (76)	28,450 (60)

[a] The positions of the bands are given in cm^{-1} with ε_{max} values in parentheses.

9.2 Term Diagrams for O_h and T_d Symmetry

9.2.1 ORGEL DIAGRAMS

Spectral terms, whose derivations were treated in Section 5.1.2, represent electronic energy states. There is not, in general, a one-to-one correspondence between microstates and spectral terms, since each spectral term represents an array of microstates. The 10 microstates for the d^1 configuration are represented by the ^2D term. In an octahedral field the d orbitals split to give e_g and t_{2g} sets, and the ^2D term splits to give ^2E_g and $^2T_{2g}$. The splitting of the d orbitals is the familiar ligand field splitting (Fig. 8.1). In the d^1 case for an infinitely strong field, there are just two discrete configurations, t_{2g}^1 and e_g^1. There are six microstates for t_{2g}^1, corresponding to $^2T_{2g}$, and four microstates for e_g^1, corresponding to ^2E_g. For the one-electron case, there is a direct correspondence between the electron configuration (d^1, t_{2g}^1, or e_g^1) and the energy state because d^1 gives only ^2D and t_{2g}^1 and e_g^1 give only $^2T_{2g}$ and ^2E_g, respectively. The splitting of the energy states is obtained by plotting the energies of the spectral terms (only ^2D here), as obtained from atomic spectroscopy, on one vertical axis (zero field or $Dq = 0$) and the energies of the infinitely strong field configurations along the other vertical axis (Fig. 9.1). The relative energies of these configurations can be expressed as the ligand field stabilization energy (LFSE), $-4Dq$ for t_{2g}^1 and $+6Dq$ for e_g^1. As we impose a gradually increasing $\mathbf{O}_h$ field on the ^2D term, the energies of ^2E_g and $^2T_{2g}$ diverge. From the other extreme, as we relax the infinitely strong field the energy separation of the states arising from e_g^1 and t_{2g}^1 decreases. Obviously, since the same states arise from the two limiting cases, they must correlate. If we draw lines, called tie lines, connecting the states, then we obtain a

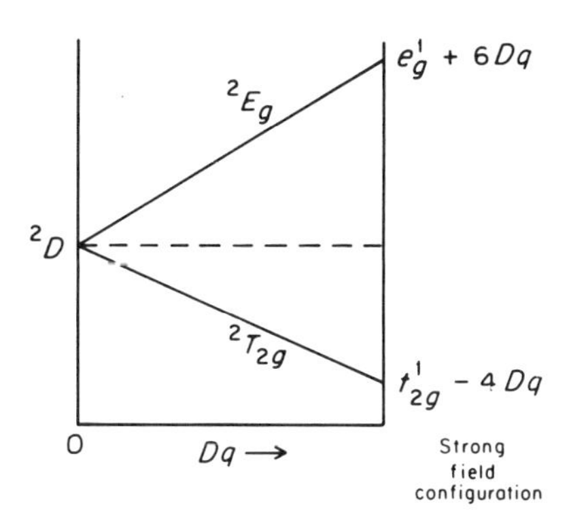

Fig. 9.1 Correlation diagram for the d^1 configuration in an octahedral field.

correlation diagram for d^1, or an energy-level splitting diagram. The energy separation between $^2T_{2g}$ and ^2E_g is $10Dq$.

For the high-spin d^6 case, the only quintet is ^5D, which splits in an $\mathbf{O}_h$ field to give ^5E_g and $^5T_{2g}$. The high-spin strong field configuration $t_{2g}^4e_g^2$ gives only $^5T_{2g}$ and high-spin $t_{2g}^3e_g^3$ gives only ^5E_g. These correspond exactly to the d^1 case, except these are quintets. The same correlation diagram can be used. There is only one transition possible for the high-spin d^6 case because the five electrons of one spin set are fixed and only the one of opposite spin is promoted.

We know that the d^9 configuration gives only a ^2D term, the one-hole case. The strong field configurations and the corresponding energy states are $t_{2g}^6e_g^3, {}^2E_g$ (lower energy) and $t_{2g}^5e_g^4, {}^2T_{2g}$. These correspond to those for d^1, except the energies are inverted. The correlation diagram is obtained by projecting the lines beyond the zero field axis (Fig. 9.2). Similarly, d^4 is the hole counterpart of d^6, and the only high-spin term is ^5D; the $\mathbf{O}_h$ states are ^5E_g (from $t_{2g}^3e_g^1$) and $^5T_{2g}$ (from $t_{2g}^2e_g^2$). Since ^5E_g is lower in energy, the splitting is opposite that of the d^6 case and the same as for d^9. All four high-spin cases are represented in the splitting diagram in Fig. 9.2. In this diagram, known as an Orgel diagram, the strong field configurations are omitted. It is obtained from correlation diagrams such as in Fig. 9.1.

We have seen that the orbital splitting diagram for $\mathbf{T}_d$ (Fig. 8.1) is opposite that for $\mathbf{O}_h$. For d^1, e^1 (^2E) is lower in energy than t_2^1 ($^2T_2$). The correlation diagram corresponds to the left side of Fig. 9.2. Similarly, we find that d^9 and high-spin d^4 are represented by the right side for $\mathbf{T}_d$ and high-spin d^6 by the left side. Since g and u have no meaning for T_d, they are omitted from the figure and should be added for the $\mathbf{O}_h$ case.

The case for d^2 in an octahedral field is handled similarly except that here there are two terms of highest spin multiplicity, ^3F and ^3P, and three strong

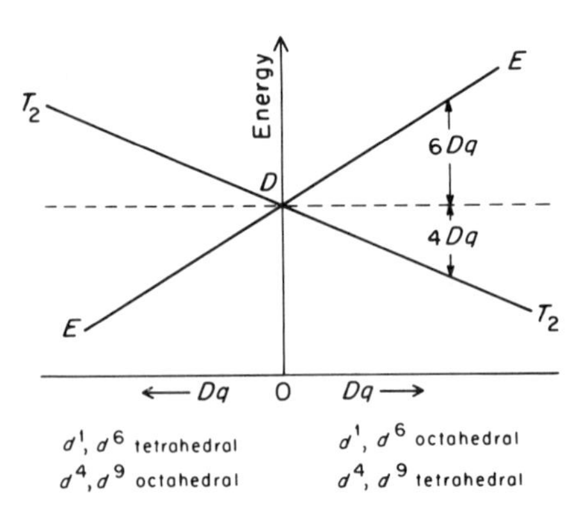

Fig. 9.2 Orgel term-splitting diagram for d^1, d^9, and high-spin d^4 and d^6 configurations.

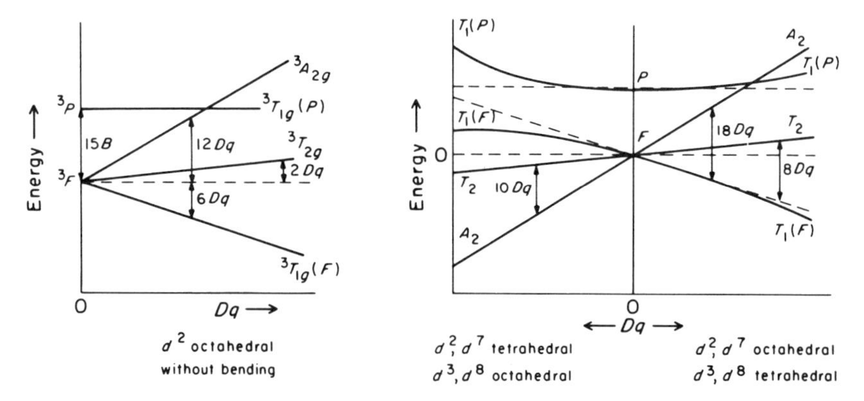

Fig. 9.3 Orgel term-splitting diagrams for d^2, d^3, d^7 (high-spin), and d^8 cases in octahedral and tetrahedral fields. (Adapted with permission from L. E. Orgel, *J. Chem. Phys.* 1955, **23,** 1004.)

field configurations increasing in energy in the order $t_{2g}^2 < t_{2g}^1 e_g^1 < e_g^2$. The ^3P term becomes $^3T_{1g}$ in an octahedral field, with no splitting. The ^3F term splits to give $^3T_{1g}$, $^3T_{2g}$, and $^3A_{2g}$ (see Section 5.2). We obtain the order of energies for these states from the correlation diagram. To construct the diagram we must sort out the singlets and triplets arising from the strong field configurations. One method is essentially trial and error. We derive the states without spin designations from each strong field configuration. Then we connect the states on the left derived from all spectral terms (these states have the same spin multiplicities as the spectral terms from which they are derived) with those on the right, avoiding the crossing of tie lines connecting states of the same symmetry species and spin multiplicity. This is the *noncrossing rule,* which prohibits the crossing of tie lines for states of the same labels because of configuration interaction.[2] We first connect those states that occur only as singlets or triplets and then others. The singlets and triplets can be sorted out more systematically using the *method of descending symmetry*, where we select a lower symmetry subgroup that removes degeneracies or *spin-factoring* (see Section 9.4).

For the present, let us just take the results for the d^2 case as shown in the Orgel diagram (Fig. 9.3). As we have seen for the d^1 and d^6 cases, the diagram applies also to the high-spin d^7 (d^{5+2}) case; the reverse splitting applies for d^3 (two holes in the half-filled set) and for d^8 (two holes). Also, the diagram applies to tetrahedral complexes as shown. For the general diagram the spin multiplicities and g and u subscripts are dropped. In the diagram for d^3 and d^8 ($\mathbf{O}_h$) and d^2 and d^7 ($\mathbf{T}_d$), we find that the line for T_1 of F parentage, $T_1(F)$,

[2] R. E. Jotham, Why Do Energy Levels Repel One Another? *J. Chem. Educ.* 1975, **52,** 377.

increases in energy with increasing Dq while that for $T_1(P)$ does not change with Dq. The extrapolation of the lines would cross; but crossing is prohibited. As Dq increases, they bend away from one another because of configuration interaction. On the right side of the diagram, the lines for the two T_1 states also bend away from one another at high field strength. In this diagram we consider only the states with the same spin multiplicity as the ground state, ignoring, for the present, states of higher multiplicities.

From the correlation diagram (which we will derive in Section 9.4.4), we find the following correlations for triplets from d^2.

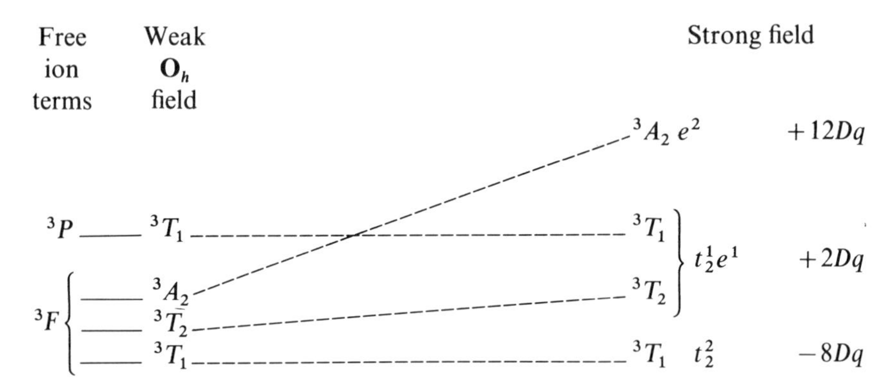

Where a state occurs only once ($^3T_2$ and $^3A_2$) we can obtain the relative energy of the state from the energy of the configuration from which it is derived, for example, $+2Dq$ for $^3T_2$ from $t_2^1e^1$ $(-4Dq + 6Dq)$ and $+12Dq$ for $^3A_2$ from e^2 $(2 \times 6Dq)$. The energy of the third state ($^3T_1$) from ^3F is derived from the unchanged "baricenter" rule, making use of the orbital degeneracy (the dimensionality of the representations, 1 for A and 3 for T):

$$\underset{A_2}{1 \times 12Dq} + \underset{T_2}{3 \times 2Dq} + \underset{T_1(F)}{3 \times xDq} = 0,$$

$$x = -6Dq \qquad \text{for} \qquad {}^3T_1(F).$$

In the Orgel diagram these splittings are relative to the "unsplit" energy (dashed horizontal line). The energy separations shown are relative to the ground state, without allowing for bending.

The energy separation between F and P for d^2, d^3, d^7, and d^8 is 15B, where B is the Racah interelectron repulsion parameter.[3] B is evaluated for the free ion. The value of B in complexes (B′) is always less than the value for the free ion because of the decreased interelectron repulsion resulting from electron

[3] The energies of spectral terms can be expressed in terms of Slater–Condon F_k parameters or Racah parameters A, B, and C. The Racah parameters are more commonly used by chemists. The energy separation between terms of maximum multiplicity is a function of the parameter B alone.

delocalization.[4] B′ is usually about 0.7B to 0.9B, depending on the amount of covalent interaction. A reasonable value of B′ can be estimated from spectra that have been analyzed for similar complexes, or lacking such information, we start with $B' = 0.8B$ and adjust the value for internal consistency.

For a d^2 ion [e.g., $V(H_2O)_6{}^{3+}$] the ground state is ${}^3T_{1g}$, and the following transitions are expected:

$$\begin{aligned} {}^3T_{1g} &\rightarrow {}^3T_{2g} & \nu_1 &= 8Dq + c, \\ {}^3T_{1g} &\rightarrow {}^3A_{2g} & \nu_2 &= 18Dq + c, \\ {}^3T_{1g} &\rightarrow {}^3T_{1g}(P) & \nu_3 &= 15\text{B}' + 6Dq + 2c. \end{aligned} \tag{9.10}$$

This is the sequence for a weak field ligand, such as H_2O. For a stronger field, ${}^3T_{1g}(P)$ is at lower energy than ${}^3A_{2g}$. The energy estimates include a parameter c for bending of ${}^3T_{1g}(F)$ [decrease in energy with increasing field strength because of configuration interaction with $T_{1g}(P)$] and, in the case of ν_3, $2c$ for bending of both ${}^3T_{1g}$ lines. The states ${}^3T_{2g}$ and ${}^3A_{2g}$ are not involved in configuration interaction, so their separation, $\nu_2 - \nu_1 = 10Dq$, is independent of bending and gives a better value of Dq.

The $[Cr(H_2O_6)]^{3+}$ (d^3) complex is expected to give three bands also:

$$\begin{aligned} {}^4A_{2g} &\rightarrow {}^4T_{2g} & \nu_1 &= 10Dq, \\ {}^4A_{2g} &\rightarrow {}^4T_{1g}(F) & \nu_2 &= 18Dq - c, \\ {}^4A_{2g} &\rightarrow {}^4T_{1g}(P) & \nu_3 &= 12Dq + 15\text{B}' + c. \end{aligned} \tag{9.11}$$

Here ν_1 is independent of configuration interaction and gives a good value of $10Dq$. The ν_3 transition is always highest in energy because the ${}^4T_{1g}$ lines cannot cross.

The parameters apply in the same way for tetrahedral complexes, but the value of Dq is smaller for the tetrahedral case.

The Orgel diagrams deal with spin-allowed transitions for all d configurations except d^0, d^5, and d^{10}. Of course, there are no $d \rightarrow d$ bands for d^0 or d^{10}, and there are no spin-allowed transitions for high-spin d^5. For the d^1-type diagram, only one transition is expected, and we have two empirical parameters (Dq and B′) so we need additional information, or we must estimate B′. For the d^2-type diagram, three transitions are expected, and there are three parameters (Dq, B′, and c). If all three bands are observed, we can evaluate the parameters. However, if Dq is near the value for the crossing of ${}^3T_{1g}(P)$ and ${}^3A_{2g}$, the bands are not likely to be resolved. In some cases the energy of ν_3 is high enough so that the band is covered by much more intense, allowed charge-transfer or ligand absorption bands.

[4] The decrease in interelectron repulsion for complexes is called the *nephelauxetic* ("cloud expanding") *effect* by Jørgensen. The greater the delocalization of the d electrons onto the ligands (covalent interaction, particularly π bonding), the greater the decrease in B′.

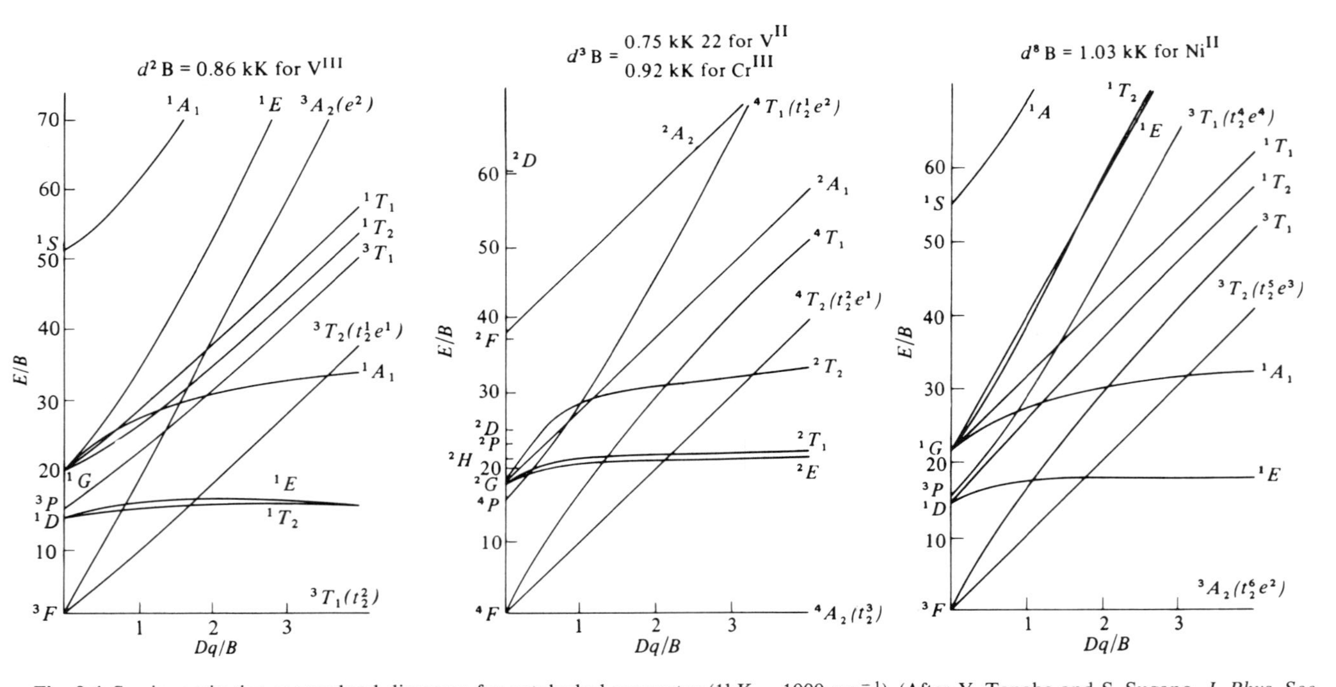

Fig. 9.4 Semiquantitative energy-level diagrams for octahedral symmetry (1kK = 1000 cm^{-1}). (After Y. Tanabe and S. Sugano, *J. Phys. Soc. Japan* 1954, **9,** 753. © 1954 Physical Society of Japan.)

9.2.2 TANABE–SUGANO DIAGRAMS

One disadvantage of the Orgel diagrams is that the reference or ground state decreases in energy with increasing field strength. In Tanabe–Sugano (T–S) diagrams for octahedral fields, the ground state is represented by a horizontal line. The diagrams are specific for a particular d^n configuration and include the appropriate bending of lines for excited states relative to the fixed ground state. States of high spin multiplicity are included, making it possible to consider spin-forbidden transitions. Tanabe–Sugano diagrams are given in Fig. 9.4 for d^2, d^3, and d^8. The axes are in dimensionless units of E/B and Dq/B (really B′ in each case) so they can be used for various ions with the same d^n configuration and various ligands. Changing the ion or the ligands affects the value of B′.

The T–S diagrams for d^4, d^5, d^6, and d^7 are shown in Figs. 9.5 and 9.6. Here, each diagram is divided into two parts separated by a vertical line. The left side

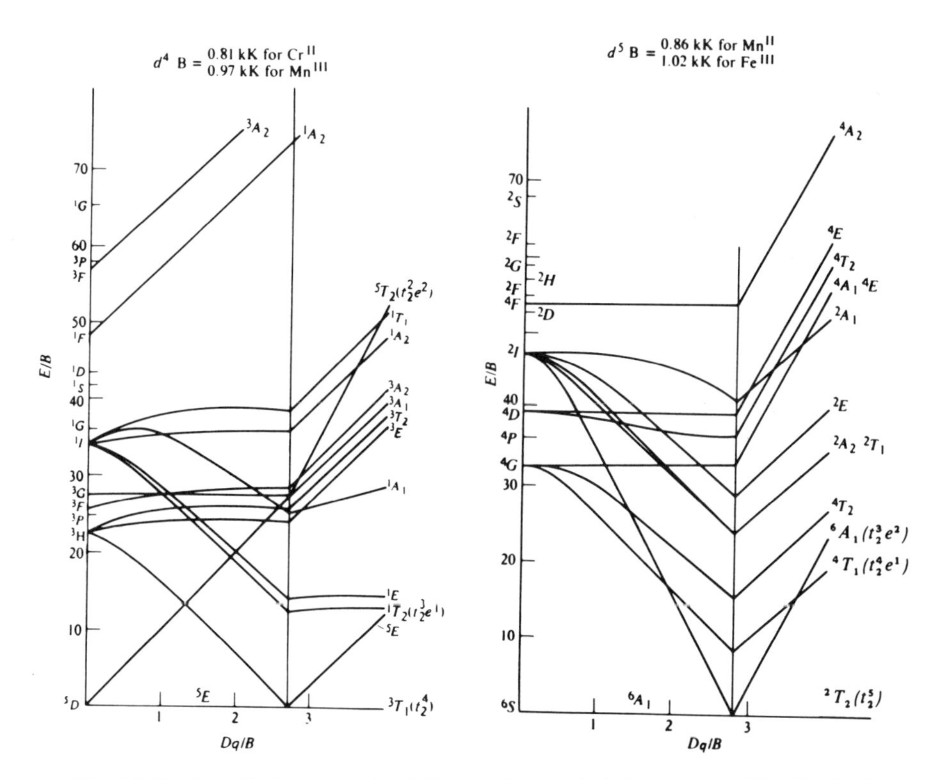

Fig. 9.5 Semiquantitative energy-level diagrams for octahedral symmetry. (After Y. Tanabe and S. Sugano, *J. Phys. Soc. Japan* 1954, **9**, 753. © 1954 Physical Society of Japan.)

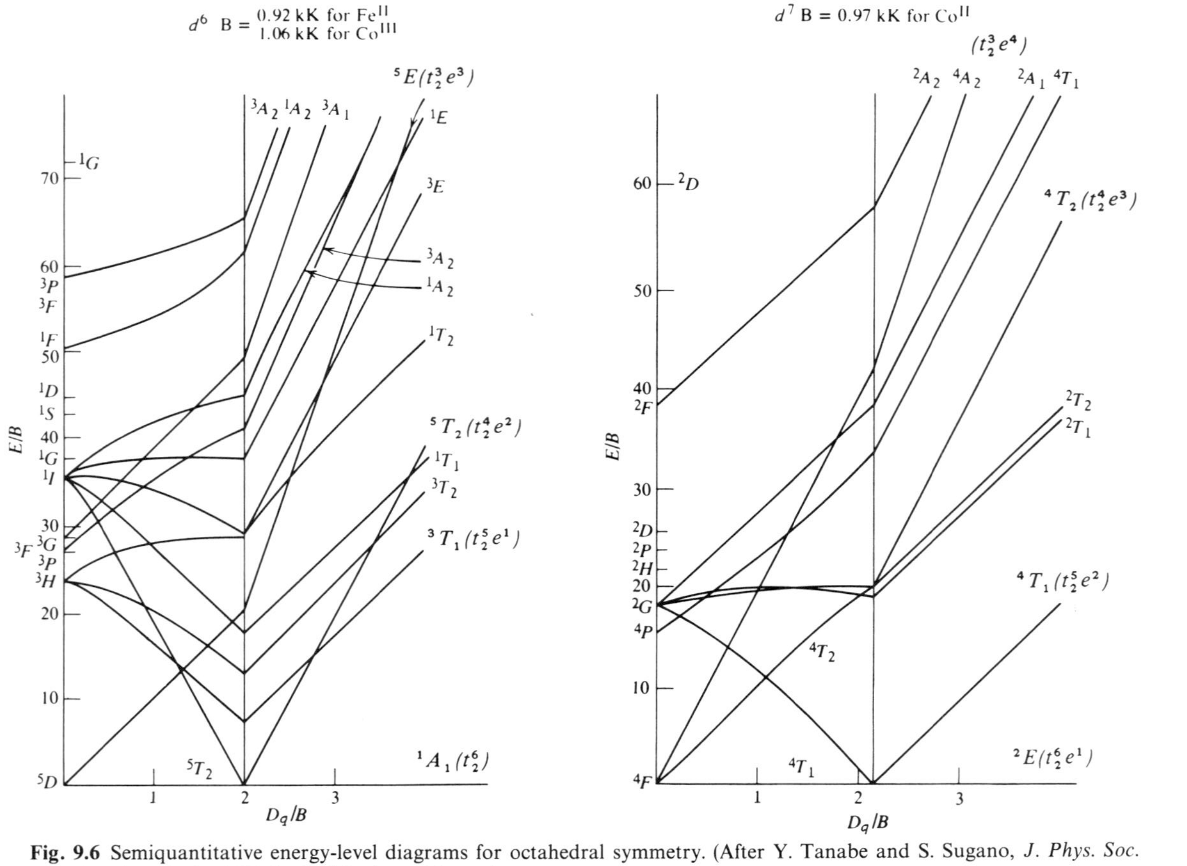

Fig. 9.6 Semiquantitative energy-level diagrams for octahedral symmetry. (After Y. Tanabe and S. Sugano, *J. Phys. Soc. Japan* 1954, **9**, 753. © 1954 Physical Society of Japan.)

applies to the high-spin case, corresponding to the appropriate Orgel diagram for the states of maximum multiplicity. The right side applies to the low-spin case with a different ground state. There are really two separate diagrams that share the same energy states but in different order except at Dq/B equal to the energy required for spin pairing. Because there are many lines, the labels, without g subscripts, are given only at the end of each line. Those at the left are the free ion terms for which the energy scale is from atomic spectra. The labels at the right are for $\mathbf{O}_h$ symmetry. For d^4, the ground state free ion term is ${}^5D$, and for the weak-field case, this gives ${}^5E$ as the ground state and ${}^5T_2$ as the only quintet excited state. Beyond $Dq/\mathrm{B} \cong 2.8$, the field is strong enough to bring about spin pairing; ${}^3T_1$ becomes the ground state, and ${}^5E$ increases in energy as an excited state. For the strong field case, the spin-allowed transitions are from ${}^3T_1$ to the other triplet states, and there are several of them from ${}^3H$, ${}^3P$, ${}^3F$, ${}^3G$, and even high energy ${}^3F$ and ${}^3P$ states. Not all lines are shown.

On the right of the d^2 T–S diagram, the ${}^3T_1$ is identified as derived from the t_2^2 strong field configuration. The next three singlets (${}^1T_2$, ${}^1E$, and ${}^1A_1$) are also from t_2^2. As a check, t_2^2 has $6 \times \frac{5}{2} = 15$ microstates and the total degeneracies of the states are the sums of the products of the spin multiplicites and the orbital degeneracies:

$$\underset{{}^3T_1}{3 \times 3} + \underset{{}^1T_2}{1 \times 3} + \underset{{}^1E}{1 \times 2} + \underset{{}^1A_1}{1 \times 1} = 15.$$

The next higher energy configuration, $t_2^1e^1$ ($6 \times 4 = 24$ microstates), gives ${}^3T_2$ (3×3), ${}^3T_1$ (3×3), ${}^1T_2$ (1×3), and ${}^1T_1$ (1×3). The highest energy states ${}^3A_2$ (3×1), ${}^1E$ (1×2), and ${}^1A_1$ (1×1) are derived from e^2 ($4 \times \frac{3}{2} = 6$ microstates). The strong field parentage is given on the right for other diagrams, but when there are many states, some of those of little importance are omitted.

9.3 Spectra of Octahedral and Tetrahedral Complexes

9.3.1 d^1 AND d^9 IONS

Only one band (${}^1T_{2g} \rightarrow {}^1E_g$ for d^1 and ${}^1E_g \rightarrow {}^1T_{2g}$ for d^9) is expected for octahedral complexes for the one-electron or one-hole case. The energy of the transition is $10Dq$. The simple Orgel diagram shows that there is no bending and there are no spin-forbidden $d \rightarrow d$ transitions. The absorption band for $[Ti(H_2O)_6]^{3+}$ shows a shoulder that is interpreted as the result of Jahn–Teller distortion for the excited state configuration e_g^1. That is, the double degeneracy of the excited state is removed, giving two closely spaced excited states. The six-coordinate Cu^{2+} (d^9) complexes are usually distorted from a regular octahedron. The single absorption band is usually broadened considerably by the distortion, and spin–orbit coupling (see Section 9.7) can contribute to the broadening. Detailed analysis might require the use of lower symmetry, probably $\mathbf{D}_{4h}$. Tetrahedral complexes are uncommon for d^1 ions. The color of species such as MnO_4^- results from charge transfer (O → Mn). The $CuCl_4^{2-}$ ion (d^9) is distorted from a regular tetrahedron to give $\mathbf{D}_{2d}$ symmetry with appreciable splitting of the energy levels.[5]

9.3.2 d^2 AND d^7 IONS

The same Orgel diagram applies to the d^2 and d^7 configurations, and the T–S diagrams for ^xF and ^xP ($x = 3$ for d^2 and $x = 4$ for d^3) are the same for d^2 and high-spin d^7. The T–S diagrams for the two configurations differ for the states of lower spin multiplicity. The spectra for octahedral complexes of d^2 (V^{3+}) and d^7 (Co^{2+}) ions are expected to show three spin-allowed bands, although often one is too high in energy to be observed because it is covered by more intense bands in the ultraviolet region. Any spin-forbidden transitions must be assigned using the particular T–S diagram for d^2 or d^7.

EXAMPLE 9.4 For $[V(H_2O)_6]^{3+}$ in alum ($KA1(SO_4)_2 \cdot 12H_2O$) crystals, bands are observed at 17,800 cm^{-1} ($\varepsilon = 3.5$) and at 25,700 cm^{-1} ($\varepsilon = 6.6$). Assign the bands and evaluate the parameters Dq, c, and B′ [1 kK (kiloKayser) = 1,000 cm^{-1}].

SOLUTION Assuming the first band is ν_1 [Eq. (9.10)],

$$17{,}800 = 8Dq + c.$$

If c were zero, Dq would be about 2200 cm^{-1} and ν_2 would be above 39,000 cm^{-1}, since bending must be included for it also ($18Dq + c$). Obviously this is a case where $^3T_{1g}(P)$ is lower than $^3A_{2g}$, and the second band must be ν_3, as designated in Eq. (9.10). Using the T–S diagram (Fig. 9.4) and a rough value of B′ 800 cm^{-1} (allowing for some lowering of B = 860 cm^{-1}), E/B′ = 17,800/800 = 22.3 for the first band, corresponding to Dq/B′ ≅ 2.4 and $Dq \cong 2.4 \times 800 \cong 1900$. The second spin-allowed band should be at E/B′ ≅ 35 or E ≅ 28,000 cm^{-1} for $^3T_1 \rightarrow {}^3T_1(P)$, with $^3A_2$ much higher in energy. Using Eqs. (9.10) for the calculation of the parameters, we must deal with two equations (for ν_1 and ν_3) and three parameters. The estimate of B′ above (800 cm^{-1}) gave reasonable fits. Using this value, we obtain Dq = 2200 cm^{-1} and c = 200 cm^{-1}. From these values the energy of ν_2 should be about 39,800 cm^{-1}.

[5] A. B. P. Lever, *Inorganic Electronic Spectroscopy*, Elsevier, New York, 1968, pp. 153–155. A new edition (1984) is available.

Low-spin d^7 complexes are not encountered for first transition series metals. Simple salts of Co are those of Co^{2+} (d^7), but if the ligand field is strong enough to bring about spin pairing, one electron is raised to the very high-energy e_g orbital and the complex is oxidized easily to Co^{III}. The spectra of octahedral low-spin d^7 complexes would be expected to be very simple with one band corresponding to the unresolved transitions $^2E_g \rightarrow {}^2T_{1g}$ and $^2T_{2g}$ and other spin-allowed transitions at very high energy.

Tetrahedral complexes of V^{3+} are rare. $[VCl_4]^-$ exists in solid $CsAlCl_4$ as host. Cobalt(II) (d^7) gives numerous tetrahedral complexes. The absorption bands of the tetrahedral complexes are usually considerably more intense than those of octahedral Co(II) complexes.

9.3.3 d^3 AND d^8 IONS

Many octahedral complexes of Cr(III) (d^3) and Ni(II) (d^8) have been studied. They share the same Orgel diagram and their T–S diagrams differ only for the states of lower spin multiplicity. There is no low-spin octahedral case to be considered. Three spin-allowed bands are expected, although one is often too high in energy to be observed. We see from the T–S diagrams that this $^xT_{1g}$ state is derived from a configuration ($^4T_{1g}$ from $t_{2g}^1e_g^2$ and $^3T_{1g}$ from $t_{2g}^4e_g^4$) that involves the promotion of two electrons from the t_{2g} orbital. Since the simultaneous promotion of two electrons has low probability, the observed bands might be expected to have low intensity. Sometimes, however, the intensity of this band is about the same as that of the other spin-allowed bands, perhaps through mixing with allowed transitions having the same symmetry. This enhancement of intensity is called intensity borrowing.

The octahedral stabilization energy is so great for the d^3 configuration that tetrahedral coordination is rare for Cr(III). Nickel(II) does form many tetrahedral complexes, such as $[NiX_4]^{2-}$ ($X = Cl^-$, Br^-, or I^-). The intensities of the absorption bands of the tetrahedral complexes are closer to those of the octahedral complexes for Ni(II), as compared with Co(II). Square planar coordination of Ni(II) is particularly favorable (d^8 is the most favorable case); the $\mathbf{D}_{4h}$ case is dealt with as a case of lower symmetry (Section 9.4).

EXAMPLE 9.5 For $[Cr(H_2O)_6]^{3+}$, bands are observed at 17,400 cm^{-1} ($\varepsilon = 13$), 24,600 cm^{-1} ($\varepsilon = 15$), and 37,800 cm^{-1} ($\varepsilon = 4$). Assign the bands and evaluate the parameters Dq, B′, and the bending parameter c.

SOLUTION The first band, ν_1 ($^4A_{2g} \rightarrow {}^4T_{2g}$, $10Dq$), gives $Dq = 1740\ cm^{-1}$. Using Eqs. (9.11) ν_2 gives $c = 6{,}700\ cm^{-1}$ and ν_3 gives $B' = 680\ cm^{-1}$. The value of B′ is a bit lower than that expected for about 0.8B (B is 920 cm^{-1} in the T–S diagram) although H_2O is not expected to give complexes involving much covalent bonding.

9.3.4 d^4 AND d^6 IONS

High-spin d^4 and d^6 ions have a ^5D ground state as the only quintet and give only one spin-allowed band. The high-spin octahedral complexes of Cr(II) and Mn(III) (d^4) and Fe(II) (d^6) usually show appreciable splitting of the main band as a result of splitting of the E_g state (Jahn–Teller splitting). Distorted tetrahedral complexes are encountered for $[CrCl_4]^{2-}$, $[FeCl_4]^{2-}$, $[FeBr_4]^{2-}$, $[Fe(NCS)_4]^{2-}$, and others. The band intensities for the tetrahedral Fe(II) complexes are much greater than for the octahedral complexes. Spin pairing occurs at $Dq/B = 2.0$ for d^6, lower than for any other d configuration. Consequently octahedral Fe(II) complexes with strong field ligands, such as CN^-, have low spin. Since Dq increases with increasing charge on M^{n+}, Co(III) complexes generally have low spin. The only well-characterized high-spin octahedral complex of Co(III) is $[CoF_6]^{3-}$. The spin-allowed transitions for octahedral low-spin d^6 complexes are $^1A_{1g} \rightarrow {}^1T_{1g}$ and $^1T_{2g}$. The other singlets are too high in energy to be observed. The absorption spectra of complexes such as $[Co(en)_3]^{3+}$, $[Co(en)_2(NH_3)_2]^{3+}$, and $[Co(en)(NH_3)_4]^{3+}$ are very similar to the spectrum of $[Co(NH_3)_6]^{3+}$. As long as the ligand atoms are the same (or very close in the spectrochemical series), the absorption spectra can be treated using effective $\mathbf{O}_h$ symmetry.

EXAMPLE 9.6 $[Fe(H_2O)_6]^{2+}$ shows one broad absorption band consisting of a main peak at 10,400 cm^{-1} and a pronounced shoulder at 8,300 cm^{-1}. $[Co(H_2O)_6]^{3+}$ shows two symmetrical bands at 16,500 and 24,700 cm^{-1} with *very* weak peaks at 8000 and 12,500 cm^{-1}. Account for the bands.

SOLUTION The one band for $[Fe(H_2O)_6]^{2+}$ indicates that this is the high-spin case. The $^5T_{2g} \rightarrow {}^5E_g$ band is split by Jahn–Teller distortion. Because of the higher ionic charge for Co(III), Dq is larger for the same ligand, and the low-spin case is more likely. The presence of two symmetrical bands, $^1A_{1g} \rightarrow {}^1T_{1g}$ and $^1A_{1g} \rightarrow {}^1T_{2g}$, are expected for low-spin Co(III). The Orgel diagrams do not apply for the low-spin cases. From the T–S diagram (Fig. 9.6), using $B' \cong$ 850 cm^{-1} and the energy of the first band as 16,500 cm^{-1}, we get $E/B' = 19.4$. This is just above $Dq/B' = 2$, the field strength required for spin pairing. The E/B' for $^1A_1 \rightarrow {}^1T_2$ is about 30 or $E \cong 25{,}500$ cm^{-1}. For this Dq/B', we expect spin-forbidden transitions at about 8000 ($^1A_1 \rightarrow {}^3T_1$) and 13,000 cm^{-1} ($^1A_1 \rightarrow {}^3T_2$). These agree with the positions of the very weak peaks.

9.3.5 d^5 IONS

The d^5 case is unique. It is its own hole counterpart, and for the high-spin case, there is only one sextet, $^6A_{1g}$, and hence, no spin-allowed transitions. The commonly encountered d^5 ions are Mn^{2+} and Fe^{3+}. Since iron is a common constituent contributing to the color of glass, the glass can be decolorized by adding some MnO_2. The Fe^{2+} is oxidized to Fe^{3+} and MnO_2 is reduced to Mn^{2+}. The salts of Fe^{3+} and Mn^{2+} are faintly colored because the transitions are spin-forbidden. The weak spin-forbidden transitions expected are to quartet states since it is even less probable that the spins of two electrons would be reversed to give a state of even lower spin multiplicity. Nevertheless, there are many quartet states, so the spectrum of $[Mn(H_2O)_6]^{2+}$ (see Fig. 9.7) consists of many very weak peaks. We note that the unresolved $^4A_{1g}$, ^4E_g peak is very sharp and that the line representing the energies of these states (high-spin case) in the T–S diagram (Fig. 9.5) is nearly horizontal, that is, the energy is independent of Dq. One broadening effect of the $d \to d$ absorption bands is the fact that the vibrational motion of the ligands varies the M–L distance and essentially sweeps a range of Dq values. Usually the energy levels are very sensitive to changes in M–L distance and Dq. In this case ($^4A_{1g}$, ^4E_g), since the energy is independent of Dq, sharp lines result. The appearance of sharp lines can be a great aid in making spectral assignments. If possible we would seek verification by examining the spectra of complexes formed with other ligands. In these cases (and also for ^1E_g and $^1T_{2g}$ for d^2) bands appear at nearly the same energy for ligands of greatly different field strength. Bands that are insensitive to changes in Dq result from transitions within an orbital set, that is, they involve spin reversal within a configuration, not promotion from one orbital set to another.

Because of the high pairing energy for d^5 (the low-spin case arises at $Dq/B \cong 2.8$), only very strong field ligands (e.g., CN^-) bring about spin pairing for Fe^{3+}. The ground state becomes $^2T_{2g}$ for low-spin d^5. Only high-spin complexes occur for Mn^{2+} since Dq values are lower because of the lower charge on the metal ion. The octahedral arrangement is not favored by LFSE for high-spin d^5, since there is no LFSE for any geometry for d^5. Consequently,

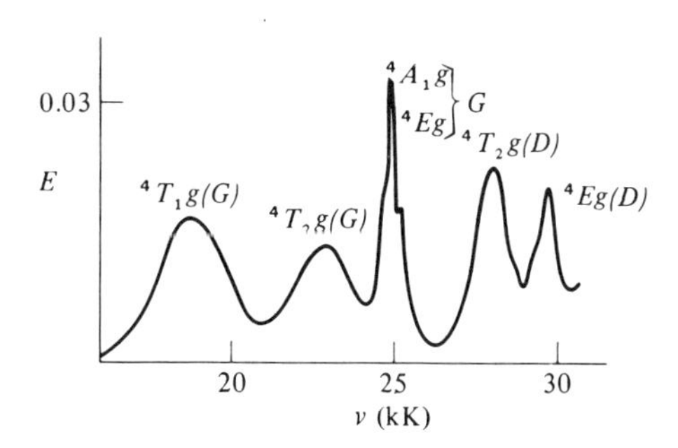

Fig. 9.7 Absorption spectrum of $[Mn(H_2O)_6]^{2+}$. (From C. K. Jørgensen, *Acta Chem. Scand.* 1954, **8,** 1502.)

whenever such factors as ligand–ligand repulsion are important, tetrahedral complexes result, for example, $MnCl_4^{2-}$ and $FeCl_4^-$.

9.3.6 CHARGE-TRANSFER AND LIGAND BANDS

The absorption bands for metal complexes resulting from $d \rightarrow d$ transitions are often referred to as ligand-field (LF) bands because their energies shift with the positions of the ligands in the spectrochemical series. These spin-allowed bands usually have relatively low intensity ($\varepsilon \approx 100$), but they are much more intense than spin-forbidden bands. Complexes of π donor ligands (e.g., Cl^-, Br^-, and I^-) often show absorption bands of such high intensity ($\varepsilon > 10^3$) that they must correspond to allowed transitions. These are charge-transfer transitions, corresponding to transitions from filled ligand orbitals to empty metal orbitals (ligand → metal). In Fig. 9.8 for $[Co(NH_3)_5F]^{2+}$, we see the two ligand field bands with minor effects caused by lowering of symmetry from $\mathbf{O}_h$. The intense charge-transfer band occurs beyond 40 kK. For $[Co(NH_3)_5Cl]^{2+}$, the charge transfer occurs above 30 kK and shows two

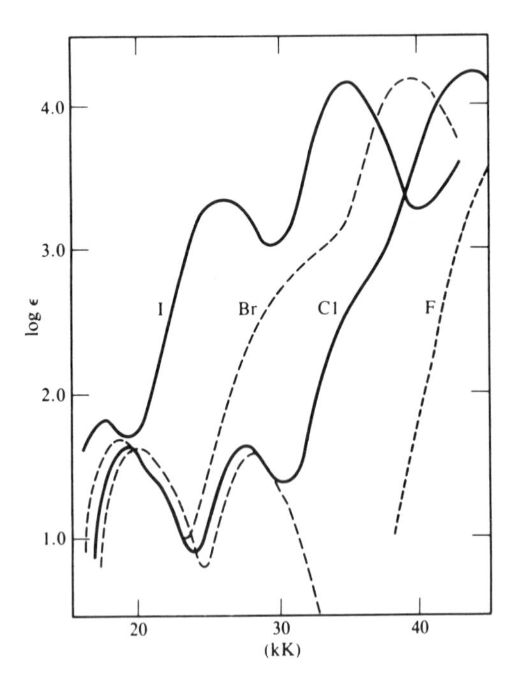

Fig. 9.8 The spectra of the $[Co(NH_3)_5X]^{2+}$ ions, where X is a halide ion. (After M. Linhard and M. Weigel, *Z. Anorg. Chem.* 1951, **266,** 49.)

components. The charge-transfer bands occur at lower energy for $[Co(NH_3)_5Br]^{2+}$ and at still lower energy for $[Co(NH_3)_5I]^{2+}$, overlapping the LF bands and obscuring the higher energy LF band.

We see that the LF bands shift slightly according to the field strength of X, but the charge-transfer bands show very large shifts. The shifts in the energies of the $X^- \rightarrow$ Co(III) transitions correspond to the changes in ease of electron removal (oxidation) of X^-. Although the transition corresponds to the transfer of an electron from X^- to Co(III), no net oxidation–reduction occurs because of the very short lifetime of the excited state. This does, however, provide a mechanism for photochemical decomposition that occurs for many complexes stored in strong light.

Metal-to-ligand charge-transfer transitions occur for complexes containing metal ions having filled, or nearly filled, t_{2g} orbitals and ligands having low-lying empty orbitals. Commonly the ligand orbitals are π^* orbitals in complexes such as those of pyridine, bipyridine, 1,10-phenanthroline, CN^-, CO, and NO. These charge-transfer bands are also very intense, with the band position depending on the relative energies of the metal and ligand orbitals involved.

Commonly the absorption spectrum of a transition metal complex of a colorless ligand that is not a good π donor or acceptor will show the LF bands in the visible region and then very intense bands in the ultraviolet region. The latter are transitions involving the ligand orbitals themselves (ligand bands). The energies of the bands might be perturbed, compared to the free ligand, because of bonding to the metal ion. The intensities of the ligand bands are determined by the selection rules applied to the ligand orbitals involved. The ones usually observed are the intense electric-dipole-allowed bands.

9.4 Correlation Diagrams for Octahedral Fields and Fields of Lower Symmetry

9.4.1 THE METHOD OF DESCENDING SYMMETRY

The trial-and-error procedure for assigning spin multiplicities to states from strong field configurations by correlation with states from spectral terms is practical only for very simple cases. One systematic way to assign spin multiplicities is to use the method of descending symmetry introduced by Bethe. The method depends on the removal of degeneracies on lowering symmetry. Usually spins can be assigned unambiguously for nondegenerate orbitals sets.

The method of descending symmetry is well illustrated by examining the d^2 case in an octahedral field. The strong field configurations are t_{2g}^2, $t_{2g}^1e_g^1$, and e_g^2. The energy states are derived from the direct products of the orbitals occupied for each configuration. For $t_{2g}^1e_g^1$, the product $(t_{2g} \times e_g)$ reduces to $T_{1g} + T_{2g}$. There is no problem with multiplicities here since the electrons are independent. We get ${}^1T_{1g} + {}^1T_{2g}$ if they have opposite spin (paired) and ${}^3T_{1g} + {}^3T_{2g}$ if they have the same spin. The degeneracy of $t_{2g}^1e_g^1$ is $6 \times 4 = 24$ microstates. The degeneracies of the states are $3 + 3 = 6$ for the singlets and $(3 \times 3) + (3 \times 3) = 18$ for the triplets, giving 24 as the total degeneracy.

For t_{2g}^2 (degeneracy $6 \times \frac{5}{2} = 15$), the direct product $(t_{2g} \times t_{2g})$ reduces to $T_{1g} + T_{2g} + E_g + A_{1g}$. We can get total degeneracy equal to 15 for these states only if there is one ${}^3T$ state and all other states are singlets or with only E_g and A_{1g} as triplets. More information is needed to choose among the three possible combinations. We can assign the multiplicities by choosing a subgroup of $\mathbf{O}_h$ that removes the orbital degeneracy of t_{2g}, preferably to give *different* nondegenerate orbitals. From Appendix 3, we see that in $\mathbf{C}_{2v}$, $A_1 + B_1 + B_2$ are derived from T_{2g} in $\mathbf{O}_h$. The direct product in $\mathbf{O}_h$ for t_{2g}^2 is $t_{2g} \times t_{2g}$. On lowering the symmetry to $\mathbf{C}_{2v}$, the two electrons can be in any combination of orbitals, or the direct product becomes $(a_1 + b_1 + b_2)(a_1 + b_1 + b_2)$, giving the following.

Product	Orbital configuration	Multiplicity
$a_1 \times a_1 = A_1$	a_1^2	${}^1A_1$
$b_1 \times a_1 = a_1 \times b_1 = B_1$	$a_1^1b_1^1$	${}^1B_1, {}^3B_1$
$b_2 \times a_1 = a_1 \times b_2 = B_2$	$a_1^1b_2^1$	${}^1B_2, {}^3B_2$
$b_1 \times b_1 = A_1$	b_1^2	${}^1A_1$
$b_2 \times b_1 = b_1 \times b_2 = A_2$	$b_1^1b_2^1$	${}^1A_2, {}^3A_2$
$b_2 \times b_2 = A_1$	b_2^2	${}^1A_1$

The A_1 states, involving two electrons in one orbital, must be singlets. Each other configuration gives a singlet and a triplet. The $\mathbf{O}_h$ states correlate with the $\mathbf{C}_{2v}$ states as follows (from correlation table).

$\mathbf{O}_h$	$\mathbf{C}_{2v}$
A_{1g}	${}^1A_1$
E_g	${}^1A_1 + A_2$
T_{1g}	$A_2 + B_1 + B_2$
T_{2g}	${}^1A_1 + B_1 + B_1$

Since all A_1 states are singlets, the corresponding $\mathbf{O}_h$ states must be singlets ($^1A_{1g}$, ^1E_g, and $^1T_{2g}$) also, and the other $\mathbf{C}_{2v}$ states derived from these $\mathbf{O}_h$ states must be singlets. The triplets must be $^3A_2 + {}^3B_1 + {}^3B_2$ derived from $^3T_{1g}$ ($\mathbf{O}_h$). This is the one triplet from t_{2g}^2 ($\mathbf{O}_h$).

The symmetry species derived from two electrons in the e_g orbitals ($4 \times \frac{3}{2} = 6$ microstates) can be obtained from the direct product $e_g \times e_g = A_{1g} + A_{2g} + E_g$. Since the total degeneracy (number of microstates) is six, the triplet cannot be ^3E_g, since the degeneracy for this state alone is $3 \times 2 = 6$ and there must be A_{1g} and A_{2g} states. We cannot tell whether A_{1g} or A_{2g} is the triplet. To assign the spin multiplicities, we choose a subgroup of $\mathbf{O}_h$ that removes the degeneracy of E_g. The $\mathbf{C}_{2v}$ group does this, but let us use a different subgroup, $\mathbf{C}_{4v}$, just to illustrate that the choice is independent of that made for t_{2g}^2. E_g ($\mathbf{O}_h$) becomes $A_1 + B_1$ ($\mathbf{C}_{4v}$). The direct product in $\mathbf{C}_{4v}$ is $(a_1 + b_1)(a_1 + b_1)$.

Product	Orbital configuration	Multiplicity
$a_1 \times a_1 = A_1$	a_1^2	$^1A_1$
$a_1 \times b_1 = B_1$	$a_1^1 b_1^1$	$^1B_1, {}^3B_1$
$b_1 \times a_1 = B_1$	$b_1^1 a_1^1$	
$b_1 \times b_1 = A_1$	b_1^2	$^1A_1$

The correlation of states gives the following.

$\mathbf{O}_h$	$\mathbf{C}_{4v}$
$^1A_{1g}$	$^1A_1$
A_{2g}	B_1
^1E_g	$^1A_1 + {}^1B_1$

Since A_1 occurs only as a singlet, the singlets are $^1A_{1g}$ and ^1E_g, and the only triplet is $^3A_{2g}$ ($\mathbf{O}_h$) corresponding to $^3B_1$ ($\mathbf{C}_{4v}$). Actually, if we just examine the possible configurations in $\mathbf{C}_{4v}$, then we see that we get $a_1^2 \rightarrow {}^1A_1$, $b_1^2 \rightarrow {}^1A_1$, and $a_1^1 b_1^1 \rightarrow {}^1B_1 + {}^3B_1$, or only the configuration involving single occupancy of orbitals gives a triplet.

The method of descending symmetry depends upon the fact that for nondegenerate orbitals we can identify the configurations (and states) requiring spin pairing. The lowering of symmetry removes orbital degeneracy, but it does not alter spin multiplicity. For the two-electron case, if we choose a subgroup giving orbitals of *different* symmetry species, we can see that only single occupancy of each orbital gives a triplet.

9.4.2 SPIN FACTORING

We can find the energy states (symmetry species) derived from a spectral term for any rotational point group by examining the effects of rotations on L (see Section 5.2). Character tables generally give the information for p and d orbitals. Those in Appendix 1 include the information needed for f orbitals. The symmetry species derived from d electron configurations are all g (for a centrosymmetric group). The u subscripts for p and f orbitals are changed to g for states derived from d^n configurations. Some discrepancies arise in noncentrosymmetric groups, as we saw in Example 5.3. Since the symmetry species are identified for $\mathbf{O}_h$ in Orgel and Tanabe–Sugano diagrams, we can use the $\mathbf{O}_h$ species and correlation tables to obtain the symmetry species for the desired lower symmetry (Table 9.2 and Appendix 3). The spin multiplicity is unchanged by lowering the symmetry,

$$^3T_{1g}\,(\mathbf{O}_h) \rightarrow {}^3A_{2g} + {}^3E_g\,(\mathbf{D}_{4h}).$$

The symmetry species, with the spin multiplicities assigned, is the information we need for the left side of a correlation diagram. On the right side we have the strong field configurations. We can get the energies of the d orbitals by the Krishnamurthy–Schaap approach (Section 8.1.4). The energies of the strong field configurations are the sums of the energies of the electrons populating the orbitals, allowing for pairing energy P where necessary. For quantitative results, configuration interaction (neglected here) would have to be included.

The method of descent of symmetry is applied easily for the $\mathbf{O}_h$ case. There are more problems if we begin with low symmetry, and direct products become cumbersome for the many-electron case. In fact, not all of the symmetry species generated by the direct product are necessarily real because some of them violate the Pauli exclusion principle. A simple, direct, and general approach developed by McDaniel[6] is available: the use of spin factoring. It has the advantage of generating from the strong field configurations the symmetry species with the *spin multiplicities assigned*. There are

TABLE 9.2
$\mathbf{O}_h$ Correlation Table

$\mathbf{O}_h$	$\mathbf{D}_{4h}$	$\mathbf{C}_{4v}$	$\mathbf{D}_3$
A_{1g}	A_{1g}	A_1	A_1
E_g	$A_{1g} + B_{1g}$	$A_1 + B_1$	E
T_{1g}	$A_{2g} + E_g$	$A_2 + E$	$A_2 + E$
T_{2g}	$B_{2g} + E_g$	$B_2 + E$	$A_1 + E$

[6] D. H. McDaniel, *J. Chem. Educ.* 1977, **54**, 147.

no spurious states to be eliminated because the Pauli exclusion principle is obeyed throughout, and it handles the many-electron case and low-symmetry cases more easily than other methods.

We have seen how spin-factoring can simplify the derivation of atomic spectral terms (Section 5.1.2). The electrons of the two spin sets (α and β) are considered separately. Any empty orbital is totally symmetric—empty atomic orbitals are associated with an S term and empty a_1, a_2, e_1, t_1, t_2, etc., orbitals are A_1 (or the totally symmetric species for the group). One electron in an atomic orbital gives a spectral term with the same capital letter designation ($s^1 \rightarrow S, p^1 \rightarrow P, f^1 \rightarrow F$, etc.), and the same is true for states in fields of any symmetry, $a_1^1 \rightarrow A_1, e^1 \rightarrow E, t_1^1 \rightarrow T_1, t_2^1 \rightarrow T_2$, etc. This provides much of the information in Table 9.3, even though we do not need to refer to a table for these cases.

TABLE 9.3
Partial States from α (or β) Electrons of One Spin Set (α or β) in Various Field Symmetries

	Under **O** or $\mathbf{T}_d$			Under $\mathbf{D}_4$	
Orbital	Electron occupancy	Partial state	Orbital	Electron occupancy	Partial state
a_1	0	A_1	a_1	0	A_1
	1	A_1		1	A_1
a_2	0	A_1	a_2	0	A_1
	1	A_2		1	A_2
e	0	A_1	b_1	0	A_1
	1	E		1	B_1
	2	A_2	b_2	0	A_1
t_1	0	A_1		1	B_2
	1	T_1	e	0	A_1
	2	T_1		1	E
	3	A_1		2	A_2
t_2	0	A_1		Under $\mathbf{D}_\infty$	
	1	T_2			
	2	T_1	$\sigma(a_1)$	0	$\Sigma^+(A_1)$
	3	A_2		1	Σ^+
	Under $\mathbf{D}_3$		$\sigma^-(a_2)$	0	Σ^+
				1	$\Sigma^-(A_2)$
a_1	0	A_1	$\pi(e_1)$	0	$\Sigma^+(A_1)$
	1	A_1		1	$\Pi(E_1)$
a_2	0	A_1		2	$\Sigma^-(A_2)$
	1	A_2	$\delta(e_2)$	0	$\Sigma^+(A_1)$
e	0	A_1		1	$\Delta(E_2)$
	1	E	$\phi(e_3)$	0	$\Sigma^+(A_1)$
	2	A_2		1	$\Phi(E_3)$
				2	$\Sigma^-(A_2)$

Where there is a correspondence to atomic orbitals, we can get the partial states from the corresponding partial atomic spectral terms; for example, for the **O** group:

$$\begin{aligned} p^0 &\to S, & t_1^0 &\to A_1,\\ p_\alpha^1 &\to P, & t_1^1 &\to T_1,\\ p_\alpha^2 &\to P, & t_1^2 &\to T_1,\\ p_\alpha^3 &\to S, & t_1^3 &\to A_1. \end{aligned}$$

The t_2 and e orbitals individually do not belong to any complete set of atomic orbitals, but they combine to give atomic orbital sets. We know that any empty, filled, or half-filled atomic orbital set is totally symmetric (e.g., p_α^3 or t_1^3 above), giving S or A_1. Thus, f_α^7 gives the S free ion term or A_1 (**O**). The individual f orbitals in **O** symmetry are a_2, t_1, and t_2, so with these filled with one spin set,

$$f_\alpha^7 \to \underset{A_2}{\underset{\downarrow}{a_2^1}}\,\underset{A_1}{\underset{\downarrow}{t_1^3}}\,\underset{\Gamma}{\underset{\downarrow}{t_2^3}}\,(\mathbf{O}) \to A_1\ (\mathbf{O})$$

or $A_2 \times A_1 \times \Gamma(t_2^3) = A_1$, giving $\Gamma(t_2^3) = A_2$. For the d orbitals filled with one spin set,

$$d_\alpha^5 \to S \to A_1(\mathbf{O}),$$

$$d_\alpha^5 \to (e^2)_\alpha(t_2^3)_\alpha(\mathbf{O}) \to A_1\ (\mathbf{O}).$$

Since $(t_2^3)_\alpha$ is A_2, $(e^2)_\alpha$ must be A_2 also ($A_2 \times A_2 = A_1$). The only missing configuration is $(t_2^2)_\alpha$. The hole formalism applies to symmetry orbitals as well as atomic orbitals, so t_2^5 (1 hole) is T_2 (**O**). The t_2^5 configuration can be written as $(t_2^2)_\alpha(t_2^3)_\beta$ and, since $(t_2^3)_{\alpha \text{ or } \beta}$ is A_2,

$$(t_2^2)_\alpha(t_2^3)_\beta \to T_2\ (\mathbf{O})$$

or $\Gamma_{(t_2^2)_\alpha} \times A_2 = T_2$ and $\Gamma_{(t_2^2)_\alpha} = T_1$ since $T_1 \times A_2 = T_2$. These results for **O** are tabulated in Table 9.3 along with the partial states for some other groups.

9.4.3 GROUND STATES

We can use the partial states to get full states as for atomic spectral terms. Although a full state does not correspond, in general, to one microstate, but rather to an array of microstates, we can obtain the states for the ground state configuration for $\mathbf{O}_h$ (or any symmetry) from the partial states for that configuration.

The lowest energy configuration has the maximum number of electrons of one spin set (maximum multiplicity) and has maximum occupancy of the lower-energy orbitals. For $\mathbf{O}_h$ symmetry, the ground states up to three d electrons are obtained from Table 9.3. d^1 $(t_{2g}^1)_\alpha$ is ${}^2T_{2g}$; d^2 $(t_{2g}^2)_\alpha$ is ${}^3T_{1g}$; and d^3 $(t_{2g}^3)_\alpha$ is ${}^4A_{2g}$. Since the partial state for $(d^5)_\alpha$ is A_{1g} (the full state is ${}^6A_{1g}$), the ground state (high spin) for d^6 $(d_\alpha^5 d_\beta^1)$ is ${}^5T_{2g}$ $(A_{1g} \times T_{2g})$, for d^7 $(d_\alpha^5 d_\beta^2)$ it is ${}^4T_{1g}$ $(A_{1g} \times T_{1g})$, and for d^8 $(d_\alpha^5 d_\beta^3)$ it is ${}^3A_{2g}$ $(A_{1g} \times A_{2g})$. The ground state configuration for d^4 is $(t_{2g}^3)_\alpha(e_g^1)_\alpha$ giving E $(A_{2g} \times E_g)$. The d^{10} configuration is ${}^1A_{1g}$, and we can get d^9 $(t_{2g}^6)(e_g^3)$ from the one hole in e_g, giving ${}^2E_g$ or as the product of the partial states $(t_{2g}^3)_\alpha(t_{2g}^3)_\beta(e_g^2)_\alpha(e_g^1)_\beta$,

$$A_{2g} \times A_{2g} \times A_{2g} \times E_g = E_g.$$

9.4.4 STATES FOR $\mathbf{O}_h$ STRONG FIELD CONFIGURATIONS AND CORRELATION DIAGRAMS

For a complete strong field electron configuration, we can obtain all states, with spin multiplicities assigned, by considering individually all possible combinations of spin sets. Let us reexamine the d^2 case $(\mathbf{O}_h)$ for which the strong field configurations are t_{2g}^2, $t_{2g}^1 e_g^1$, and e_g^2. For t_{2g}^2 $(6 \times \frac{5}{2} = 15$ microstates) we get

$$t_{2g}^2 \quad (t_{2g}^2)_{\alpha \text{ or } \beta} \rightarrow {}^3T_{1g}, \qquad 3 \times 3 = 9 \text{ degeneracy},$$

$$(t_{2g}^1)_\alpha(t_{2g}^1)_\beta \qquad \begin{matrix} T_{2g} \times T_{2g} \\ \text{degeneracy:} \end{matrix} = \begin{matrix} {}^1A_{1g} & + & {}^1E_g & + & ({}^1T_{1g}) & + & {}^1T_{2g} \\ 1 & & 2 & & (3) & & 3 \end{matrix}.$$

However, we must recognize that the state of highest spin multiplicity $({}^3T_{1g})$ with $S = 1$ represents an array with $M_S = \pm 1, 0$. By taking the direct product $T_{2g} \times T_{2g}$ for $(t_{2g}^1)_\alpha(t_{2g}^1)_\beta$, we get a ${}^1T_{1g}$ state that must be dropped as part of the array (with $M_S = 0$) for ${}^3T_{1g}$. We begin by examining the configuration of highest spin because *we can expect the same symmetry species to be replicated with all possible lower spin multiplicities*, but these really represent the expected lower M_S values. A quintet such as ${}^5T_{2g}$ would be replicated as ${}^3T_{2g}$ and ${}^1T_{2g}$, and a quartet such as ${}^4T_{2g}$ would be replicated as ${}^2T_{2g}$. We see that ${}^3T_{1g}$, ${}^1A_{1g}$, and ${}^1T_{2g}$ account for the expected degeneracy (15).

For $(t_{2g}^1)_{\alpha \text{ or } \beta}$ $(e_g^1)_{\alpha \text{ or } \beta}$ there are $6 \times 4 = 24$ microstates. The states are

$$T_{2g} \times E_g = T_{1g} + T_{2g}.$$

Here, all possible spin combinations exist, so we get ${}^3T_{1g}$, ${}^3T_{2g}$ and ${}^1T_{1g}$, ${}^1T_{2g}$ for a total degeneracy of 24. The same result is obtained by taking the product for $(t_{2g}^1)_\alpha(e_g^1)_\alpha$ (or all β) *and* for $(t_{2g}^1)_\alpha(e_g^1)_\beta$ (or reverse α and β). Here the ${}^1T_{1g}$ and ${}^1T_{2g}$ are not dropped since they arise from *independent* configurations. Comparing the number of microstates with the total degeneracy serves as a check.

For $(e_g^2)_{\alpha \text{ or } \beta}$, there are 6 microstates, and we get $(e_g^2)_{\alpha \text{ or } \beta} \rightarrow {}^3A_{2g}$ (degeneracy 3) and $(e_g^1)_\alpha(e_g^1)_\beta \rightarrow E_g \times E_g = {}^1A_{1g} + ({}^1A_{2g}) + {}^1E_g$. As before (for t_{2g}^2), considering all spin combinations in *one set* of orbitals, we expect the ${}^3A_{2g}$ to be replicated as ${}^1A_{2g}$, so this state is dropped. The states ${}^3A_{2g}$, ${}^1A_{1g}$, ${}^1E_g$ account for the total degeneracy (6).

We now have all states from the strong field $\mathbf{O}_h$ configurations *with spin multiplicities* assigned. This is the information we need for the right side of the $\mathbf{O}_h$ correlation diagram for d^2 (Fig. 9.9). The ordering for parallel lines

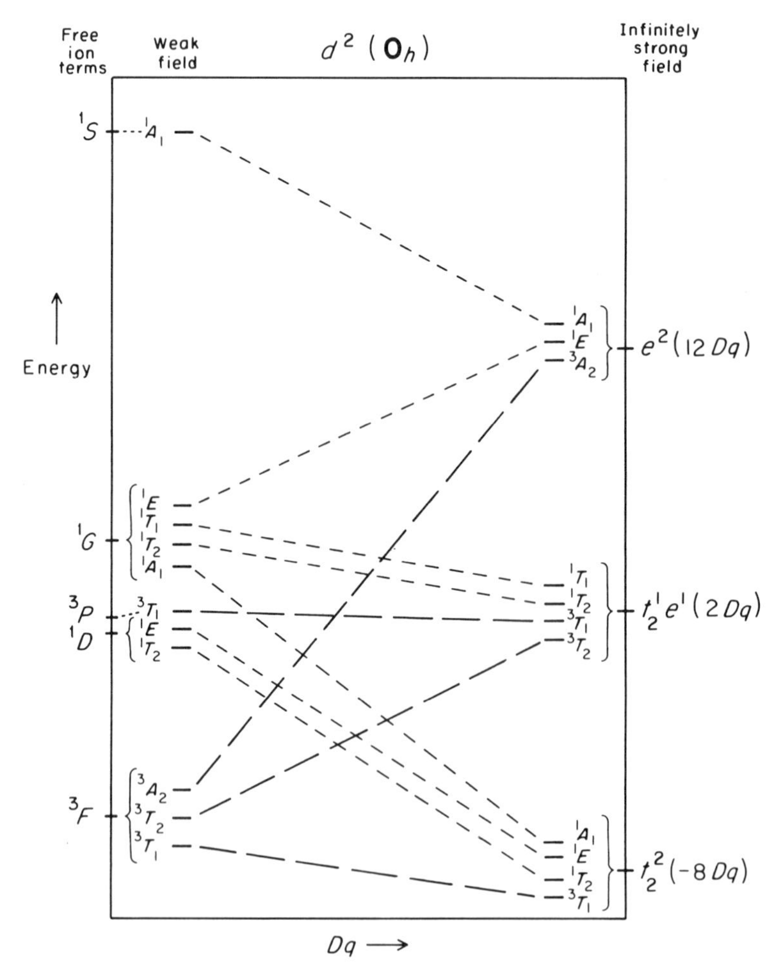

Fig. 9.9 Correlation diagram for a d^2 ion in $\mathbf{O}_h$ symmetry. Since all energy states and orbitals have g symmetry, the g subscripts are omitted.

($^1T_1, {}^1T_2(G)$ and $^1E, {}^1T_2(D)$) is arbitrary in this treatment. We minimize crossing, even for different symmetry species. Thus, $^3A_2$ (^3F) is higher in energy than $^3T_2$ (^3F), because $^3A_2$ correlates with the higher energy configuration. Note that the solid lines, connecting the triplets, constitute an Orgel diagram.

Now let us obtain the correlation diagram for d^3 ($\mathbf{O}_h$) in more direct fashion, as we would handle such cases.

Strong field configurations:	t_{2g}^3	$t_{2g}^2e_g^1$	$t_{2g}^1e_g^2$	e_g^3
Number of microstates:	20	60	36	4

States t_{2g}^3 $\quad (t_2^3)_{\alpha \text{ or } \beta} \rightarrow \boxed{^4A_2}$

$(t_2^2)_\alpha(t_2^1)_\beta$ [and $(t_2^2)_\beta(t_2^1)_\alpha$]

$T_1 \times T_2 = (^2A_2) + \boxed{^2E + {}^2T_1 + {}^2T_2}$

($^2A_2$ is the $M_S = \pm\frac{1}{2}$ component of $^4A_2$)

States $\quad \boxed{^4A_{2g} + {}^2E_g + {}^2T_{1g} + {}^2T_{2g}}$

$\quad 4 + 4 + 6 + 6 = 20$ (degeneracy)

$e_g^3 \quad (e^3)_{\alpha \text{ or } \beta} \rightarrow {}^2E \qquad$ (1 hole case)

state $\boxed{^2E_g} \qquad$ degeneracy = 4

$t_{2g}^2e_g^1 \quad (t_2^2)_{\alpha \text{ or } \beta} \rightarrow {}^3T_1 \qquad$ (partial state)

$(t_2^1)_\alpha(t_2^1)_\beta \rightarrow T_2 \times T_2 = {}^1A_1 + {}^1E + {}^1T_2 + (^1T_1)$

($^1T_1$ is the $M_S = 0$ component of $^3T_1$)

$(t_2^2)_{\alpha \text{ or } \beta}(e^1)_{\alpha \text{ or } \beta} \rightarrow {}^3T_1 \times {}^2E \rightarrow \boxed{^4T_1 + {}^4T_2 + {}^2T_1 + {}^2T_2}$

We retain the doublets since the electrons can be all of one spin set or of both sets.

$(t_{2\alpha}^1t_{2\beta}^1)(e^1)_{\alpha \text{ or } \beta} \rightarrow {}^1(T_2 \times T_2)(^2E) = (^1A_1 + {}^1E + {}^1T_2)(^2E)$

$= \boxed{^2E + {}^2A_1 + {}^2A_2 + {}^2E + {}^2T_1 + {}^2T_2}$

States $\quad \boxed{^4T_{1g} + {}^4T_{2g} + 2\,{}^2T_{1g} + 2\,{}^2T_{2g} + 2\,{}^2E_g + {}^2A_{1g} + {}^2A_{2g}}$

Degeneracy $\quad 12 + 12 + 12 + 12 + 8 + 2 + 2 = 60$

$t_{2g}^1e_g^2 \quad t_{2g}^1 \rightarrow {}^2T_{2g} \qquad (e_{\alpha \text{ or } \beta}^2) \rightarrow {}^3A_2$

$(e_\alpha^1e_\beta^1) \quad E \times E = {}^1A_1 + {}^1E + (^1A_2) \qquad$ ($M_S = 0$ component of $^3A_2$)

$(t_2^1)(e^2)_{\alpha \text{ or } \beta} \qquad {}^2T_2 \times {}^3A_2 = \boxed{^4T_1 + {}^2T_1} \qquad$ (retain both)

$(t_2^1)(e_\alpha^1e_\beta^1) \quad {}^2T_2 \times (^1A_1 + {}^1E) = \boxed{^2T_1 + 2\,{}^2T_2}$

States $\quad \boxed{^4T_{1g} + 2\,{}^2T_{1g} + 2\,{}^2T_{2g}}$

Degeneracy $\quad 12 + 12 + 12 = 36$

d^3 configuration $\qquad 10 \times \frac{9}{2} \times \frac{8}{3} = 120 \quad$ total degeneracy

A qualitative correlation diagram is given in Fig. 9.10 for d^3 ($\mathbf{O}_h$). Not all tie lines for doublets are shown. The Orgel diagrams are such correlation diagrams considering only the states of highest spin multiplicity. Other d configurations can be handled similarly. The cases involving more than two electrons are very cumbersome to handle by other methods.

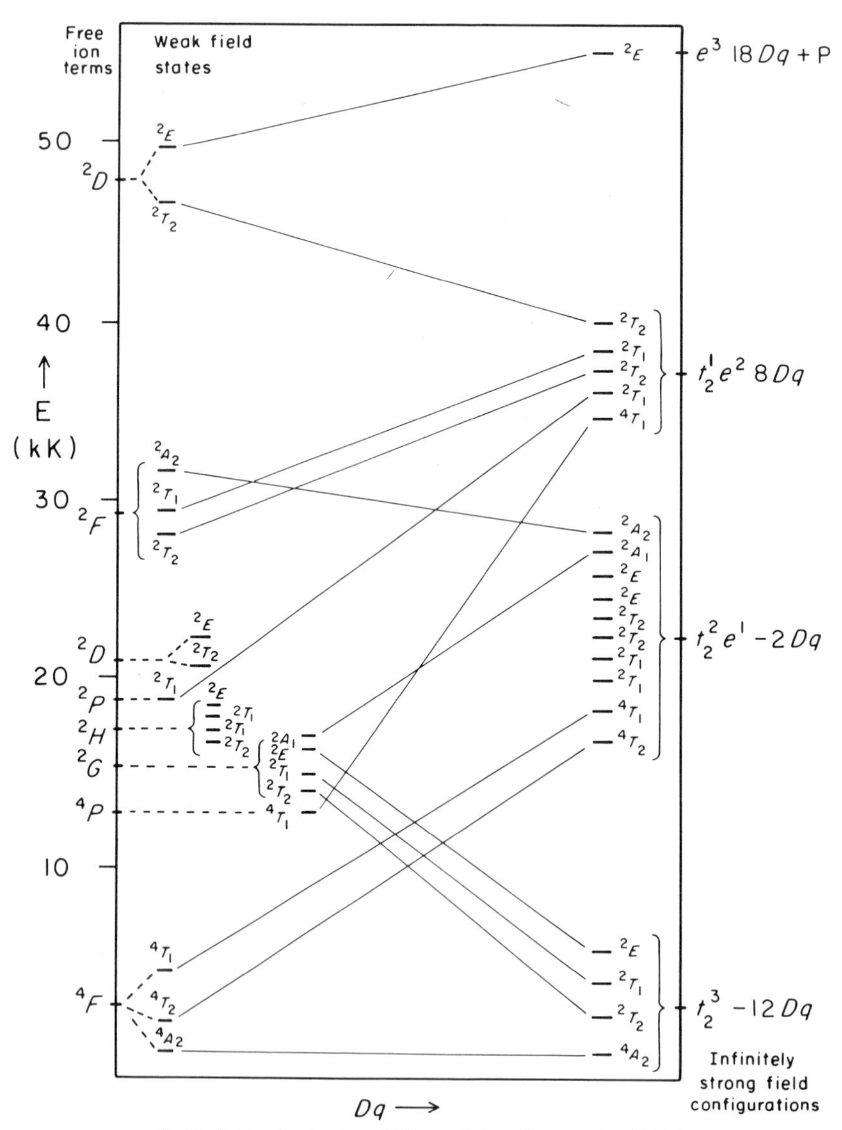

Fig. 9.10 Qualitative (partial) correlation diagram for d^3 ($\mathbf{O}_h$).

9.4.5 CORRELATION DIAGRAMS FOR FIELDS OF LOWER SYMMETRY

Now we can construct a correlation diagram for any electron configuration and any symmetry. The symmetry species for the *d* orbitals are available from character tables. Their relative energies are available from the Krishnamurthy–Schaap treatment (Section 8.1.4). McDaniel applied spin factoring to obtain the information necessary for a correlation diagram for d^2 in a trigonal bipyramidal ($\mathbf{D}_{3h}$) field. Let us examine the square pyramidal ($\mathbf{C}_{4v}$) and square planar ($\mathbf{D}_{4h}$) cases.

EXAMPLE 9.7 Obtain a qualitative correlation diagram for a d^3 ion in a square pyramidal field ($\mathbf{C}_{4v}$).

SOLUTION From Section 8.14, we obtain the energies of the *d* orbitals as given in Table 9.4. The spectral terms for d^3 are listed also. The corresponding energy states in $\mathbf{O}_h$ and those for $\mathbf{C}_{4v}$ obtained from a correlation table (Table 9.2) are

TABLE 9.4
Energy States for d^3 in $\mathbf{C}_{4v}$ Symmetry

Free ion terms	$\mathbf{O}_h$	$\mathbf{C}_{4v}$
^4F	$^4A_{2g}$	$^4B_1$
	$^4T_{1g}$	$^4A_2 + {}^4E$
	$^4T_{2g}$	$^4B_2 + {}^4E$
^4P	$^4T_{1g}$	$^4A_2 + {}^4E$
	^2E_g	$^2A_1 + {}^2B_1$
^2H	$2\,^2T_{1g}$	$2\,^2A_2 + 2\,^2E$
	$^2T_{2g}$	$^2B_2 + {}^2E$
^2G	$^2A_{1g}$	$^2A_1$
	^2E_g	$^2A_1 + {}^2B_1$
	$^2T_{1g}$	$^2A_2 + {}^2E$
	$^2T_{2g}$	$^2B_2 + {}^2E$
^2F	$^2A_{2g}$	$^2B_1$
	$^2T_{1g}$	$^2A_2 + {}^2E$
	$^2T_{2g}$	$^2B_2 + {}^2E$
$2\,^2P$	$2\,^2E_g$	$2\,^2A_1 + 2\,^2B$
	$2\,^2T_{2g}$	$2\,^2B_2 + 2\,^2E$
^2P	$^2T_{1g}$	$^2A_2 + {}^2E$
$10 \times \frac{9}{2} \times \frac{8}{3}$ = Degeneracy: 120	120	120

Relative energies for $\mathbf{C}_{4v}$

b_1 — $+9.14Dq\,(d_{x^2-y^2})$

Energy ↑ a_1 — $+0.86\,(d_{z^2})$

b_2 — $-0.86\,(d_{xy})$

e — $-4.57\,(d_{xz}, d_{yz})$

TABLE 9.5
Configuration Energies and the Corresponding Energy States for d^3 in C_{4v} Symmetry

Configuration	Degeneracy	Energy (*Dq*)	States in C_{4v}
$a_1^1b_1^2$	2	19.14 + P[a]	$^2A_1$
$b_2^1b_1^2$	2	17.42 + P	$^2B_2$
$e^1b_1^2$	4	13.71 + P	^2E
$a_1^2b_1^1$	2	10.86 + P	$^2B_1$
$b_2^1a_1^1b_1^1\,(2\times 2\times 2)$	8	9.14	$^4A_1 2^2A_2$
$b_2^2b_1^1$	2	7.42 + P	$^2B_1$
$e^1b_1^1a_1^1$	16	5.43	^4E, 2^2E
$e^1b_2^1b_1^1$	16	3.71	^4E, 2^2E
$b_2^1a_1^2$	2	0.86 + P	$^2B_2$
$e^2b_1^1$	12	0	$^4B_2, {}^2A_1, {}^2A_2, {}^2B_1, {}^2B_2$
$b_2^2a_1^1$	2	−0.86 + P	$^2A_1$
$e^1a_1^2$	4	−2.85 + P	^2E
$e^1b_2^1a_1^1\,(4\times 2\times 2)$	16	−4.57	^4E, 2^2E
$e^1b_2^2\,(4\times 1)$	4	−6.29 + P	^2E
$e^2a_1^1\,(6\times 2)$	12	−8.28	$^4A_2, {}^2A_1, {}^2A_2, {}^2B_1, {}^2B_2$
$e^2b_2^1$	12	−10	$^4B_1, {}^2A_1, {}^2A_2, {}^2B_1, {}^2B_2$
e^3	4	−13.71 + P	^2E
Degeneracy	120		120

[a] P is the pairing energy.

given. Table 9.5 lists all configurations for the three-electron case. The energies are the sums of the orbital energies for the three electrons plus the pairing energy (P) for low-spin configurations. The energy states correspond to the orbital labels for one-electron occupancy (a_1^1, b_1^1, b_2^1, and e^1) and for e^3. The configurations a_1^2, b_1^2, and b_2^2 (filled orbitals) are A_1. The e^2 configuration includes $e_\alpha^1 e_\beta^1$ ($E\times E$) and $e^2_{\alpha\text{ or }\beta}$, which is A_2 (the high-spin d^5 configuration $e^2b_2^1a_1^1b_1^1$ is A_1, and $B_2\times A_1\times B_1 = A_2$, therefore $e^2_{\alpha\text{ or }\beta}$ must be A_2 since $A_2\times A_2 = A_1$). The states in C_{4v} are the products of the representations of the occupied orbitals. The correlation diagram is shown in Fig. 9.11. The energy of $^4B_2$ is the same as that of $e^2b_1^1$ ($0Dq$) and that of $^4B_1$ is $-10Dq$ from $e^2b_2^1$. These are the only states that occur just once. For others, configuration interaction must be considered.

EXAMPLE 9.8 Obtain a qualitative correlation diagram for d^8 (low spin) in a square planar (D_{4h}) complex.

SOLUTION The relative energies of the *d* orbitals in D_{4h} and the energy states are given in Table 9.6. The energies of the configurations (neglecting pairing energies and configuration interaction) are given in Table 9.7, along with the correspond-

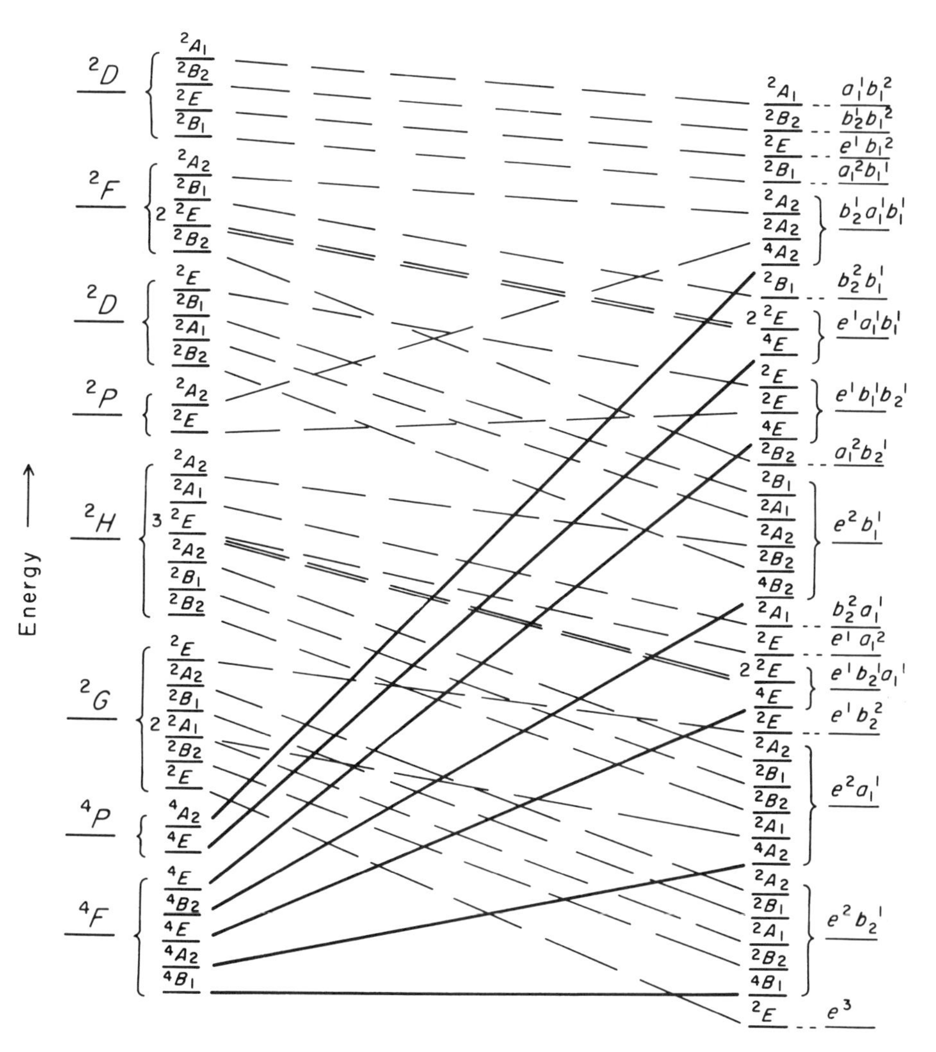

Fig. 9.11 Correlation diagram for d^3 in $\mathbf{C}_{4v}$ symmetry.

ing energy states. The correlation diagram for d^8 in Fig. 9.12 combines the weak field $\mathbf{D}_{4h}$ states derived from the free ion terms and the states derived from the strong field configurations. The spin-factoring approach allows us to consider only spin-allowed transition (singlets), omitting the states derived from ^3F and ^3P. Square planar Ni(II) complexes commonly show two major bands, sometimes preceded by a very weak lower-energy band. Commonly these bands are $^1A_{1g} \rightarrow {}^3A_{2g}$ (weak), $^1A_{1g} \rightarrow {}^1A_{2g}$, and $^1A_{1g} \rightarrow {}^1B_{1g}$.

TABLE 9.6
Relative Energies of d Orbitals in D_{4h} and the Energy States

Energy of Orbitals
b_{1g}— $12.28Dq(d_{x^2-y^2})$
b_{2g}— $2.28\,(d_{xy})$
a_{1g}— $-4.28\,(d_{z^2})$
e_g—— $-5.14\,(d_{xz}, d_{yz})$
degeneracy for d^8 (or d^2) = 45

Free ion terms	O_h	D_{4h}
^3F	$^3A_{2g}$	$^3B_{1g}$
	$^3T_{1g}$	$^3A_{2g} + {}^3E_g$
	$^3T_{2g}$	$^3B_{2g} + {}^3E_g$
^1D	^1E_g	$^1A_{1g} + {}^1B_{1g}$
	$^1T_{2g}$	$^1B_{2g} + {}^1E_g$
^3P	$^3T_{1g}$	$^3A_{2g} + {}^3E_g$
^1G	$^1A_{1g}$	$^1A_{1g}$
	^1E_g	$^1A_{1g}, {}^1B_{1g}$
	$^1T_{1g}$	$^1A_{2g} + {}^1E_g$
	$^1T_{2g}$	$^1B_{2g} + {}^1E_g$
^1S	$^1A_{1g}$	$^1A_{1g}$

TABLE 9.7
Energies for d^8 Electron Configuration in D_{4h} Symmetry with the Corresponding Energy States without Pairing or Configuration Interaction[a]

Energy	Configuration	Degeneracy	States
10.28	$e^2a_1^2b_2^2b_1^2$	6	$e^2_{\alpha\text{ or }\beta} \rightarrow {}^3A_2$
			$e^1_\alpha e^1_\beta \rightarrow {}^1A_1 + ({}^1A_2) + {}^1B_1 + {}^1B_2$
9.42	$e^3a_1^1b_2^2b_1^2$	8	$^3E, {}^1E$
8.56 + P	$e^4b_2^2b_1^2$	1	$^1A_1$
2.86	$e^3a_1^2b_2^1b_1^2$	8	$^3E, {}^1E$
2.00	$e^4a_1^1b_2^1b_1^2$	4	$A_1 \times A_1 \times B_2 \times A_1 = {}^1B_2, {}^3B_2$
−4.56 + P	$e^4a_1^2b_1^2$	1	$A_1 \times A_1 \times A_1 = {}^1A_1$
−7.14	$e^3a_1^2b_2^2b_1^1$	8	$E \times A_1 \times A_1 \times B_1 = {}^3E, {}^1E$
−8.00	$e^4a_1^1b_2^2b_1^1$	4	$A_1 \times A_1 \times A_1 \times B_1 = {}^1B_1, {}^3B_1$
−14.56	$e^4a_1^2b_2^1b_1^1$	4	$A_1 \times A_1 \times B_2 \times B_1 = {}^1A_2, {}^3A_2$
−24.56 + P	$e^4a_1^2b_2^2$	1	$^1A_1$
		45	

[a] g subscripts are omitted.

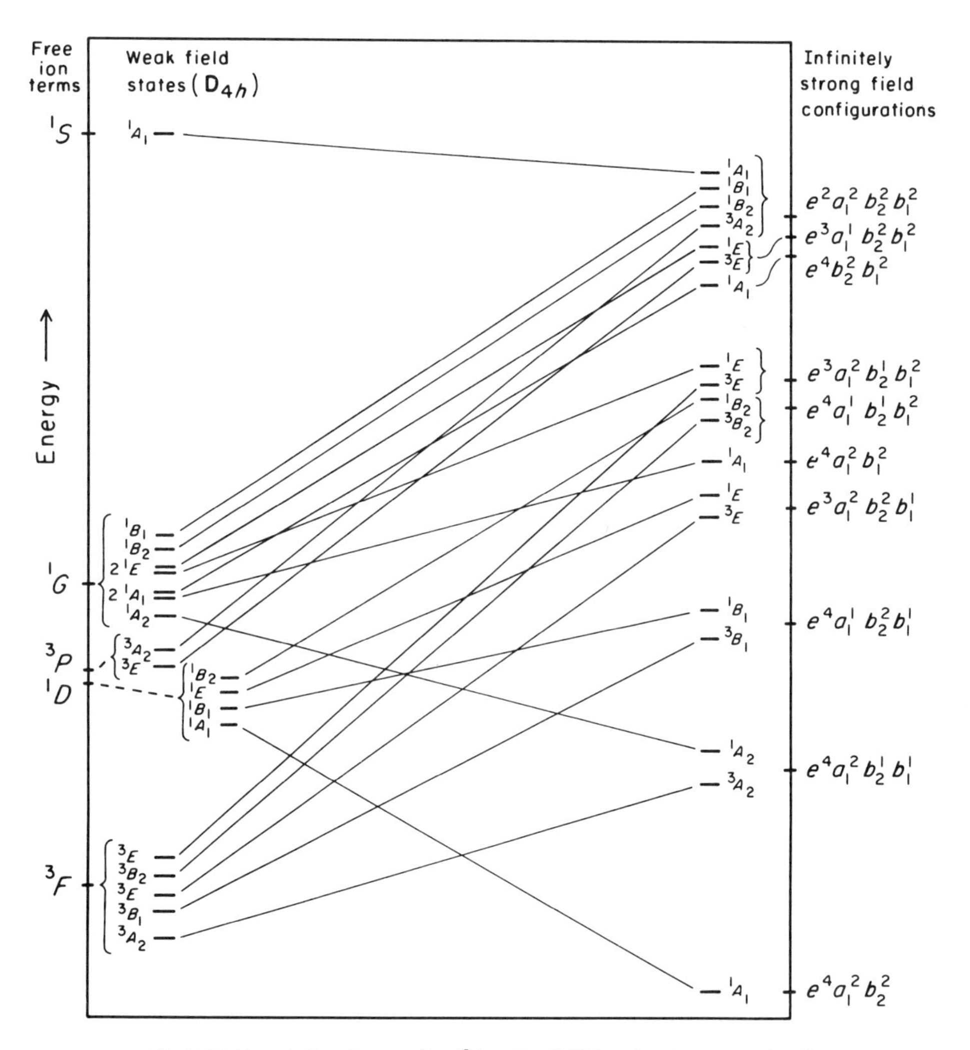

Fig. 9.12 Correlation diagram for d^8 in a $\mathbf{D}_{4h}$ field (g subscripts are omitted).

9.5 Spectra of Six-Coordinate Complexes of Lower Symmetry

9.5.1 GENERAL FEATURES

As we have seen for $[Co(NH_3)_6]^{3+}$ and $[Co(en)_3]^{3+}$ (Section 9.1.2), the addition of chelate rings usually does not change the *effective* symmetry for absorption spectra. The absorption spectra of $[Co(en)(NH_3)_4]^{3+}$,

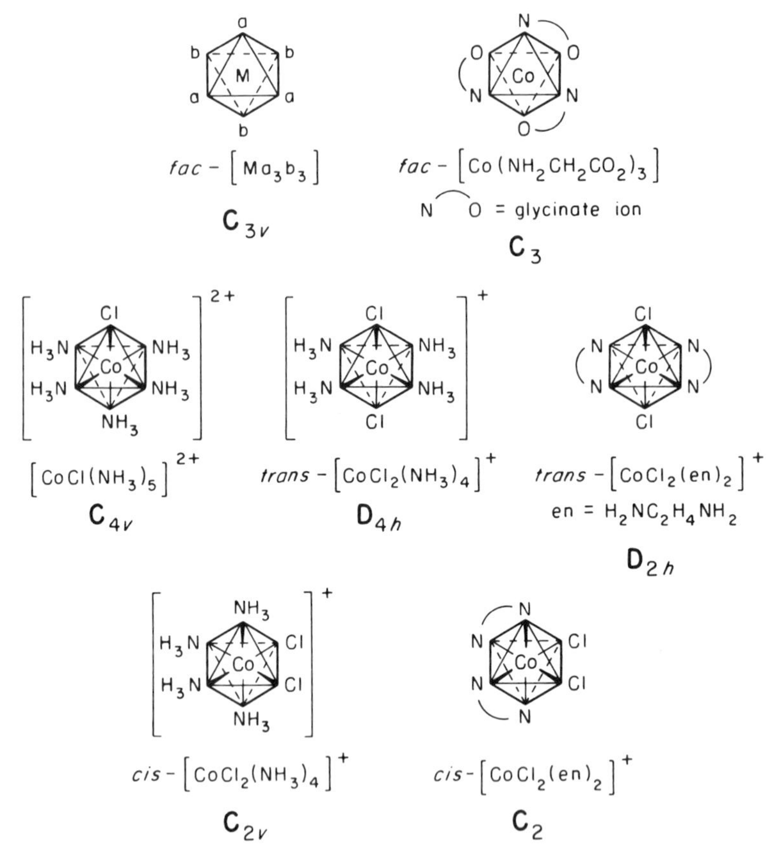

Fig. 9.13 Structures of some cobalt(III) complexes.

$[Co(en)_2(NH_3)_2]^{3+}$ (2 isomers), $[Co(en)_3]^{3+}$, $[Co(dien)_2]^{3+}$ $[Co(trien)(en)]^{3+}$, etc.[7] can all be treated using effective $\mathbf{O}_h$ symmetry.

The *trans*-$[CoCl_2(NH_3)_4]^+$ complex is bright green and the *cis* isomer is violet. The arrangement of *different* ligand atoms does have a great influence on the spectra. The *trans* isomer has $\mathbf{D}_{4h}$ symmetry and the *cis* isomer is $\mathbf{C}_{2v}$ (see Fig. 9.13). The corresponding complexes of ethylenediamine $[CoCl_2(en)_2]^+$ have $\mathbf{D}_{2h}$ (*trans*) and $\mathbf{C}_2$ (*cis*) symmetry, but the spectra are so similar that the higher effective symmetry of the NH_3 complexes can be used.

The replacement of one or more of the ligands of an octahedral complex, ML_6, by other ligands can be treated as a perturbation or minor distortion of the octahedron. Since the *d* orbitals are centrosymmetric, there is an averaging effect for ligands in *trans* positions. The *fac*-$[a_3b_3]$ isomer (e.g., *fac*-

[7]dien = $NH_2C_2H_2NHC_4H_4NH_2$, diethylenetriamine,
trien = $NH_2C_2H_4NHC_2H_4NHC_2H_4NH_2$, triethylenetetraamine.

$[Co(NH_2CH_2CO_2)_3]$; see Fig. 9.13) shows two symmetrical absorption bands as expected for $\mathbf{O}_h$ symmetry. Here, along *each* axis we have *trans*-(N, O). The spectrum is that expected for octahedral $[CoX_6]$, where the field strength of X is the average of that for N and O $[Dq_X = (Dq_N + Dq_O)/2]$. Since the effective (averaged) fields are the same along x, y, and z, the effective symmetry is $\mathbf{O}_h$ (again the chelate rings can be neglected for absorption spectra, but see Section 9.7 for optical activity). This is known as the cubic case because there is no apparent splitting.

Because of the averaging effect for *trans* ligands, the splitting pattern for $[CoCl(NH_3)_5]^{2+}$ ($\mathbf{C}_{4v}$) is the same as that expected for *trans*-$[Co(NH_3)_4X_2]^{n+}$, where $Dq_X = (Dq_N + Dq_{Cl})/2$. However, the splitting for *trans*-$[CoCl_2(NH_3)_4]^+$ is *twice* as great as for $[CoCl(NH_3)_5]^{2+}$ because along the unique axis we have the effect of two Cl^- replacing NH_3 instead of one. Both of these are treated as having effective $\mathbf{D}_{4h}$ symmetry, and this is referred to as the tetragonal case because there is one unique axis. The Krishnamurthy–Schaap treatment (Section 8.1.4) verifies that the splitting of the energies of the d orbitals is twice as great for *trans*-$[MX_4Y_2]$ as for $[MX_5Y]$.

The *cis*-$[CoCl_2(NH_3)_4]^+$ has one unique axis (2 NH_3 along z) and two equivalent axes (NH_3 and Cl^- *trans* along x and y). This also can be treated as a tetragonal case with the strong field along z (2 NH_3) and the weaker field (average field strength for NH_3 and Cl^-) along x and y. Once again, we can use effective $\mathbf{D}_{4h}$ symmetry for *cis*-$[CoCl_2(NH_3)_4]^+$ and *cis*-$[CoCl_2(en)_2]^+$ even though the symmetries are $\mathbf{C}_{2v}$ and $\mathbf{C}_2$, respectively. Again Krishnamurthy–Schaap treatment verifies that the splitting for *cis*-$[MX_4Y_2]$ is half as great as for *trans*-$[MX_4Y_2]$ and in the *opposite direction* (strong field along unique axis for the *cis* isomer and weak field along unique axis for the *trans* isomer).

The much more pronounced splitting of the $T_{1g}(\mathbf{O}_h)$ band for *trans*-$[Co(en)_2F_2]^+$ compared to *cis*-$[Co(en)_2F_2]^+$ is shown in Fig. 9.14. The

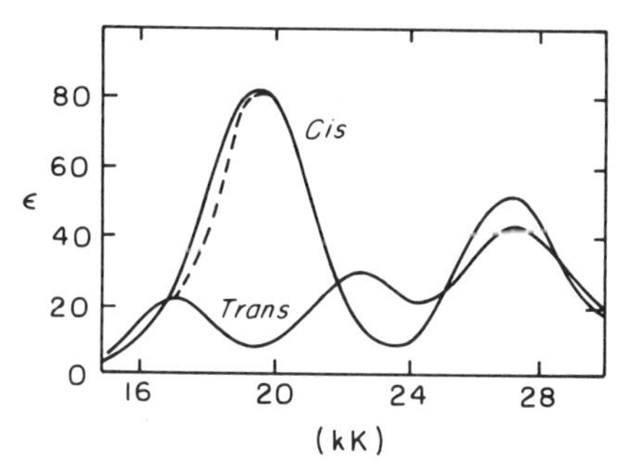

Fig. 9.14 The absorption spectra of *cis*- and *trans*-$[Co(en)_2F_2]NO_3$. The dotted line outlines the main Gaussian band. (After F. Basolo, C. J. Ballhausen, and J. Bjerrum, *Acta Chem. Scand.* 1955, **9**, 810.)

splitting of the high-energy $T_{2g}(\mathbf{O}_h)$ band is usually very slight or not apparent. The full range of splitting patterns is shown for the complexes in the series $[Cr(H_2O)_6]^{3+} \rightarrow [Cr(H_2O)_{6-x}(CN)_X]^{3-x} \rightarrow [Cr(CN)_6]^{3-}$ in Fig. 9.15. The positions expected are shown for the two *trans* isomers, although spectra were not available.

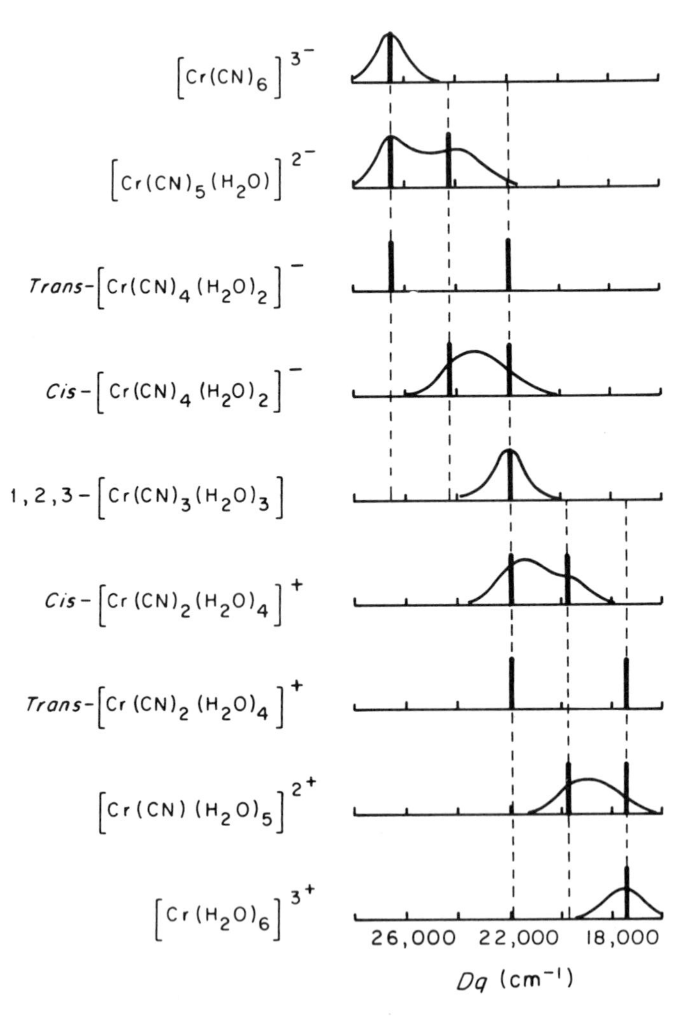

Fig. 9.15 Comparison of observed frequencies and splittings of the first *d-d* bands in aquacyanochromium(III) complexes with values calculated neglecting configuration interaction. ($Dq_{CN} = 2650\ cm^{-1}$; $Dq_{H_2O} = 1740\ cm^{-1}$; splitting $= \frac{35}{4}\ Dt$, where $Dt = \mp\frac{2}{7}(Dq_{CN} - Dq_{H_2O})$ for single substitution. (From W. B. Schaap, R. Krishnamurthy, D. K. Wakefield, and W. F. Coleman, in *Coordination Chemistry* (S. Kirschner, ed.), Plenum, New York, 1969, p. 177.)

9.5.2 LIGAND FIELD PARAMETERS FOR FIELDS OF LOWER SYMMETRY

Crystal field theory (CFT) for an octahedral complex treats the six ligands as charges interacting with electrons in the d orbitals, ignoring covalent bonding. The e_g orbitals are higher in energy than the t_{2g} orbitals by $10Dq$. We can modify this approach as the ligand field theory (LFT) by recognizing that covalent bonding is usually important and treating $10Dq$ as an empirical parameter including covalent interaction. Still only one parameter ($10Dq$) is required for the octahedral d^1 case. An increase in σ-bonding interaction (involving e_g orbitals only) increases $10Dq$ and π bonding (involving t_{2g} orbitals only) can increase or decrease $10Dq$.

Distorted octahedral complexes can be treated as perturbations of the $\mathbf{O}_h$ case using LFT. Covalent bonding is ignored since the participation of the various d orbitals in π bonding can be much more complex than for the $\mathbf{O}_h$ case. Nevertheless, as we shall see later, the LF parameters added for cases of lower symmetry can be related to σ and π interactions using the angular overlap model and earlier treatments of Yamatera and McClure (see Section 9.5.3).

We shall examine the case of tetragonally distorted complexes, such as $[CoCl(NH_3)_5]^{2+}$ ($\mathbf{C}_{4v}$) and *trans*-$[CoCl_2(NH_3)_4]^+$ ($\mathbf{D}_{4h}$), where we retain a C_4 axis. In addition to Dq, two other parameters, Ds and Dt, are required:

$$Dq = \frac{1}{6}\frac{Ze^2}{a^5}\langle r^4\rangle, \tag{9.12}$$

$$Ds = \frac{2}{7}Ze^2\langle r^2\rangle\left[\frac{1}{a_z^3} - \frac{1}{b_{xy}^3}\right], \tag{9.13}$$

$$Dt = \frac{2}{7}Ze^2\langle r^4\rangle\left[\frac{1}{a_z^5} - \frac{1}{b_{xy}^5}\right], \tag{9.14}$$

where Z is the charge on the central ion, e the charge on the electron, $\langle r^n\rangle$ is an integral dependent on the nth power of the mean radius of the d electron considered, and a is the metal–ligand distance. For the tetragonal complexes, a_{xy} is the distance to the a ligands and b_z is the distance to the b ligand(s) along z. Ds describes the effects of the distortion on the $\langle r^2\rangle$ term. Dt describes the differences in Dq values for the ligands along z and those in the xy plane. Ds and Dt for the $\mathbf{D}_{4h}$ case, *trans*-$[Ma_4b_2]$, are twice as great as for the $\mathbf{C}_{4v}$ case, $[Ma_5b]$ because there are two b ligands along z for $\mathbf{D}_{4h}$,

$$Dt = \tfrac{2}{7}[Dq(z) - Dq(xy)] \quad \text{for } \mathbf{C}_{4v} \text{ case}, \tag{9.15}$$

$$Dt = \tfrac{4}{7}[Dq(z) - Dq(xy)] \quad \text{for } \mathbf{D}_{4h} \text{ case}. \tag{9.16}$$

The signs of Ds and Dt are those of Perumareddi (*J. Phys. Chem.* 1967, **71**, 3144, 3155) and opposite those introduced by Moffitt and Ballhausen and used by Wentworth and Piper. In recent work, computer fits of the spectral data and the energy matrices (and signs) of Perumareddi have been used.

Following Wentworth and Piper[8], we assume that Dq for $[Co(NH_3)_6]^{3+}$ can be used for Dq for the four NH_3 in the xy plane[9] for $[CoCl(NH_3)_5]^{2+}$ and *trans*-$[CoCl_2(NH_3)_4]^+$. For $[Co(NH_3)_6]^{3+}$, the energies[8] for the spin-allowed transitions and the first spin-forbidden transition in terms of Dq and the Racah electron repulsion parameters (B and C)[10] are:

I $^1A_{1g} \rightarrow {}^1T_{1g}$	$10Dq - C$	21.20 kK
II $^1A_{1g} \rightarrow {}^1T_{2g}$	$10Dq + 16B - C$	29.55 kK
III $^1A_{1g} \rightarrow {}^3T_{1g}$	$10Dq - 3C$	13.40 kK

$$\text{II} - \text{I} = 16B = 8.35, \qquad B = 522 \text{ cm}^{-1}$$

$$\text{I} - \text{III} = 2C = 7.80, \qquad C = 3.90 \text{ kK}$$

$$\text{I } 21.20 = 10Dq - C = 10Dq - 3.90$$

$$10Dq = 25.10 \text{ kK.}$$

The symmetry species for the singlet energy states are the same for low-spin d^6 $\mathbf{C}_{4v}$ and $\mathbf{D}_{4h}$ complexes except for adding the g subscripts for $\mathbf{D}_{4h}$. The energies of the transitions are as follows.

	$[CoCl(NH_3)_5]^{2+}$	*trans*-$[CoCl_2(NH_3)_4]^+$
$^1A_1 \rightarrow {}^1E_a$ $10Dq + \frac{35}{4}Dt - C$	18.72 kK	15.90 kK
$^1A_1 \rightarrow {}^1A_2$ $10Dq - C$	21.35 kK	21.00 kK
$^1A_1 \rightarrow {}^1B_2$ $10Dq + 4Ds + 5Dt + 16B - C$ and $^1A_1 \rightarrow {}^1E_b$ $10Dq - 2Ds + \frac{25}{4}Dt + 16B - C$	27.50 kK	24.94 kK

There is an obvious problem since we have three experimental energies in each case and five parameters! We see that the energy of $^1A_1 \rightarrow {}^1A_2$ is the same as that of the $^1A_{1g} \rightarrow {}^1T_{1g}$ ($\mathbf{O}_h$) transition. Using the same value of C (3.90) and the relative constancy of $^1A_2$, we have a check on $10Dq$ (25.25 for $\mathbf{C}_{4v}$ and 25.00

[8] R. A. D. Wentworth and T. S. Piper, *Inorg. Chem.* 1965, **4**, 709. Data are presented for a series of Co(III) and Cr(III) complexes.

[9] D. A. Rowley and R. S. Drago (*Inorg. Chem.* 1968, **7**, 795) found that for tetragonal nickel(II) complexes, the value of Dq for ligands along z were not constant with changes in ligands in the xy plane. Wentworth and Piper found that the values of Dq for particular ligands to be quite constant for the complexes they studied.

[10] The Racah parameter A, B, and C are defined in terms of the Slater–Condon F_k parameters. Chemists find them convenient since transitions involving states of maximum multiplicity require B $(F_2 - 5F_4)$ only. Here we need C $(35F_2)$ also since these are not states of maximum multiplicity.

for $\mathbf{D}_{4h}$). Using these $10Dq$ values (although the value for the parent $\mathbf{O}_h$ complex is used often) for ${}^1E_a$ we get

$$\mathbf{C}_{4v} \quad 18.72 = 25.25 + \tfrac{35}{4}Dt - 3.90,$$

$$Dt = -0.30 \quad \text{kK},$$

$$\mathbf{D}_{4h} \quad 15.90 = 25.00 + \tfrac{35}{4}Dt - 3.90,$$

$$Dt = -0.59 \quad \text{kK}.$$

Thus, we see that the splitting parameter Dt is very nearly twice as great for the $\mathbf{D}_{4h}$ case.

As is usually the case, the ${}^1A_{1g} \rightarrow {}^1T_{2g}$ ($\mathbf{O}_h$) band is unsplit for the tetragonally distorted complexes. Hence, we are unable to evaluate Ds. Wentworth and Piper estimated $Ds < \frac{5}{3}Dt$. The energy of band II decreases from $[Co(NH_3)_6]^{3+}$ to $[CoCl(NH_3)_5]^{2+}$ to *trans*-$[CoCl_2(NH_3)_4]^+$ because of the substantial positive Dt term for each tetragonal component ($+5Dt$ or $+6\frac{1}{4}Dt$).

The case for evaluation of tetragonal parameters is more favorable for complexes of Cr(III) (d^3) since more transitions can be observed. The energies of the spin-allowed ligand field transitions for d^3 were given in Eq. (9.11). Tetragonal distortion to give a $\mathbf{D}_{4h}$ complex gives the following energy states for which the energies relative to octahedral terms as zero are given by Lever[11] (but using Perumareddi's sign convention here):

d^3 (and d^8)	$\mathbf{O}_h$	$\mathbf{D}_{4h}$	Energy[a]	Separation
	${}^4A_{2g}$	${}^4B_{1g}$	$-7Dt$	
	${}^4T_{2g}$	${}^4B_{2g}$	$-7Dt$	$\frac{-35}{4}Dt$
		${}^4E_g$	$+\frac{7}{4}Dt$	
	${}^4T_{1g}(F)$	${}^4A_{2g}$	$+4Ds - -2Dt$	$6Ds - \frac{5}{4}Dt$
		${}^4E_g$	$-2Ds - \frac{3}{4}Dt$	
	${}^4T_{1g}(P)$	${}^4A_{2g}$	$-2Ds + 8Dt$	$-3Ds + 5Dt$
		${}^4E_g$	$+Ds + 3Dt$	

(9.17)

[a] Relative to $\mathbf{O}_h$ states as zero.

Configuration interaction between the two ${}^4T_{1g}$ states increases their separation but reduces the splitting of each of these states, giving separations of $4.9Ds - 0.5Dt$ for the ${}^4T_{1g}(F)$ and $-1.9Ds + 4.35Dt$ for ${}^4T_{1g}(P)$. For complexes of the type $[Cra_5b]$ and *trans*-$[Cra_4b_2]$, the ${}^4T_{1g}$ ($\mathbf{O}_h$) bands show little or no splitting, so we can only evaluate $10Dq$ and Dt from the splitting of the first (${}^4T_{2g}$) band. If Dq is greater for the ligands in the xy plane than for those along z, then Dt is negative and ${}^4B_{1g} \rightarrow {}^4E_g$ ($10Dq + \frac{35}{4}Dt$) is lower in

[11] A. B. P. Lever, *Inorganic Electron Spectroscopy*, Elsevier, Amsterdam, 1968, p. 193.

energy than ${}^4B_{1g} \rightarrow {}^4B_{2g}$ ($10Dq_{xy}$). In some cases[12] curve analysis of the second band [${}^4B_{1g} \rightarrow {}^4T_{1g}(F)$] reveals components that can be used to evaluate Ds.

The Orgel diagram for d^3 is the same as that for d^8 [Ni(II)] and likewise the same tetragonal splitting parameters apply. For d^3 complexes, the highest energy spin-allowed band is often not observed because it is covered by a more intense ligand or charge-transfer band. The *d–d* bands for Ni(II) occur at lower energy than those of Cr(III) and usually all spin-allowed bands are observed for Ni(II). However, in both cases lowering of symmetry causes significant splitting for low-energy bands only. Since the d^2 case is the hole counterpart of d^8, the energies of the tetragonal components for d^2 are the same as those for d^8, but with the signs of the Ds and Dt terms reversed.

EXAMPLE 9.9 The low-temperature single-crystal polarized spectrum of *trans*-[$Cr(en)_2F_2$]ClO_4 reveals the following bands:[13]

Band position (polarization)		Assignment
18,800 cm^{-1}	(z)	${}^4B_{1g} \rightarrow {}^4E_g({}^4T_{2g})$
21,700	(xy)	$\rightarrow {}^4B_{2g}$
25,000	(z)	$\rightarrow {}^4E_g[{}^4T_{1g}(F)]$
29,300	(xy)	$\rightarrow {}^4A_{2g}$
41,000 (shoulder)	(xy)	$\rightarrow {}^4A_{2g}[{}^4T_{1g}(P)]$

Calculate the parameters Dq, Dt, Ds, and **B** [allowing for configuration interaction between the T_{1g} states, see Eq. (9.11)].

SOLUTION We can treat the complex with D_{2h} symmetry as effectively $\mathbf{D}_{4h}$ because we have seen that chelate rings have little effect on the local symmetry at the metal ion. Here $10Dq$ is the energy of the transition to ${}^4B_{2g}$ or $Dq_{en} = 2170\ cm^{-1}$. From the ${}^4B_{1g} \rightarrow {}^4E_g({}^4T_{2g})$ transition we get[14]

$$18{,}800 = 10Dq_{en} + \tfrac{35}{4}Dt,$$

$$Dt = -331 \quad cm^{-1}$$

From the separation of the ${}^4T_{1g}(F)$ components we get

$${}^4A_{2g} - {}^4E_g = 4300 = 6Ds - \tfrac{5}{4}Dt,$$

$$Ds = 648.$$

[12] W. A. Baker, Jr. and M. G. Phillips, *Inorg. Chem.* 1966, **5**, 1042.

[13] L. Dubicki, M. A Hitchman, and P. Day, *Inorg. Chem.* 1970, **9**, 188.

[14] Dubicki *et al.* report Dt to be negative using the matrices given by Perumareddi (*J. Phys. Chem.* 1967, **71**, 3144). D. A. Rowley (*Inorg. Chem.* 1971, **10**, 397), using Dubicki's data and the opposite signs of Dt and Ds, found $Dt = 362$ and $Ds = -741$ using a computer-fit program. As defined in Eq. (9.17), if ${}^4T_{2g}$ splits to give ${}^4E_a$ as the lower energy component, Dt is negative.

Since here the Ds term determines the order of the splitting ($6Ds > \frac{5}{4}Dt$) and $^4A_{2g}$ is higher in energy, Ds must be positive [see Eq. (9.17)]. We can evaluate the extent of configuration interaction (bending parameter c) for the two T_{1g} ($\mathbf{O}_h$) states, using Eq. (9.11) to get

$$^4B_{1g} \rightarrow {}^4E_g(T_{1g}) \qquad 18Dq - c - 2Ds - \tfrac{3}{4}Dt$$

$$= 39{,}060 - 1300 + 248 - c = 25{,}000$$

$$c = 13{,}000 \quad \text{cm}^{-1}.$$

We see that configuration interaction is very great. Using Eqs. (9.11) and (9.17) and the energy of the $^4A_{2g}[^4T_{1g}(P)]$ transition, we calculate B',

$$^4B_{1g} \rightarrow {}^4A_{2g} \qquad 12Dq + 15\text{B}' + c - 2Ds + 8Dt = 41{,}000 \quad \text{cm}^{-1},$$

$$\text{B}' = 400 \quad \text{cm}^{-1}.$$

In Example 9.9 we calculated $10Dq$ from the energy of the $^4B_{1g} \rightarrow {}^4B_{2g}$ transition. This transition is not involved in configuration interaction since this is the only $^4B_{2g}$ state. Dt was calculated from the $^4B_{1g} \rightarrow {}^4E_g(^4T_{2g})$ transition. The $^4\text{T}_{2g}$ ($\mathbf{O}_h$) state is not involved in configuration interaction, but the ^4E_g ($\mathbf{D}_{4h}$) state will interact with the other ^4E_g states, causing an error in Dt. Ds was calculated from the splitting of the $^4T_{1g}(F)$ band assuming that the splitting is unaffected by configuration interaction. As we have noted, configuration interaction decreases the splittings. The Dt and Ds values were used to calculate the bending parameter c assuming that the important splitting is between the $^4T_{1g}$ ($\mathbf{O}_h$) states. This, of course, is an oversimplification. The two $^4A_{2g}$ states interact and all three ^4E_g states interact. The matrices for the energies of these states are given by Lever.[15] The 2×2 matrix for $^4A_{2g}$ leads to a quadratic equation and the 3×3 matrix for ^4E_g leads to a cubic equation. For research applications computer programs are used to solve for the parameters using the spectral data rather than estimated peak positions.[16] For *trans*-$[Cr(en)_2F_2]^+$ treated in Example 9.9, the results (in cm^{-1}) are $Dq = 2170$, $Dt = -320$, $Ds = 800$, and $B = 625$.

9.5.3 ANGULAR OVERLAP MODEL PARAMETERS FOR FIELDS OF LOWER SYMMETRY

Yamatera[17] dealt with the splitting patterns of distorted octahedral cobalt(III) complexes using a molecular orbital approach. He used parameters to separate the σ and π contributions. McClure's treatment[18] is similar.

[15] A. B. P. Lever, *Inorganic Electronic Spectroscopy*, Elsevier, Amsterdam, 1968, Appendix 5.
[16] L. Dubicki and P. Day, *Inorg. Chem.* 1971, **10**, 2043.
[17] H. Yamatera, *Bull. Chem. Soc. Japan* 1958, **31**, 95
[18] D. S. McClure, *Advances in the Chemistry of the Coordination Compounds*, (S. Kirchner, ed.), Macmillan, New York, 1961, p. 498.

Results of application of the angular overlap model (AOM) to the energy levels of complexes of lower symmetry are essentially equivalent to the pioneering work of Yamatera and McClure. The AOM approach fits their treatments into a useful broader scheme (see Section 8.3.2).

We are interested in the effects of σ and π interaction of a ligand on the energies of the d orbitals involved in ligand field transitions, where $3e_\sigma - 4e_\pi = 10Dq$. Let us limit our consideration to linearly bonded ligands[19] (J), neglecting δ interaction, and defining the splitting parameters

$$\Delta_{\sigma J} = 3e_{\sigma J}, \qquad \Delta_{\pi J} = 4e_{\pi J}. \tag{9.18}$$

The term $\Delta_{\sigma J}$ is positive because $e_{\sigma J}$ represents the σ-antibonding energy. We are interested in transitions to the empty σ^* orbitals. The term $\Delta_{\pi J}$ is positive for π-donor ligands, but it can be negative if the π-acceptor role is more important. The overall splitting of the d orbitals is determined by the number of ligands and their Δ values. Since the e_g ($\mathbf{O}_h$) orbitals are involved in σ bonding only and the t_{2g} orbitals are involved in π bonding only, the splittings for *trans*-$[Cr(NH_3)_4X_2]^+$ are given by

$$\begin{aligned}
\Delta(d) &= \tfrac{1}{3}\Delta_X + \tfrac{2}{3}\Delta_N = \tfrac{1}{3}(\Delta_{\sigma X} - \Delta_{\pi X}) + \tfrac{2}{3}(\Delta_{\sigma N} - \Delta_{\pi N}),\\
\Delta(e_g) &= -\tfrac{2}{3}(\Delta_{\sigma X} - \Delta_{\sigma N}),\\
\Delta(t_{2g}) &= -\tfrac{1}{2}(\Delta_{\pi X} - \Delta_{\pi N}).
\end{aligned} \tag{9.19}$$

To reduce the number of parameters to be evaluated, we follow Schäffer *et al.*[20] in defining new parameters relative to $\Delta_{\pi N}$, since $\Delta_{\pi N}$ should vanish for NH_3. The new parameters are

$$\begin{aligned}
\Delta'_{\sigma N} &\equiv \Delta_N,\\
\Delta'_{\sigma X} &= \Delta_{\sigma X} - \Delta_{\pi N}\\
\Delta'_{\pi X} &= \Delta_{\pi X} - \Delta_{\pi N}.
\end{aligned} \tag{9.20}$$

Equations (9.19) become

$$\begin{aligned}
\Delta(d) &= \tfrac{1}{3}\Delta'_{\sigma X} - \tfrac{1}{3}\Delta'_{\pi X} + \tfrac{2}{3}\Delta_N,\\
\Delta(e_g) &= -\tfrac{2}{3}\Delta'_{\sigma X} + \tfrac{2}{3}\Delta_N,\\
\Delta(t_{2g}) &= -\tfrac{1}{2}\Delta'_{\pi X}.
\end{aligned} \tag{9.21}$$

The paper by Glerup *et al.* presents a thorough treatment of a series of *trans*-$[Cr(NH_3)_4X_2]^+$, *trans*-$[Cr(NH_3)_4XY]^+$, *trans*-$[Cr(py)_4X_2]^+$, and *trans*-

[19] For a linearly bonded ligand, the M–J bonding has $\mathbf{C}_{\infty v}$ symmetry. NH_3 can be considered to be a linealry bonded ligand, but NO_2^- is not because of oriented π bonding.

[20] J. Glerup, O. Mønsted, and C. E. Schäffer, *Inorg. Chem.* 1976, **15**, 1399.

$[Cr(py)_4XY]^+$ complexes. The spectral parameters and the angular overlap parameters were evaluated by computer programs. They established the following "two-dimensional" spectrochemical series for tetragonal complexes considering the σ and π effects separately:

$$\Delta_J: \quad Br^- < Cl^- < F^- < H_2O < py < NH_3 \quad \text{(overall)},$$

$$\Delta'_{\sigma J}: \quad Br^- < Cl^- < py < H_2O(?) < NH_3 < F^- < OH^-(?)$$

$$\Delta'_{\pi J}: \quad py < NH_3 < Br^- < Cl^- < H_2O(?) < F^- < OH^-(?).$$

The Δ_J value of F^- is lower than those of NH_3, py, and H_2O because of the π-antibonding contribution of F^-. The positions of H_2O and OH^- are shown as doubtful because of nonlinear π interaction. Not enough is known about these nonlinear contributions to treat them adequately. The $\Delta'_{\pi J}$ value of py is less than that of NH_3 because the results indicate that py serves as a π acceptor to some extent.

EXAMPLE 9.10 Spectral data for *trans*-$[Cr(en)_2F_2]^+$ are given in Example 9.9. The value of $\Delta(e_g)$ is $-1600\ cm^{-1}$ and $\Delta(t_{2g})$ is $-4000\ cm^{-1}$. Calculate Δ_N, $\Delta'_{\sigma F}$, $\Delta'_{\pi F}$, and $\Delta(d)$.

SOLUTION The transition ${}^4B_{1g} \rightarrow {}^4B_{2g}$ is unaffected by the ligands added along z, so the energy of this transition is Δ_N ($21{,}700\ cm^{-1}$).

$$\Delta(t_{2g}) = -4000 = -\tfrac{1}{2}\Delta'_{\pi F},$$

$$\Delta'_{\pi F} = 8000 \quad cm^{-1},$$

$$\Delta(e_g) = -1600 = -\tfrac{2}{3}\Delta'_{\sigma F} + \tfrac{2}{3}(21{,}700),$$

$$\Delta'_{\sigma F} = 24{,}100 \quad cm^{-1}$$

$$\Delta(d) = \tfrac{1}{3}(24{,}100) - \tfrac{1}{3}(8000) + \tfrac{2}{3}(21{,}700)$$

$$= 19{,}800\ cm^{-1}.$$

9.6 Magnetic-Dipole Transitions

The rare earth metal compounds give absorption spectra that are characterized by sharp peaks. These are magnetic-dipole transitions that involve a change in magnetic-dipole moment. The integral to be examined is of the same form as that for electric-dipole transitions [Eq. (9.6)] except that the operator

μ_m is the magnetic-dipole moment operator

$$\int \psi_e^{gr} \mu_m \psi_e^{ex} \, d\tau. \tag{9.22}$$

The components of μ_m belong to the same symmetry species as rotations about the x, y, and z axes, R_x, R_y, R_z. Since these rotations are g, the symmetry selection rule requires that ψ^{gr} and ψ^{ex} both be g or both u: $g \times g \times g = g$ and $u \times g \times u = g$, but $g \times g \times u = u$. Hence $d \to d$ and $f \to f$ transitions are magnetic-dipole allowed. The more detailed application of the selection rule is similar to that for electric-dipole transitions—the integral must be totally symmetric (A_{1g}, A_1, etc.). However, now the symmetry of the transition ($\Gamma_{\psi^{gr}} \times \Gamma_{\psi^{ex}}$) must be the same as that of R_x, R_y, or R_z.

9.7 Optical Activity

An optically active substance rotates the plane of polarization of plane polarized light passing through it. A plot of the observed angle of rotation (the molar rotation, on a molar basis) versus wavelength (or frequency) is an *optical rotatory dispersion* (ORD) curve. It was shown by Fresnel in 1825 that the angle of rotation can be related to the difference in the indices of refraction for the components of right and left circularly polarized light, which combine to give plane polarized light (see Fig. 9.16 and 9.17),

$$\alpha = (n_l - n_r)\pi/\lambda \qquad \text{(Fresnel).} \tag{9.23}$$

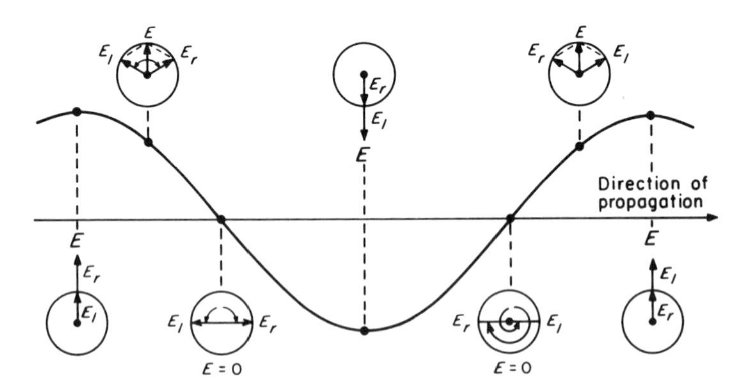

Fig. 9.16 A beam of polarized light viewed from the side (sine wave) and along the direction of propagation at specific times (circles). The resultant vector **E** and the circularly polarized components $\mathbf{E}_l$ and $\mathbf{E}_r$ are shown.

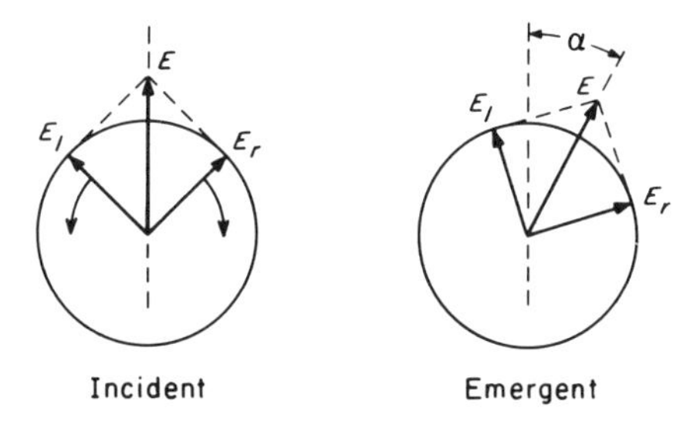

Fig. 9.17 Plane polarized light before entering and after emerging from an optically active substance.

Since one circularly polarized component is slowed down more than the other ($n_l \neq n_r$), the resultant vector **E** is rotated through an angle α.

In the region of an absorption band the right and left circularly polarized components are absorbed to different extents. If plane polarized light enters a sample, the light that emerges is elliptically polarized (Fig. 9.18). The

$$\varepsilon_l \neq \varepsilon_r, \qquad \Delta\varepsilon = \varepsilon_l - \varepsilon_r = \text{CD} \tag{9.24}$$

differential absorption ($\Delta\varepsilon$) is the *circular dichroism* (CD). Since $\varepsilon_l - \varepsilon_r$ defines the minor axis of the ellipse, $\varepsilon_l + \varepsilon_r$ defines the major axis, and $\tan\theta = (\varepsilon_l - \varepsilon_r)/(\varepsilon_l + \varepsilon_r)$, $\tan\theta$ (or θ itself) defines the eccentricity of the ellipse also. The CD is sometimes expressed as the ellipticity defined by θ,

$$[\theta] \quad (\text{degrees}) \cong 3298\ (\Delta\varepsilon). \tag{9.25}$$

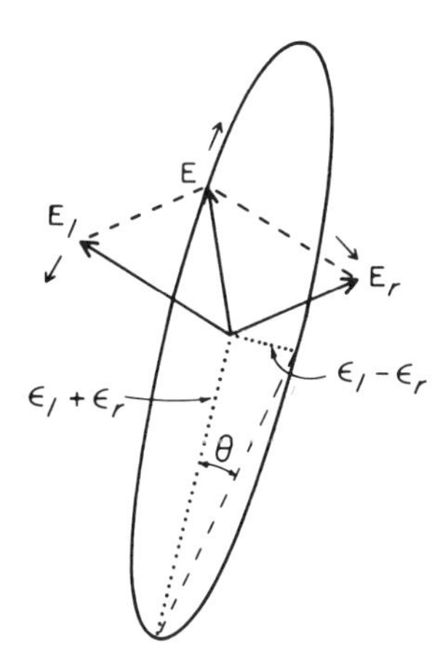

Fig. 9.18 Elliptically polarized light which emerges from a sample of an absorbing optically active substance, $\varepsilon_l \neq \varepsilon_r$ and $n_l \neq n_r$. The incident light was plane polarized.

The CD and ORD are two manifestations of the interaction of polarized light with an optically active substance. The phenomena together are known as the *Cotton effect* (after A. Cotton, the discoverer). Figure 9.19 shows the ORD and CD in relation to an absorption band for one of a pair of optical isomers. It is not necessary, generally, to measure both ORD and CD, since the ORD curve is the derivative of the CD curve.

The CD curves have more simple shape (treated as Gaussian) and have the great advantage that the intensity drops to zero not far from the peak maximum. The observed rotation for a colorless substance, such as tartaric acid, in the visible region is a composite of the tails of all of the electronic transitions that satisfy the selection rules—and these are located far away in the ultraviolet region.

Selection Rules for Optical Activity. The area of a CD curve is related to the rotational strength of the transition, corresponding to the dipole strength for electronic absorption. The rotational strength R is equal to the imaginary part of the product of the integrals for the electric-dipole and magnetic-dipole transitions.

$$R = \mathrm{Im}\left(\int_{-\infty}^{\infty} \Psi_{\mathrm{e}}^{\mathrm{gr}} \mu_{\mathrm{e}} \Psi_{\mathrm{e}}^{\mathrm{ex}}\, d\tau\right)\left(\int_{-\infty}^{\infty} \Psi_{\mathrm{e}}^{\mathrm{gr}} \mu_{\mathrm{m}} \Psi_{\mathrm{e}}^{\mathrm{ex}}\, d\tau\right) \tag{9.26}$$

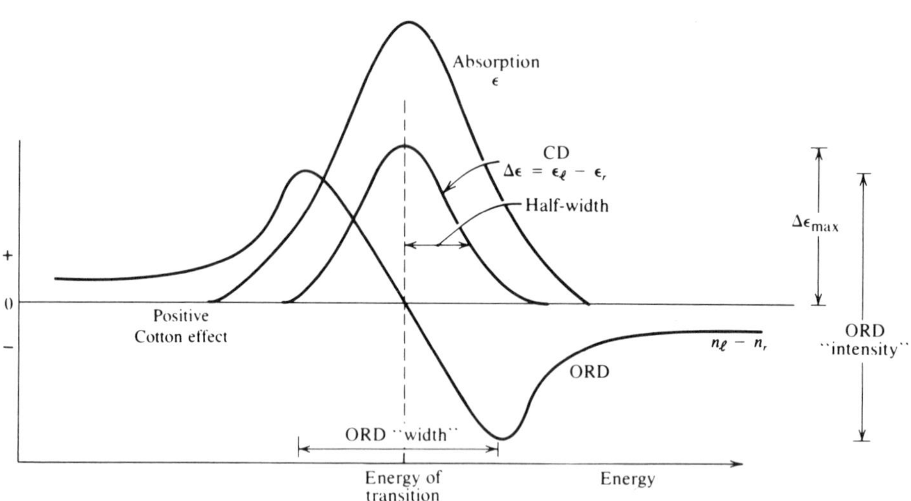

Fig. 9.19 Theoretical curves for absorption, circular dichroism, and optical rotatory dispersion for a single electronic transition of one optical isomer. For the other optical isomer the Cotton effect will be negative for this transition, the CD curve will be negative, and the ORD curve will be the mirror image of the one shown, with the trough at lower energy than the peak. (From B. E. Douglas, D. H. McDaniel, and J. J. Alexander, *Concepts and Models of Inorganic Chemistry*, 2nd ed., Wiley, New York, 1983, p. 315. © 1983 John Wiley & Sons, Inc. Reproduced by permission.)

Since we saw that the first integral prohibits $g \rightarrow g$ or $u \rightarrow u$ transitons and the second prohibits $g \rightarrow u$ or $u \rightarrow g$ transitions, they are mutually exclusive for centrosymmetric point groups. We can see why a substance cannot be optically active if it possesses a center of symmetry ($i = S_2$), but, in fact, it must lack *any* S_n axis. Consequently, only molecules belonging to $\mathbf{C}_n$ and $\mathbf{D}_n$ groups are optically active (there are no real examples of **O**, **T**, or **I**).

In Fig. 9.20, we see the absorption, CD, and ORD curves for $[Co(en)_3]^{3+}$. A striking feature is that in the region of the first absorption band ($A_{1g} \rightarrow T_{1g}$ for $\mathbf{O}_h$) there are *two* CD peaks of opposite sign. These are the two components expected for $\mathbf{D}_3$ symmetry. Since the separation between the components is small and the absorption bands are broad, the absorption components merge into one broad band. We see two CD peaks within this manifold because they are not so broad, and it helps that they are of opposite sign. The $\mathbf{D}_3$ components of T_{2g} are E and A_1, but the $A_1 \rightarrow A_1$ transition is electric- and magnetic-dipole forbidden (see Example 9.11) and does not show up in the CD spectrum. The presence of two components in the first band region is not so apparent for the ORD curve. This is one of the disadvantages of ORD, resulting partly from the more complex curve shape.

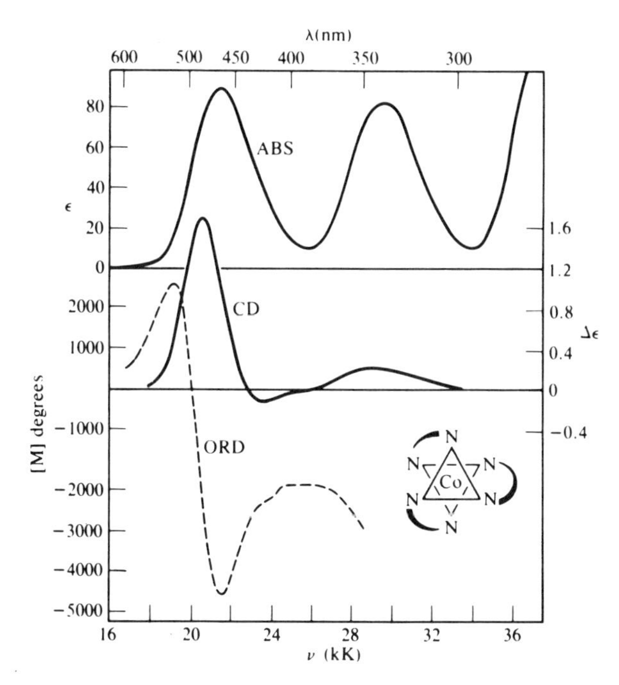

Fig. 9.20 Absorption, ORD, and CD curves of Λ-(+)-$[Co(en)_3]Cl_3$. (From B. E. Douglas, D. H. McDaniel, and J. J. Alexander, *Concepts and Models of Inorganic Chemistry*, 2nd ed., Wiley, New York, 1983, p. 316. © 1983 John Wiley & Sons, Inc. Reproduced by permission.)

EXAMPLE 9.11 Examine the selection rules for optical activity for $A_1 \rightarrow A_1$ and $A_1 \rightarrow A_2$ for $[Co(en)_3]^{3+}$.

SOLUTION Since the ground state is A_1, the transition symmetries are the same as for the excited states, A_1 and A_2. The operators z and R_z belong to A_2 and (x, y) and (R_x, R_y) belong to E. Therefore the A_2 transition is electric- and magnetic-dipole allowed and A_1 is electric- and magnetic-dipole forbidden. A transition is allowed only if *both* integrals are nonzero.

9.8 Polarized Crystal Spectra

We see from the equations for symmetry selection rules [Eq. (9.6)] that the operation of the selection rule for electric-dipole transitions depends on the direction of the change in electric-dipole moment—along x, y, or z. (The selection rules for magnetic-dipole transitions are treated similarly.) Thus, in Example 9.1 for $[Cr(en)_3]^{3+}$ ($\mathbf{D}_3$), we saw that ${}^1A_2 \rightarrow {}^1A_1$ was electric-dipole allowed using the operator μ_z. The ${}^1A_2 \rightarrow {}^1E$ transition is allowed using the operator $\mu_{x,y}$. This means that if the electric dipole changes along z (the C_3 axis), the ${}^1A_2 \rightarrow {}^1A_1$ transition is allowed, and if it changes in the xy plane, ${}^1A_2 \rightarrow {}^1E$ is allowed. Experimentally this means that if the electric vector of a beam of light is parallel to $z(C_3)$ for the complex ion, then the absorption band ${}^1A_2 \rightarrow {}^1A_1$ would be observed, and if it is in the xy plane, then ${}^1A_2 \rightarrow {}^1E$ is observed. Both bands should appear using unpolarized light (all orientations of the electric vector) or for random orientations of $[Cr(en)_3]^{3+}$ as occur in solution, regardless of whether or not the light is polarized. Actually this is not a practical case, since the $\mathbf{D}_3$ splittings of the $\mathbf{O}_h$ states (T_{2g} and T_{1g}) are very small, and since the absorption bands are very broad, the small splitting is obscured and the peaks merge into one broad envelope showing no obvious splitting (see Fig. 9.20 for $[Co(en)_3]^{3+}$ for which the absorption spectrum is very similar). The existence of two $\mathbf{D}_3$ components can be seen in the first LF band region in the CD spectrum of $[Co(en)_3]^{3+}$ (Fig. 9.20). Polarized CD studies of crystals[21] of $[Co(en)_3]^{3+}$ were used to identify the E and A_2 components from the T_{1g} ($\mathbf{O}_h$) band for $[Co(en)_3]^{3+}$.

In crystals, molecules or ions frequently show preferred orientation. We might expect square planar molecules or ions to stack like pancakes—all with the thin direction parallel, but not necessarily in neat stacks on top of one another. Since the crystal symmetry depends on the manner of stacking of the molecules or ions, a unique crystallographic axis is likely to coincide with a unique molecular symmetry axis. In a crystal of $[Cr(en)_3]^{3+}$ in which the $C_3(z)$

[21] R. E. Ballard, A. J. McCaffery, and S. F. Mason, *Proc. Chem. Soc.* 1962, 331.

symmetry axis is parallel to the unique crystallographic axis, the $^4A_2 \rightarrow {}^4A_1$ transition would be allowed with the crystal oriented such that the electric vector of a beam of linearly polarized light is parallel to the unique axis of the crystal. The $^4A_2 \rightarrow {}^4E$ transition is forbidden for this orientation, but it is allowed if the crystal is rotated 90° so that the electric vector of the polarized light is perpendicular to z, or is the xy plane.

Polarized crystal spectra provide a powerful technique for interpreting spectra in cases where the components can be resolved. Often, assignments based on solution spectra are uncertain because the sequence in terms of energy can be difficult to determine. It helps if the x-ray crystal structure is known. However, striking differences in the results of application of the selection rules along different crystallographic directions means that highly ordered stacking exists. If the assignments agree with those expected for the matching of crystallographic and molecular symmetry axes, then that is the likely stacking pattern. Some crystals are strikingly dichroic, showing different colors when viewed by light transmitted in different crystallographic directions.

The cases where the LF bands are electric-dipole allowed are generally those of noncentrosymmetric (or rotational) groups, such as $\mathbf{D}_3$. In these cases the splittings are usually very small, as noted for the $[M(en)_3]^{3+}$ case. In dealing with absorption spectra we use the highest effective symmetry that is consistent with the number of components observed, and this usually corresponds to a centrosymmetric group. The forbidden LF bands appear in centrosymmetric groups because of vibronic coupling (Section 9.1.2). A vibronically-allowed transition is one for which the product of the representations of the transition symmetry and the electric dipole moment operator is the same representation as one of the vibrational modes for the complex [Eqs. (9.8) and (9.9)]. Hence, if a transition is not formally electric-dipole allowed it is vibronically allowed if the product corresponds to the representation of a vibrational mode. We can consider polarized crystal spectra of vibronic bands by adding this step.

One of the best examples showing the value of polarized crystal spectra is *trans*-$[CoCl_2(en)_2]^+$. The bright green crystals of *trans*-$[CoCl_2(en)_2]Cl \cdot 2HCl \cdot 2H_2O$ and *trans*-$[CoCl_2(en)_2]ClO_4$ were studied by Yamada and co-workers.[22] The molecular symmetry is $\mathbf{D}_{2h}$, but, as is often the case, the absorption spectra do not reveal the effect of the two chelate rings, so the effective symmetry is $\mathbf{D}_{4h}$. This is the microsymmetry or the symmetry of the *trans*-$[CoCl_2(N)_4]^+$ chromophore. In $\mathbf{D}_{4h}$, the transitions are $A_{1g} \rightarrow {}^1E_g$, $^1A_{2g}$ (from $^1A_{1g} \rightarrow {}^1T_{1g}$, $\mathbf{O}_h$), and $^1A_{1g} \rightarrow {}^1E_g$, $^1B_{2g}$ (from $^1A_{1g} \rightarrow {}^1T_{2g}$, $\mathbf{O}_h$).

[22] S. Yamada and R. Tsuchida, *Bull. Chem. Soc. Japan* 1952, **25**, 127; S. Yamada, A. Nakahara, Y. Shimura, and R. Tsuchida, *Bull. Chem. Soc. Japan* 1955, **28**, 222.

The electric-dipole moment operators are $A_{2u}(\mu_z)$ and $E_u(\mu_{x,y})$. Applying the Laporte selection rule we get the following.

Transition	Symmetry of transition[a]	Direct product with operator A_{2u}	Direct product with operator E_u	Polarization[b]
${}^1A_{1g} \to {}^1A_{2g}$	A_{2g}	A_{1u}	E_u	$\perp$
${}^1A_{1g} \to {}^1B_{2g}$	B_{2g}	B_{1u}	E_u	$\perp$
${}^1A_{1g} \to {}^1E_g$	E_g	E_u	$A_{1u} + A_{2u}$ $B_{1u} + B_{2u}$	$\|^c, \perp$

[a] Product of electronic wave functions.
[b] Refers to xy polarization or polarization perpendicular to the principal axis (C_4).
[c] Refers to z polarization or polarization parallel to C_4.

Since the operators are odd (u) and both electronic wave functions are even (g), the product is necessarily odd so only an odd vibration can make a transition vibronically allowed. The odd vibrations of *trans*-[ML_4X_2] ($\mathbf{D}_{4h}$) are $2A_{2u} + B_{2u} + 3E_u$. Since there are no A_{1u} or B_{1u} modes, ${}^1A_{1g} \to {}^1A_{2g}$ and ${}^1A_{1g} \to {}^1B_{2g}$ are vibronically allowed only with $xy\,(\perp)$ polarization. Since there are modes belonging to the representations of the products with each operator for ${}^1A_{1g} \to {}^1E_g$, this transition is vibronically allowed with either polarization.

The polarized crystal absorption spectra of $[CoCl_2(en)_2]^+$ for both polarizations are shown in Fig. 9.21. In the region of the first LF band T_{1g} of $\mathbf{O}_h$ parentage, one component is seen for $\|$ polarization, so this should be

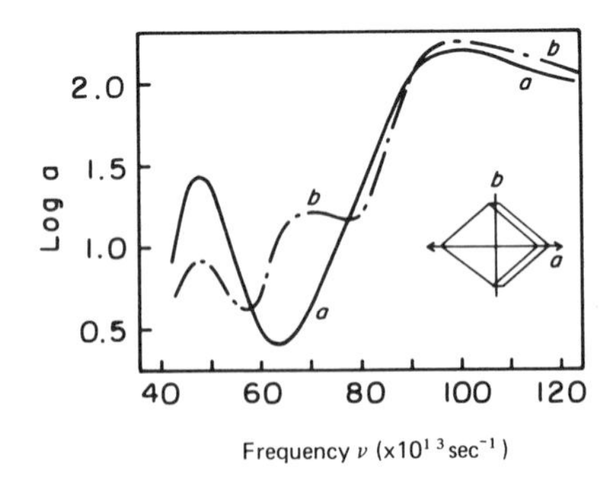

Fig. 9.21 Polarized absorption spectra of *trans*-[$CoCl_2(en)_2$]ClO_4. (From S. Yamada, A. Nakahara, Y. Shimura, and R. Tsuchida, *Bull. Chem. Soc. Japan* 1955, **28**, 222.)

$^1A_g \rightarrow {^1E_g}$, and two components are seen for $\perp$ polarization, $^1A_{1g} \rightarrow {^1E_g}$ (lower energy) and $^1A_{1g} \rightarrow {^1A_{2g}}$.

In the higher-energy region ($^1T_{2g}$, $\mathbf{O}_h$ parentage), one broad band is observed for either polarization. This should be $^1A_{1g} \rightarrow {^1E_g}$. Since we do not see the expected additional $^1A_{1g} \rightarrow {^1B_{2g}}$ component for $\perp$ polarization, this is presumed to be unresolved. On lowering symmetry the splitting of the T_{2g} ($\mathbf{O}_h$) band is usually considerably smaller than that of T_{1g} ($\mathbf{O}_h$), so it is expected that the two components would be likely to be unresolved.

9.9 Spin–Orbit Coupling and Double Groups

In the derivation of spectral terms (Section 5.1.2) and in dealing with spectra in this chapter, we have assumed that the quantum numbers L (giving terms such as S, P, D, F) and S (giving the spin multiplicity) could be dealt with separately. This is the Russell–Saunders coupling scheme, the l values combine vectorially to give L, s values give S, and $J = L + S$. It is also possible for l and s values to combine to give j. The j values combine to give J. In this spin–orbit coupling scheme L and S have lost their significance. The total angular momentum is given by J.

In Sections 3.13.2 and 5.2, we saw that the representation for an orbital or state wave function could be obtained from the angular momentum quantum number l or L. The character of the representation for rotation by the angle α is given by

$$\chi(\alpha) = \frac{\sin(l + \frac{1}{2})\alpha}{\sin \alpha/2}. \tag{9.27}$$

This equation can be used to obtain the representations for integral values of S and J also. However, S and J can also have half-integral values, and there the equation does not apply without modification of the group under consideration. The difficulty can be seen if we consider rotation by 2π (360°). This should be the equivalent of the identity operation. For integral quantum numbers $\chi(\alpha) = \chi(\alpha + 2\pi)$ but not for half-integral quantum numbers,

$$\chi(\alpha + 2\pi) = \frac{\sin[(J + \frac{1}{2})(\alpha + 2\pi)]}{\sin[(\alpha + 2\pi)/2]} = \frac{\sin[(J + \frac{1}{2})\alpha + \pi(2J + 1)]}{\sin(\alpha/2 + \pi)}, \tag{9.28}$$

and since $\pi(2J + 1)$ is a multiple of 2π for any half-integral value of J,

$$\pi(2J + 1) = 2\pi, \quad \text{if} \quad J = \tfrac{1}{2},$$

$$\pi(2J + 1) = 4\pi, \quad \text{if} \quad J = \tfrac{3}{2},$$

and $\sin(\alpha/2 + \pi) = -\sin \alpha/2$,

$$\chi(\alpha + 2\pi) = \frac{\sin(J + \frac{1}{2})\alpha}{-\sin(\alpha/2)} \quad \text{or} \quad \chi(\alpha + 2\pi) = -\chi(\alpha) = (-1)^{2J}(\alpha). \quad (9.29)$$

We cannot identify the representation for half-integral quantum numbers in this way, since the character changes sign on rotation by 2π.

Bethe provided a way of handling half-integral quantum numbers by considering rotation by 2π as an independent symmetry operation. Since we introduce a new symmetry operation, designated R, we must consider all products of R with other operations of the group. (The order of multiplication does not matter since rotations about the axis commute, $C_n^m R = RC_n^m$.) The new group is called a *double group* since there are twice as many symmetry operations. There will be new classes for the double group, and new representations, but not twice as many.

In evaluating the characters, we note that Eq. (9.27) reduces to $0/0$ for $\alpha = 0$ or 2π. The indeterminant can be evaluated using l'Hospital's rule (see Example 3.18) to give

$$\begin{aligned} \chi(0) &= 2J + 1, \\ \chi(2\pi) &= 2J + 1 \qquad \text{for integral } J \text{ values,} \\ &= -(2J + 1) \qquad \text{for half-integral } J \text{ values.} \end{aligned}$$

It follows that for half-integral values of J,

$$\chi(RC_n) = -\chi(C_n),$$

and RC_n and C_n must be in different classes, except for the case of C_2, for which

$$\chi(RC_2) = \chi(C_2) = 0.$$

C_2 is a special case of a more general result. It can be shown (see Problem 3.13) that for a rotation by $(m/n)(2\pi)$.

$$\chi\left[\frac{m}{n}(2\pi) + 2\pi\right] = \chi\left[\left(\frac{n - m}{n}\right)2\pi\right]$$

and so

$$\chi[RC_n^m] = \chi[C_n^{n-m}].$$

Thus, RC_n^m and C_n^{n-m} can be in the same class.

The characters are worked out for each rotational operation, and then the operations are grouped into classes. For a rotational group, the new classes are R and C_nR (other than C_2R). For the dihedral groups ($\mathbf{D}_n$ other than $\mathbf{D}_2$), C_n and C_n^{n-1} are in the same class. For the corresponding double groups (designated $\mathbf{D}'_n$), they are in separate classes, the classes being grouped as follows

$$\begin{array}{ccccccc} & & C_n & C_n^{n-1} & C_2 & C'_2 & \text{etc.} \\ E & R & C_n^{n-1}R & C_nR & C_2R & C'_2R & \end{array}$$

Let us obtain the character table for the double group $\mathbf{D}'_4$. For $\mathbf{D}_4$, there are eight operations, five classes, and five representations. For $\mathbf{D}'_4$, there are 16 operations and two new classes $-R$ and C_4R. We can fill in the table for all of the representations from the $\mathbf{D}_4$ group, since for these the characters are the same for E as for R and for C_4 as for C_4^3. Since there are two new classes, there must be two new representations. These must be two-dimensional representations for the sum of the squares of the dimensions to give the order of the group [Eq. (3.75)];

$$(1)^2 + (1)^2 + (1)^2 + (1)^2 + (2)^2 + (2)^2 + (2)^2 = 16.$$

The new two-dimensional representations have $+2$ characters for E, -2 for R, and 0 for any C_2 operation. This leaves only the characters for C_4 and C_4R to be evaluated. For $J = \frac{1}{2}$,

$$\chi(C_4) = \frac{\sin(\frac{1}{2} + \frac{1}{2})90^\circ}{\sin 45^\circ} = \sqrt{2},$$

$$\chi(C_4R) = \frac{\sin 450^\circ}{\sin 225} = -\sqrt{2}.$$

These are the characters for one of the new representations, E'_2. The other new two-dimensional representation has the same characters as E'_2 for all operations except C_4 and C_4R. For it to be orthogonal to E'_2, the characters for C_4 and C_4R must be the same with signs reversed. The resulting table is as follows.

	$\mathbf{D}'_4$	E	R	C_4 C_4^3R	C_4^3 C_4R	C_2 C_2R	$2C'_2$ $2C'_2R$	$2C''_2$ $2C''_2R$
Γ_1	A'_1	1	1	1	1	1	1	1
Γ_2	A'_2	1	1	1	1	1	-1	-1
Γ_3	B'_1	1	1	-1	-1	1	1	-1
Γ_4	B'_2	1	1	-1	-1	1	-1	1
Γ_5	E'_1	2	2	0	0	-2	0	0
Γ_6	E'_2	2	-2	$\sqrt{2}$	$-\sqrt{2}$	0	0	0
Γ_7	E'_3	2	-2	$-\sqrt{2}$	$\sqrt{2}$	0	0	0

The character tables for the common double groups involving improper rotations can be obtained by deriving the character table for the isomorphic rotational double group. Thus, the groups $\mathbf{D}_3$ ($\mathbf{C}_3 \wedge \mathbf{C}_2$) and $\mathbf{C}_{3v}$ ($\mathbf{C}_3 \wedge \mathbf{C}_s$) are isomorphic (see Sections 2.2 and 2.7.3), and the character tables are the same for the double groups $\mathbf{D}'_3$ and $\mathbf{C}'_{3v}$. Similarly, $\mathbf{D}'_{2d}$ and $\mathbf{C}'_{4v}$ are isomorphic with $\mathbf{D}'_4$. The character tables in Appendix 2 group important double groups as isomorphic groups with the corresponding symmetry operations identified.

The representations in the double group are designated using the Mulliken symbols with primes added or using Bethe's notation as Γ_1, Γ_2, etc. Both sets are used and are given here.

Let us consider the application to a one-electron case, the Ti^{3+} ion in an octahedral field. The free ion term for d^1 is ^2D with $J = \frac{3}{2}$ or $\frac{5}{2}$. Neglecting spin–orbit coupling, the energy states in an $\mathbf{O}_h$ field are ^2E_g and $^2T_{2g}$. If the effect of spin–orbit coupling is much greater than that of the ligand field, then the energy states are $^2D_{3/2}$ and $^2D_{5/2}$. So far we have considered cases where only the ligand field effect is important. Now we want to consider the intermediate case where both effects are important. We can approach this intermediate case from either extreme. If the ligand field effect is more important, then spin–orbit coupling can be considered a perturbation of the ^2E_g and $^2T_{2g}$ states. If spin–orbit coupling is more important, then we would consider the perturbation of the $^2D_{3/2}$ and $^2D_{5/2}$ levels by the $\mathbf{O}_h$ field. We choose the approach that is closer to the actual case.

For light elements, such as Ti, spin–orbit coupling is relatively small, so we might begin with the LF states, ^2E_g and $^2T_{2g}$. These are the states from a D term in an $\mathbf{O}_h$ field, that is, considering only the effect of the $\mathbf{O}_h$ field on the orbital quantum number. The representations for $L = 2$ are E_g and T_{2g} in $\mathbf{O}_h$. Here the spin is $\frac{1}{2}$, so we need to find the representation for $S = \frac{1}{2}$ using the double group $\mathbf{O}'$ (Appendix 2).

$\mathbf{O}'$	E	R	$8C_3$	$8C_3R$	$6C_2$	$6C_4$	$6C_4R$	$12C'_2$
$S = \frac{1}{2}$	2	-2	1	-1	0	$\sqrt{2}$	$-\sqrt{2}$	0
or								
$\Gamma_{(S=1/2)} = \Gamma_6(E'_2)$.								

Using the Bethe notation, E_g ($\mathbf{O}_h$) is $\Gamma_3(E'_1)$ in $\mathbf{O}'$ and T_{2g} ($\mathbf{O}_h$) is $\Gamma_5(T'_2)$ in $\mathbf{O}'$. The spin–orbit direct products of the representation for $S = \frac{1}{2}$ (Γ_6) and $L = 2$ (Γ_3 and Γ_5) using the $\mathbf{O}'$ table are

$$\Gamma_6 \times \Gamma_3 = \Gamma_8 \qquad \Gamma_6 \times \Gamma_5 = \Gamma_7 + \Gamma_8.$$

The product $\Gamma_6 \times \Gamma_5$ is a six-dimensional, reducible representation, that reduces to $\Gamma_7 + \Gamma_8$. These results tell us that the effect of spin–orbit coupling is to convert the LF state E_g ($\mathbf{O}_h$) into Γ_8 and to split T_{2g} ($\mathbf{O}_h$) into Γ_7 and Γ_8.

If we were dealing with a heavy element with large spin–orbit coupling, then we would begin with the J values $\frac{3}{2}$ and $\frac{5}{2}$. In an octahedral field, we can find the representations for $J = \frac{3}{2}$ and $\frac{5}{2}$ by using Eq. (9.27) to obtain the characters for each of the operations of the $\mathbf{O}'$ group.

$\mathbf{O}'$	E	R	$8C_3$	$8C_3R$	$6C_2$	$6C_4$	$6C_4R$	$12C_2'$
$J = \frac{3}{2}$	4	4	-1	1	0	0	0	0
$J = \frac{5}{2}$	6	-6	0	0	0	$-\sqrt{2}$	$\sqrt{2}$	0

From the $\mathbf{O}'$ table, we see that $\Gamma_{3/2}$ is Γ_8 and $\Gamma_{5/2}$ reduces to $\Gamma_7 + \Gamma_8$, or the $J = \frac{5}{2}$ level is split by the octahedral field to give two spin–orbit states Γ_7 and Γ_8. These are necessarily the same spin–orbit states we obtained starting with the LF states E_g and T_{2g}. Now let us draw a correlation diagram starting with the information that for d^1, the energy of $^2T_{2g}$ is lower than that for ^2E_g, and $D_{3/2}$ is lower in energy than $D_{5/2}$. The result is shown in Fig. 9.22. The separation of

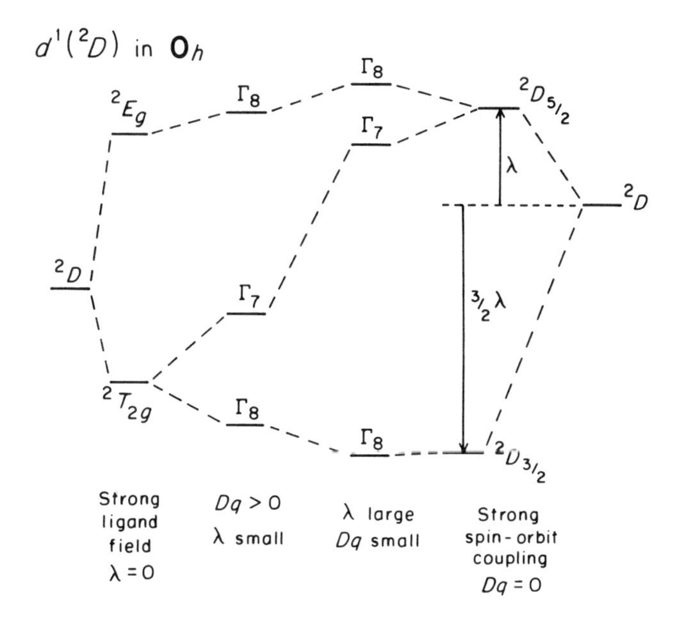

Fig. 9.22 The effects of spin–orbit coupling, starting with the ligand field states ($^2T_{2g}$ and ^2E_g on the left) and with the J levels ($\frac{3}{2}$ and $\frac{5}{2}$ on the right).

the J levels is shown in terms of λ, the spin–orbit coupling constant. The interval (separation between the J levels) is $\lambda \times J$ (of higher term) or $\frac{5}{2}\lambda$.

The d^9 case for Cu^{2+} is similar except for the inversion of the splitting of LF terms and J values. Also spin–orbit coupling is more important for Cu than for Ti. In the case of the Ag^{2+} (also d^9), spin–orbit coupling is still more important, and it would be more realistic to begin with the J values and examine the effect of imposing the ligand field on these. The correlation diagram for a square planar complex ($\mathbf{D}_{4h}$) of Ag^{2+} is shown in Fig. 9.23. We use the $\mathbf{D}'_4$ double group so we require characters for rotations only.

The J levels are $\frac{5}{2}$ (lower in energy for d^9) and $\frac{3}{2}$. The characters for the operations in $\mathbf{D}_4$ are obtained for these J values using Eq. (9.2.8), and the representations are determined. From the ligand field side we obtain the representations by operating on $L = 2$ or using the $\mathbf{O}_h$ LF states (the lower energy ^2E_g and $^2T_{2g}$) and a correlation table to get the corresponding representations in $\mathbf{D}_{4h}$ and in $\mathbf{D}'_4$ (drop the gs and primes). Then we obtain the representation for $S = \frac{1}{2}$ in $\mathbf{D}'_4$ (as for the $\mathbf{O}'$ case) and take the direct products of this representation (Γ_6) with the $\mathbf{D}'_4$ representations derived from the ligand field terms. We obtain the same set of representations, using each approach.

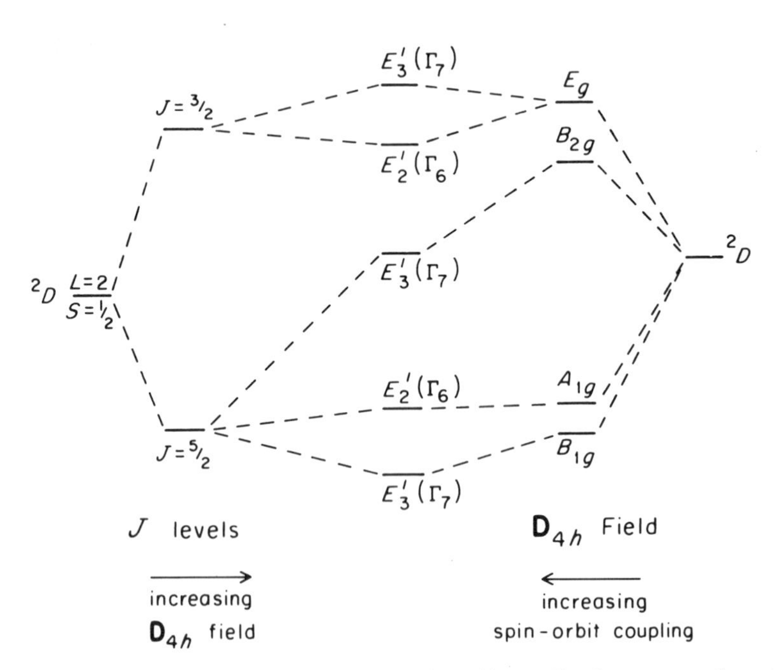

Fig. 9.23 Qualitative correlation diagram for spin–orbit coupling for a square planar ($\mathbf{D}_{4h}$) complex of Ag^{2+} (d^9) starting with the J levels (left) and the ligand field states (right).

The spin-selection rule applies strictly to the model represented by Russell–Saunders coupling. The effect of spin–orbit coupling is to mix spin states so that S is no longer a valid quantum number. This is the mechanism for the appearance of spin-forbidden bands. Generally the intensities of spin-forbidden bands increase as the extent of spin–orbit coupling increases.

9.10 Notation for Types of Electronic Spectra

We have dealt with electronic spectra using the symmetry species for the ground and excited states. This has the advantage that selection rules can be examined directly and designation of classification is superfluous if all bands considered are of the same type. In the more general case, it is often helpful to designate transitions by type.

(1) *Ligand field.* In this chapter, we have been concerned with the $d \rightarrow d$ transitions of metal complexes. These are usually referred to as ligand field (LF) bands. In general, these bands have low intensities ($\varepsilon \simeq 10\text{–}10^2$) because they are symmetry- (Laporte-) forbidden (see Section 9.1.2). If the bands are also spin-forbidden, involving a change in the number of unpaired electrons, the intensities are lowered by about two orders of magnitude. Even though the ligand field bands have relatively low intensities, they account for the beautiful varied colors of most transition metal complexes. The bands occur in the visible region because of the small energy separation of the d orbitals.

Vibronic. In the cases we have considered, the vibronic bands are LF bands that gain intensity because of coupling of the electronic transitions with vibrational transitions.

(2) *Change transfer.* Charge-transfer bands, in simplest terms, involve the transfer of an electron from one atom to another. We have seen examples (see Section 9.3.6) in the case of metal complexes of halide ions where an electron in the ground state of the molecule is transferred from a halide ion to an empty orbital that is largely localized on the metal ion. Metal-to-ligand charge transfer occurs where the metal has filled orbitals and the ligand has low-energy empty orbitals, such as the π^* orbitals of CN^-, CO, or bipyridine. Many such complexes are intensely colored because the transitions are symmetry- (and spin-) allowed, and the energy separations are small enough to correspond to the visible region of the spectrum. Charge-transfer bands can occur in organic and other inorganic compounds also. They usually occur in the ultraviolet region.

(3) $\pi \rightarrow \pi^*$. For compounds for which π bonding is important, the frontier orbitals are likely to be the bonding π orbitals (or the higher-energy,

predominantly bonding π orbitals), the HOMO, and the π^* orbitals (or the lower-energy, predominantly π^* orbitals), the LUMO. The $\pi \rightarrow \pi^*$ bands are intense when the symmetry species involved obey the selection rule. These are the important low energy transitions of aromatic compounds such as benzene.

(4) $\sigma \rightarrow \sigma^*$. Because the σ bonding orbitals are usually lowest in energy and the σ^* antibonding orbitals are highest in energy, $\sigma \rightarrow \sigma^*$ transitions usually occur at very high energy, probably corresponding to the far ultraviolet (vacuum ultraviolet) region of the spectrum.

(5) $n \rightarrow \pi^*$. Transitions from a nonbonding orbital, in many cases a lone pair, to a π^* orbital are designated as $n \rightarrow \pi^*$. We can examine the selection rule using the symmetry species involved for particular orbitals, but in some cases a physical picture provides useful information. For example, in the case of a heterocyclic compound such as pyridine, the lone pair on N occupies an orbital in the plane of the molecule and the π and π^* orbitals are perpendicular to this plane. Here the $n \rightarrow \pi^*$ transitions are expected to have low intensities, since they are orthogonal, unless there is some orbital mixing in either the ground or excited states.

(6) $n \rightarrow \sigma^*$. Transitions from a nonbonding orbital to a σ^* orbital are more likely to give net positive overlap and, hence, to obey selection rules than the $n \rightarrow \pi^*$ transitions for heterocyclic compounds, but they are expected to occur at higher energies.

Additional Reading

C. J. Ballhausen, *Introduction to Ligand Field Theory*, McGraw-Hill, New York, 1962. An excellent early source.

B. Douglas, D. H. McDaniel, and J. J. Alexander, *Concepts and Models of Inorganic Chemistry*, 2nd ed., Wiley, New York, 1983. The text covers background assumed in this book.

B. N. Figgis, *Introduction to Ligand Fields*, Wiley, New York, 1966.

M. Gerloch and R. C. Slade, *Ligand Field Parameters*, Cambridge Univ. Press, New York, 1961. A high-level treatment. The notation often differs from that used by most chemists.

G. Herzberg, *Spectra of Diatomic Molecules*, 2nd ed., Van Nostrand Reinhold, New York, 1950.

G. Herzberg, *Electronic Spectra of Polyatomic Molecules*, Van Nonstrand Reinhold, New York, 1966.

H. H. Jaffé and M. Orchin, *Theory and Applications of Ultraviolet Spectroscopy*, Wiley, New York, 1962.

A. B. P. Lever, *Inorganic Electronic Spectroscopy*, 2nd ed. Elsevier, Amsterdam, 1984. Excellent for theory and spectral data.

L. E. Orgel, *An Introduction to Transition-Metal Chemistry: Ligand Field Theory*, Methuen, London, 1960. A summary of Orgel's work that made chemists aware of LF theory.

H. L. Schlafer and G. Glieman, *Basic Principles of Ligand Field Theory*, Wiley (Interscience), New York, 1969.

Problems

9.1 (a) Identify the ground and excited states of the same spin multiplicity for $[Ni(NH_3)_6]^{2+}$ and for $[Ni(en)_3]^{2+}$. Use the full symmetry of each ion.
(b) Which spin-allowed transitions are electric-dipole allowed and with what polarization?
(c) Which are magnetic-dipole allowed and with what polarization?

9.2 (a) For Fe^{3+} complexes, what main features would you expect to distinguish spectra for high-spin $\mathbf{O}_h$ complexes from those for low-spin complexes?
(b) What are the spin-allowed transitions for low-spin $\mathbf{O}_h$ complexes of Fe^{3+}? Are any of these electric-dipole allowed?

9.3 Why might *d–d* absorption bands for tetrahedral complexes be expected to be more intense than those for octahedral complexes for the same metal ions?

9.4 For the $[Ni(NH_3)_6]^{2+}$ transitions (Problem 9.1), which are vibronically allowed? The M–N vibrations for $\mathbf{O}_h$ belong to the symmetry species A_{1g}, E_g, T_{2g}, T_{1u}, and T_{2u}.

9.5 Identify the less intense band for each of the following pairs and indicate the reason for your choice. Each pair represents a *different* reason for differences in intensity.
(a) For $[Ni(NH_3)_6]^{2+}$: ${}^3A_{2g} \rightarrow {}^3T_{1g}$ and ${}^3A_{2g} \rightarrow {}^1T_{1g}$.
(b) For $[Co(H_2O)_6]^{2+}$: ${}^4A_{2g} \rightarrow {}^4T_{1g}$ and $[CoCl_4]^{2-}$ ${}^4A_2 \rightarrow {}^4T_1$.
(c) The most intense band in the visible or near ultraviolet region for $[Cr(NH_3)_6]^{3+}$ and for $[CrCl(NH_3)_5]^{2+}$.
(d) $[VF_6]^{3-}$ (d^2): ${}^3T_{1g} \rightarrow {}^3T_{2g}$ or ${}^3T_{1g} \rightarrow {}^3A_{2g}$. (Check the correlation diagram, Fig. 9.9.)

9.6 The spin-allowed transitions of $[Cr(NH_3)_6]^{3+}$ and $[Cr(en)_3]^{3+}$ have similar intensities. Is this expected from the selection rules for the full symmetry of the ions? Why is it true?

9.7 The *d–d* electronic transitions, forbidden by the symmetry selection rule for centrosymmetric point groups, can become partially allowed through mixing between *d* and *p* orbitals. However, *d–p* mixing is prohibited for some noncentrosymmetric point groups. Identify two of these groups.

9.8 It is possible for symmetry forbidden *d–d* transitions to gain intensity from *d–f* mixing for most noncentrosymmetric groups. For which noncentrosymmetric groups is *d–f* mixing prohibited? What other consideration makes d–f mixing less important than d–p mixing, at least for the first transition series?

9.9 The complex $[VF_6]^{3-}$ shows bands at 14,800 and 23,000 cm^{-1}. Assign the bands, calculate Dq and B′, and account for any missing bands. Can the bending parameter be evaluated from these data?

9.10 Ni^{2+} in MgO as host lattice (assume octahedral Ni(II)O_6 as the chromophore) shows bands at 8600, 13,700, 24,500, and a weak band at 21,500 cm^{-1}. Assign the bands and calculate Dq, B′, and c.

9.11 The tetrahedral $[CoCl_4]^{2-}$ ion shows two absorption bands centered at 5460 cm^{-1} (1830 nm, $\varepsilon = 140$) and 14,700 cm^{-1} (680 nm, $\varepsilon = 1400$). Assign the bands considering the likelihood of detecting a third band in the region of the spectrum involved and the application of selection rules. Vibrations for $\mathbf{T}_d$ belong to A_1, E, and $2T_2$ symmetry species.

9.12 Use spin factoring to derive the states in $\mathbf{O}_h$ for d^7. Draw the correlation diagram.

9.13 Derive the correlation diagram for d^8 in $\mathbf{C}_{4v}$.

9.14 Apply the selection rules to determine which transitions are allowed for d^8 (high spin) in $\mathbf{C}_{4v}$.

9.15 Derive the correlation diagram for d^2 in $\mathbf{D}_{3h}$.

9.16 Compare the splitting of the spin-allowed $\mathbf{O}_h$ absorption bands of $[Cr(NH_3)_6]^{3+}$ for *cis*-$[CrCl_2(NH_3)_4]^+$, *trans*-$[CrCl_2(NH_3)_4]^+$, and *fac*-$[CrCl_3(NH_3)_3]$.

9.17 For the complex *trans*-$[Co(en)_2(H_2O)_2]^{3+}$ (assume $\mathbf{D}_{4h}$ symmetry), one absorption band shows two components, at 18.2 and 22.5 kK, and a second band appears at 29.0 kK. The two spin-allowed bands for $[Co(en)_3]^{3+}$ occur at 21.47 and 29.50 kK and the ${}^1A_{1g} \rightarrow {}^3T_{1g}$ ($\mathbf{O}_h$) occurs at 13.8 kK. Calculate Dq for en, and for H_2O calculate C and Dt.

9.18 The absorption spectrum of *trans*-$[Cr(en)_2(H_2O)_2]Cl_3 \cdot H_2O$ shows components of band I at 20.0 (z polarized) and 23.5 kK, and components of band II at 27.5 (z polarized) and 29.3 kK. The second component in each case is unpolarized. Assign the bands and calculate the LF parameters Dq(en), $Dq(H_2O)$, Dt, Ds, and c. Odd vibrations for $\mathbf{D}_{4h}$ are A_{2u}, B_{2u}, and E_u.

9.19 In solid Cs_2CuCl_4 the symmetry of $[CuCl_4]^{2-}$ is $\mathbf{D}_{2d}$ with possible slight distortion by the $\mathbf{C}_s$ site symmetry. The polarized crystal spectrum shows a band centered at about 5000 cm^{-1} with (x, y) polarization, a very weak band with z polarization at 8000 cm^{-1}, and a band at 9000 cm^{-1} with z polarization. Account for the bands assuming ${}^2B_2$ as the ground state and ${}^2B_1$ lower than ${}^2A_1$.

9.20 For *trans*-$[CrCl_2(NH_3)_4]^+$ the components of band I occur at 16.94 (^4E_g) and 21.11 kK and those of band II at 24.88 and 25.66 (E_g) kK. Assign the other components. The value of $\Delta(e_g)$ is -0.66 kK and that of $\Delta(t_{2g})$ is -3.30 kK. Calculate Δ_N, $\Delta'_{\sigma Cl}$, $\Delta'_{\pi Cl}$, and $\Delta(d)$.

9.21 To which point groups must optically active molecules belong?

9.22 (a) Which of the spin-allowed transitions of $Ni(en)_3]^{2+}$ ($\mathbf{D}_3$) might be expected to appear in the circular dichroism spectrum based on selection rules? (Actually the complex is too labile to be resolved.)

(b) What is the polarization of the allowed CD transitions?

9.23 For a $[CuCl_5]^{3-}$ complex with $\mathbf{C}_{4v}$ symmetry, obtain the spin–orbit states starting with the $\mathbf{C}_{4v}$ ligand field states and starting with the J levels. Draw a correlation diagram relating these two extremes.

10

Vibrational Spectroscopy

The normal mode analysis is introduced for linear molecules. Normal modes are described pictorially for H_2O. The symmetry species for the normal modes of H_2O are derived from the nine degrees of freedom and then by a simple procedure that is applied to other molecules, including cage structures. Vibrational spectra of functional groups, coordinated ligands, and ionic solids are considered briefly.

10.1 Introduction

Electronic absorption spectra commonly occur in the visible or ultraviolet region. The electronic excited states differ from the ground states in the population of energy levels. Although the time span for the transition to the excited state is normally too short for atom rearrangement, the most favorable molecular geometry for the electronic excited states often differ from that for the ground state. For any given electronic state, the atoms are in constant motion. The motion increases with increasing temperature, but some motion persists even at absolute zero. The atomic motions correspond to molecular vibrations and molecular rotations in addition to translations. Just as the electronic energy states correspond to representations of the point group, the molecular vibrations also can be resolved into normal modes, each belonging to an irreducible representation of the group. The rotations belong to the representations corresponding to R_x, R_y, and R_z in the character table.

It is instructive to consider the vibrations of a diatomic molecule. The stretching of the bond can be treated as a harmonic oscillator, although it becomes increasingly more anharmonic with increasing amplitude. The molecule differs from the harmonic oscillator in that the repulsion caused by interpenetration of electron clouds becomes more important for very short distances, and the bond becomes weak at long distances. The potential energy of the harmonic oscillator as a function of d is given by the parabola in Fig. 10.1. The potential energy curve for a diatomic molecule deviates from the parabola at distances much shorter or longer than d_0. Just as with electronic energy states, the energy is quantized. The lowest energy state, the *vibrational ground state*, corresponds to v_0, and higher vibrational states are designated as 1, 2, 3,.... If enough vibrational energy is added, then the amplitude of vibration becomes so great that the molecule dissociates: this is the dissociation energy of the molecule D. As the temperature increases, internal molecular motion increases, corresponding to larger numbers of molecules in vibrational excited states. For an electronic excited state, there is another potential energy curve. Electronic transitions can occur from the v_0 of the ground state to v_0 of the electronic excited state—the $0 \rightarrow 0$ transition. Or the transition can occur from any occupied vibrational level of the electronic ground state to any vibrational level (within the limitations of the selection rules) of the electronic excited state. This corresponds to the coupling of vibrational and electronic transitions—*vibronic* transitions (see Section 9.1.2). Pure vibrational transitions correspond to absorption in the infrared (IR) region ($\sim 10^2$–10^4 cm^{-1}). Pure rotational spectra occur in the microwave

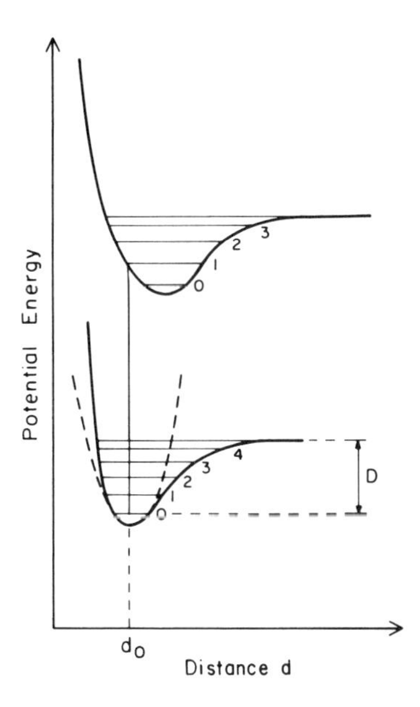

Fig. 10.1 Potential energy curves for the electronic ground state (parabola shown by the dotted line) and an electronic excited state for X_2.

(0.1–10 cm^{-1}) or far infrared region (~100 cm^{-1}). Rotational and vibrational transitions couple to give vibrational–rotational transitions.

10.1.1 FORCE CONSTANTS

If we treat two atoms A and B of a diatomic molecule, connected by a spring, as a simple harmonic oscillator, Hooke's law gives the restoring force for displacement by x (from d_0 to $d_0 + x$) as

$$\text{restoring force} = f = -kx. \tag{10.1}$$

The proportionally constant k is called the *force constant*. The potential energy (V) is given by

$$V = \frac{kx^2}{2}. \tag{10.2}$$

For a molecule with only discrete (quantized) vibrational energies allowed, the energy levels (E_v) are characterized by vibrational quantum numbers v ($v = 0, 1, 2, 3, \ldots$) and frequencies ν,

$$\begin{aligned} E_v &= (v + \tfrac{1}{2})h\nu \\ &= (v + \tfrac{1}{2})hc\bar{\nu} \qquad (\bar{\nu} \text{ in cm}^{-1}), \end{aligned} \tag{10.3}$$

where

$$\nu = \frac{1}{2\pi}\left(\frac{k}{\mu}\right)^{1/2} \tag{10.4}$$

or

$$\bar{\nu} = \frac{1}{2\pi c}\left(\frac{k}{\mu}\right)^{1/2}.$$

Here μ is the *reduced mass* of the molecule and c is the velocity of light.

$$\mu = \frac{m_1 m_2}{m_1 + m_2}. \tag{10.5}$$

For a diatomic molecule as a simple harmonic oscillator, the vibrational levels are spaced equally. The energy of the transition from the ground state ($v = 0$) to the first excited state ($v = 1$) (a fundamental transition) is $\Delta E = h\nu$. From the observed frequency, we can calculate the force constant for the A—B bond.

A better fit for the trough of the potential energy curve (Fig. 10.1) is obtained by including an anharmonic correction. Equation (10.3) with energies in wave numbers becomes

$$\bar{E}_v = \bar{\nu}_e(v + \tfrac{1}{2}) - \bar{\nu}_e x_e(v + \tfrac{1}{2})^2, \tag{10.6}$$

where $\bar{\nu}_e x_e$ is a small anharmonic correction and $\bar{\nu}_e$ is the corrected stretching frequency, differing only slightly from the observed fundamental $\bar{\nu}_{0-1}$. The observed frequency for CO is 2141 cm^{-1} and $\bar{\nu}_e$ is 2170 cm^{-1}.

We can calculate force constants from Eq. (10.4), but better values (k_e) are obtained using $\bar{\nu}_e$ to give

$$k_e = (200\pi c \bar{\nu}_e)^2 \mu. \tag{10.7}$$

A factor of 100 is included since k_e is in newtons per meter and $\bar{\nu}$ is in reciprocal centimeters. We use $c = 2.9979 \times 10^8$ m/sec and convert μ in amu to kilograms by dividing by 6.022×10^{26} amu/kg (Avogadro's number based on the kilogram–mole). Equation (10.7) (with μ in amu) reduces to

$$k_e = 5.8916 \times 10^{-5} (\bar{\nu}_e)^2 \mu. \tag{10.8}$$

Force constants in Table 10.1 were calculated using Eq. (10.8).

From the data in Table 10.1 we see for H_2 that the stretching frequencies greatly depend upon the masses of the isotopes involved, but the force constants do not. It is tempting to relate the force constants to bond strengths because of some obvious parallels. Very strong bonds, such as those of N_2 and

TABLE 10.1
Vibrational Constants of Diatomic Molecules[a]

	μ (amu)	$\bar{\nu}_e$ (cm^{-1})	$\bar{\nu}_e x_e$ (cm^{-1})	k_e ($N\,m^{-1}$)	D (kJ $mole^{-1}$)	d_0 (pm)
$^1H_2$	0.50391	4401.21	121.34	575.1	432.11	74.14
$^2H_2$	1.00705	3315.50	61.82	575.9	439.66	74.15
$^{(2,3)}H_2$	1.20764	2845.5	51.38	576.1	441.25	74.14
$^7Li_2$	3.508	351.43	2.61	25.5	100.9	267.29
$^{133}Cs_2$	66.4527	42.02	0.08	6.9	154.6	447
$^{24}Mg_2$	11.9925	51.12	1.65	1.8	4.83	389
$^{(129,132)}Xe_2$	65.19361	21.12	0.65	1.7	2.22	436
F_2	9.49920	916.64	11.24	470.2	154.6	141.19
$^{35}Cl_2$	17.48443	559.7	2.67	322.7	239.25	198.79
$^{79}Br_2$	39.4592	325.32	1.08	246.0	190.16	228.10
$^{127}I_2$	63.45224	214.50	0.61	172.0	148.84	266.6
$^{16}O_2$	7.997458	1580.19	11.98	1176.5	493.63	120.75
$^{14}N_2$	7.001537	2358.57	14.32	2294.7	941.74	109.77
$^{12}C^{16}O$	6.856209	2169.81	13.29	1901.8	1070.3	112.83
HF	0.95706	4138.32	89.88	965.6	566.3	91.68
$H^{35}Cl$	0.979593	2990.95	52.82	516.3	427.86	127.46
$H^{81}Br$	0.995427	2648.98	45.22	411.5	362.6	141.44
$H^{127}I$	0.99985	2309.01	39.64	314.1	294.7	160.92

[a] Data from K. P. Huber and G. Herzberg, *Molecular Spectra and Molecular Structure IV. Constants of Diatomic Molecules*, Van Nostrand-Reinhold, New York, 1979.

CO, have large force constants, and very weak bonds, such as those of Mg_2 and Xe_2, have very low force constants. Force constants are in the range 300–600 N m^{-1} for many diatomic molecules involving "normal" single bonds (e.g., H_2, F_2, Cl_2, HCl, and HBr), about twice that for doubly bonded molecules (e.g., O_2) and about three times as great for triply bonded molecules (e.g., N_2 and CO). Also, the force constants decrease with decreasing bond energies in the series HF > HCl > HBr > HI and $Li_2 > Cs_2$. However, we see that the force constants follow a regular trend for the halogens, $F_2 > Cl_2 > Br_2 > I_2$, but the bond dissociation energies do not. There is no simple relationship between k and bond energy. The force constant depends on the stiffness of the spring of our harmonic oscillator or the curvature of the well at the bottom of the potential energy curve (Fig. 10.1). The dissociation energy of the molecule is the depth of the well.

In spite of the tenuous relationship between force constants and bond strengths, worthwhile correlations have been applied within series of closely related compounds such as some metal carbonyls. The Cotton–Kraihanzel[1] method has been applied rather widely within series of similar metal carbonyls.

Since we are interested in the spectral applications where symmetry is most helpful, we will focus attention on the vibrational spectra of polyatomic molecules, other than diatomics.

10.1.2 NUMBER OF NORMAL MODES

A molecule with n atoms has a total of $3n$ degrees of freedom—three degrees of freedom per atom. Three of these degrees of freedom can be taken to be those of the position of the center of mass and therefore correspond to translational degrees of freedom. A nonlinear molecule has three rotational degrees of freedom, and a linear molecule has two rotational degrees of freedom. The remaining degrees of freedom are vibrational. Thus, for a nonlinear (linear) molecule, there are $3n - 6$ ($3n - 5$) vibrational degrees of freedom.

10.2 Vibrational and Normal Modes for Linear Molecules

The various possible modes of vibration of a polyatomic molecule are in general difficult to visualize. However, if we make the assumption (which is an approximation for real molecules) that all relative displacements of the atoms

[1] F. A. Cotton and C. S. Kraihanzel, *J. Am. Chem. Soc.* 1962, **84**, 4432; *Inorg. Chem.* 1963, **2**, 533; F. A. Cotton, A. Musco, and G. Yagupsky, *Inorg. Chem.* 1967, **6**, 1357.

are related to the corresponding forces by Hooke's law, then a mathematical description of the vibrations becomes fairly simple. According to this model, there is for each vibrational degree of freedom a *normal mode* that behaves as an independent harmonic oscillator. A general vibrational motion will be a mixture of these normal modes. The entity that undergoes the harmonic motion in a normal mode is called the normal coordinate. It is a vector in the $3n$-dimensional space of the degrees of freedom of the molecule. (Actually, as we shall see, it is confined to the vibrational subspace of this $3n$-dimensional space.) The oscillatory motion depends upon an effective force constant (a function of the various interatomic force constants) and upon an effective mass (a function of the various atomic masses). A mathematical procedure called *normal mode analysis* gives these functions.

The calculation of normal coordinates and their frequencies is usually not a trivial matter. Fortunately, many important properties of the normal modes (and of the corresponding vibrational spectra) can be obtained from a consideration of effects of molecular symmetry alone, without any normal mode calculations. The primary purpose of this chapter is the delineation of these symmetry considerations. However, such a discussion might be followed more easily if we first describe the *results* of the normal mode analysis of a simple example. This should help the reader to visualize the various mathematical entities involved.

We pick for our example the linear XY_2 molecule and the nine-dimensional Cartesian coordinate system labeled as shown in Fig. 10.2. The nine-dimensional vectors in this space will be expressed as

$$(x_1, y_1, z_1; x_2, y_2, z_2; x_3, y_3, z_3),$$

where the semicolons are used for convenience in keeping track of the degrees of freedom of the individual atoms.

A normal mode analysis gives the following results. The frequencies can be expressed as

$$\begin{aligned}
\nu_1 = \nu_2 &= \frac{1}{2\pi}\sqrt{(2k/m_Y)(1 + (2m_Y/m_X))}, \\
\nu_3 &= \frac{1}{2\pi}\sqrt{k'/m_Y}, \\
\nu_4 &= \frac{1}{2\pi}\sqrt{(k''/m_Y)(1 + (2m_Y/m_X))}, \\
\nu_5 = \nu_6 = \nu_7 = \nu_8 = \nu_9 &= 0,
\end{aligned} \tag{10.9}$$

where k, k', and k'' are force constants; m_X and m_Y are the atomic masses. The

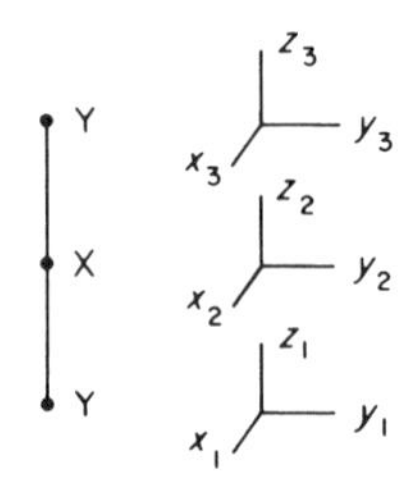

Fig. 10.2 Coordinate system for linear XY_2.

zero frequencies ($\nu_5, \ldots, \nu_9$) correspond to the translational and rotational degrees of freedom.

The normal coordinates can be expressed as

$$
\begin{aligned}
\mathbf{Q}_1 &= \frac{1}{\sqrt{2[1 + 2(m_Y/m_X)^2]}}\left(1,0,0; \frac{-2m_Y}{m_X}, 0,0; 1,0,0\right), \\
\mathbf{Q}_2 &= \frac{1}{\sqrt{2[1 + 2(m_Y/m_X)^2]}}\left(0,1,0; \frac{-2m_Y}{m_X}, 0,0; 0,1,0\right), \\
\mathbf{Q}_3 &= \frac{1}{\sqrt{2}}(0,0,1; 0,0,0; 0,0,-1), \\
\mathbf{Q}_4 &= \frac{1}{\sqrt{2[1 + 2(m_Y/m_X)^2]}}\left(0,0,1; 0,0, \frac{-2m_Y}{m_X}; 0,0,1\right), \\
\mathbf{Q}_5 &= \frac{1}{\sqrt{2}}(1,0,0; 0,0,0; -1,0,0) \qquad \text{rotation about } y \text{ axis}, \\
\mathbf{Q}_6 &= \frac{1}{\sqrt{2}}(0,1,0; 0,0,0; 0,-1,0) \qquad \text{rotation about } x \text{ axis}, \\
\mathbf{Q}_7 &= \frac{1}{\sqrt{3}}(1,0,0; 1,0,0; 1,0,0) \qquad \text{translation in } x \text{ direction}, \\
\mathbf{Q}_8 &= \frac{1}{\sqrt{3}}(0,1,0; 0,1,0; 0,1,0) \qquad \text{translation in } y \text{ direction}, \\
\mathbf{Q}_9 &= \frac{1}{\sqrt{3}}(0,0,1; 0,0,1; 0,0,1) \qquad \text{translation in } z \text{ direction}.
\end{aligned}
\tag{10.10}
$$

Rotation about the z axis (the line of centers) would have the zero vector as normal coordinate (i.e., there would be no change in any coordinate) and so does not exist as a degree of freedom. There are several things to be noted about these normal coordinates. First, they are orthonormal. They are normalized because

$$\mathbf{Q}_i \cdot \mathbf{Q}_i = 1,$$

and they are orthogonal because

$$\mathbf{Q}_i \cdot \mathbf{Q}_j = 0 \qquad \text{if} \quad i \neq j.$$

The four vibrational vectors, therefore, span a four-dimensional subspace that can be considered independently of the five-dimensional translational–rotational subspace. For the vibrational subspace, we can pick a coordinate system with axes along the normal coordinates. This will give a new four-dimensional Cartesian system in which the normal coordinates can be expressed as

$$\begin{aligned} \mathbf{Q}_1 &= (1, 0, 0, 0), \\ \mathbf{Q}_2 &= (0, 1, 0, 0), \\ \mathbf{Q}_3 &= (0, 0, 1, 0), \\ \mathbf{Q}_4 &= (0, 0, 0, 1). \end{aligned} \tag{10.11}$$

However, to picture the normal modes we go back to the original nine-dimensional representation and look at the changes in the nine coordinates that are represented by the nine-dimensional vectors. This gives the results shown in Fig. 10.3, where relative magnitudes of the displacements of X and Y

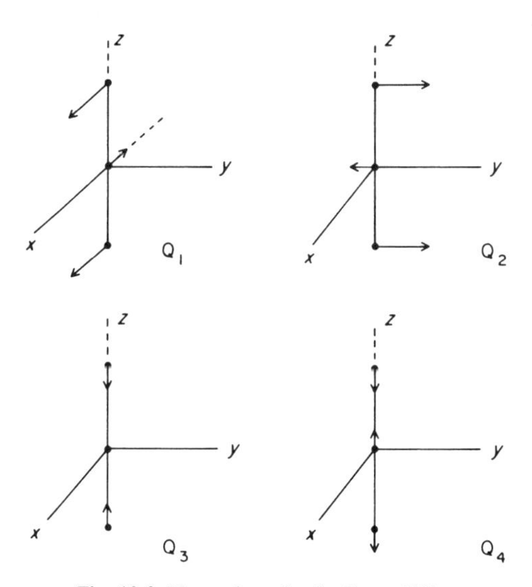

Fig. 10.3 Normal modes for linear XY_2.

depend upon the ratio m_Y/m_X. [We emphasize here that the $\mathbf{Q}_i$ are nine- (or four-) dimensional vectors, and their orthogonality has meaning only in those higher-dimensional spaces. Obviously, in the ordinary three-dimensional space of the molecule there could not be four (or more) vectors orthogonal to one another.]

These vectors $\mathbf{Q}_i$ are constant. They represent the *direction* of the motion in the nine- (or four-) dimensional space. The motion is oscillatory along these constant directions.

It will be noted from Fig. 10.3 that the two degenerate (they have the same frequency) modes $\mathbf{Q}_1$ and $\mathbf{Q}_2$ are pure bending modes and that $\mathbf{Q}_3$ and $\mathbf{Q}_4$ are pure stretching modes—symmetric and antisymmetric, respectively. This is not the usual situation. Generally, a normal mode will be a mixture of stretching and bending modes, and correspondingly, a stretching or bending mode will be a mixture of normal modes. As we shall see, symmetry considerations alone will tell us whether a normal mode must be pure stretching or bending, or whether it can be a mixture. Symmetry considerations alone will not give us information about the composition of the mixture when it can exist.

Since molecular vibrations are described by quantum mechanics, we need a Ψ function for each of the normal modes. Each normal mode has the Ψ function of a harmonic oscillator, which can be expressed as[2]

$$\Psi_i(v) = N_v H_v \exp\left[\frac{-\pi v_i}{h} \mathbf{Q}_i \cdot \mathbf{Q}_i\right], \tag{10.12}$$

where N_v is a normalizing constant, v is the vibrational quantum number, and H_v is the Hermite polynomial of degree v. H_v is an even polynomial (containing powers of $\mathbf{Q}_i \cdot \mathbf{Q}_i$) when v is even, and H_v is an odd polynomial with terms such as $(\mathbf{Q}_i \cdot \mathbf{Q}_i)^{2s}\mathbf{Q}_i$ $(s = 0, 1, 2, \ldots)$ when v is odd. Since $\mathbf{Q}_i \cdot \mathbf{Q}_i$ is invariant under all orthogonal transformations (a representation times itself is totally symmetric), all $\Psi_i(v)$ with even v will be invariant under all symmetry operations, that is, will belong to the completely symmetric representation of the symmetry group. On the other hand, all $\Psi_i(v)$ with odd v will transform in the same way as the corresponding $\mathbf{Q}_i$. Each normal coordinate $\mathbf{Q}_i$ belongs to some *irreducible* representation of the symmetry group of the molecule, while stretching and bending modes might belong to reducible representations, corresponding to combinations of the normal modes.

It is the fact that normal coordinates belong to irreducible representations of the molecular group that makes it possible to use symmetry operations in the ordinary three-dimensional space to investigate the symmetry properties

[2] The vector $\mathbf{Q}_i$ in Eq. (10.4) is not normalized but has a magnitude that corresponds to the oscillatory motion.

of normal modes, even though the normal coordinates belong to another vector space. In most of the applications to be considered in this chapter, the vector space of the normal coordinates never enters into the considerations explicity.

10.3 Symmetry Species for Vibrational Modes of H_2O

We can resolve the normal modes into stretching and bending modes to use the symmetry of the molecule to advantage. Let us consider a simple case, the H_2O molecule ($\mathbf{C}_{2v}$). The $(3 \times 3) - 6 = 3$ bending and stretching modes are easy to visualize—we can have the O—H bonds stretch symmetrically or unsymmetrically and the H atoms can move away from one another, increasing the bond angle. These are described as the symmetric and antisymmetric stretching modes and the bending mode. In each case, motion of the H atoms requires smaller displacement of the heavier O to retain the same center of gravity. If we examine the transformation properties of the vectors in Fig. 10.4 using the character table for $\mathbf{C}_{2v}$, we see that the vectors describing the atom displacements of the symmetric stretch and the bending mode are left unchanged by all of the symmetry operations of the group. They are totally symmetric, belonging to the A_1 representation. The displacement vectors for the antisymmetric stretching vibration are unchanged by the E and σ_{yz} operations (character $+1$), and they are reversed in sign (direction) by C_2 and σ_{xz} (character -1). Thus, the antisymmetric stretching vibration transforms as B_2. The three vibrational modes are $2A_1 + B_2$. The two *normal modes*

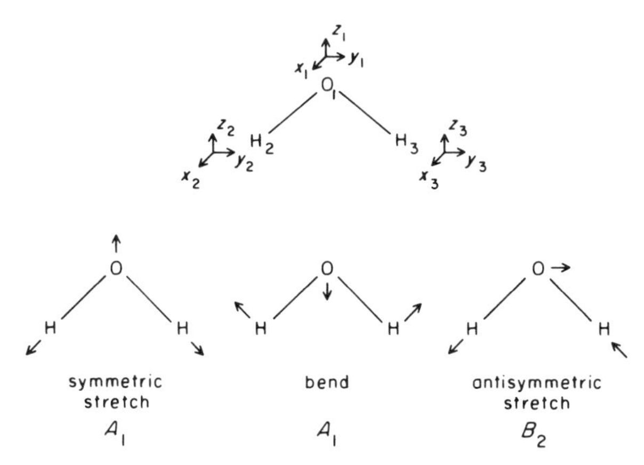

Fig. 10.4 Coordinates and vibrational modes for H_2O.

TABLE 10.2
Transformation Matrices for Atoms of the H_2O Molecule

E	x_1	y_1	z_1	x_2	y_2	z_2	x_3	y_3	z_3
x'_1	1	0	0						
y'_1	0	1	0		**0**			**0**	
z'_1	0	0	1						
x'_2				1	0	0			
y'_2		**0**		0	1	0		**0**	
z'_2				0	0	1			
x'_3							1	0	0
y'_3		**0**			**0**		0	1	0
z'_3							0	0	1

$\chi(E) = 9$

$(C_2)_z$	x_1	y_1	z_1	x_2	y_2	z_2	x_3	y_3	z_3
	−1	0	0						
	0	−1	0		**0**			**0**	
	0	0	1						
							−1	0	0
		0			**0**		0	−1	0
							0	0	1
				−1	0	0			
		0		0	−1	0		**0**	
				0	0	1			

$\chi(C_2) = -1$

for σ_{xz} only O is unshifted

σ_{xz}	x_1	y_1	z_1
x'_1	1	0	0
y'_1	0	−1	0
z'_1	0	0	1

$\chi(\sigma_{xz}) = 1$

All atoms are unshifted by σ_{yz}

σ_{yz}	x	y	z
x'	−1	0	0
y'	0	1	0
z'	0	0	1

$\chi(\sigma_{yz}) = 1$ (for each atom)

of A_1 symmetry are each a mixture of the stretching and bending modes of A_1 symmetry. The *normal mode* of B_2 symmetry is pure stretching.

We need a more systematic approach to deal with more complex cases.

Let us go through a detailed analysis for H_2O and then see how we can simplify the treatment greatly. The coordinates are defined and the atoms are numbered as in Fig. 10.4. The matrices for the identity operation and for C_2 are given in Table 10.2. The character of the E operation is 9 and that for C_2 is −1. We see that only those atoms unshifted by an operation can contribute to the character for the full matrix, and only these need be considered. Atoms that are shifted contribute to the off-diagonal elements of the matrix. For σ_{xz}, only O is unshifted. The character is +1 from the matrix in Table 10.2. For σ_{yz}, all atoms are unshifted with the same transformation of coordinates for each atom as given in the table. The character is 1 for each atom or 3 for all atoms. The characters for this set of vectors (Table 10.3) describe a reducible representation of the group. We could reduce the nine-dimensional representation directly, or we can subtract the representations for translations and rotations for simplification. The translations transform as x, y, and z and the rotations as R_x, R_y, and R_z. Subtracting the characters for these representations gives the representations for the vibrational modes Γ_{vib}, Table 10.3).

TABLE 10.3
Γ_{vib} from Γ_{total}

	E	C_2	σ_{xz}	σ_{yz}	
Γ_{total}	9	−1	1	3	
Characters for translations and rotations					
A_1 (z)	1	1	1	1	
A_2 (R_z)	1	1	−1	−1	
$2B_1$ (x, R_y)	2	−2	2	−2	
$2B_2$ (y, R_x)	2	−2	−2	2	
Totals					
Γ_{trans} } Γ_{rot} }	6	−2	0	0	(subtract)
Γ_{vib}	3	1	1	3	

A simpler approach is to write down the sum of the characters of the representations corresponding to $\Gamma_{translation}$ $[B_1\,(x), B_2\,(y)$, and $A_1\,(z)]$ from the character table (or from the transformation of the x, y, z vectors), the number of atoms unmoved by each operation of the group, and take the direct product of these representations to give Γ_{total} (Table 10.4—the same nine-dimensional representation obtained above). We subtract the characters for the translations (already tabulated as Γ_{trans}) and rotations (Γ_{rot}) to give Γ_{vib}. This

TABLE 10.4
Γ_{vib} and $\Gamma_{stretch}$

	E	C_2	σ_{xz}	σ_{yz}	
Γ_{trans}	3	−1	1	1	
Unmoved atoms	3	1	1	3	
Γ_{total}	9	−1	1	3	(product)
Γ_{trans}	3	−1	1	1	(subtract)
	6	0	0	2	
Γ_{rot}	3	−1	−1	−1	(subtract)
Γ_{vib}	3	1	1	3	
Unmoved atoms	3	1	1	3	
	1	1	1	1	(subtract)
$\Gamma_{stretch}$	2	0	0	2	$= A_1 + B_2$

three-dimensional representation is reduced easily by inspection to give $\Gamma_{vib} = 2A_1 + B_2$.

We began by examining the sketches of the vibrational modes. Normally, we would get these following the procedure just used. The task of drawing the vibrational modes is often simplified if we know which representations are stretches and which are bending modes. We can find the representation for the stretching modes ($\Gamma_{stretch}$) by subtracting 1 from each character for the representation for the unmoved atoms (Table 10.3).[3] The result ($\Gamma_{stretch}$) is $A_1 + B_2$, so the other A_1 must be a bending mode. Stretching modes involve the H atoms moving along the bond directions with the necessary compensating motion of O. The symmetric stretch (A_1) must involve the same motion for each H, and the direction of one vector along the bond must be reversed for B_2 (unsymmetrical stretch). Since the bending vibration is A_1, the outward motion of both H atoms must be represented by equivalent vectors that are interchanged by C_2 and by σ_{xz} (scissoring motion).

Although the vibrations for any angular XY_2 molecule are $2A_1 + B_2$, the relative motions of the atoms (consistent with the symmetry) depend very much on the masses of the atoms X and Y. For H_2O, the H atoms are displaced much more than the heavier O atom. The relative displacements are much different for SO_2 and OCl_2 (Fig. 10.5).

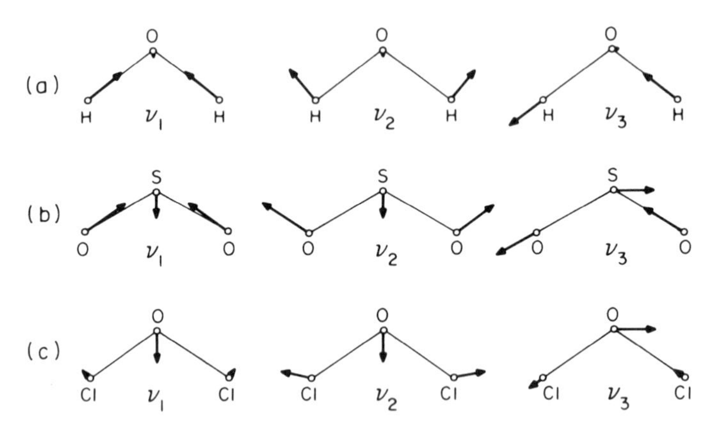

Fig. 10.5 Actual form of the normal vibrations of H_2O, SO_2, and Cl_2O. The scale of amplitudes is much larger than the scale of internuclear distances. The scales differ for different diagrams. (From G. Herzberg, *Molecular Spectra and Molecular Structure II. Infrared and Raman Spectra*, Van Nostrand, New York, 1945, p. 171.)

[3] The vectors for stretching vibrations of an MX_n molecule (M at the origin) coincide with the σ bonds. If we subtract one from the number of atoms unmoved by each operation, we eliminate the atom at the origin and obtain the number of unshifted vectors along the σ bonds.

10.4 Selection Rules for Infrared and Raman Spectra

The absorption of a photon with the correct energy raises a molecule from one vibrational level to the next. Since the energy differences are small, the absorption occurs in the infrared region. All wave functions for normal vibrational modes in their ground states belong to the totally symmetric representation of the point group of the molecule, describing the undistorted molecule. Thus, all ground state normal modes are A_{1g}, A_1, A', etc. The symmetry species of the first vibrationally excited state (fundamental) is determined by the symmetry of the sets of vectors representing the atomic motions, as we have seen for H_2O. Excitation to the first vibrational state above the ground state for a particular mode is a *fundamental* transition. Double excitation corresponds to the first overtone, with twice the frequency. Here we are concerned primarily with fundamental transitions.

The selection rule for infrared absorption corresponds to that for electric dipole transitions, but it is simplified in that the ground state (Ψ_v^0) is always A_{1g} (A_1, A', etc.). The transition is allowed if the integral in Eq. (10.13) is totally symmetric. This requires that the wave function

$$\int \Psi_v^0 \mu_e \Psi_v^{ex} \, d\tau \tag{10.13}$$

for the vibrationally excited state must belong to the *same representation* as μ_e. The operator μ_e transforms as x, y, or z, corresponding to the electric vector along x, y, or z.[4] In physical terms, a vibration is infrared active only if it causes a change in the dipole moment of the molecule. Of course, the dipole moment fluctuates during the vibration.[5]

[4] The symmetry selection rule [Eq. (10.13)] has the same form as Eq. (9.5) for electric dipole transitions, and the selection rules depend on polarization along x, y, and z in the same way. Since infrared spectra are usually obtained for mulls, pellets, solutions, or neat liquids, polarized spectra are used infrequently.

[5] If it is assumed that Ψ_v^0 and Ψ_v^{ex} in Eq. (10.13) are for harmonic motion and that μ_e is a linear function of the normal coordinates, then the integral is zero unless $\Delta v = 1$ for one normal mode and $\Delta v = 0$ for all others. That is, all overtones and all *combinations* (where more than one normal mode is excited) would be forbidden. Only fundamentals would be allowed. The fact that overtones and combinations are sometimes observed experimentally means that in such cases one (or both) of the aforementioned assumptions is not a sufficiently good approximation. Selection rules apply in the usual way to combination bands (take the direct product of the symmetry species combined) and overtones (Γ_i^2 for the first overtone, Γ_i^3 for the second, etc.). Thus, the first overtone of a nondegenerate vibration is always totally symmetric and is forbidden for a centrosymmetric group. For this case, the second overtone ($\Gamma_i^3 = \Gamma_i$) is allowed if the fundamental is allowed.

When light passes through a sample, some of it might be absorbed, but most of the light passes through undeflected. A small amount of the light (with the same energy) is scattered in all directions. In 1928, Raman observed that some of the scattered light can differ in energy from the incident light. The differences correspond to the energy spacing of the vibrational levels. Thus, if a photon is scattered inelastically, the molecule interacting might be left in a higher vibrational level, so that the energy of the scattered photon is $\nu_{\text{incident}} - \Delta\nu$. This is the case for Stokes radiation—the scattered light has less energy than the incident light because some of the energy went into vibrational excitation. If initially some of the molecules of the sample are in vibrationally excited states (the higher vibrational states are more highly populated with increasing temperature), then after inelastic scattering of a photon, the molecule might be left in a lower vibrational state. In this case, the photon emitted has *greater* energy than the incident light, $\nu_{\text{incident}} + \nu_{\Delta\nu}$—anti-Stokes radiation.

Infrared absorption is measured directly by the decrease in intensity of the infrared radiation passing through the sample. In a Raman experiment, we measure the radiation scattered by the sample at 90° to the incident beam. Monochromatic light of any wavelength, but higher in energy than infrared light is used. Commonly, monochromatic light in the visible range is used. Only a small fraction of the light ($\sim\frac{1}{1000}$) is scattered over all directions. Consequently, an intense source is needed, and lasers are ideally suited. Most of the scattered light has the same frequency as the incident light, but the Raman lines correspond to the energy of the incident light minus or plus the energy of vibrational excitation. The lines lowered in energy (Stokes) are more intense than those raised in energy, as might be expected from probability.

Infrared absorption occurs only if a vibration causes a change in the dipole moment of a molecule. The symmetrical expansion (breathing mode) of a molecule causes a change in the polarizability of the molecule since the expanded molecule is more polarizable than the contracted molecule. The oscillating electric field of light induces a dipole moment in the molecule, and the magnitude of the induced dipole moment depends upon the polarizability of the molecule. A vibration is Raman active if there is a change in the polarizability of the molecule during the vibration. Symmetric breathing vibrations for highly symmetric molecules are Raman active even though they are not infrared active. For Raman scattering, the integral corresponding to Eq. (10.13) involves the polarizability tensor **P** as the operator,

$$\int \Psi_v^0 \mathbf{P} \Psi_v^{\text{ex}} \, d\tau. \tag{10.14}$$

The components of the polarizability tensor transform as the quadratic functions of x, y, or z. In the character tables, the representations are those corresponding to the functions x^2, y^2, z^2, xy, xz, yz, $x^2 - y^2$, $x^2 + y^2$, etc.

For a centrosymmetric molecule (ground state A_{1g}) the Raman and IR selection rules are mutually exclusive. Since μ_e belongs to a u representation, only odd (u) vibrations are IR active, and since **P** belongs to a g representation, only even (g) vibrations are Raman active. In such cases, the two types of vibrational spectroscopy complement one another Vibrations of noncentrosymmetric molecules can be both IR and Raman active. For H_2O, all three vibrational modes are IR and Raman active.

Polarized Raman spectra are useful even for studies in solution as a means of distinguishing transitions to totally symmetric vibrational excited states from those of lower symmetry. Polarized light is used, and the depolarization ratio ($I_{\perp}/I_{||}$) of the intensities of scattered light perpendicular ($I_{\perp}$) and parallel ($I_{||}$) to the direction of polarization of incident light are measured. Polarized bands (low $I_{\perp}/I_{||}$) identify transitions to totally symmetric excited states.

10.5 Square Planar Molecules (D_{4h})

There are $(3 \times 5) - 6 = 9$ vibrational modes for XeF_4 or $[PtCl_4]^{2-}$. We

can write down the number of atoms left unmoved by each symmetry operation and multiply by the corresponding characters for $\Gamma_{x,y,z}$ to give Γ_{total} (Table 10.5). Subtracting Γ_{trans} ($\Gamma_{x,y,z}$) and Γ_{rot} gives the nine-dimensional representation for Γ_{vib}. Subtracting one from the number of atoms left unmoved by each operation gives the four-dimensional representation for the stretching modes. We get the representation for the bending modes by subtracting that for the stretching modes from Γ_{vib}. These can be reduced by inspection or by using Eq. (3.79). It is instructive to compare these results with the results of the LGO for σ and π bonding for square planar complexes. In that case, we dealt with π bonding parallel to C_4 [$\Gamma_\pi(||)$] and perpendicular to C_4 [$\Gamma_\pi(\perp)$].

$$\Gamma_{\text{stretch}} = A_{1g} + B_{1g} + E_u, \qquad \Gamma_\sigma = A_{1g} + B_{1g} + E_u,$$

$$\Gamma_{\text{bend}} = A_{2u} + B_{2u} + B_{2g} + E_u, \qquad \Gamma_\pi(||) = A_{2u} + B_{2u} + E_g$$

$$\Gamma_\pi(\perp) = A_{2g} + B_{2g} + E_u.$$

TABLE 10.5
Stretching and Bending Modes of XeF_4

$\mathbf{D}_{4h}$	E	C_4	C_2	C'_2	C''_2	i	S_4	σ_h	σ_v	σ_d	
Unmoved atoms	5	1	1	3	1	1	1	5	3	1	
$\Gamma_{x,y,z}$	3	1	−1	−1	−1	−3	−1	1	1	1	
Γ_{total}	15	1	−1	−3	−1	−3	−1	5	3	1	(product)
Γ_{trans}	3	1	−1	−1	−1	−3	−1	1	1	1	(subtract)
Γ_{rot}	3	1	−1	−1	−1	3	1	−1	−1	−1	(subtract)
Γ_{vib}	9	−1	1	−1	1	−3	−1	5	3	1	
Unmoved atoms	5	1	1	3	1	1	1	5	3	1	
	1	1	1	1	1	1	1	1	1	1	(subtract)
$\Gamma_{stretch}$	4	0	0	2	0	0	0	4	2	0	$= A_{1g} + B_{1g} + E_u$
Γ_{bend}	5	−1	1	−3	1	−3	−1	1	1	1	$= A_{2u} + B_{2u} + B_{2g} + E_u$

It should not be a surprise that there is an exact correspondence between Γ_{stretch} and Γ_σ since they correspond to the transformation of the same vectors directed along the bonds. Why are there two more representations (A_{2g} and E_g) for Γ_π than for Γ_{bend}? Here we can deal with the same set of vectors on the 4F (of XeF_4) to get Γ_{bend} as were used to obtain Γ_π (Fig. 10.6).[6]

The in-plane A_{2g} combination is $x_1 + x_2 + x_3 + x_4$, corresponding to rotation about z. The out-of-plane $y_1 - y_3$ corresponds to rotation about y, and $y_2 - y_4$ corresponds to rotation about x. Together these belong to E_g. The character table confirms that the representations for rotations to be eliminated are A_{2g} (R_z) and E_g (R_x, R_y).

From the comparison with Γ_π ($\perp$ and $\|$), we can identify A_{2u} and B_{2u} as out-of-plane and B_{2g} and E_u as in-plane bending vibrations. The combinations of vectors needed to draw the bending and stretching vibrations are the same as the LGOs we generated earlier (Fig. 6.11) using the projection operator method. If we did not have these results available, we could sketch the vibrational modes knowing their symmetry properties, which are stretches, and which are in-plane and out-of-plane bending modes. This is a particularly favorable case, because of the high symmetry and the matching of the bond directions with the coordinate axes and the usual descriptions of the atomic orbitals. In fact, we can use the atomic orbitals as guides (templates) for obtaining the sign combinations for the vectors (Fig. 10.6). The A_{1g} stretching mode is totally symmetrical (s) and corresponds to the symmetrical lengthening and shortening of the bonds—a breathing vibration. The B_{1g} stretch can be drawn by analogy with the sign combination of $d_{x^2-y^2}$ and the E_u stretches match the signs of p_x and p_y. The Xe moves in the case of E_u but not for A_{1g} and B_{1g}. The out-of-plane A_{2u} matches p_z—all vectors on the four fluorine atoms point in the same direction, being symmetric with respect to C_4, σ_v, and σ_d. There is no orbital on the central atom with B_{2u} symmetry, so the LGO is nonbonding. However, we see that B_{2u} changes sign for the C_4 operation and for σ_d but not for C_2 or for σ_v, so the signs alternate around the plane. The remaining bending vibrations involve the in-plane vectors. The B_{2g} mode matches the alternating signs of d_{xy} and the signs of the two E_u modes match those of p_x and p_y. Here the vibrations are pure bending or pure stretching vibrations except for E_u. Since we found an E_u for stretching and E_u for

[6] The coordinate axes are chosen in Fig. 10.6 so that z for each F is along the Xe—F bond and y is parallel to the C_4 axis. The choice of axes is made to show the correspondence between the vibrational modes and Γ_σ and Γ_π. There are other approaches to the analysis of vibrational modes. We can treat the Xe—F bending modes directly. However, this leads to four in-plane bends, but
  |
  F

there are only three. That is, we can allow three of the four angles to bend independently, but then the fourth is fixed. This analysis requires the elimination of the "spurious" mode.

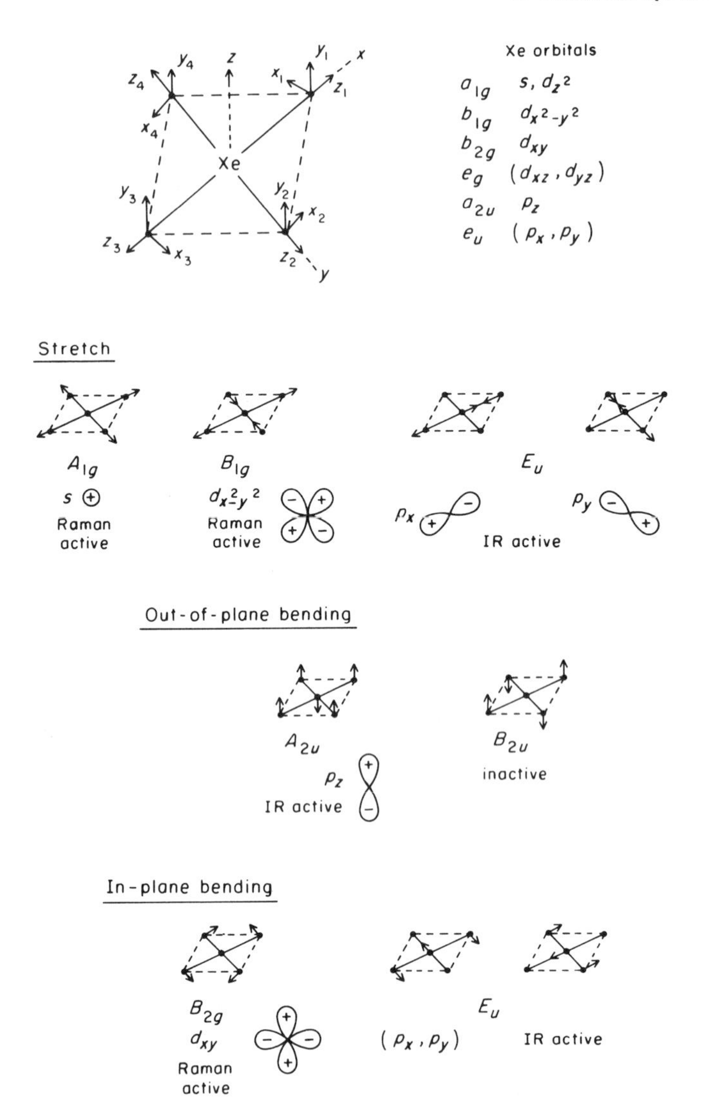

Fig. 10.6 Vibrational modes of XeF_4 (square planar, $\mathbf{D}_{4h}$).

bending, they will mix. The two E_u normal modes will involve mixtures of stretching and bending.

The representations for $\Gamma_{x,y,z}$ are $A_{2u} + E_u$, so these modes are IR active. From the character table we see that the A_{1g}, B_{1g}, and B_{2g} modes are Raman active since they correspond to quadratic functions of x, y, or z. The B_{2u} bending mode is inactive in IR and Raman.

EXAMPLE 10.1 A great deal of structural information can be gained from vibrational spectra without complete analysis. The number of C—O stretching frequencies and their IR and Raman activity can often be used to distinguish among possible geometries of $M(CO)_n$ compounds. Determine the representations for the C—O stretches and their IR and Raman activity for $Cr(CO)_6$ ($\mathbf{O}_h$). The Cr—C≡O bonds are linear.

SOLUTION The C—O stretching frequencies transform as vectors along the bonds and hence along the coordinate axes. They are the same as the Cr—C stretches and the same as the Cr—C σ bonds or the σ bonds of any octahedral MX_6 compound. We get the representations as we did for stretching modes for XeF_4 or as we did for σ bonds for MX_6 (Section 8.2.1).

$\mathbf{O}_h$	E	C_3	C_2'	C_4	C_2	i	S_4	S_6	σ_h	σ_d
Γ_σ	6	0	0	2	2	0	0	0	4	2

$$\Gamma_\sigma = A_{1g} + E_g + T_{1u}$$

A_{1g} and E_g are Raman active

T_{1u} is IR active

EXAMPLE 10.2 Identify the symmetry species for the stretching and bending vibrations of CO_3^{2-} ($\mathbf{D}_{3h}$). Which are IR and Raman active?

SOLUTION See Table 10.6. There will be mixing of the stretching and bending modes of E' symmetry. Thus, the normal mode of A_1' is pure stretching and A_2'' is pure bending. The two normal modes of E' symmetry are mixtures

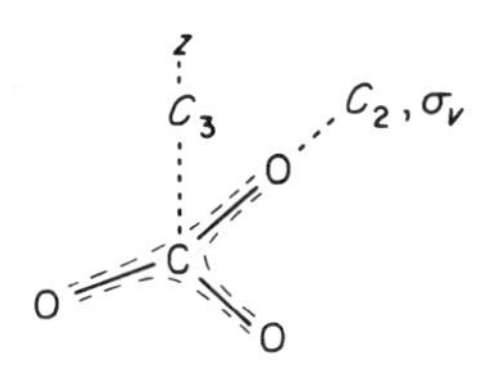

TABLE 10.6
Solution of Example 10.2

$\mathbf{D}_{3h}$	E	C_3	C_2	σ_h	S_3	σ_v	
$\Gamma_{x,y,z}$	3	0	−1	1	−2	1	
Unmoved atoms	4	1	2	4	1	2	
Γ_{total}	12	0	−2	4	−2	2	(product)
Γ_{trans}	3	0	−1	1	−2	1	(subtract)
Γ_{rot}	3	0	−1	−1	2	−1	(subtract)
Γ_{vib}	6	0	0	4	−2	2	
Unmoved atoms	4	1	2	4	1	2	
	1	1	1	1	1	1	(subtract)
Γ_{stretch}	3	0	1	3	0	1	(subtract from Γ_{vib})
Γ_{bend}	3	0	−1	1	−2	1	

$\Gamma_{\text{stretch}} = A_1' + E'$ $\Gamma_{\text{bend}} = A_2'' + E'$
IR active A_2'' and E' Raman active A_1' and E'

10.6 Trigonal Bipyramidal Molecules ($\mathbf{D}_{3h}$)

Let us examine in detail the trigonal bipyramidal molecule PCl_5 ($\mathbf{D}_{3h}$) (see Table 10.7).

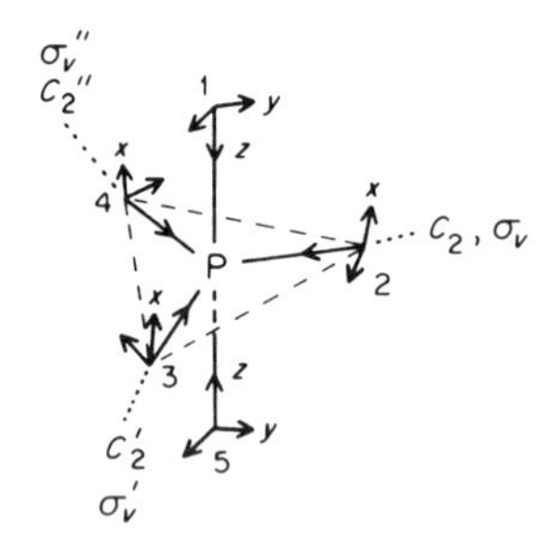

The A_1' vibrations are pure stretching vibrations and E'' is pure bending, but the bending and stretching modes of A_2'' symmetry and of E' symmetry mix.

In sketching the vibrational modes it helps if we recall that for the trigonal bipyramid (TPB), the vectors in the equatorial plane are in a separate set from those in the axial positions. With the z vectors directed toward P as the stretching (or σ) vectors, x and y are involved in bending (or π) vectors. The x and y vectors in the axial positions are treated as a set since C_2 interchanges the axial positions and C_3 mixes x and y. The y vectors on the equatorial Cl can be

TABLE 10.7
Symmetry Species for Vibrations of PCl_5

$\mathbf{D}_{3h}$	E	C_3	C_2	σ_h	S_3	σ_v	
$\Gamma_{x,y,z}$	3	0	−1	1	−2	1	
Unmoved atoms	6	3	2	4	1	4	
Γ_{total}	18	0	−2	4	−2	4	(product)
Γ_{trans}	3	0	−1	1	−2	1	(subtract)
Γ_{rot}	3	0	−1	−1	2	−1	(subtract)
Γ_{vib}	12	0	0	4	−2	4	
Unmoved atoms	6	3	2	4	1	4	
	1	1	1	1	1	1	(subtract)
$\Gamma_{stretch}$	5	2	1	3	0	3	(subtract from Γ_{vib})
Γ_{bend}	7	−2	−1	1	−2	1	

$\Gamma_{stretch} = 2A_1' + A_2'' + E'$ $\Gamma_{bend} = A_2'' + 2E' + E''$
IR active A_2'' and E' Raman active A_1', E', and E''.

treated as one set of in-plane vectors and the corresponding vectors form a set of out-of-plane vectors. We saw from Example 10.2 that the stretches for CO_3^{2-} (also $\mathbf{D}_{3h}$) are A_1' and E'. These are the same as the equatorial stretches for the TBP, so the axial stretches of the TBP are A_1' and A_2''. This provides enough information to sketch these vibrations. The E' stretching vibrations correspond to the E' LGOs (σ) for trigonal planar or trigonal bipyramidal molecules, each with a single nodal plane.

Since the bending vibrations for CO_3^{2-} are A_2'' and E' (Example 10.2), the remaining bending vibrations E' and E'' must involve the axial atoms. For each of the axial atoms the vectors x and y combine as E'—these combine as $E' + E'$ and $E' - E'$, or looking at the vectors individually, $x_1 + x_5$ and $y_1 + y_5$, and $x_1 - x_5$ and $y_1 - y_5$. The sum is still E' and the difference is E''. The E' is the straightforward bending mode (although it mixes with the E' stretch), but we will have to reexamine the E'' case in connection with $(E'')_{eq}$.

We could sort out descriptions of the vibrations involving equatorial vectors pictorially, but let us proceed more systematically using the projection operator method. The results of carrying out all symmetry operations of the group for the in-plane (y) and out-of-plane (x) vectors are given in Table 10.8. A_2'' and E'' vanish for y, and E' vanishes for x, giving

$$A_2'' = 4x_2 + 4x_3 + 4x_4 \quad \text{or} \quad x_2 + x_3 + x_4,$$

$$E' = 4y_2 - 2y_3 - 2y_4 \quad \text{or} \quad 2y_2 - y_3 - y_4,$$

$$E'' = 4x_2 - 2x_3 - 2x_4 \quad \text{or} \quad 2x_2 - x_3 - x_4,$$

TABLE 10.8
Application of the Projection Operator for D_{3h}

D_{3h}	E	C_3	C_3'	C_2	C_2'	C_2''	σ_h	S_3	S_3'	σ_v	σ_v'	σ_v''
y_2	y_2	y_3	y_4	$-y_2$	$-y_4$	$-y_3$	y_2	y_3	y_4	$-y_2$	$-y_4$	$-y_3$
x_2	x_2	x_3	x_4	$-x_2$	$-x_4$	$-x_3$	$-x_2$	$-x_3$	$-x_4$	x_2	x_4	x_3
a_2''	y_2	y_3	y_4	y_2	y_4	y_3	$-y_2$	$-y_3$	$-y_4$	$-y_2$	$-y_4$	$-y_3$
e'	$2y_2$	$-y_3$	$-y_4$	0	0	0	$2y_2$	$-y_3$	$-y_4$	0	0	0
e''	$2y_2$	$-y_3$	$-y_4$	0	0	0	$-2y_2$	y_3	y_4	0	0	0
a_2''	x_2	x_3	x_4	x_2	x_4	x_3	x_2	x_3	x_4	x_2	x_4	x_3
e'	$2x_2$	$-x_3$	$-x_4$	0	0	0	$-2x_2$	x_3	x_4	0	0	0
e''	$2x_2$	$-x_3$	$-x_4$	0	0	0	$2x_2$	$-x_3$	$-x_4$	0	0	0

For each of the E representations, we have obtained one of the pair. We can go through the process of obtaining the other linear combinations as we did for σ and π bonding, but the pattern is now familiar for the doubly degenerate case with C_3 as the major axis. In each case, there is one nodal plane, and the other equivalent combination has a nodal plane perpendicular to this one, through the P—Cl bond, giving $x_3 - x_4$ (E'') and $y_3 - y_4$ (E').

There are two A_1' stretching vibrations, one derived for the axial fluorines and one for the equatorial fluorines. They can combine as a sum and a difference. The $(A_2'')_{ax}$ is an unsymmetrical stretch and $(A_2'')_{eq}$ is a bending vibration. The combination of these with all vectors, including that on P, in one direction is translation along z. There are two A_2'' vibrations, independent of the translations. The translations along x and y are E'—the respective combinations of $+x$ on all atoms and $+y$ on all atoms. All three of the E' vibrations mix. Two E'' combinations are shown for axial and equatorial atoms. There is only one E'' bending mode. We note that the rotations for $\mathbf{D}_{3h}$ are A_2' and E''. The A_2' is $y_2 + y_3 + y_4$ to give rotation about the C_3 axis. Looking at the vectors in Fig. 10.7, we see that the combination of $(x_1 - x_5)$ with $(-x_3 + x_4)$ involves rotation (tumbling toward the viewer). Combination of $(y_1 - y_5)$ with $(-2x_2 + x_3 + x_4)$ (the reverse of the sketch) involves rotation to the right. These are the E'' rotations. The other combinations, $(x_1 - x_5) + (x_3 - x_4)$ and $(y_1 - y_5)$ with $(2x_2 - x_3 - x_4)$ are the pair of E'' bending modes, and these are purely bending.

10.7 Framework and Cage Molecules

When there is no atom at the origin, there is no clear symmetry distinction between stretching and bending vibrations. (The procedure for obtaining the reducible representation for the stretching vibrations by subtracting one from

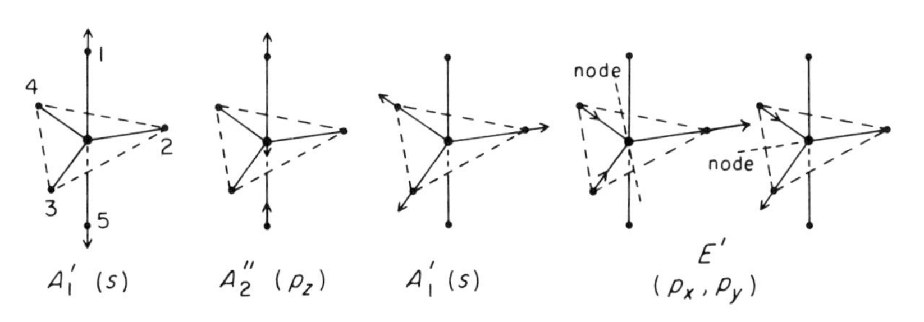

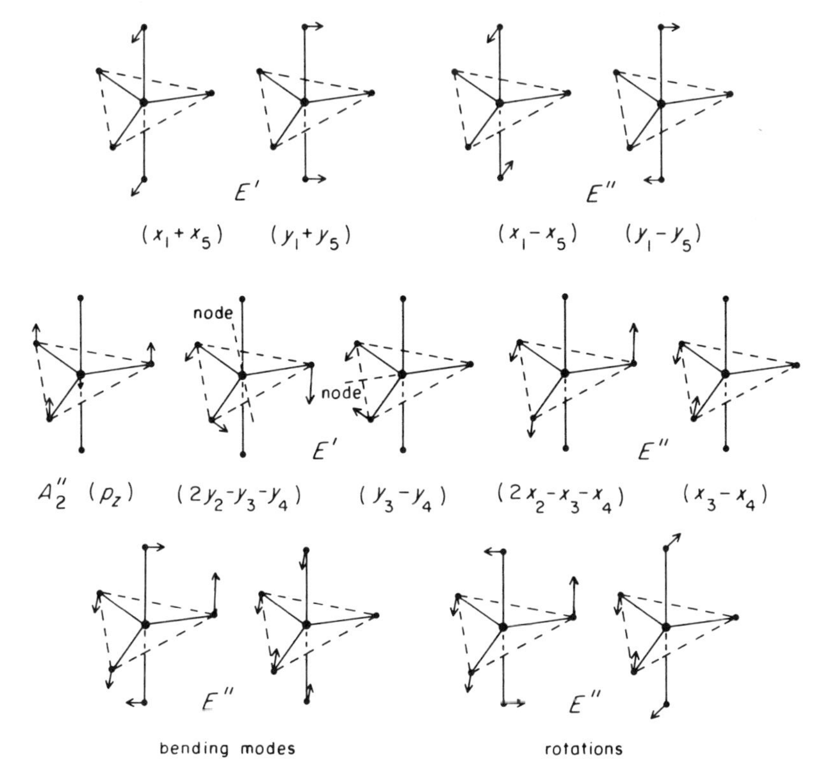

Fig. 10.7 Vibrational modes for PCl_5 ($\mathbf{D}_{3h}$).

the number of atoms unmoved by each operation of the group does not apply.) As examples of simple framework molecules, let us consider B_2H_6 and the closely related C_2H_4 (both $\mathbf{D}_{2h}$).

B_2H_6

y axis perpendicular to B_2H_4 plane

$\mathbf{D}_{2h}$	E	$(C_2)_z$	$(C_2)_y$	$(C_2)_x$	i	σ_{xy}	σ_{xz}	σ_{yz}	
$\Gamma_{x,y,z}$	3	−1	−1	−1	−3	1	1	1	
Unmoved atoms	8	2	2	0	0	2	6	4	
Γ_{total}	24	−2	−2	0	0	2	6	4	(product)
Γ_{trans}	3	−1	−1	−1	−3	1	1	1	(subtract)
Γ_{rot}	3	−1	−1	−1	3	−1	−1	−1	(subtract)
Γ_{vib}	18	0	0	2	0	2	6	4	

$$\Gamma_{\text{vib}} = 4A_g + B_{1g} + 2B_{2g} + 2B_{3g} + A_u + 3B_{1u} + 2B_{2u} + 3B_{3u}$$

Here the B_2H_6 molecule is oriented with the borons and terminal hydrogen atoms in the xz plane. Interchanging the x and z axes (arbitrary for $\mathbf{D}_{2h}$) would give $2B_{1g}$ and $1B_{3g}$. Sketches of all of the vibrations (interchanging the x and z axes) are given by Harris and Bertolucci[7] with descriptions as twisting, wagging, and rocking modes.

The ethylene molecule is treated similarly, using the same choice of axes. The resulting vibrational modes correspond to the framework vibrations of B_2H_6. The six missing vibrations of B_2H_6 ($A_g + B_{1g} + B_{3g} + B_{1u} + B_{2u} + B_{3u}$) are those involving the bridging hydrogens.

$\mathbf{D}_{2h}$	E	$(C_2)_z$	$(C_2)_y$	$(C_2)_x$	i	σ_{xy}	σ_{xz}	σ_{yz}	
$\Gamma_{x,y,z}$	3	−1	−1	−1	−3	1	1	1	
Unmoved atoms	6	2	0	0	0	0	6	2	
Γ_{total}	18	−2	0	0	0	0	6	2	(product)
Γ_{trans}	3	−1	−1	−1	−3	1	1	1	(subtract)
Γ_{rot}	3	−1	−1	−1	3	−1	−1	−1	(subtract)
Γ_{vib}	12	0	2	2	0	0	6	2	

$$\Gamma_{\text{vib}} = 3A_g + 2B_{2g} + B_{3g} + A_u + 2B_{1u} + B_{2u} + 2B_{3u}$$

Note that the C_2H_4 molecule is used frequently as an example, and the subscripts of some of the representations are changed for a different choice of axes.

[7] D. C. Harris and M. D. Bertolucci, *Symmetry and Spectroscopy*, Oxford University Press, New York, p. 149.

The octahedral $B_6H_6^{2-}$ ion relates the vibrational modes of a cage compound to the familiar bonding description. We orient the vectors as for the σ- and π-bonding treatment of octahedral complexes. The vectors are duplicated for B and H (each can move along the bond). Although H cannot form π bonds, it can move in planes perpendicular to the B—H bond (it has three degrees of freedom).

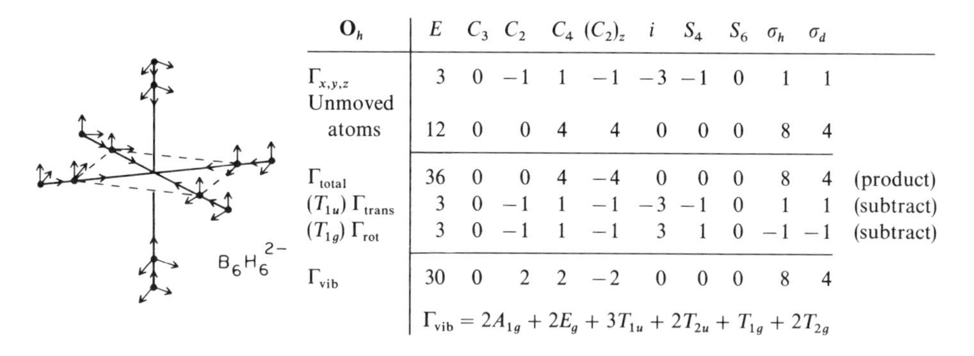

$\mathbf{O}_h$	E	C_3	C_2	C_4	$(C_2)_z$	i	S_4	S_6	σ_h	σ_d	
$\Gamma_{x,y,z}$	3	0	−1	1	−1	−3	−1	0	1	1	
Unmoved atoms	12	0	0	4	4	0	0	0	8	4	
Γ_{total}	36	0	0	4	−4	0	0	0	8	4	(product)
$(T_{1u})\ \Gamma_{\text{trans}}$	3	0	−1	1	−1	−3	−1	0	1	1	(subtract)
$(T_{1g})\ \Gamma_{\text{rot}}$	3	0	−1	1	−1	3	1	0	−1	−1	(subtract)
Γ_{vib}	30	0	2	2	−2	0	0	0	8	4	

$\Gamma_{\text{vib}} = 2A_{1g} + 2E_g + 3T_{1u} + 2T_{2u} + T_{1g} + 2T_{2g}$

For octahedral ML_6,

$$\Gamma_\sigma = A_{1g} + E_g + T_{1u},$$

$$\Gamma_\pi = T_{1g} + T_{2g} + T_{1u} + T_{2u}.$$

Combining the B and H vectors along the σ bonds we get $2A_{1g} + 2E_g + 2T_{1u}$, but we must subtract one T_{1u} for translations. Combining the B and H vectors perpendicular to the σ bond direction we get $2T_{1g} + 2T_{2g} + 2T_{1u} + 2T_{2u}$, but we must subtract one T_{1g} for rotations. We obtain the same representations for the vectors on H as those on B. The sums of the $(T_{1u})_B + (T_{1u})_H$ vectors along the directions for σ bonds in octahedral ML_6 in combination with appropriate T_{1u} vectors of π type correspond to translations (see Fig. 10.8). The differences $(T_{1u})_B - (T_{1u})_H$ mix stretching (along σ-bond directions) and bending vibrations. The differences $(A_{1g})_B - (A_{1g})_H$ and $(E_g)_B - (E_g)_H$ correspond to B—H stretching vibrations. The sums $(A_{1g})_B + (A_{1g})_H$ and $(E_g)_B + (E_g)_H$ are vibrations of the B—H groups along the axial (radial) directions. The sums and differences of the B and H vectors perpendicular to the radial B—H bonds [each sum and each difference has the same symmetry as the representations combined, e.g., $(T_{1g})_B \pm (T_{1g})_H = T_{1g}$] are vibrational modes except for $(T_{1g})_B + (T_{1g})_H$, which corresponds to the rotations (T_{1g}). Thus, the representations for vibrations obtained from the σ- and π-type vectors are reduced to those generated from the numbers of atoms unshifted by the operations of the group.

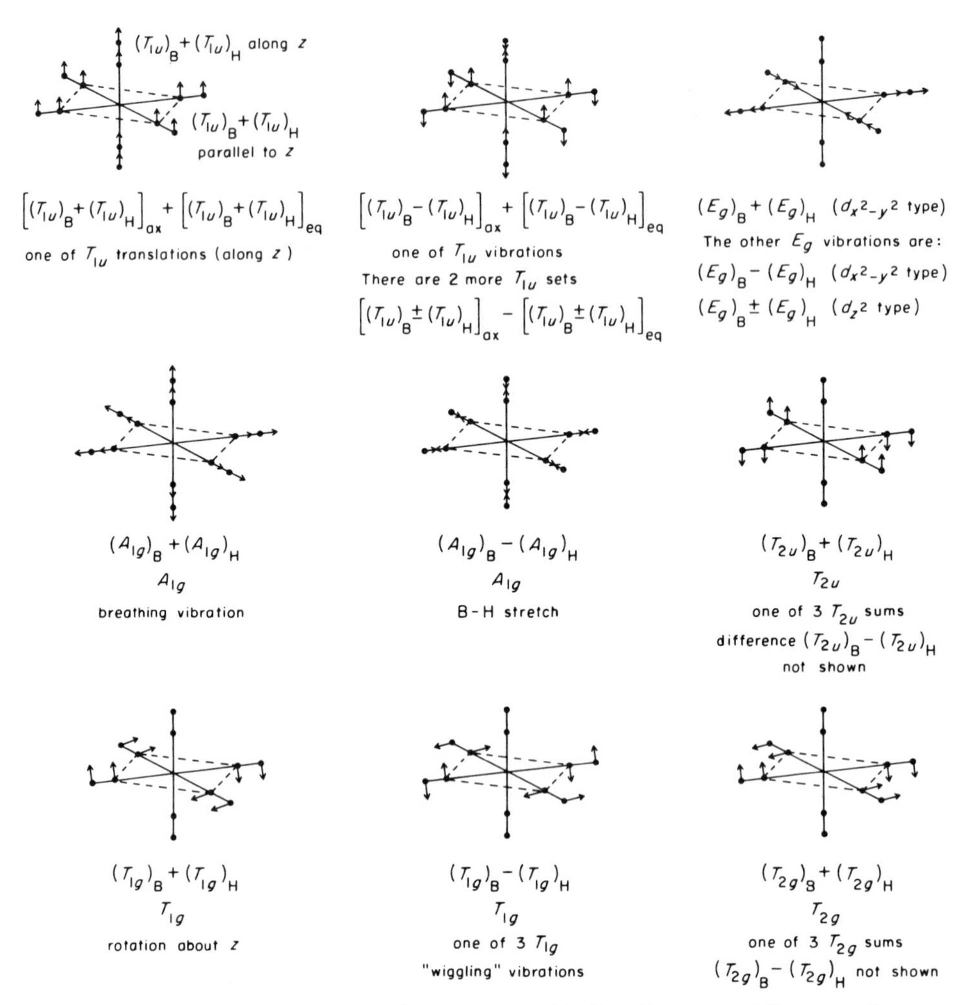

Fig. 10.8 Vibration modes for $B_6H_6^{2-}$. The motion of the light H atoms should be greater than that of B.

10.8 Jahn–Teller Distortions

We have seen that the unequal population of the e_g orbitals of an octahedral complex, as for six-coordinate Cu^{2+} (d^9) complexes (Section 8.1.3), leads to distortion of the octahedron because of the Jahn–Teller effect. The potential energy (E) in terms of small displacements (Q) of the atoms, is given by Eq. (10.15) (U is the nuclear–nuclear and nuclear–electronic potential energy).

$$E = E_0 + Q\left\langle \Psi_0 \left| \frac{\partial U}{\partial Q} \right| \Psi_0 \right\rangle + \frac{Q^2}{2}\left\langle \Psi_0 \left| \frac{\partial^2 U}{\partial Q^2} \right| \Psi_0 \right\rangle + Q^2 \sum_k \frac{\left[\left\langle \Psi_0 \left| \frac{\partial U}{\partial Q} \right| \Psi_k \right\rangle\right]^2}{E_0 - E_k}. \quad (10.15)$$

The second term (linear in Q) applies to degenerate ground states [^2E_g ($\mathbf{O}_h$) for a d^9 complex] leading to the distortion resulting from the *first-order Jahn–Teller* (FOJT) *effect*. The Hamiltonian must be totally symmetric, so $|\partial U/\partial Q|$ must belong to the same representation as Q, and this representation (a normal mode for the molecule) must be contained in the direct product $\Psi_0 \times \Psi_0$ ($E_g \times E_g = A_{1g} + A_{2g} + E_g$). The normal modes for $\mathbf{O}_h$ include A_{1g} and E_g, but since A_{1g} causes no distortion, only E_g will be effective in removing the degeneracy of the degenerate state and of the unequally occupied orbitals.

In the case of a hypothetical low-spin d^8 octahedral complex, the ground state should be $^1A_{1g}$ (one electron in each e_g orbital), and there is no FOJT effect. However, the configuration with both electrons in one e_g orbital should be only slightly higher in energy, giving the ^1E_g state. For the fourth term in Eq. (10.15), Q^2 is totally symmetric and the integral is totally symmetric if $\partial U/\partial Q$ (and Q) belongs to E_g, since $\Gamma_{\Psi_0} \times \Gamma_{\Psi_k} = A_{1g} \times E_g = E_g$. This situation involving two energy states (derived from an incompletely filled shell, but with a nondegenerate ground state) differing only slightly in energy has been called the *pseudo Jahn–Teller effect*. Even though this effect is governed by the quadratic terms of Eq. (10.15), the magnitude of the effect is generally small, as for the FOJT effect. Pearson,[8] who clarified the distinctions among Jahn–Teller effects, groups the pseudo Jahn–Teller effect with the FOJT effect.

The potential energy of a molecule with filled shells can be lowered also from the second quadratic term in cases where there is a small energy separation between the highest filled orbital and the lowest empty orbital. It has been suggested[8] that only this situation be designated as the *second-order Jahn–Teller* (SOJT) *effect*. As for the pseudo Jahn–Teller effect, the direct product $\Gamma_{\Psi_0} \times \Gamma_{\Psi_k}$ must belong to the same representation as Q. As a result of the SOJT distortion, forming a new point group, the HOMO and LUMO must belong to the *same* representation in the new point group. The consequence is that these states can continue to mix, lowering one and raising the other, leading to much larger distortions than encountered for FOJT and pseudo Jahn–Teller effects.

Pearson[8] describes CH_4^+ as an example illustrating FOJT and SOJT distortions. The correlation diagram for CH_4^+ for $\mathbf{T}_d$, $\mathbf{D}_{2d}$, and $\mathbf{D}_{4h}$ are shown in Fig. 10.9. If we assume CH_4^+ is formed from CH_4 by removing an electron, we expect $a_1^2t_2^5$ as the configuration in $\mathbf{T}_d$ symmetry. The degenerate

[8] R. G. Pearson, *Proc. Nat. Acad. Sci. U.S.A.* 1975, **72,** 2104.

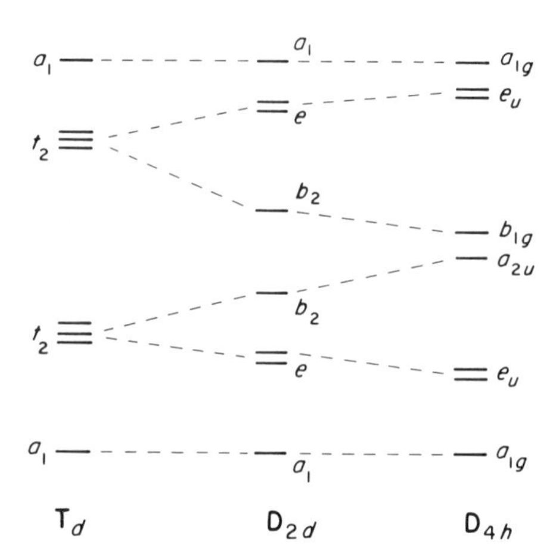

Fig. 10.9 Molecular orbital correlation diagram for CH_4^+ in tetrahedral ($\mathbf{T}_d$), flattened tetrahedral ($\mathbf{D}_{2d}$), and square planar ($\mathbf{D}_{4h}$) geometries. (From R. G. Pearson, *Proc. Nat. Acad. Sci. USA* 1975, **72,** 2104.)

ground state ${}^2T_2$ is subject to FOJT distortion. For the linear term of (10.15), $(T_2 \times T_2) = A_1 + E + T_2$, so an E or T_2 mode could cause the distortion. The E mode leads to a $\mathbf{D}_{2d}$ structure and T_2 leads to a $\mathbf{C}_{3v}$ structure, expected to be higher in energy. From the correlation diagram for $\mathbf{D}_{2d}$, we expect an $a_1^2e^4b_2^1$ configuration, without further interaction among the species.

If we assume CH_4^+ to be planar ($\mathbf{D}_{4h}$), then the configuration is $a_{1g}^2e_u^4a_{2u}^1$ or $a_{1g}^2e_u^4b_{1g}^1$, depending in the relative energies of a_{2u} and b_{1g}, which are expected to be very close in energy. The direct product of the HOMO and LUMO is $B_{1g} \times A_{2u} = B_{2u}$ for either case. There is an out-of-plane B_{2u} vibration that can lead to the $\mathbf{D}_{2d}$ point group. As we see from the correlation diagram (Fig. 10.9), both b_{1g} and a_{2u} become b_2 in $\mathbf{D}_{2d}$, so they can continue to interact.

Either initial symmetry ($\mathbf{T}_d$ or $\mathbf{D}_{4h}$) leads to the actual symmetry ($\mathbf{D}_{2d}$) of CH_4^+.

An interesting example of the SOJT effect was reported by Cotton and Fang.[9] The compound $W_4(OEt)_{16}$ with only eight electrons for the metal cluster was found to have low symmetry ($\mathbf{C}_i$),[10] rather than the $\mathbf{C}_{2h}$ symmetry expected if 10 electrons (5 M—M bonds in M_4^{14+}) are available for the cluster. Figure 10.10 shows a correlation diagram for $\mathbf{C}_{2h}$ (Mo_4^{14+}) and $\mathbf{C}_i$ (Mo_4^{16+}), using the Mo compounds as models for the W compound. Since there are no degeneracies, no FOJT effect is expected. Starting with $\mathbf{C}_{2h}$ symmetry, we see that the five low-energy orbitals are filled by 10 electrons, but with eight electrons, $1b_g$ (HOMO) is filled, but $2a_g$ (LUMO), of slightly higher

[9] F. A. Cotton and A. Fang, *J. Am. Chem. Soc.* 1982, **104,** 113.
[10] M. H. Chisholm, J. C. Huffman, and J. Leonelli, *J. Chem. Soc., Chem. Commun.* 1981, 270.

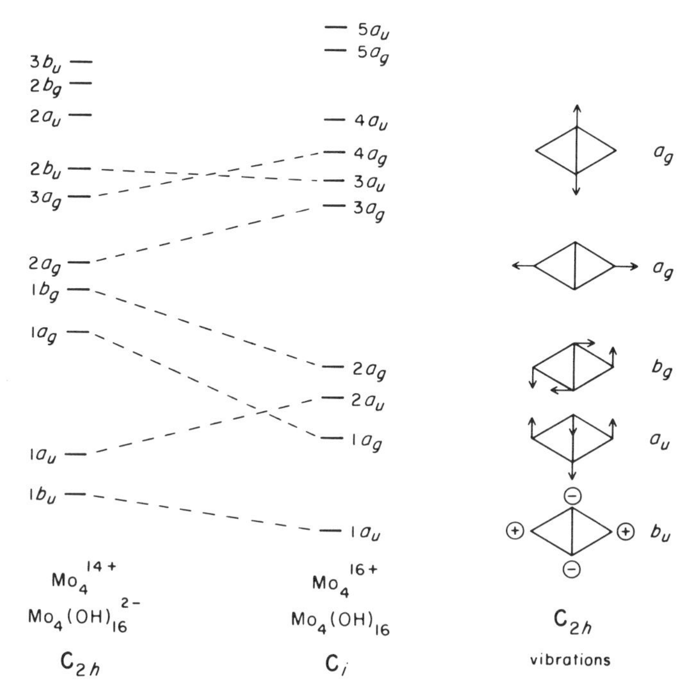

Fig. 10.10 Correlation diagram for $\mathbf{C}_{2h} \rightarrow \mathbf{C}_i$ for a Mo_4 cluster. Vibrations for $\mathbf{C}_{2h}$ are sketched at right. (Adapted with permission from F. A. Cotton and A. Fang, *J. Am. Chem. Soc.* 1982, **104,** 113. Copyright © 1982, American Chemical Society.)

energy, is empty. The product ($B_g \times A_g = B_g$) corresponds to a vibration (sketched in Fig. 10.10) causing distortion leading to $\mathbf{C}_i$ symmetry. Both a_g and b_g orbitals become a_g in the new point group, increasing their energy separation because of configuration interaction.

10.9 Practical Applications

10.9.1 GROUP FREQUENCIES

Many organic functional groups can be identified by characteristic group frequencies. The carbonyl group $>$C=O has a strong characteristic absorption peak at about 1700 cm^{-1}. The frequency is shifted only slightly in going from aldehydes to ketones to carboxylic acids.

The carbonyl frequency has been a valuable aid in the investigation of complexes of ethylenediaminetetraacetic acid (edta), which can function as a sexidentate ligand. It is not always easy to determine if the ligand is attached through both nitrogens and all four carboxylate groups. The carbonyl band occurs at about 1730 cm^{-1} in aliphatic carboxylic acids, but the absorption is shifted to about 1600 cm^{-1} in the carboxylate anions. The large shift is attributed to the lower bond order of the two equivalent carbon–oxygen bonds in RCO_2^-. The bond order is two in $R\overset{\overset{O}{\|}}{C}OH$, but it should be about 1.5 in $(R{-}C\langle^{O}_{O})^-$. The carbonyl band in edta complexes of most divalent metal ions is very near 1600 cm^{-1}, indicating that the carboxylate group remains essentially unchanged. In a complex such as that of cobalt(III), K[Co(edta)], the carbonyl band occurs at about 1650 cm^{-1}, indicating that the C—O bond order has increased (relative to RCO_2^-) because of more localization of the π-electron pair but not to as great an extent as for a carboxylic acid. One free carboxylic acid group is present in complexes such as Na[Co(Hedta)Cl] and a second band appears at 1750 cm^{-1}. This complex can be converted to Na_2[Co(edta)Cl] in which there is an uncoordinated carboxylate ion that absorbs at 1600 cm^{-1}. The extent of the shift of the coordinated carbonyl band in edta complexes has been interpreted as a measure of the extent of the covalency of the M—O bond.

Group frequencies are useful for identification in "fingerprint" IR spectra. Characteristic group frequencies are tabulated[11] for easy identification of observed bands. It is helpful in following the course of a reaction to look for the appearance of bands characteristic of a functional group expected in the product or the disappearance of bands characteristic of functional groups in the starting materials.

10.9.2 METAL CARBONYLS

Infrared spectroscopy has been useful in elucidating structures of metal carbonyls. The stretching vibration of carbon monoxide is at 2141 cm^{-1}, but in $Ni(CO)_4$ ($\mathbf{T}_d$) the stretching vibration occurs at 2057 cm^{-1}. This has been interpreted as an indication of lowering of the bond order for CO because of Ni—C π bonding (back bonding). Some efforts have been made to correlate

[11] L. J. Bellamy, *The Infrared Spectra of Complex Molecules*, 3rd ed., Wiley, New York, 1975, pp. 5–9; R. S. Drago, *Physical Methods in Chemistry*, Saunders, Philadelphia, 1977, pp. 164–69 (a good selection of inorganic examples).

the lowering of the C—O stretching frequency with the lowering of the C—O bond order.

The most common manner of coordination involves coordination of CO as a terminal ligand with CO stretching frequencies in the range 2000–2100 cm^{-1}. Bridging CO ligands (M—C(=O)—M) give much lower frequencies (1800–1900 cm^{-1}), corresponding to the lower bond order of the CO bond. Vibrational spectroscopy has been very helpful in the study of the rather complex structures of metal carbonyls.[12]

A good illustration of the application concerns the choice of the structure of $Fe(CO)_5$. Bridged structures can be eliminated since there are no bands as low as 1900 cm^{-1}. The two likely structures are the trigonal bipyramid (TBP, $\mathbf{D}_{3h}$; see Section 10.6) and the square pyramid (SP, $\mathbf{C}_{4v}$). Looking at the transformation of the vectors along the bonds (Γ_σ) for the two cases, we obtain the following.

TBP $\mathbf{D}_{3h}$	E	C_3	C_2	σ_h	S_3	σ_v	
Γ_σ	5	2	1	3	0	3	$= 2A'_1 + A''_2 + E'$

SP $\mathbf{C}_{4v}$	E	C_4	C_2	σ_v	σ_d	
Γ_σ	5	1	1	3	1	$= 2A_1 + B_1 + E$

Applying the selection rules, we expect the following.

	TBP	SP	Observed C—O stretches
IR	A''_2, E'	$2A_1, E$	2028, 1994 cm^{-1}
Raman	$2A'_1, E'$	$2A_1, B_1, E$	2114, 2031, 1984 cm^{-1}

The two IR bands and three Raman bands are consistent with $\mathbf{D}_{3h}$ symmetry and not with $\mathbf{C}_{4v}$ symmetry.

10.9.3 OTHER COORDINATED LIGANDS

The carbonate ion ($\mathbf{D}_{3h}$, Example 10.2) has four normal modes (A'_1, A''_2, and $2E'$), with the A'_1 IR inactive. As expected, three IR bands are observed for Na_2CO_3. The local symmetry at the carbonate ion is reduced to $\mathbf{C}_s$ if it serves as a unidentate ligand, and it is reduced to $\mathbf{C}_{2v}$ for CO_3^{2-} as a bidentate ligand (see Fig. 10.11). In each of these cases of lower symmetry, ν_1 becomes IR active

[12] S. F. A. Kettle, *Top. Curr. Chem.* 1977, **71,** 111.

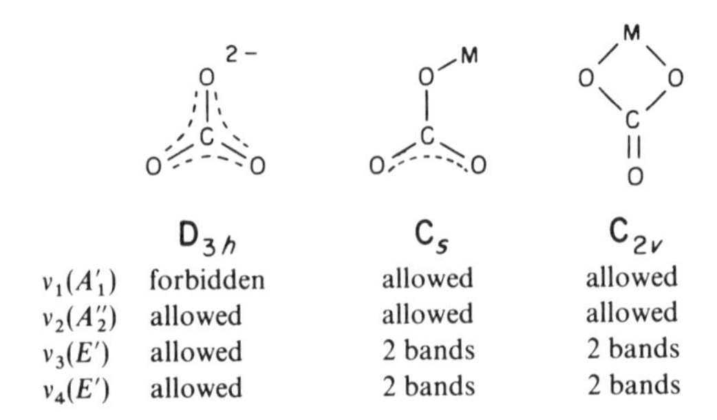

2−
O
C
O
O
M
O
C
O
O
M
O
O
C
O
D_{3h}
C_s
C_{2v}
$\nu_1(A_1')$ forbidden allowed allowed
$\nu_2(A_2'')$ allowed allowed allowed
$\nu_3(E')$ allowed 2 bands 2 bands
$\nu_4(E')$ allowed 2 bands 2 bands

Fig. 10.11 Infrared active bands for CO_3^{2-} and its complexes.

and the degeneracy of E' is removed, resulting in two IR bands from each E'. The splitting of ν_3 is much greater for the $\mathbf{C}_{2v}$ case (see Fig. 10.12).

Similarly, infrared spectra can be useful in distinguishing the manner of coordination of sulfate ion. The tetrahedral ($\mathbf{T}_d$) SO_4^{2-} ion has four normal modes of vibration (Fig. 10.13), only two of which (ν_3 and ν_4) are IR active. If the SO_4^{2-} ion functions as a unidentate ligand, the symmetry is lowered to

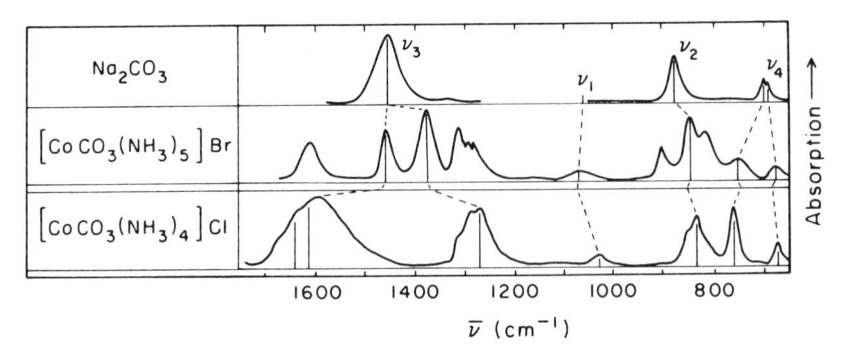

Fig. 10.12 Infrared spectra of sodium carbonate, unidentate, and bidentate Co(III) carbonato complexes. (Adapted from K. Nakamoto, *Infrared Spectra of Inorganic and Coordination Compounds*, New York, 1963, p. 160. © 1963 John Wiley & Sons, Inc. Reproduced by permission.)

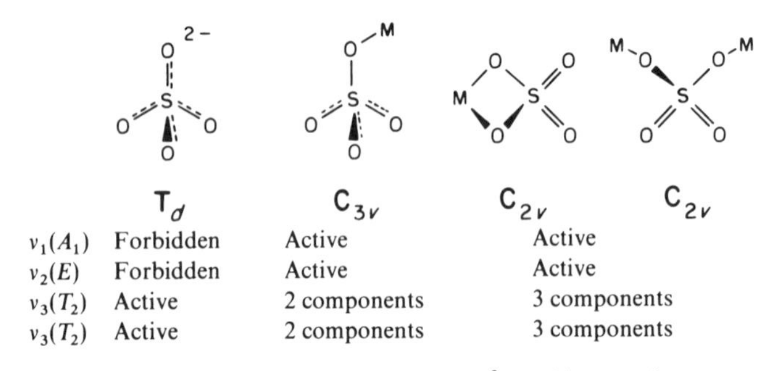

Fig. 10.13 Infrared active bands for SO_4^{2-} and its complexes.

C_{3v}, causing ν_1 and ν_2 to become IR active and ν_3 and ν_4 to split into two components each. If the SO_4^{2-} ion is bidentate, or if it serves as a bidentate bridging ligand, then the symmetry is lowered to C_{2v}, increasing the number of components of ν_3 and ν_4 to three each. All four situations have been observed. Simple sulfates show only two IR bands (ν_3 and ν_4). The complex $[Co(NH_3)_6]_2(SO_4)_3 \cdot 5H_2O$ shows the ν_3 and ν_4 bands and a very weak ν_1 band which presumably appears because of perturbations in the crystal. The complex $[Co(NH_3)_5SO_4]Br$ (unidentate SO_4^{2-}) shows the expected six peaks

and

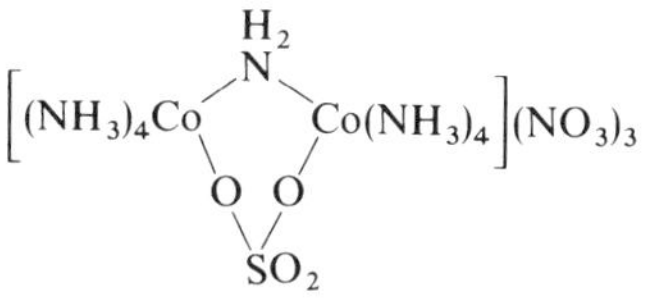

$$\left[(NH_3)_4Co\overset{H_2}{\underset{O\quad O}{\overset{N}{\diagup\quad\diagdown}}}Co(NH_3)_4\right](NO_3)_3 \quad (\text{O, O bridged by } SO_2)$$

shows the expected eight peaks.

Infrared spectra can be useful in distinguishing linkage isomers. We can determine whether SCN^- is uncoordinated, coordinated through the S (thiocyanato or *S*-thiocyanato complex), or coordinated through N (isothiocyanato or *N*-thiocyanato complex), by examining the C—N and C—S IR stretching frequencies. The C—S stretching frequencies, relative to those of KSCN, are lower for complexes involving M—SCN bonding and higher for M—NCS complexes (see Table 10.9). The shifts in the C—N stretching

TABLE 10.9
Infrared Frequencies (cm^{-1}) of the Thiocyanate Ligand[a]

Compound	C—N stretch[b]	C—S stretch[b]
HNCS	2038	848
KSCN	2049	748
$K_2Pt(SCN)_6$	2126(s), 2118 (m), 2076(w)	694(w)
$K_2Pt(SCN)_4$	2128(s), 2099(s), 2077(sh)	696(w)
$K_2Hg(SCN)_4$	2101(s), 2086(s)	718
$K_3Cr(NCS)_6$	2098(vs), 2058(vs)	820(vw)
$(Et_4N)_2Co(NCS)_4$	2062(s)	837(w)
$[Ir(NH_3)_5NCS]^{2+}$	2140(s)	825
$[Ir(NH_3)_5SCN]^{2+}$	2110(s)	700
$[Mn(CO)_5NCS]^{2+}$	2113	813
$[Mn(CO)_5SCN]^{2+}$	2160	676
Bridged complexes $Zn[Hg(SCN)_4]$	2157, 2146	786
$Hg_3[Cr(NCS)_6]_2$	2134	801

[a] Data selected from R. A. Bailey, S. L. Kozak, T. W. Michelsen, and W. N. Mills, *Coord. Chem. Rev.* 1971, **6,** 407 and J. L. Burmeister, *Coord. Chem. Rev.* 1968, **3,** 225.

[b] Intensities are described as (s), strong; (m), medium; (w), weak; (vs), very strong; (vw), very weak; (sh), shoulder.

frequencies are somewhat variable, and they are in the same direction for both types of bonding. The increases in frequencies, however, are often greater for M—SCN compounds. In complexes where SCN^- bridges two metals, the frequency increases for C—S are less than for the M—NCS complexes, and the C—N frequencies increase somewhat more than in either of the other cases. The correlations of the stretching frequencies with M—N or M—S coordination is complicated by the fact that there is usually mixing of the C—N, C—S, and M—N or M—S vibrations. More detailed analysis requires consideration of the force constants also.

The C—S frequency shifts for the M—NCS complexes indicate that the C—S bond order is increased relative to the free SCN^- ion, which is to be expected since electrons will tend to be displaced toward N. The lower C—S bond order indicated by the frequency shifts for the M—SCN complexes is expected also, since coordination through the sulfur should reduce the tendency of S to participate in π bonding.

10.9.4 COMBINATION BANDS AND FERMI RESONANCE

Vibrational spectra are not necessarily as easily interpreted as you might conclude. Some of the bands expected might have very low intensity or they might be shifted from the expected region because they appear as combination bands. For ethylene (see Section 10.7), all vibrations are Raman active (g species) or IR active (u species) except for A_u. By combination of A_u with B_{1u}, we get B_{1g} ($A_u \times B_{1u}$), which is Raman allowed. Combination of A_u with any of the u species (including the first overtone $A_u \times A_u = A_g$) corresponds to a Raman-allowed transition. Combination of A_u with any g species (e.g., $A_u \times B_{2g} = B_{2u}$) gives an IR-allowed transition. Not all combinations are expected to be observed. In cases where two excitations have the same symmetry, combinations or overtones can interact with a fundamental, shifting the frequency of the fundamental. This is known as *Fermi resonance*.

10.10 Ionic Solids

Suppose there are N monatomic ions per unit cell. The total number of degrees of freedom per unit cell is $3N$. Three of these degrees of freedom can be associated with the center of mass of the unit cell, and the motion of this center of mass corresponds to the motion of the unit cell as a whole, leading to three *acoustical* modes of vibration (the three acoustical branches). The acoustical modes are vibrations of the crystal as a whole and are characterized by their longer wavelengths, the crystal size determining the upper limit. The remaining degrees of freedom make up $3N - 3$ *optical* branches. They correspond to motions of the ions relative to one another in the unit cell, the center of mass remaining fixed.

If an ion in the crystal is polyatomic, containing n atoms, it has $3n - 3$ "internal" (or "intramolecular") degrees of freedom. These include the rotational degrees of freedom of the free ion and (unless the ion can rotate relatively freely in the solid) are *librational* degrees of freedom. These additional degrees of freedom correspond to optical modes.

It is our purpose to illustrate how group theory can be used to predict which of the optical modes of a crystal are IR or Raman active. We shall describe the "correlation method" by outlining two of the examples given by Fateley *et al.*[13] The symmetry restrictions on the motion at a site are those of the site symmetry. The combined motion of all the ions in the unit cell is in accordance with the crystal symmetry. A correlation between the site and crystal symmetries is involved. The process is one of going from site (subgroup) to crystal (group) symmetries—just the opposite to the process involved in descending symmetry. For polyatomic ions, we also correlate the symmetry of the free ion with the site symmetry of the ion in the crystal.

In addition to knowing the site symmetry of all the ions and the crystal symmetry, we must also know the orientations of the site symmetries relative to the crystal symmetry. The correlation depends on this. For example, in the case of a site symmetry $\mathbf{D}_{2d}$ in a crystal of class $\mathbf{D}_{4h}$, we need to know which of the C_2 axes of $\mathbf{D}_{2d}$ corresponds to the C_4 axis of $\mathbf{D}_{4h}$ in the crystal being considered (Table 10.10).

As an example for the monatomic ionic crystals, we use anatase, TiO_2 (see Section 2.9). The required crystallographic information is the following: The crystal class is $\mathbf{D}_{4h}$. There are two Ti atoms and four O atoms per unit cell. The site symmetry of the Ti atoms is $\mathbf{D}_{2d}$ with the correlation designated $C_2'' \rightarrow C_2'$. The site symmetry of the O atoms is $\mathbf{C}_{2v}$ with the correlation designated C_2, σ_v. The motion of the ions must transform as x, y, and z. The relevant symmetry species for the $\mathbf{D}_{2d}$ sites are B_2 (z), and E (x, y); those for the $\mathbf{C}_{2v}$ sites are A_1 (z), B_1 (x), and B_2 (y). The correlations of these with the symmetry species of $\mathbf{D}_{4h}$ are as follows: For Ti,

$$B_2 \rightarrow B_{1g}, A_{2u}, \qquad E \rightarrow E_g, E_u,$$

$$\Gamma_{\text{Ti}} = B_{1g} + A_{2u} + E_g + E_u \qquad \text{(six degrees of freedom).}$$

For O,

$$A_1 \rightarrow A_{1g}, B_{1g}, A_{2u}, B_{2u}, \qquad B_1 \rightarrow E_g, E_u \qquad B_2 \rightarrow E_g, E_u,$$

$$\Gamma_{\text{O}} = A_{1g} + B_{1g} + A_{2u} + B_{2u} + 2E_g + 2E_u \quad \text{(12 degrees of freedom).}$$

[13] W. G. Fateley, F. R. Dollish, N. T. McDevitt, and F. F. Bentley, *Infrared and Raman Selection Rules for Molecular and Lattice Vibrations*, Wiley (Interscience), New York, 1972.

TABLE 10.10
Some Correlations for the Examples of Vibrations in Ionic Solids[a]

$\mathbf{D}_{4h}$	$C_2' \rightarrow C_2'$ $\mathbf{D}_{2d}$	$C_2'' \rightarrow C_2'$ $\mathbf{D}_{2d}$	$\mathbf{C}_{4v}$	C_2, σ_v $\mathbf{C}_{2v}$	C_2, σ_d $\mathbf{C}_{2v}$	C_2' $\mathbf{C}_{2v}$	C_2'' $\mathbf{C}_{2v}$
A_{1g}	A_1	A_1	A_1	A_1	A_1	A_1	A_1
A_{2g}	A_2	A_2	A_2	A_2	A_2	B_1	B_1
B_{1g}	B_1	B_2	B_1	A_1	A_2	A_1	B_1
B_{2g}	B_2	B_1	B_2	A_2	A_1	B_1	A_1
E_g	E	E	E	$B_1 + B_2$	$B_1 + B_2$	$A_2 + B_2$	$A_2 + B_2$
A_{1u}	B_1	B_1	A_2	A_2	A_2	A_2	A_2
A_{2u}	B_2	B_2	A_1	A_1	A_1	A_2	B_2
B_{1u}	A_1	A_2	B_2	A_2	A_1	A_2	B_2
B_{2u}	A_2	A_1	B_1	A_1	A_2	B_2	A_2
E_u	E	E	E	$B_1 + B_2$	$B_1 + B_2$	$A_1 + B_1$	$A_1 + B_1$

$\mathbf{T}_d$	$\mathbf{T}$	$\mathbf{D}_{2d}$	$\mathbf{D}_2$	$\mathbf{C}_{2v}$
A_1	A	A_1	A	A_1
A_2	A	B_1	A	A_2
E	E	$A_1 + B_1$	$2A$	$A_1 + A_2$
T_1	T	$A_2 + E$	$B_1 + B_2 + B_3$	$A_2 + B_1 + B_2$
T_2	T	$B_2 + E$	$B_1 + B_2 + B_3$	$A_1 + B_1 + B_2$

[a] The meanings of the column designations are as follows: $C_2' \rightarrow C_2'$ means the C_2' axis of $\mathbf{D}_{4h}$ is the C_2' axis of $\mathbf{D}_{2d}$ (similarly for $C_2'' \rightarrow C_2'$); C_2, σ_v means the C_2 axis (coincident with the C_4 axis) of $\mathbf{D}_{4h}$ is the C_2 axis of $\mathbf{C}_{2v}$, and the σ_v planes of $\mathbf{D}_{4h}$ are the reflection planes of $\mathbf{C}_{2v}$ (similarly for C_2, σ_d); C_2' means the C_2' axis of $\mathbf{D}_{4h}$ is the C_2 axis of $\mathbf{C}_{2v}$.

For the unit cell we have

$$\Gamma_{\text{cryst}} = \Gamma_{\text{Ti}} + \Gamma_{\text{O}} \qquad \text{(18 degrees of freedom).}$$

These 18 degrees of freedom include the three acoustical modes. These acoustical modes must transform as x, y, and z in the symmetry $\mathbf{D}_{4h}$. They are the bases for the species A_{2u} (z) and E_u (x, y).

$$\Gamma_{\text{acoust}} = A_{2u} + E_u.$$

We must subtract Γ_{acoust} from Γ_{cryst} to obtain the representations for the optical branches (Γ_{vib}). This leaves

$$\Gamma_{\text{vib}} = \Gamma_{\text{cryst}} - \Gamma_{\text{acoust}}$$

$$= \underset{\text{(R)}}{A_{1g}} + \underset{\text{(R)}}{2B_{1g}} + \underset{\text{(IR)}}{A_{2u}} + B_{2u} + \underset{\text{(R)}}{3E_g} + \underset{\text{(IR)}}{2E_u}.$$

The species that are IR active are those that have as bases x, y, or z. These are (as noted above) A_{2u} and E_u. There will be one A_{2u} vibration and two doubly degenerate E_u vibrations that are IR active.

The species that are Raman active are those in $\mathbf{D}_{4h}$ that have the polarizability components, or their combinations, as bases. These are A_{1g}, B_{1g}, and E_g. The Raman active vibrations are as follows: one A_{1g}, two B_{1g}, and three doubly degenerate E_g.

As an example of a crystal containing a polyatomic ion, we take NH_4I. The crystal class is $\mathbf{D}_{4h}$ with two ions of each type per unit cell. The site symmetry of the ammonium ions is $\mathbf{D}_{2d}$ (correlation $C_2'' \to C_2'$). The site symmetry of the iodide ions is $\mathbf{C}_{4v}$ (only one correlation possible for this).

For the analysis of the interionic vibrations (i.e., with NH_4^+ being treated as a rigid body, without rotation) the results obtained for $\Gamma_{NH_4^+\,(\mathrm{rigid})}$ are the same as those we already obtained for Γ_{Ti},

$$\Gamma_{NH_4^+\,(\mathrm{rigid})} = B_{1g} + A_{2u} + E_g + E_u.$$

For the iodide ions, the relevant species in $\mathbf{C}_{4v}$ are A_1 (z) and E (x, y). The correlations are

$$A_1 \to A_{1g}, A_{2u}, \qquad E \to E_g, E_u,$$

$$\Gamma_{\mathrm{I}^-} = A_{1g} + A_{2u} + E_g + E_u.$$

Γ_{acoust} is the same as for TiO_2, so we have

$$\begin{aligned} \Gamma_{NH_4I(\mathrm{ext\text{-}vib})} &= \Gamma_{NH_4^+\,(\mathrm{rigid})} + \Gamma_{\mathrm{I}^-} - \Gamma_{\mathrm{acoust}} \\ &= \underset{(\mathrm{R})}{A_{1g}} + \underset{(\mathrm{R})}{B_{1g}} + \underset{(\mathrm{IR})}{A_{2u}} + \underset{(\mathrm{R})}{2E_g} + \underset{(\mathrm{IR})}{E_u}. \end{aligned}$$

The IR active species are A_{2u} and E_u, and the Raman active species are A_{1g}, B_{1g}, and $2E_g$.

We now analyze the internal vibrations and rotations of NH_4^+ (the rotations become librations in the crystal). We first analyze the NH_4^+ as a free ion of symmetry $\mathbf{T}_d$. The 12-dimensional representation with the coordinates of the four hydrogens as basis has the following characters in $\mathbf{T}_d$ (Problem 10.3c).

E	$8C_3$	$3C_2$	$6S_4$	$6\sigma_d$
12	0	0	0	2

This resolves as follows:

$$\Gamma_{NH_4^+\,(\mathrm{free})} = \underset{(\mathrm{R})}{A_1} + \underset{(\mathrm{R})}{E} + T_1 + \underset{(\mathrm{R,IR})}{2T_2}.$$

One of the T_2 species corresponds to the rotations. (For future comparison, we note that for the free ion, only species T_2 is IR active, while A_1, E, and T_2 are Raman-active species. The species T_1 corresponds to the inactive rotations)

The next step is to correlate these $\mathbf{T}_d$ species with the site symmetry $\mathbf{D}_{2d}$. These are for the vibrations

$$A_1 \to A_1,$$
$$E \to A_1 + B_1,$$
$$2T_2 \to 2B_2 + 2E,$$

and for the librations

$$T_1 \to A_2 + E.$$

Now we must correlate these $\mathbf{D}_{2d}$ species with the crystal symmetry $\mathbf{D}_{4h}$. For the vibrations we obtain

$$2A_1 \to 2A_{1g}, 2B_{2u},$$
$$B_1 \to B_{2g}, A_{1u},$$
$$2B_2 \to 2B_{1g}, 2A_{2u},$$
$$2E \to 2E_g, 2E_u,$$

and for the librations

$$A_2 \to A_{2g}, B_{1u},$$
$$E \to E_g, E_u.$$

These give

$$\Gamma_{\mathrm{NH_4^+\,(int\text{-}vib)}} = \underset{(\mathrm{R})}{2A_{1g}} + A_{1u} + \underset{(\mathrm{IR})}{2A_{2u}} + \underset{(\mathrm{R})}{2B_{1g}} + \underset{(\mathrm{R})}{B_{2g}} + 2B_{2u} + \underset{(\mathrm{R})}{2E_g} + \underset{(\mathrm{IR})}{2E_u}$$
(18 degrees of freedom),

$$\Gamma_{\mathrm{NH_4^+\,(lib)}} = A_{2g} + B_{1u} + \underset{(\mathrm{R})}{E_g} + \underset{(\mathrm{IR})}{E_u} \qquad \text{(6 degrees of freedom).}$$

You may wish to compare these results with the species of $\Gamma_{\mathrm{NH_4^+\,(free)}}$.

Additional Reading

D. M. Adams, *Metal–Ligand and Related Vibrations*, St. Martin's Press, New York, 1968.

N. L. Alpert, W. E. Keiser, and H. A. Szymanski, *Theory and Practice of Infrared Spectroscopy*, Plenum, New York, 1970.

L. J. Bellamy, *The Infrared Spectra of Complex Molecules*, 3rd ed., Wiley (Halsted), New York, 1975.

N. B. Colthup, L. H. Daly, and S. E. Wiberley, *Introduction to Infrared and Raman Spectroscopy*, 2nd ed., Academic Press, New York, 1975.
R. T. Conley, *Infrared Spectroscopy*, Allyn and Bacon, Boston, 1972.
G. Herzberg, *Infrared and Raman Spectra of Polyatomic Molecules*, Van Nostrand-Reinhold, New York, 1945. Still a useful source.
K. Nakamoto, *Infrared and Raman Spectra of Inorganic and Coordination Compounds*, 3rd ed., Wiley (Interscience), New York, 1978.
Sadtler Research Laboratories, *Sadtler Standard Spectra*, Philadelphia. An extensive collection of spectra.
E. B. Wilson, Jr., J. C. Decius, and P. C. Cross, *Molecular Vibrations*, McGraw-Hill, New York, 1955. A classic.

Problems

10.1 (a) From Eq. (10.4) relating vibrational frequency to force constants and the tenuous relationship between force constants and bond strengths, arrange the following molecules in the order of increasing vibrational energies for $v_0 \rightarrow v_1$: H_2, Mg_2, F_2, O_2, and N_2.
(b) Qualitatively, how do force constants and vibrational frequencies vary for H_2 compared to D_2?

10.2 Obtain the representations for the normal modes for linear C_2H_2, H—C≡C—H. Sketch them.

10.3 Determine the number of normal modes and their representations for the following molecules. Identify the stretching modes. (a) NH_3, $\mathbf{C}_{3v}$, (b) BF_3, $\mathbf{D}_{3h}$, (c) CH_4, $\mathbf{T}_d$, (d) SF_4, $\mathbf{C}_{2v}$, (e) BrF_5, $\mathbf{C}_{4v}$, (f) SF_6, $\mathbf{O}_h$.

10.4 Determine which of the normal modes in Problem 10.3 are infrared (IR) active and which are Raman active.

10.5 (a) Determine the symmetry species for the normal modes of ClF_3 ($\mathbf{C}_{2v}$).
(b) Identify the stretching modes.
(c) Which vibrations are Raman allowed and which are IR allowed?

10.6 For ethylene (Section 10.7) we might have chosen the coordinate axes differently. What would be the symmetry species for the vibrations with the molecule in the xy plane and the x axis through the C atoms?

10.7 (a) Find the symmetry species for the normal modes of *trans* (planar) N_2F_2.
(b) Which are IR and which are Raman active?
(c) Which are in plane?
(d) Sketch the modes.

10.8 For N_2F_2 in Problem 10.7, what would be the IR and Raman activity of (a) each first overtone, (b) each combination of fundamentals?

10.9 Identify the symmetry species of the normal modes of a pyramidal molecule sketched. Only one of a pair of doubly degenerate modes is sketched.

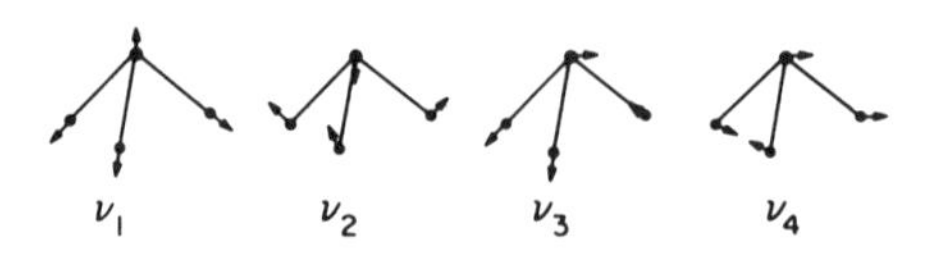

10.10 (a) What would be the symmetry species for the C—N stretching vibrations for square planar $[Pt(CN)_4]^{2-}$?
(b) Which are Raman active? and IR active?
(c) What would be the symmetry species of the normal modes if the complex were tetrahedral? (Pt—C≡N bonds are linear.)

10.11 For $Fe_2(CO)_9$, infrared absorption for the C—O stretching vibrations occurs at 2080, 2034, and 1828 cm^{-1}.
(a) Are there both terminal and bridging COs present?
(b) What would be the highest plausible symmetry with a C_3 axis?
(c) Obtain the symmetry species for the C—O stretches. Which are in-plane?
(d) Is the structure in (b) consistent with the number of infrared bands for the ~2050 and ~1850 cm^{-1} ranges?
(e) Which C—O stretches should be Raman active?

10.12 (a) For the hypothetical planar ($\mathbf{D}_{3h}$) B_3Cl_3, what are the symmetry species for the normal modes?
(b) Which are in-plane vibrations?
(c) Which are Raman active and which are IR active?
(d) Which are Raman polarized?

10.13 (a) Determine the number of and the symmetry species for the normal modes of benzene.
(b) Determine which are in-plane vibrations.
(c) Which are IR active? and Raman active?

10.14 Consider the nondegenerate normal modes of XeF_4 (Fig. 10.6). Which of these would be expected to be IR and Raman active as
(a) first overtones?
(b) second overtones?
(c) Which combination bands of the nondegenerate normal modes would be IR active?
(d) Which normal modes would be expected to be Raman polarized?

10.15 (a) Obtain the symmetry species for the normal modes of $CHCl_3$ and determine which are IR active and which are Raman active. Bands are observed in the infrared and Raman at about the same frequencies: ν_1, 3019; ν_2, 668; ν_3, 366; ν_4, 1216; ν_5, 761; and ν_6, 262 cm^{-1}. Bands ν_1, ν_2, and ν_3 are Raman polarized. For $CDCl_3$ the frequencies of bands ν_1 and ν_4 are lowered greatly and those of ν_3 and ν_6 are essentially unchanged.

(b) Assign the bands to the proper symmetry species.

(c) Sketch the nondegenerate modes and assign observed bands to them.

11

Symmetry-Controlled Reactions

Bonding interactions are determined by the symmetry of orbitals available. Since many chemical reactions break some bonds and form new bonds, it seems obvious that reaction mechanisms involving interactions of orbitals of like symmetry would be favored (allowed). This is a beautiful example where elegant theory and experimental observations come together in an appealing way. Symmetry rules can be expressed in terms of orbital symmetry, state symmetry, or bond symmetry.

11.1 Simple Bimolecular Reactions

The reaction of H_2 with I_2 to form HI was considered to be a typical example of a bimolecular reaction until 1967 when it was shown to be more complex. It seemed entirely reasonable that the broadside collision of H_2 with I_2 could break the H—H and I—I bonds while the H—I bonds were formed. As we shall see, such reactions are symmetry-forbidden.

Let us begin with the simplest example of a reaction between diatomic molecules, the H–D exchange reaction,

$$H_2 + D_2 \rightleftarrows 2\,HD.$$

If the molecules collide, new bonds cannot be formed by the interaction between the two filled bonding orbitals or between the two empty antibonding orbitals. The new H—D bonds can be formed only by electron transfer from the filled σ orbital of one molecule to the empty σ^* orbital of the other. The combinations are equivalent, so let us consider the filled σ orbital to belong to H_2.

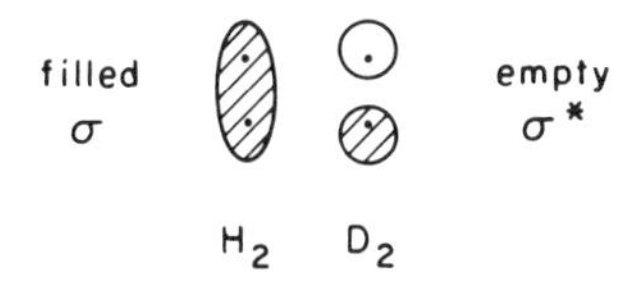

The net orbital overlap is zero. Since the orbitals that might effect electron transfer differ in symmetry so that the net overlap is zero, the reaction is *symmetry-forbidden.*

Now we consider a similar reaction that is more favorable thermodynamically and involves different orbitals:

$$H_2 + F_2 \longrightarrow 2\,HF. \tag{11.1}$$

The σ bond in F_2 involves overlap of one p orbital of each F. For each F, the much lower energy s orbital and the other two filled p orbitals are nonbonding. The *highest (energy) occupied molecular orbital* (HOMO) of F_2 is σ and the *lowest (energy) unoccupied molecular orbital* (LUMO) is σ^*. The occupied orbitals of F_2 are lower in energy than those of H_2 corresponding to the greater electronegativity of fluorine. Electron transfer to form H—F bonds must occur from the HOMO (σ) of H_2 to the LUMO (σ^*) of F_2.

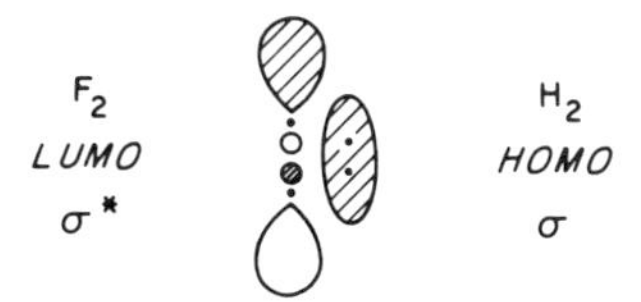

Obviously the net overlap is zero. This reaction is forbidden also. Of course, it is this *mechanism* that is symmetry-forbidden. The reaction occurs by a free radical mechanism.

When an active metal such as Na (a free radical) reacts with an active nonmetal such as F_2, we must consider transfer of an electron from the highest occupied orbital of Na (3s) to the LUMO of F_2. Addition of an electron to the antibonding orbital causes the F—F bond to break (F_2^- is unstable), giving $Na^+ + F^- + \cdot F$, a symmetry-allowed reaction.

$$Na^+ \ \ F{-}F \longrightarrow Na^+ + F^- + \cdot F \tag{11.2}$$

The reaction between Na and F produces ions. In cases involving smaller differences in electronegativities, the electrons transferred are shared, whether these are oxidation–reduction or acid–base reactions. For molecules, the electrons from the Lewis base (or the reductant) in the HOMO are transferred to the LUMO of the Lewis acid (or the oxidant).

Proton affinities parallel ionization energies for simple molecules in the gas phase. In solution, relative acidities are affected by solvation effects to a great extent. Relative base strengths toward metal ions that are hard Lewis acids parallel the base strengths relative to the proton. Highly polarizing cations are soft Lewis acids. Base strengths of soft bases are enhanced toward soft acids in comparison to the proton.

Carbon monoxide shows no significant basicity toward the proton, but it is among the strongest field ligands toward transition metal ions. We expect oxygen, with its lone pairs, to be basic, and few carbon compounds serve as bases with carbon as the donor. Nevertheless, in metal carbonyls carbon is the donor. The HOMO of CO is the nonbonding orbital with electron density localized on C, the carbon lone pair. Of course, this does not explain the extraordinarily high field strength of CO in transition metal carbonyls. In these cases the base strength of CO is enhanced tremendously with metals late in each transition series, particularly those in low oxidation states. These metals have filled, or nearly filled, d orbitals of suitable symmetry for π donation from the metal to empty π^* orbitals of CO. This is called "back donation" because CO is a σ donor and a π acceptor. As we shall see later, such interaction of metal ions with olefins is important in bringing about reactions that are symmetry-forbidden without the intervention of a transition metal ion.

BF_3 is a good Lewis acid, combining with F^- to form BF_4^-. The LUMO of BF_3 is a_2'', a π^* orbital that has the character of p_z on boron predominately. One of the lone pairs on F^- in the HOMO (a p orbital) is donated to the LUMO of BF_3 to form a new σ bond. The reaction is symmetry-allowed. Since BF_3 is a hard acid, it combines readily with the hard base F^-. The soft acid BH_3 (from B_2H_6) combines more readily with the soft bases H^- or CO.

11.2 Electrocyclic Reactions

11.2.1 BUTADIENE–CYCLOBUTENE CONVERSION

Ring closure of polyolefins and the reverse ring opening process are referred to as *electrocyclic reactions*. Orbitals used for π bonding in the polyolefin are used for σ bonding in the cyclic compound. These are of particular interest because the reactions are stereospecific. Woodward and Hoffmann[1] showed

[1] R. B. Woodward and R. Hoffmann, *Angew. Chem. Int. Edit.* 1969, **8,** 781; also available as *The Conservation of Orbital Symmetry*, Verlag Chemie, GmbH, Weinheim/Bergstrasse, 1970. Their earlier papers are cited.

that the products depend on the number of π electrons in the polyolefin, and the products differ for thermal and photochemical reactions. Thus, heating *cis*-3,4-dimethylcyclobutene gives *cis-trans*-2,4-hexadiene, while heating *trans*-3,4-dimethylcyclobutene gives *trans-trans*-2,4-hexadiene. The photochemically induced cyclization reactions yield the *opposite* isomers of dimethylcyclobutene.

These are concerted reactions—the bond breaking and bond making occur continuously without the formation of intermediates or the interaction of a catalyst. The principle of microscopic reversibility requires that the changes in stereochemistry and orbital interactions can be traced reversibly from reactant to product. We see from Fig. 11.1 that conrotatory motion is required to convert the *cis-trans*-butadiene into *cis*-cyclobutene (thermal route) and disrotatory motion is required to obtain *trans*-cyclobutene (photochemical route). These are the required changes to match the experimental

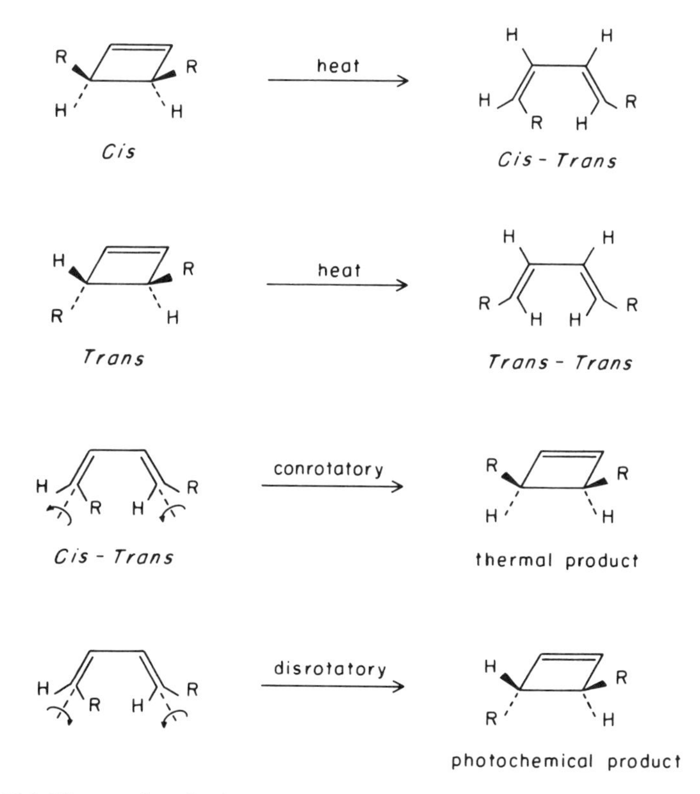

Fig. 11.1 Ring opening of substituted cyclobutenes and the reverse photochemical reaction. Conrotatory and disrotatory motions are illustrated.

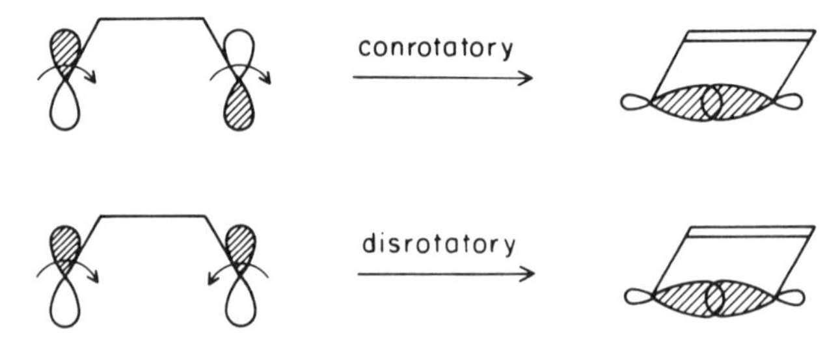

Fig. 11.2 Sign combinations of orbitals required for σ bond formation by conrotatory and disrotatory motions.

observations. If we examine the sign patterns required for the p orbitals to form σ bonds by the two routes, we see the result shown in Fig. 11.2.

Figure 11.3 shows the LCAOs for the π MO's for butadiene and for the corresponding orbitals in cyclobutene. We see that for conrotatory motion, σ

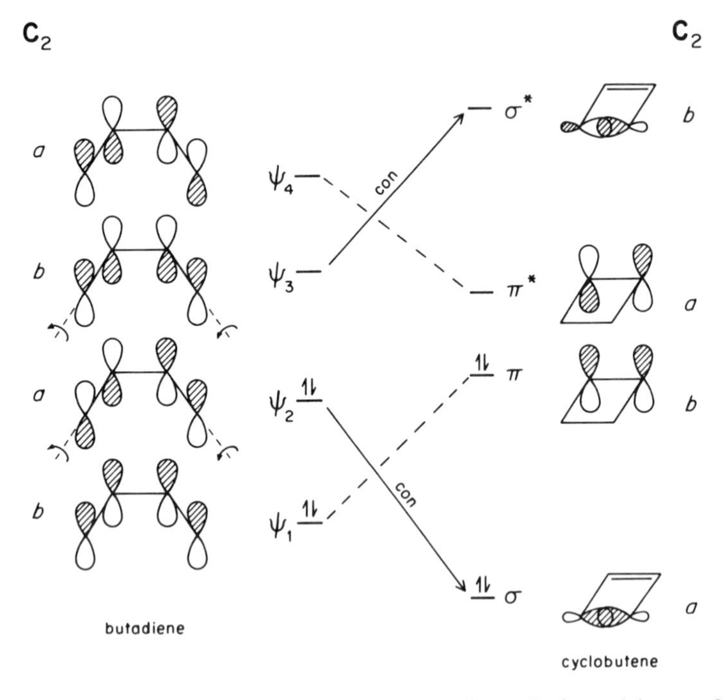

Fig. 11.3 The πMOs for butadiene and the corresponding MOs for cyclobutene. Orbitals interconverted by conrotatory motion are connected. The symmetry species are identified for $\mathbf{C}_2$ symmetry.

is derived from Ψ_2 and σ^* from Ψ_3. The sign patterns for the corresponding p orbitals are the same for Ψ_1 and π and for Ψ_4 and π^*. The symmetry that persists from reactant to product and for all intermediate orientations of p orbitals is $\mathbf{C}_2$, so we can correlate the two sets of orbitals under $\mathbf{C}_2$ symmetry. We see that our tie lines connect orbitals belonging to the same (a or b) symmetry species. The conrotatory process is *symmetry-allowed* because bonding orbitals correlate with bonding orbitals of the same symmetry.

The results for the disrotatory process are shown in Fig. 11.4. Here the reactant, product, and intermediate orientations belong to the $\mathbf{C}_s$ group. The symmetry patterns for the arrows representing disrotatory motion establish the symmetry for intermediate orientations. The symmetry species are identified for each MO in the figure. We see that the bonding Ψ_2 MO correlates with the antibonding π^* orbital and the bonding π orbital of cyclobutene correlates with the antibonding Ψ_3 MO. The disrotatory process is *symmetry-forbidden.*

For each of these processes we see that the lower-energy bonding orbital correlates with a bonding orbital for ring opening or closing by a conrotatory or disrotatory process. It is the HOMO that remains a bonding MO only for

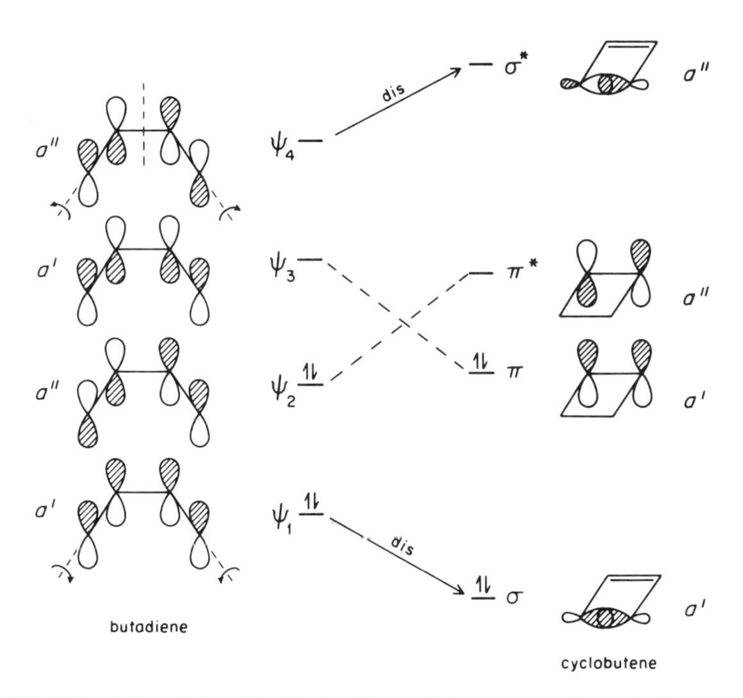

Fig. 11.4 Correlation of the πMOs for butadiene and the corresponding MOs for cyclobutene by disrotatory motion. The symmetry species are identified for $\mathbf{C}_s$ symmetry.

the symmetry-allowed reaction. It is to be expected that the HOMOs will be intimately involved in the reaction and will restrict the course of the reaction. If we excite an electron from Ψ_2 to Ψ_3 of butadiene photochemically, the HOMO becomes Ψ_3. The disrotatory process (Fig. 11.4) would give an excited state of cyclobutene ($\sigma^2\pi^1\pi^{*1}$), and Ψ_3 correlates with the π bonding orbital. Thus, the photochemical reaction is symmetry-allowed for disrotatory motion. It is symmetry-forbidden for conrotatory motion (Fig. 11.3), since here Ψ_3 correlates with the highest-energy σ^* orbital.

The symmetry used for the conrotatory process is $\mathbf{C}_2$, and for the disrotatory process it is $\mathbf{C}_s$. The symmetry with respect to the plane through all carbon atoms does not help, since all of the π and π^* orbitals change sign with respect to this plane. Also, we wish to examine the transformation of p orbitals perpendicular to this plane (Ψ_2 or Ψ_1) into σ bonding orbitals contained in this plane. For the conrotatory process, the $\mathbf{C}_2$ group describes the symmetry of the orbitals and the molecular motion. For the disrotatory process, the $\mathbf{C}_s$ group is appropriate since the C_2 axis is destroyed. Substituents alter the symmetry of cyclobutene and butadiene, but the stereospecificity of the electrocyclic reactions remains the same. The course of the reaction is determined by the symmetry of the *orbitals* involved.

11.2.2 CYCLIZATION OF THE ALLYL CATION AND ALLYL ANION

Let us consider the allyl system where we can examine $H_2CCHCH_2^+$ with two π electrons and $H_2CCHCH_2^-$ with four π electrons. The correlation diagram for ring closure of the allyl cation is shown in Fig. 11.5. Let us use $\mathbf{C}_2$ symmetry to identify the symmetry species for conrotatory motion and $\mathbf{C}_s$ for disrotatory motion (taking the rotational vectors into account). We see that the reaction is symmetry-allowed for the disrotatory process since the bonding π orbital (a') correlates with the bonding σ orbital (a') of the cyclopropyl cation. For the $H_2CCHCH_2^-$ anion, disrotatory motion correlates the HOMO (a'') with σ^* of the cyclic product. For disrotatory ring opening of the cylic anion, the HOMO (a') correlates with π^* of the allyl anion. For the conrotatory process ($\mathbf{C}_2$ symmetry), the HOMO of $H_2CCHCH_2^-$ (a) correlates with the σ orbital of the cyclic anion, and the HOMO of the cyclic anion (b) correlates with the π bonding orbital of $H_2CCHCH_2^-$. The symmetry-allowed process is conrotatory for the anion containing four π electrons.

Thermal electrocyclic reactions are symmetry-allowed for conrotatory motion for $4n$ ($n = 0, 1, 2, \ldots$) π electrons and for disrotatory motion for $(4n + 2)$ π electrons. For photochemical reactions, the opposite rotatory motion is symmetry-allowed.

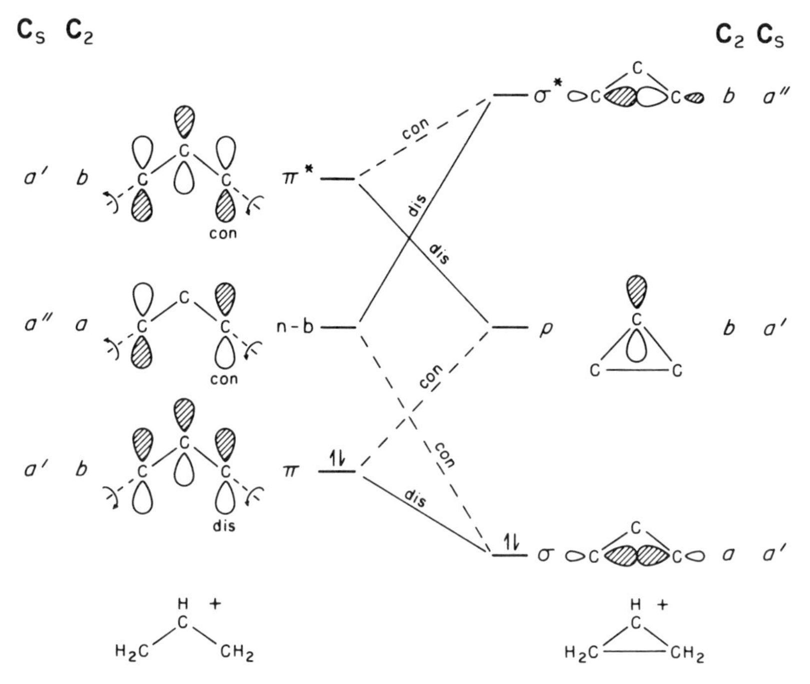

Fig. 11.5 Correlation diagram for conrotatory ($\mathbf{C}_2$ group) and disrotatory ($\mathbf{C}_s$ group) motion in the cyclization of $H_2CCHCH_2^+$.

11.2.3 FRONTIER ORBITAL APPROACH

We have used the Woodward–Hoffmann[1] correlation-diagram approach to determine the symmetry-allowed electrocyclic reactions. Fukui has developed a parallel approach using frontier orbitals.[2] This approach has been developed further by Pearson.[3] For ring closure, the rotatory motion of the p orbitals of the terminal carbon atoms of the HOMO to create a σ bonding orbital is obvious. It is conrotatory for butadiene (Fig. 11.3), disrotatory for $H_2CCHCH_2^+$ (Fig. 11.5), and conrotatory for $H_2CCHCH_2^-$. For ring opening, the symmetry-allowed direction of rotation is not always obvious without the full correlation diagram.

For a bond to be broken, we must promote electrons from the HOMO to remove electrons from the bonding orbital or add electrons to the corresponding antibonding orbital. The direct product of the symmetry species involved

[2] K. Fukui, *Acc. Chem. Res.* 1971, **4,** 57; *Angew. Chem. Int. Edit.* 1982, **11,** 801.

[3] R. G. Pearson, *Chem. and Engr. News* Sept. 28, 1970, p. 66; *Acc. Chem. Res.* 1971, **4,** 1952; *Symmetry Rules for Chemical Reactions*, Wiley (Interscience), New York, 1976.

in the promotion must belong to the same symmetry species as the reaction coordinate. The symmetry of the reaction coordinate describes the symmetry of the molecular motion involved. Using $\mathbf{C}_2$ symmetry for the cyclopropyl cation, we see that conrotatory motion is A and disrotatory motion is B.

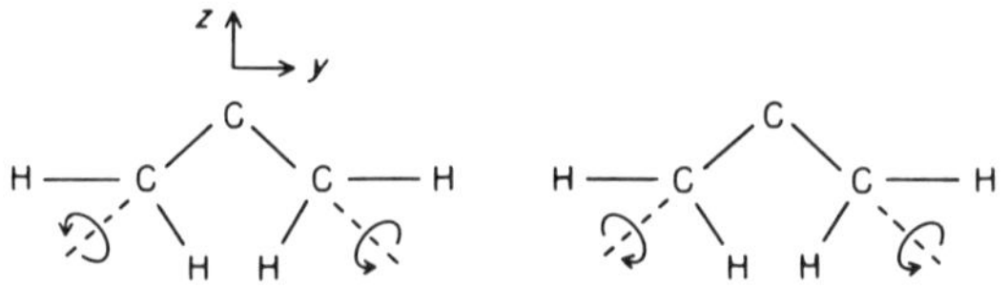

con, A dis, B

For the cyclopropyl cation (Fig. 11.5), electrons must be promoted from the HOMO (σ, a) to the LUMO (π, b). The direct product $A \times B = B$ corresponds to the disrotatory process. For the cyclopropyl anion, promotion from the HOMO (π, b) to σ^* (b) would break the σ bond. The direct product $B \times B = A$ gives the symmetry of conrotatory motion.

Checking the ring closure process in the same way, for $H_2CCHCH_2{}^+$, promotion of electrons from the π orbital (b) to the nonbonding a orbital would break the π bond and put electrons into the orbital that is transformed into the σ orbital in the product. The product $B \times A = B$ corresponds to a disrotatory process. For $H_2CCHCH_2{}^-$, HOMO–LUMO promotion is not suitable. We need to retain electrons in the nonbonding a orbital since this is the orbital needed to form the σ-bonding orbital with a symmetry. The π bond is broken by $\pi \rightarrow \pi^*$ promotion. The direct product for this promotion is $B \times B = A$, corresponding to the conrotatory process. The nonbonding a orbital correlates with the σ (a) orbital and π^* correlates with the nonbonding b orbital of the product.

Let us consider the choice of symmetry. The reactant $H_2CCHCH_2{}^+$ ion in Fig. 11.5, if assumed planar, has $\mathbf{C}_{2v}$ symmetry, as does the product. However, if we examine an intermediate orientation, conrotatory motion destroys both planes of symmetry, and disrotatory motion destroys one symmetry plane and the C_2 axis. If we try to use $\mathbf{C}_{2v}$ symmetry, the nonbonding orbital (a_2) of the reactant does not have the same symmetry as the a_1 orbital of the cyclic product with which it must correlate. The direct products, under $\mathbf{C}_{2v}$, of the reactant orbitals involved are $B_1 \times A_2 = B_2$ for the cation and $B_1 \times B_1 = A_1$ for the anion. Neither of these corresponds to conrotatory (A_2) or disrotatory (B_1) motion under $\mathbf{C}_{2v}$. Under $\mathbf{C}_2$ symmetry the critical reactant and product orbitals that must correlate have the same symmetry.

For ring opening of cyclobutene (Fig. 11.3), let us use the $\mathbf{C}_2$ group. We must break the σ bond and the π bond between atoms 2 and 3. This can be accomplished by promotion from the σ (a) orbital to π^* (a), requiring A ($A \times A = A$) or conrotatory motion. Promotion from π (b) to σ^* (b)

also involves breaking the σ and π bonds and gives the same result since $B \times B = A$. In this case, we can use $\mathbf{C}_{2v}$ symmetry since the orbitals that correlate belong to the same symmetry species.

11.3 Cycloaddition Reactions

11.3.1 DIELS–ALDER REACTIONS

We have examined the π orbitals of butadiene with respect to ring closure; now let us consider the addition of an olefin to butadiene. This is the simplest form of the Diels–Alder reaction, the addition of a double bond to a 1,3-diene. For the addition to occur by a concerted process, the π orbitals of the terminal carbons of butadiene must overlap those of the olefin symmetrically so that the two σ bonds can be formed simultaneously as the π bonds are broken. The reacting molecules must approach such that they share a plane of symmetry. This symmetry plane persists through the course of the reaction. Orbitals are designated as a' for symmetric or a'' for antisymmetric with respect to the mirror plane under $\mathbf{C}_s$ symmetry in Fig. 11.6.

Since one reactant has four π electrons and one has two π electrons, this is described as a [4 + 2] cycloaddition. The new bonds can be formed from the combination of the HOMO of butadiene with the LUMO (π^*) of ethylene, both with a'' symmetry, and the LUMO of butadiene with the HOMO (π) of ethylene, both with a' symmetry. These combinations are shown in Figs. 11.6 and 11.7. The reaction is symmetry-allowed. Each combination of orbitals of the same symmetry on the two reactants results in one MO of lower energy than the HOMO involved and one MO of higher energy than the LUMO involved, resulting in stabilization of the bonding orbitals. The bonding combinations are shown in Fig. 11.6. The antibonding combinations results from reversal of the signs of the p orbitals of ethylene.

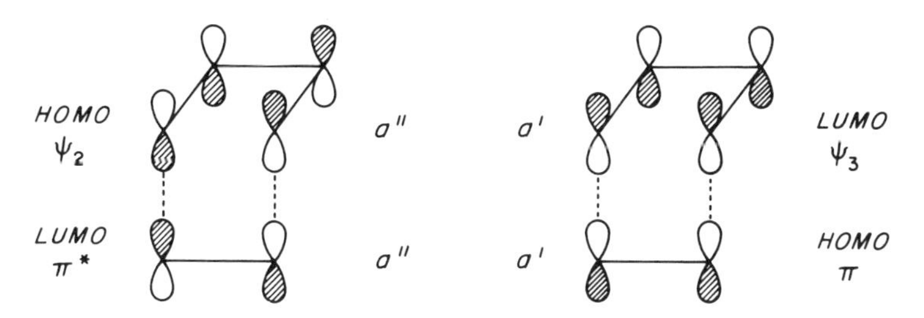

Fig. 11.6 HOMO–LUMO interactions for the [4 + 2] cycloaddition of 1,3-butadiene and ethylene.

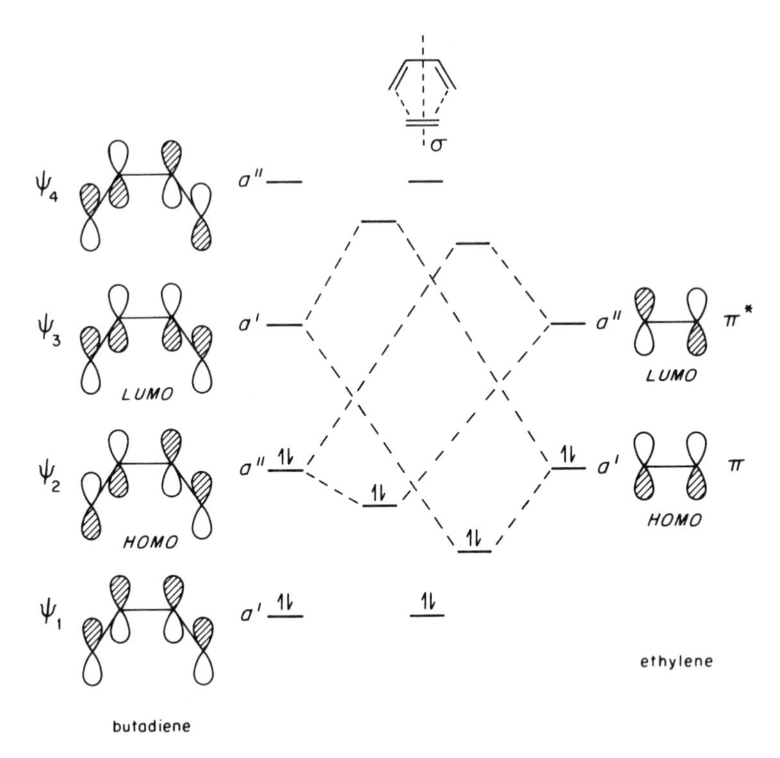

Fig. 11.7 Correlation diagram for the cycloaddition of butadiene and ethylene under $\mathbf{C}_s$ symmetry.

11.3.2 DIMERIZATION OF ETHYLENE

Now let us examine [2 + 2] cycloaddition with the dimerization of ethylene as an example. Planar cyclobutane has $\mathbf{D}_{4h}$ symmetry. As parallel ethylene molecules approach one another the symmetry is $\mathbf{D}_{2h}$. If we use $\mathbf{D}_{2h}$ symmetry to identify the interacting orbitals, the symmetry species for the product can be obtained, since $\mathbf{D}_{2h}$ is a subgroup of $\mathbf{D}_{4h}$. Our task is simplified by using the minimum number of symmetry elements required to make the necessary distinctions. Three planes of symmetry generate the full symmetry of the $\mathbf{D}_{2h}$ group since two intersecting symmetry planes generate a C_2 axis. During the course of the reaction two π bonds become σ bonds. The plane through all carbon atoms distinguishes σ orbitals from π orbitals, but it does not distinguish one σ combination from another or one π combination from another. The other two planes are sufficient to characterize the orbital combinations. Two perpendicular planes of symmetry define the $\mathbf{C}_{2v}$ group, a subgroup of $\mathbf{D}_{2h}$. We will use the symmetry species of $\mathbf{C}_{2v}$ to identify the orbital combinations during the course of the reaction.

Each ethylene has one π orbital and one π^* orbital. As the ethylenes approach, the orbitals overlap along the direction required for σ interaction. The combinations are shown in Fig 11.8. The π orbitals of the two ethylenes can interact to give a bonding combination (a_1) and a combination (b_1) with strong antibonding interaction between the ethylenes. Thd π^* orbital of one ethylene can interact with the π^* orbital of the other ethylene to give a net bonding interaction (b_2) and also a combination that is strongly antibonding for all carbons (a_2). The correlation diagram resulting from these new bonding and antibonding interactions is shown in Fig. 11.9. The σ bonds formed have lower energy than the π bonds broken because of the more favorable overlap for σ bonding.

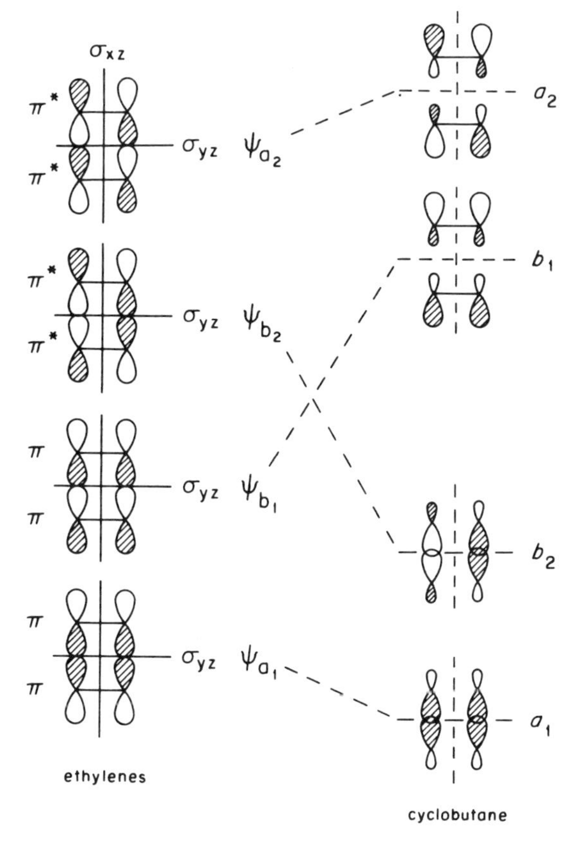

Fig. 11.8 Combinations of the π and π^* orbitals of two approaching ethylene molecules (using $\mathbf{C}_{2v}$ symmetry) to form the new MOs of cyclobutane.

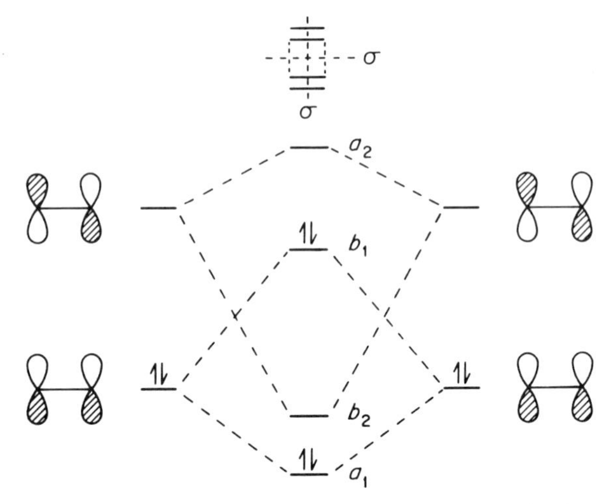

Fig. 11.9 Correlation diagram for the dimerization of ethylene under $\mathbf{C}_{2v}$ symmetry. The resulting orbital combinations are sketched in Fig. 11.8.

Since we have combinations only between HOMO–HOMO of the same symmetry and LUMO–LUMO of the same symmetry, the reaction is symmetry-forbidden. A bonding electron pair of ethylene becomes an antibonding pair upon dimerization. This represents forming the product in an excited state, or a high-energy barrier for the [2 + 2] cycloaddition by a thermal route. However, if an electron is promoted from π to π^* on one ethylene (or an electron for the interacting pair is promoted from b_1 to b_2), then an excited state of cyclobutane with the configuration $a_1^2 b_2^1 b_1^1$ is formed. There is net stabilization and the photochemical dimerization is symmetry-allowed.

11.3.3 FRONTIER ORBITAL APPROACH

For a bimolecular reaction, the symmetry requirement is that the symmetry of the bonds formed must be the same as the symmetry of the bonds broken. Net positive overlap between the HOMO and the LUMO occurs only if they belong to the same symmetry species. For a concerted reaction the point group must remain the same during the course of the reaction. If the point group of the combined product has higher symmetry than that of the combining molecules, the point group of the combining molecules must be a subgroup of the larger group. For the dimerization of ethylene, $\mathbf{D}_{2h}$ symmetry is maintained throughout, with $\mathbf{D}_{4h}$ symmetry being achieved in cyclobutane.

For the symmetry-allowed butadiene–ethylene [4 + 2] addition we have favorable overlap for the a' orbitals and for the a'' orbitals of both reactants.

The symmetry-adapted σ bonds have a' and a'' symmetry, and the new π bond is a'. The bonds broken are a' and a'' for butadiene and a' for ethylene. The bonds made in the ethylene dimerization are a_{1g} and b_{3u} in $\mathbf{D}_{2h}$ (a_1 and b_1 in $\mathbf{C}_{2v}$). The symmetries of the bonds broken are determined for the symmetry combinations for the two π-bonded ethylene molecules. These are the a_1 and b_2 ($\mathbf{C}_{2v}$) sketches in Fig. 11.8. They are a_{1g} and b_{2u} in $\mathbf{D}_{2h}$. Using either $\mathbf{C}_{2v}$ or $\mathbf{D}_{2h}$, we see that one of the bonds made ($b_{3u}, \mathbf{D}_{2h}$ or b_1, $\mathbf{C}_{2v}$) differs in symmetry from the bond broken ($b_{2u}, \mathbf{D}_{2h}$ or $b_2, \mathbf{C}_{2v}$), so the thermal reaction is symmetry-forbidden.

11.4 State-Correlation Diagrams

We have seen in the ligand field treatment that for the d^1 (or d^9) case the orbital-splitting diagram corresponds to the term diagram (an energy-state diagram). However, in the case of more than one electron (or more than one hole), because of electron repulsion, several terms (energy states) arise, and the term (or state) diagram is more complex than the orbital diagram. The term diagram contains the full information necessary to deal with electronic transitions from the ground state to excited states. Similarly the orbital-correlation diagrams we have used so far are less informative than the corresponding state diagrams. For simple cases, the orbital diagrams make correct predictions, but they can be misleading because some orbital correlations are seen to be invalid from state diagrams.

Let us first examine the state-correlation diagram for the dimerization of ethylene to form butane. For two weakly interacting ethylene molecules, the combinations of orbitals are as shown on the left in Fig. 11.8. The energy sequence is $a_1 < b_1 < b_2 < a_2$. The resulting MOs for cyclobutane (right side of Fig. 11.8) are in the sequence $a_1 < b_2 < b_1 < a_2$. We construct the electronic-state diagram by plotting the energies of the electron configurations, showing the orbital occupancy, for two interacting ethylenes on the left and butane on the right, as shown in Fig. 11.10. For each configuration, we obtain the symmetry label of the energy state by taking the direct products of the symmetry species for *each electron*. Since the direct product of any one-dimensional (A or B) species times itself is totally symmetric, the symmetry species for any doubly occupied a or b orbital is A (or A_1, A', etc.). Hence, the energy states are A_1 ($\mathbf{C}_{2v}$) for filled orbitals, and we need examine only the direct products of singly occupied orbitals to obtain the symmetry species as shown in the figure. We connect the energy states belonging to the same symmetry species. Thus, A_2 is derived from $a_1^2 b_2^1 b_1^1$ for the interacting ethylenes and for butane. The energy of A_1 derived from $a_1^2 b_2^2$ for interacting ethylenes increases, and that for A_1 from $a_1^2 b_1^2$ decreases as the interaction increases. The

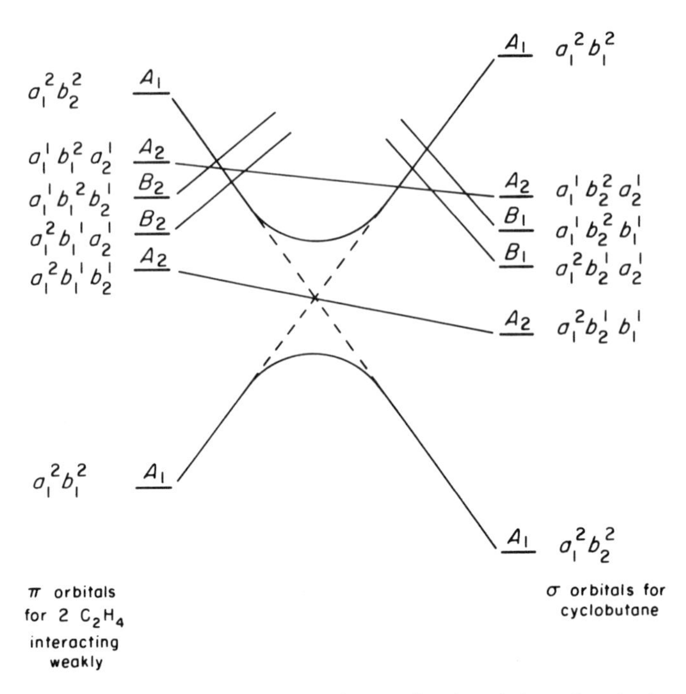

Fig. 11.10 An electronic state-correlation diagram for the ethylene dimerization reaction. Other higher-energy configurations are not shown.

tie lines would cross, but this is prohibited by the noncrossing rule because of configuration interaction. The result is as shown in Fig. 11.10. The energy of the lowest A_1 on the left increases as interaction increases, and the energy of the lowest A_1 on the right increases as the bonds in cyclobutane are elongated. The increase continues as long as bond breaking is more important than bond making. As the bond making becomes more important (in either direction), the energy of the A_1 state decreases until the bonds are formed fully. Only the lower-energy states appearing on both sides are shown since these are involved in reactions of the molecules in their ground states. The large increase in energy of the ground state (A_1) for dimerization of ethylene or dissociation of cyclobutane indicates a high activation energy for the reaction.

In a sense, the state-correlation diagram indicates that the reaction violates no symmetry rules, since the ground state and first excited state of two ethylenes and of cyclobutane correlate. However, the high activation energy results from the configuration interaction with higher-energy states of the same symmetry. Hence the conclusion is the same: the reaction does not occur thermally because of a symmetry restriction. The state diagram makes it more apparent than the orbital diagram that a pair of interacting ethylenes excited to the first excited state (A_2) can dimerize without a symmetry-imposed barrier. The photochemical reaction is allowed.

Next let us consider the state diagrams for ring closure of the allyl cation $H_2CCHCH_2^+$ by conrotatory and disrotatory processes. The configurations for the ally cation are shown in the center of Fig. 11.11. The relative energies of the two highest-energy states are not certain, but their sequence does not affect the predictions. The symmetry species are identified under $\mathbf{C}_s$ symmetry for the disrotatory process and under $\mathbf{C}_2$ for the conrotatory process. For the disrotatory process, we see that the ground states correlate without any significant activation energy. This is the preferred process as predicted from the orbital diagram (Fig. 11.5). The A' excited states are not connected, because of uncertainty about their relative energies. Spin multiplicities are not shown. Singlets arise from two electrons in one orbital, but the other configurations can give singlets and triplets. States that correlate must have the same spin multiplicity, and the noncrossing rule applies only to states with the same spin multiplicity.

For the conrotatory process, we cannot connect the A' states from the two b^2 configurations and those from the two a^2 configurations because of the noncrossing rule. Although the ground states correlate, contrary to the orbital

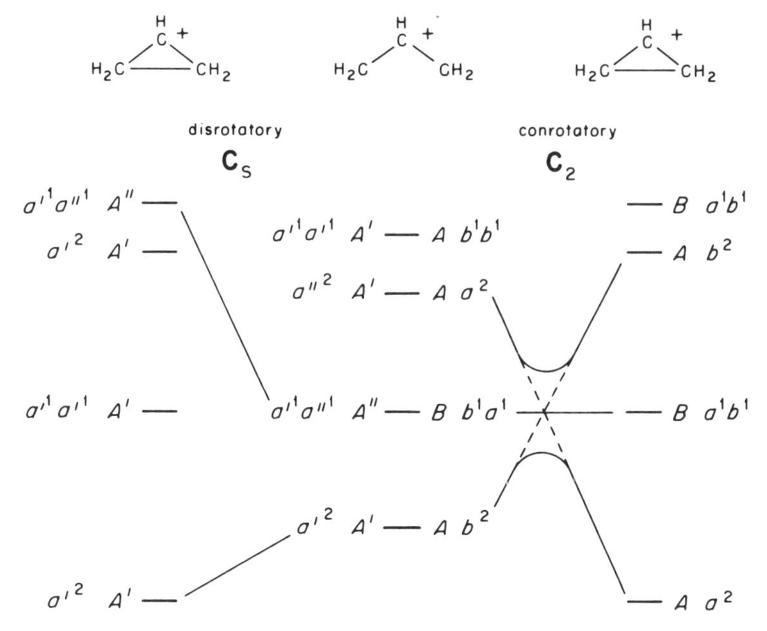

Fig. 11.11 State-correlation diagram for the cyclization of $H_2CCHCH_2^+$ by disrotatory and conrotatory processes.

correlations, there is high activation energy for ring opening or ring closure. In either case, the energy increases as the bond is broken until the formation of the new bond is more important than bond breaking.

In the cases examined, the conclusions are the same from the orbital and state-correlation diagrams. In the cases where the orbital correlations are seen to be invalid, it is because of the noncrossing rule, and this results in high activation energy, as seen from the state correlations. For these simple cases with few electrons, only a few of the lower energy states are important in providing the essential correlations. In more complex cases, we would have to have more detailed information about the relative energies of the states, and we would have to consider the spin multiplicities. The state diagrams can provide the most reliable predictions, but their construction can require the most detailed information. Orbital- and state-correlation diagrams have been applied recently to thermal substitution reactions of cobalt(III) complexes.[4]

11.5 Summary of Symmetry Rules

Symmetry rules are often called Woodward–Hoffmann rules because the work of Woodward and Hoffmann gained wide attention and acceptance in a very short time. They presented their ideas in such a way that chemists could see their value and wide applicability. Many of the applications were presented in the most impressive way, by making *predictions*, not just rationalizations of observations. Usually when important ideas become popular it is soon recognized that similar ideas had been proposed by others earlier or that similar ideas developed independently elsewhere. The important earlier contributions of C. A. Coulson, H. C. Longuet-Higgins, E. W. Abrahamson, R. F. W. Bader, K. Fukui, and others are cited by Woodward and Hoffmann. Although the work of Woodward and Hoffmann was best known, the great significance of Fukui's work with frontier orbitals was justly recognized when Fukui and Hoffmann shared the Nobel Prize in Chemistry in 1981.

Ralph Pearson has had a great influence in developing the theory of hard and soft acids and bases and in recognizing the great value of and popularizing crystal field theory and symmetry rules for chemical reactions. His lectures, reviews, and the book *Symmetry Rules for Chemical Reactions* have developed a distinctive approach and have demonstrated the applicability of symmetry rules to many inorganic reactions and to predictions of molecular structure. In the applications to follow we will follow Pearson's approach most closely. It is often easier to examine the symmetry of critical orbitals than to construct a

[4] L. G. Vanquickenborne and K. Pierloot, *Inorg. Chem.* 1984, **23,** 1471.

correlation diagram. Caution must be exercised in constructing correlation diagrams. Correlation tables are based on character tables using a particular choice of axes. The "major" axis might change for point groups involved in a correlation diagram. Thus, the usual choice of the z axis for a linear MX_2 molecule is along the bonds, but for the sidewise approach of an $X—X$ molecule under $\mathbf{C}_{2v}$ symmetry, it is between the atoms approaching and in the plane of the atoms. For a final square planar ($\mathbf{D}_{4h}$) complex, z is perpendicular to the plane of the atoms. There is, of course, a C_2 axis common to these three groups.

The basis for the Woodward–Hoffmann rules is that *orbital symmetry is conserved in concerted reactions.* This requires that from initiation to completion one or more symmetry elements must persist. The molecular orbitals (or electronic states) must belong to the same symmetry species of the point group of the reacting system during the entire course of the reaction. If filled low-energy orbitals of the reactant(s) correlate with high-energy orbitals of the product(s), then there will be high activation energy and the reaction is *symmetry-forbidden.*

Pearson's symmetry rules for a bimolecular reaction require the following.

(1) Electron density must flow from the HOMO of the donor to the LUMO of the acceptor.

(2) The HOMO and LUMO involved must belong to the same symmetry species of the point group of the reacting system. This is necessary for net positive overlap. If the HOMO and LUMO do not have the correct symmetry, then other orbitals comparable in energy to the HOMO or LUMO might be involved.

(3) The HOMO and LUMO (or the orbitals involved) must be close in energy (about 6 eV).

(4) Removing electrons from a bonding HOMO breaks the necessary bond(s), and filling a bonding LUMO must correspond to formation of the new bond(s). Removing electrons from an antibonding HOMO strengthens the bond involved, and adding electrons to an antibonding LUMO weakens or destroys the bond involved.

Another approach used by Pearson is the *bond-symmetry rule*: *a reaction is allowed if the symmetry of the bonds that are made is the same as the symmetry of the bonds broken.* The effective symmetry (point group) of the reacting system is used. Lone pairs of electrons must be considered if they become bonding or if bonding electrons become lone pairs. Thus, the $H_2 + C_2H_4$ reaction is symmetry-forbidden because the HOMO (π) of C_2H_4 is a_1 ($\mathbf{C}_{2v}$) and the LUMO (σ^*) of H_2 is b_2. Using the bond-symmetry rule, the bonds broken are a_1 (σ, H_2) and a_1 (π, C_2H_4), and those formed are symmetry-adapted a_1 and b_2, the sum and difference for the new bonds.

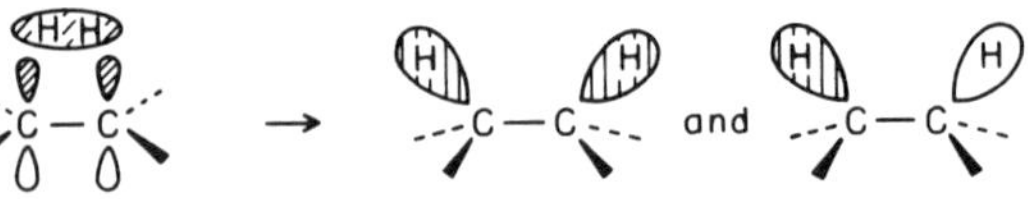

$a_1 + a_1$ → a_1 and b_2

For unimolecular reactions, the HOMO and LUMO (or suitable orbitals of similar energy to the HOMO or LUMO) belong to the same molecule. The electron transfer results in change in electron density within the molecule causing the nuclei to move into the new regions of high electron density. The HOMO and LUMO do not necessarily belong to the same symmetry species, but *their direct product must belong to the same symmetry species as that describing the nuclear motion.* The nuclear motion corresponds to one or more of the normal modes of vibration of the molecule.

Pearson has used the symmetry rules as they apply to unimolecular reactions for the prediction of stable shapes of molecules. Two possible structures are considered along with the reaction coordinate required to interconvert them. The stable structure will have a large energy gap between the appropriate HOMO and LUMO.

The cycloaddition reactions are bimolecular, and the reverse reactions are unimolecular. Both must obey the symmetry rules. If we focus attention on the combinations of the orbitals between the two halves of cyclobutane, the orbitals are the same as those in Fig. 11.9, except the bonding orbitals a_1 and b_2 are occupied. The product of the HOMO and LUMO orbitals is $B_2 \times B_1 = A_2$. The concerted splitting of cyclobutane requires the symmetrical separation of the parallel ethylene molecules, $\leftrightarrows\rightrightarrows$, giving A_1 symmetry. The reaction is symmetry-forbidden. Using the correlation diagram (Fig. 11.9), we see that the bonding b_2 orbital of cyclobutane correlates with the π^* orbital of ethylene.

11.6 Oxidative Addition Reactions

A simple example of oxidative addition is the reaction of PCl_3 with Cl_2,

$$PCl_3 + Cl_2 \longrightarrow PCl_5. \qquad (11.3)$$

The reaction can be treated by symmetry rules to examine the possibilities for axial–axial, equatorial–axial, or equatorial–equatorial addition. However, PX_5 molecules are fluxional with rapid interchange between axial and equatorial sites. If the X groups are not all the same, the less electronegative groups prefer equatorial sites. We have seen that the addition of Cl_2 to H_2 or to C_2H_4 is symmetry-forbidden. For $\mathbf{C}_{2v}$ symmetry of the reacting system, the HOMO is a_1 for H_2 and for C_2H_4, and the LUMO (σ^*) of Cl_2 is b_2.

The concerted reactions to add Cl_2 to TlCl or $PbCl_2$ are symmetry-forbidden.

$$Cl{-}Tl\,(:) + Cl{-}Cl \rightarrow Cl{-}TlCl_2 \quad (11.4)$$

s, a_1 HOMO $\quad \sigma^*, b_2$ LUMO

$$Cl_2Pb\,(:) + Cl{-}Cl \rightarrow Cl_2PbCl_2 \quad (11.5)$$

s, a_1 HOMO $\quad \sigma^*, b_2$ LUMO

The HOMOs are s lone pairs ($a_1, \mathbf{C}_{2v}$); or, considering the bond-symmetry rule, the bonds broken are a_1 for the lone pair and a_1 for the Cl—Cl bond, and those made are a_1 and b_2 (symmetry-adapted for $\mathbf{C}_{2v}$).

$$M\,(:) + Cl{-}Cl \rightarrow MCl_2 \quad (11.6)$$

a_1 $\quad \sigma, a_1$ $\quad a_1\,(+\,+)$ and $b_2\,(+\,-)$

The oxidative addition of Cl_2 to SCl_2 is symmetry-allowed considering the HOMO–LUMO match [Eq. (11.7)] or the bond symmetries [Eq. (11.8)] under $\mathbf{C}_{2v}$ symmetry.

$$Cl_2S{:}\text{---}z + Cl{-}Cl \rightarrow Cl_4S\,(:)\text{---}z \quad (11.7)$$

HOMO $b_2(p_y)$ $\quad$ LUMO b_2 $\quad$ sawhorse (see Table 6.1)

$$Cl_2S{:}\text{---}z + Cl{-}Cl \rightarrow Cl_4S\,(:)\text{---}z \quad (11.8)$$

$a_1 + b_2$ lone pairs $\quad a_1$ σ $\quad a_1$ (lone pair) $+ a_1 + b_2$ (symmetry-adapted S—Cl bonds)

Oxidative Addition to Linear ML_2 *Complexes.* Symmetry rules indicate that for the concerted addition of X_2 to a linear ML_2, *cis* addition is allowed and *trans* addition is forbidden.[5] While experimental evidence for the reaction

[5] R. G. Pearson, *Symmetry Rules for Chemical Reactions*, Wiley (Interscience), New York, 1976, p. 286.

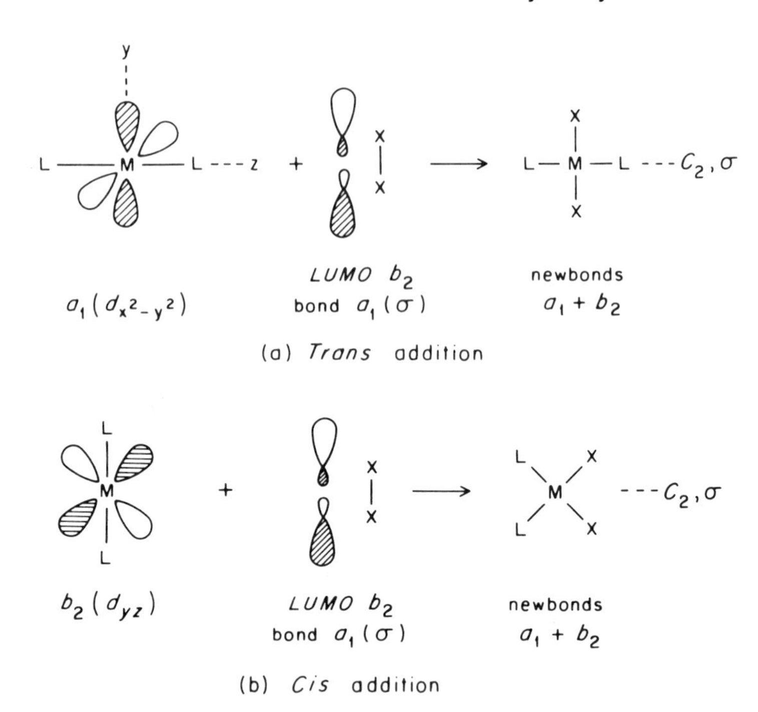

Fig. 11.12 (*a*) *Trans* addition of X_2 to a linear d^{10} ML_2 molecule. The concerted reaction is forbidden. (*b*) *Cis* addition is allowed.

of $Au(CN)_2^-$ with I_2 is not consistent with this prediction and perhaps another mechanism is operative, reductive elimination from planar gold(III) complexes supports this prediction. Exclusively *cis* elimination occurs on heating *trans*-$[Au(CH_3)_2(C_2H_5)P(C_6H_5)_3]$ to give propane and no ethane:

$$C_2H_5\text{—}\underset{\displaystyle CH_3}{\overset{\displaystyle CH_3}{\underset{|}{\overset{|}{Au}}}}\text{—}P(C_6H_5)_3 \xrightarrow{\text{heat}} C_3H_8 + \underset{\text{linear}}{CH_3\text{—}Au\text{—}P(C_6H_5)_3}. \tag{11.9}$$

Considering the reverse oxidative addition ($\mathbf{C}_{2v}$ for the reacting system), the HOMO of linear Au(I) is d_{z^2} (a_1). The major lobes of this orbital are along the L—Au—L bonds, giving poor directional overlap and *no* net positive overlap with the LUMO (π^*, b_2) of X—X. It also violates the bond-symmetry rule. The orbitals perpendicular to the L—Au—L bonds are $d_{x^2-y^2}$ (a_1) and d_{xy} (a_2). These also give no net positive overlap for *trans* addition (see Fig. 11.12). This reaction using the $d_{x^2-y^2}$ orbital is forbidden according to the bond-symmetry rule for *trans* addition because a_1 (σ) bond of Cl_2 is broken, the a_1 lone pair of ML_2 is used, and the new MX_2 symmetry-adapted bonds are a_1 and b_2. For *cis* addition, the orbital of proper symmetry (b_2 with the atoms in the yz plane) is d_{yz}. This orbital gives net positive overlap, and the directional characteristics are good for *cis* addition. Considering the a_1 (σ) bond of X_2 to be broken and the b_2 lone pair to be used, the bond-symmetry rule holds as well. The new symmetry-adapted bonds are a_1 $(+\,+)$ and b_2 $(+\,-)$. The b_2 orbital is not the HOMO but the next lower-energy orbital. When the HOMO has the wrong symmetry, orbitals not much lower in energy can be used.

Symmetry rules must apply, of course, to forward *and* reverse concerted reactions. Let us consider the *cis* elimination of a d^8 square planar complex *cis*-$[ML_2X_2]$ to give linear ML_2 (d^{10}) and X_2. If the d_{yz} orbital is the one interacting with all four ligands in the square planar complex, as formed in Fig. 11.12, it is highest in energy.[6] The d_{xy} and d_{xz} orbitals interact about equally for this orientation of ligands and orbitals, and their relative energies are uncertain. The $d_{x^2-y^2}$ and d_{z^2} orbitals interact least, but again their relative energies are uncertain. The orbital correlation for the square planar and linear complexes is shown in Fig. 11.13 using the choice of axes shown and symmetry labels for $\mathbf{C}_{2v}$.

There are two σ bonds (a_1 and b_2) in ML_2, and two more (also a_1 and b_2) are formed in ML_2X_2 using the X orbitals, a pair of electrons for the X_2 bond, and another pair of electrons from d_{yz} (b_2). The electron transfer accompanying reductive elimination is from σ (b_2) to d_{yz} (b_2). The vacated b_2 orbital becomes σ^* in X_2 and one a_1 σ bonding orbital becomes the a_1 σ bonding orbital in X_2. The symmetrical motion of the departing X atoms and the two L ligands parallel to the C_2 axis is described by $\rightrightarrows\rightrightarrows$-$C_2, \sigma$. This motion belongs to the A_1 symmetry species. The direct product for the orbitals involved in electron transfer gives $B_2 \times B_2 = A_1$. The concerted process is symmetry-allowed.

Oxidative Addition to Square Planar Complexes. This is a common reaction that can give *cis* or *trans* products. The Cl atoms add in *trans*

$$[PtCl_4]^{2-} + Cl_2 \longrightarrow [PtCl_6]^{2-} \qquad (11.10)$$

```
                                                   Cl
                                                   |
     Cl-----PPh3                              Cl---+---PPh3
    / \    /  /                               / \  |  /  /
   /   Ir    /     +  Cl2  ------->          /   Ir     /          (11.11)
  /   /   \ /                               /   /  \   /
Ph3P------CO                             Ph3P------+-CO
Vaska's compound                                   |
                                                   Cl
```

positions in the case of Vaska's compound [Eq. (11.11)]. Many X_2 and XY reagents can be added with the formation of *cis* products in some cases, and for some, *trans* products are formed. Since many organometallic compounds add H_2 to form metal hydrides (containing the H^- ligand), even H_2 causes

[6] In fact, the d_{yz} orbital will be the d orbital participating in σ bonding. Since we are focusing on the d manifold, this is really the antibonding σ orbital concentrated on the metal.

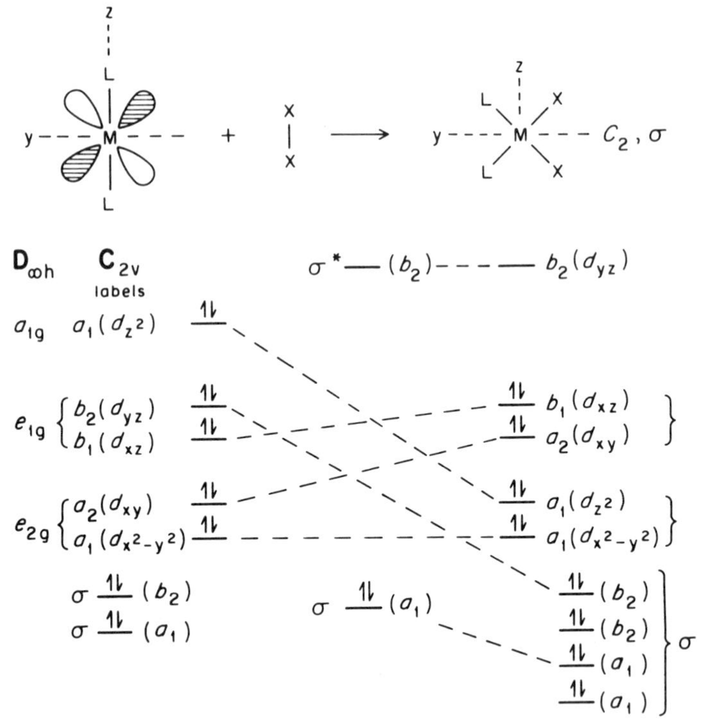

Fig. 11.13 Correlation diagram showing the effects on the d orbitals for *cis* oxidative addition to a d^{10} linear ML_2 complex or *cis* reductive elimination from a d^8 square planar ML_4 (ML_2X_2) complex.

oxidative addition. Oxidative addition to a square planar complex with the ligands in the xy plane by a concerted process involves electron transfer from the d_{xz} (or d_{yz}) orbital to the σ^* orbital of the X_2 (or XY) to give a *cis* product, similar to the *cis* addition for linear complexes in Fig. 11.12. Other reasonable two-step mechanisms lead to *trans* products.

11.7 Indirect Routes to Symmetry-Forbidden Reactions

The following reactions are forbidden:

$$N_2 + H_2 \longrightarrow N_2H_2, \tag{11.12}$$

$$H_2CCH_2 + H_2 \longrightarrow H_3CCH_3, \tag{11.13}$$

$$H_2CCH_2 + Cl_2 \longrightarrow H_2ClCCClH_2. \tag{11.14}$$

In each case the HOMO and LUMO do not have proper symmetry for the formation of the new bonds and for breaking the H_2 or Cl_2 bonds. Reaction (11.12) can be accomplished indirectly by first forming a transition metal dihydride using a transition metal with filled, or nearly filled, *d* orbitals.[7] Here the HOMO is a filled metal *d* orbital and the LUMO is the σ^* of H_2, breaking the H—H bond and forming two M—H bonds.

$$M \quad H{-}H \longrightarrow MH_2 \tag{11.15}$$

The metal hydride can combine with N_2 to form N_2H_2 by transferring electrons from the filled M—H b_1 symmetry-adapted σ orbital into the π^* orbital (b_1) of N_2, breaking the triple bond and forming two N—H bonds.

$$\sigma_1 C_2 \text{ ---- } N{\equiv}N \text{ --- } M \longrightarrow HN{=}NH + M \tag{11.16}$$

The N_2H_2 can then transfer hydrogens to C_2H_4. Electrons are donated from the filled σ orbital of appropriate symmetry to the π^* orbital of C_2H_4, breaking the double bond and forming two C—H bonds. Of course, the MH_2 could react directly with C_2H_4.

$$H_2C{=}CH_2 \quad N{=}N \longrightarrow H_3C{-}CH_3 + N{\equiv}N \tag{11.17}$$

Similarly, the symmetry-forbidden chlorination of ethylene can be accomplished using $PbCl_4$ [Eq. (11.5)] or $SbCl_5$ as a chlorinating agent:

$$PbCl_4 + C_2H_4 \longrightarrow PbCl_2 + C_2H_4Cl_2. \tag{11.18}$$

The symmetry for the reacting unit is C_{2v}. The bonds broken are a_1 and b_1 ($PbCl_4$) and a_1 (π) for C_2H_4. The new symmetry-adapted C—Cl bonds are a_1 and b_1, and a lone pair occupies an a_1 orbital on Pb. These orbitals are sketched in Fig. 11.14. Electrons are transferred from b_1 of $PbCl_4$ to b_1 (π^*) of C_2H_4 to break the π bond (a_1).

[7] R. G. Pearson, *Chem. and Engr. News* Sept. 28, 1970, 66.

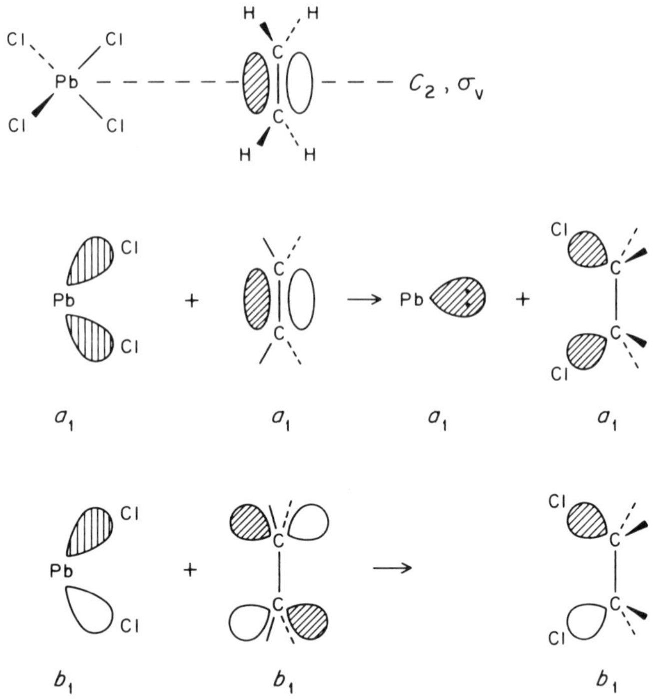

Fig. 11.14 Orbitals involved in the atom transfer for the chlorination of C_2H_4 by $PbCl_4$.

11.8 Metal Catalysis of Symmetry-Forbidden Reactions

The thermal dimerization of ethylene to form cyclobutane is symmetry-forbidden (see Figs. 11.8 and 11.9), but such a reaction occurs with the intervention of a d^8 transition metal atom, such as a complex of Fe(0) or Pd(II). Ethylene does not react, but more complex olefins form substituted cyclobutanes. We can see that this could be achieved if an empty metal d orbital accepted the electron pair from the π orbital of b_1 symmetry and a filled metal orbital donated an electron pair to the π^* orbital of b_2 symmetry to break the π bond and form a b_2 σ bond.

The metal catalysis of olefin dimerization was first interpreted in terms of orbital-symmetry rules by Mango and Schachtschneider.[8] The reaction

[8] F. D. Mango and J. H. Schachtschneider, *J. Am. Chem. Soc.* 1967, **89,** 2484; F. D. Mango, *Adv. Catalysis* 1969, **20,** 291; F. D. Mango and J. H. Schachtschneider, *Transition Metals in Homogeneous Catalysis* (G. N. Schrauzer, ed.), Marcel Dekker, New York, 1971, p. 223.

requires that two olefins bond to a metal atom or ion in proper orientation to form the new σ bonds of cyclobutane. The uncoordinated cyclobutane would leave for two more olefins to come in. The composition and stereochemistry of the metal complex must change considerably from the initial interaction with two olefins to the final release of cyclobutane. Since we do not know the stereochemistry or other bonding interactions, let us focus on the d orbitals that might accomplish the needed electron transfer. Although the two olefins might bond to the metal initially in some other orientation, let us assume that they arrive at the orientation shown in Fig. 11.15 for the beginning of the concerted reaction. The symmetry of the bis(olefin)metal unit, M(C=C)$_2$ (M bonded to two C=C units), is $\mathbf{C}_{2v}$.

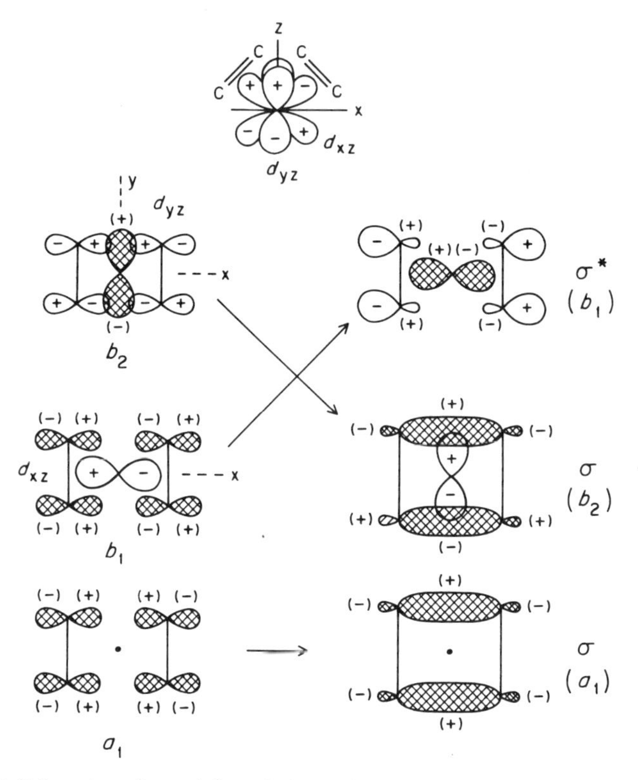

Fig. 11.15 Orientation of two olefins relative to the coordinate axes and to the metal d_{yz} and d_{xz} orbitals. The electron transfer from $b_1 \rightarrow d_{xz}$ to form σ^* and from $d_{yz} \rightarrow b_2$ to form σ (b_2) are shown. Occupied orbitals are shaded and empty orbitals are open.

The C_2 axis is along z and the C—C bonds are parallel to the yz plane. If the d_{xz} orbital is empty, it can accept the electron pair from the b_1 combination orbital ($\pi + \pi$). A filled d_{yz} orbital could transfer an electron pair to the b_2 combination orbital ($\pi^* + \pi^*$), breaking the π bonds and forming the σ orbital with b_2 symmetry. There is no symmetry barrier for the conversion of the a_1 π orbital to the a_1 σ orbital.

Figure 11.16 shows the correlation of the π orbitals of two ethylene molecules, the d_{xz} and d_{yz} orbitals of a d^8 metal atom, and the σ orbitals of cyclobutane. The net effect is the breaking of two π bonds and the formation of two σ bonds in cyclobutane. The pair of electrons originally in d_{yz} ends up in d_{xz}. This is of no consequence since the choice of x and y axes is arbitrary. We also expect considerable reorganization with the departure of cyclobutane before two more olefins are bonded. The transformation shown in Fig. 11.16 is a concerted process. The removal of cyclobutane and reentry of two more olefins oriented exactly as required surely is not a concerted process. The splitting pattern for the d orbitals must change to a great extent during this process.

The manner of participation of the d orbitals in the dimerization of olefins postulated requires an empty d orbital and at least one filled d orbital. The d^8 case is ideal. However, Ni(0), a d^{10} atom, is also effective for the dimerization process. Another mechanism must be operative.

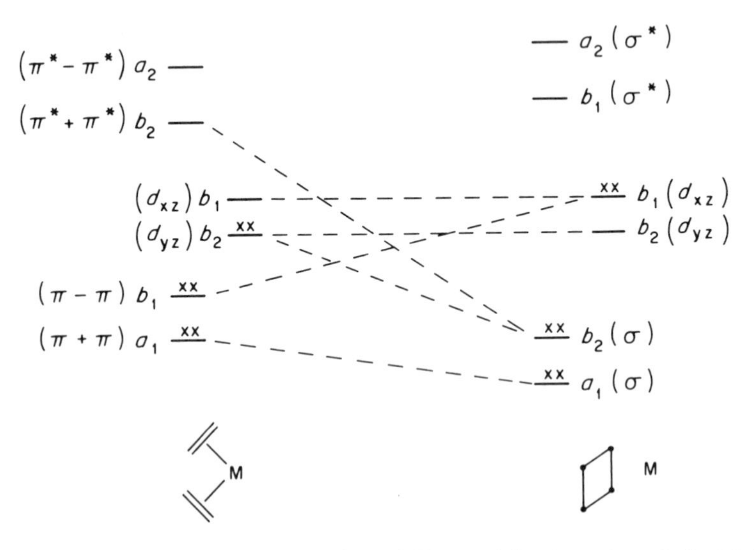

Fig. 11.16 Correlation of the π orbitals of two C_2H_4, metal d_{xz}, d_{yz}, and σ and σ^* orbitals of cyclobutane.

A reaction of two simple substituted olefins of the types

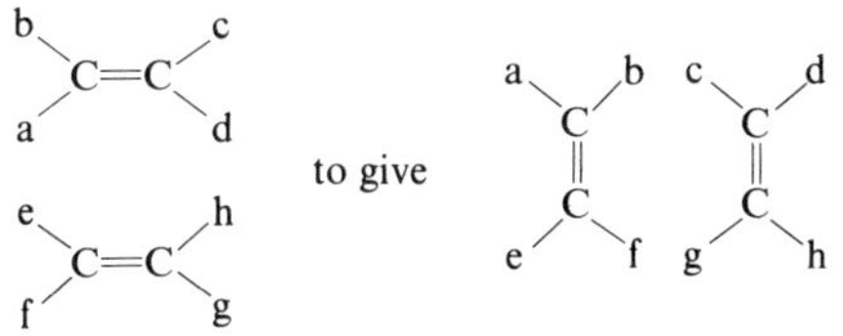

b c
C=C
a d
to give
e h
C=C
f g

a b c d
C C
‖ ‖
C C
e f g h

is catalyzed by metal atoms or ions that bring about dimerization of more complex olefins. Although these simpler olefins do not dimerize catalytically, surely the two reactions have much in common.

As we have seen, the disrotatory ring opening of cyclobutene to form butadiene is symmetry-forbidden (see Fig. 11.4). This process is catalyzed by Fe(0) (d^8) also. If the d_{z^2} orbital is the empty highest-energy orbital, it can accept an electron pair from the π orbital of butene to break this bond. An electron pair can be donated from the d_{xz} or d_{yz} orbital (depending on the choice of axes) to form the new a'' π bond of butadiene. There is no symmetry barrier for the disrotatory conversion of the σ (a') orbital to the π (a') orbital. Pearson[9] has reviewed several other metal-catalyzed reactions.

Additional Reading

K. Fukui, *Acc. Chem. Res.* 1971, **4**, 57; *Angew. Chem. Int. Edit.* 1982, **11**, 801. The frontier orbital approach.

H. C. Longuet-Higgins and E. W. Abrahamson, *J. Am. Chem. Soc.* 1965, **87**, 2045. Introduces the use of state diagrams.

R. G. Pearson, *Chem. and Engr. News*, Sept. 28, 1970, p. 66: *Acc. Chem. Res.* 1971, **4**, 1952. General introduction with a wide range of applications.

R. G. Pearson, *Symmetry Rules for Chemical Reactions*, Wiley (Interscience), New York, 1976. Many inorganic applications. Includes frontier orbitals and the bond-symmetry rules.

R. B. Woodward and R. Hoffmann, *J. Am. Chem. Soc.* 1965, **87**, 395. Their first paper on symmetry rules.

R. B. Woodward and R. Hoffmann, *The Conservation of Orbital Symmetry*, Verlag Chemie, Gmbh, Weinheim/Bergstrasse, 1970. Excellent summary with many applications.

Problems

11.1 Considering the orbital symmetries, would you expect the attack of H_2 by H^- to lead to a triangular or linear arrangement? Explain.

[9] R. G. Pearson, *Symmetry Rules for Chemical Reactions*, Wiley (Interscience), New York, 1976, pp. 398–443.

11.2 Based on relative orbital energies and charge distribution, explain why you expect one product to be preferred for each of the following reactions. Which orbitals are involved in electron transfer?

(a) $Cl^+ + Cl{-}F \longrightarrow Cl{-}Cl{-}F^+$ or $Cl{-}F{-}Cl^+$,
(b) $Cl^- + Cl{-}F \longrightarrow Cl{-}Cl{-}F^-$ or $Cl{-}F{-}Cl^-$.

11.3 Based on orbital correlations, is the cyclization of the pentadienyl cation expected to be allowed for conrotatory or disrotatory motion?

11.4 Draw a state-correlation diagram for the addition of ethylene to butadiene to determine if the reaction is symmetry-allowed.

11.5 Draw a state-correlation diagram for the ring opening of cyclobutene by conrotatory and disrotatory processes to determine whether the reaction is symmetry-allowed.

11.6 Draw state-correlation diagrams for conrotatory and disrotatory ring closing of the allyl anion to determine if either is symmetry-allowed.

11.7 Consider the addition of Cl_2 to butadiene.

(a) Is the reaction allowed based on the bond-symmetry rule? Which carbon atoms should be substituted?

(b) Draw the correlation diagram for the reaction to determine whether the reaction is symmetry-allowed.

11.8 (a) Apply the bond-symmetry rule to determine if the addition of Cl_2 to the allyl cation is symmetry-allowed.

(b) Is addition of Cl_2 to the allyl anion symmetry-allowed?

11.9 Consider the bond-symmetry rule to determine whether 1,4 addition of cyclopropanone to butadiene is allowed.

11.10 Is CO loss allowed from (a) cyclopropanone and (b) cyclopentenone (to give CO and butadiene)?

11.11 Based on the bond-symmetry rule, are the following dissociation reactions allowed?

a) (C–C ring with SO_2) $\longrightarrow$ $C_2H_4 + SO_2$

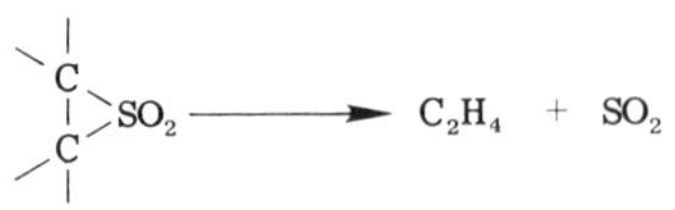

b) ($HC{=}CH$ ring with two CH_2 and SO_2) $\longrightarrow$ $HC{=}CH_2$ / $HC{=}CH_2$ + SO_2

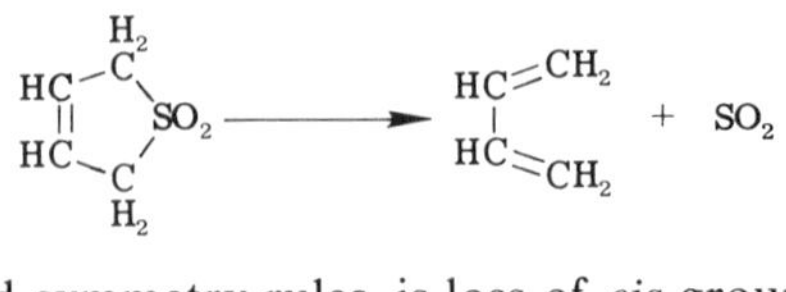

11.12 Based on bond-symmetry rules, is loss of *cis* groups or *trans* groups allowed for removal of F_2 from XeF_4? Use $\mathbf{C}_2$ symmetry and maintain the C_2 axis perpendicular to the departing pair of F atoms.

11.13 Consider the sidewise approach of F_2 toward Br of BrF to form BrF_3, maintaining $\mathbf{C}_{2v}$ symmetry. Using the bond-symmetry rule, is the reaction allowed? Neglect the unchanged Br—F bond and the lone pairs on the F atoms. Consider the lone pairs on the reacting Br to be in s, p_x, and p_y orbitals and the lone pairs in the central Br of BrF_3 to be in a pair of delocalized orbitals (consider the pair combinations) above and below the plane of the "bent T" BrF_3 molecule (a trigonal bipyramid with equatorial lone pairs).

11.14 Consider the oxidative addition of an X_2 molecule to linear ML_2 to form tetrahedral ML_2X_2. Is the reaction allowed for any d^n configuration? Use $\mathbf{C}_{2v}$ labels and the relative energies for the d orbitals in linear ML_2 given in Fig. 11.13 and the usual small tetrahedral splitting of the d orbitals (but with $\mathbf{C}_{2v}$ labels) to construct a correlation table.

Appendix 1

Character Tables

THE NONAXIAL GROUPS

$\mathbf{C}_1$	E
A	1

$\mathbf{C}_s$	E	σ_h			
A'	1	1	x, y, R_z	x^2, y^2, z^2, xy	$xz^2, yz^2, x(x^2 - 3y^2), y(3x^2 - y^2)$
A''	1	-1	z, R_x, R_y	yz, xz	$z^3, xyz, z(x^2 - y^2)$

$\mathbf{C}_i$	E	i			
A_g	1	1	R_x, R_y, R_z	$x^2, y^2, z^2, xy, xz, yz$	
A_u	1	-1	x, y, z		all cubic functions

THE AXIAL GROUPS

The $\mathbf{C}_n$ groups

$\mathbf{C}_2$	E	C_2			
A	1	1	z, R_z	x^2, y^2, z^2, xy	$z^3, xyz, z(x^2 - y^2)$
B	1	-1	x, y, R_x, R_y	yz, xz	$xz^2, yz^2, x(x^2 - 3y^2), y(3x^2 - y^2)$

$\varepsilon = \exp(2\pi i/3)$

$\mathbf{C}_3$	E	C_3	C_3^2			
A	1	1	1	z, R_z	$x^2 + y^2, z^2$	$z^3, x(x^2 - 3y^2), y(3x^2 - y^2)$
E	$\left\{\begin{matrix} 1 \\ 1 \end{matrix}\right.$	$\begin{matrix} \varepsilon \\ \varepsilon^* \end{matrix}$	$\left.\begin{matrix} \varepsilon^* \\ \varepsilon \end{matrix}\right\}$	$(x, y), (R_x, R_y)$	$(x^2 - y^2, xy), (yz, xz)$	$(xz^2, yz^2), [xyz, z(x^2 - y^2)]$

$\mathbf{C}_4$	E	C_4	C_2	C_4^3			
A	1	1	1	1	z, R_z	$x^2 + y^2, z^2$	z^3
B	1	-1	1	-1		$x^2 - y^2, xy$	$xyz, z(x^2 - y^2)$
E	$\left\{\begin{matrix} 1 \\ 1 \end{matrix}\right.$	$\begin{matrix} i \\ -i \end{matrix}$	$\begin{matrix} -1 \\ -1 \end{matrix}$	$\left.\begin{matrix} -i \\ i \end{matrix}\right\}$	$(x, y), (R_x, R_y)$	(xz, yz)	$(xz^2, yz^2), [x(x^2 - 3y^2), y(3x^2 - y^2)]$

$\mathbf{C}_5$	E	C_5	C_5^2	C_5^3	C_5^4			$\varepsilon = \exp(2\pi i/5)$
A	1	1	1	1	1	z, R_z	$x^2 + y^2, z^2$	z^3
E_1	1	ε	ε^2	ε^{2*}	ε^*	$(x, y), (R_x, R_y)$	(yz, xz)	(xz^2, yz^2)
	1	ε^*	ε^{2*}	ε^2	ε			
E_2	1	ε^2	ε^*	ε	ε^{2*}		$(x^2 - y^2, xy)$	$[xyz, z(x^2 - y^2)], [x(x^2 - 3y^2), y(3x^2 - y^2)]$
	1	ε^{2*}	ε	ε^*	ε^2			

$\mathbf{C}_6$	E	C_6	C_3	C_2	C_3^2	C_6^5	$\mathbf{C}_6 = \mathbf{C}_3 \times \mathbf{C}_2$		$\varepsilon = \exp(2\pi i/6)$
A	1	1	1	1	1	1	z, R_z	$x^2 + y^2, z^2$	z^3
B	1	-1	1	-1	1	-1			$x(x^2 - 3y^2), y(3x^2 - y^2)$
E_1	1	ε	$-\varepsilon^*$	-1	$-\varepsilon$	ε^*	(x, y), (R_x, R_y)	(xz, yz)	(xz^2, yz^2)
	1	ε^*	$-\varepsilon$	-1	$-\varepsilon^*$	ε			
E_2	1	$-\varepsilon^*$	$-\varepsilon$	1	$-\varepsilon^*$	$-\varepsilon$		$(x^2 - y^2, xy)$	$[xyz, z(x^2 - y^2)]$
	1	$-\varepsilon$	$-\varepsilon^*$	1	$-\varepsilon$	$-\varepsilon^*$			

The $\mathbf{C}_n$ groups (*continued*)

$\mathbf{C}_7$	E	C_7	C_7^2	C_7^3	C_7^4	C_7^5	C_7^6	$\varepsilon = \exp(2\pi i/7)$		
A	1	1	1	1	1	1	1	z, R_z	$x^2 + y^2, z^2$	z^3
E_1	1	ε	ε^2	ε^3	ε^{3*}	ε^{2*}	ε^*	(x, y), (R_x, R_y)	(xz, yz)	(xz^2, yz^2)
	1	ε^*	ε^{2*}	ε^{3*}	ε^3	ε^2	ε			
E_2	1	ε^2	ε^{3*}	ε^*	ε	ε^3	ε^{2*}		$(x^2 - y^2, xy)$	$[xyz, z(x^2 - y^2)]$
	1	ε^{2*}	ε^3	ε	ε^*	ε^{3*}	ε^2			
E_3	1	ε^3	ε^*	ε^2	ε^{2*}	ε	ε^{3*}			$[x(x^2 - 3y^2), y(3x^2 - y^2)]$
	1	ε^{3*}	ε	ε^{2*}	ε^2	ε^*	ε^3			

$\mathbf{C}_8$	E	C_8	C_4	C_2	C_4^3	C_8^3	C_8^5	C_8^7	$\mathbf{C}_8 = \mathbf{C}_4 \times \mathbf{C}_2$	$\varepsilon = \exp(2\pi i/8)$	
A	1	1	1	1	1	1	1	1	z, R_z	$x^2 + y^2, z^2$	z^3
B	1	-1	1	1	1	-1	-1	-1			
E_1	1	ε	i	-1	$-i$	$-\varepsilon^*$	$-\varepsilon$	ε^*	(x, y), (R_x, R_y)	(xz, yz)	(xz^2, yz^2)
	1	ε^*	$-i$	-1	i	$-\varepsilon$	$-\varepsilon^*$	ε			
E_2	1	i	-1	1	-1	$-i$	i	$-i$		$(x^2 - y^2, xy)$	$[xyz, z(x^2 - y^2)]$
	1	$-i$	-1	1	-1	i	$-i$	i			
E_3	1	$-\varepsilon$	i	-1	$-i$	ε^*	ε	$-\varepsilon^*$			$[x(x^2 - 3y^2), y(3x^2 - y^2)]$
	1	$-\varepsilon^*$	$-i$	-1	i	ε	ε^*	$-\varepsilon$			

The $\mathbf{S}_n$ groups

$\mathbf{S}_4$	E	S_4	C_2	S_4^3			
A	1	1	1	1	R_z	x^2+y^2, z^2	$xyz, z(x^2-y^2)$
B	1	−1	1	−1	z	x^2-y^2, xy	z^3
E	$\begin{Bmatrix}1 \\ 1\end{Bmatrix}$	$\begin{matrix}i \\ -i\end{matrix}$	$\begin{matrix}-1 \\ -1\end{matrix}$	$\begin{matrix}-i \\ i\end{matrix}$	$(x, y), (R_x, R_y)$	(xz, yz)	$(xz^2, yz^2), [x(x^2-3y^2), y(3x^2-y^2)]$

$\mathbf{S}_6$	E	C_3	C_3^2	i	S_6^5	S_6	$\mathbf{S}_6 = \mathbf{C}_3 \times \mathbf{C}_i$		$\varepsilon = \exp(2\pi i/3)$
A_g	1	1	1	1	1	1	R_z	x^2+y^2, z^2	
E_g	$\begin{Bmatrix}1 \\ 1\end{Bmatrix}$	$\begin{matrix}\varepsilon \\ \varepsilon^*\end{matrix}$	$\begin{matrix}\varepsilon^* \\ \varepsilon\end{matrix}$	$\begin{matrix}1 \\ 1\end{matrix}$	$\begin{matrix}\varepsilon \\ \varepsilon^*\end{matrix}$	$\begin{matrix}\varepsilon^* \\ \varepsilon\end{matrix}$	(R_x, R_y)	(x^2-y^2, xy), (xz, yz)	
A_u	1	1	1	−1	−1	−1	z		$z^3, x(x^2-3y^2), y(3x^2-y^2)$
E_u	$\begin{Bmatrix}1 \\ 1\end{Bmatrix}$	$\begin{matrix}\varepsilon \\ \varepsilon^*\end{matrix}$	$\begin{matrix}\varepsilon^* \\ \varepsilon\end{matrix}$	$\begin{matrix}-1 \\ -1\end{matrix}$	$\begin{matrix}-\varepsilon \\ -\varepsilon^*\end{matrix}$	$\begin{matrix}-\varepsilon^* \\ -\varepsilon\end{matrix}$	(x, y)		$(xz^2, yz^2), [xyz, z(x^2-y^2)]$

$\mathbf{S}_8$	E	S_8	C_4	S_8^3	C_2	S_8^5	C_4^3	S_8^7		$\varepsilon = \exp(2\pi i/8)$	
A	1	1	1	1	1	1	1	1	R_z	x^2+y^2, z^2	
B	1	−1	1	−1	1	−1	1	−1	z		z^3
E_1	$\begin{Bmatrix}1 \\ 1\end{Bmatrix}$	$\begin{matrix}\varepsilon \\ \varepsilon^*\end{matrix}$	$\begin{matrix}i \\ -i\end{matrix}$	$\begin{matrix}-\varepsilon^* \\ -\varepsilon\end{matrix}$	$\begin{matrix}-1 \\ -1\end{matrix}$	$\begin{matrix}-\varepsilon \\ -\varepsilon^*\end{matrix}$	$\begin{matrix}-i \\ i\end{matrix}$	$\begin{matrix}\varepsilon^* \\ \varepsilon\end{matrix}$	(x, y), (R_x, R_y)		(xz^2, yz^2)
E_2	$\begin{Bmatrix}1 \\ 1\end{Bmatrix}$	$\begin{matrix}i \\ -i\end{matrix}$	$\begin{matrix}-1 \\ -1\end{matrix}$	$\begin{matrix}-i \\ i\end{matrix}$	$\begin{matrix}1 \\ 1\end{matrix}$	$\begin{matrix}i \\ -i\end{matrix}$	$\begin{matrix}-1 \\ -1\end{matrix}$	$\begin{matrix}-i \\ i\end{matrix}$		(x^2-y^2, xy)	$[xyz, z(x^2-y^2)]$
E_3	$\begin{Bmatrix}1 \\ 1\end{Bmatrix}$	$\begin{matrix}-\varepsilon^* \\ -\varepsilon\end{matrix}$	$\begin{matrix}-i \\ i\end{matrix}$	$\begin{matrix}\varepsilon \\ \varepsilon^*\end{matrix}$	$\begin{matrix}-1 \\ -1\end{matrix}$	$\begin{matrix}\varepsilon^* \\ \varepsilon\end{matrix}$	$\begin{matrix}i \\ -i\end{matrix}$	$\begin{matrix}-\varepsilon \\ -\varepsilon^*\end{matrix}$		(xz, yz)	$[x(x^2-3y^2), y(3x^2-y^2)]$

The $\mathbf{C}_{nv}$ groups $\mathbf{C}_{nv} = \mathbf{C}_n \wedge \mathbf{C}_s$

$\mathbf{C}_{2v}$	E	C_2	$\sigma_v(xz)$	$\sigma_v(yz)$			
A_1	1	1	1	1	z	x^2, y^2, z^2	$z^3, z(x^2 - y^2)$
A_2	1	1	-1	-1	R_z	xy	xyz
B_1	1	-1	1	-1	x, R_y	xz	$xz^2, x(x^2 - 3y^2)$
B_2	1	-1	-1	1	y, R_x	yz	$yz^2, y(3x^2 - y^2)$

$\mathbf{C}_{3v}$	E	$2C_3$	$3\sigma_v$			
A_1	1	1	1	z	$x^2 + y^2, z^2$	$z^3, x(x^2 - 3y^2)$
A_2	1	1	-1	R_z		$y(3x^2 - y^2)$
E	2	-1	0	$(x, y), (R_x, R_y)$	$(x^2 - y^2, xy), (xz, yz)$	$(xz^2, yz^2), [xyz, z(x^2 - y^2)]$

$\mathbf{C}_{4v}$	E	$2C_4$	C_2	$2\sigma_v$	$2\sigma_d$			
A_1	1	1	1	1	1	z	$x^2 + y^2, z^2$	z^3
A_2	1	1	1	-1	-1	R_z		
B_1	1	-1	1	1	-1		$x^2 - y^2$	$z(x^2 - y^2)$
B_2	1	-1	1	-1	1		xy	xyz
E	2	0	-2	0	0	$(x, y), (R_x, R_y)$	(xz, yz)	$(xz^2, yz^2), [x(x^2 - 3y^2), y(3x^2 - y^2)]$

$\mathbf{C}_{5v}$	E	$2C_5$	$2C_5^2$	$5\sigma_v$			
A_1	1	1	1	1	z	$x^2 + y^2, z^2$	z^3
A_2	1	1	1	−1	R_z		
E_1	2	2 cos 72°	2 cos 144°	0	$(x, y), (R_x, R_y)$	(xz, yz)	(xz^2, yz^2)
E_2	2	2 cos 144°	2 cos 72°	0		$(x^2 - y^2, xy)$	$[xyz, z(x^2 - y^2)], [x(x^2 - 3y^2), y(3x^2 - y^2)]$

$\mathbf{C}_{6v}$	E	$2C_6$	$2C_3$	C_2	$3\sigma_v$	$3\sigma_d$			
A_1	1	1	1	1	1	1	z	$x^2 + y^2, z^2$	z^3
A_2	1	1	1	1	−1	−1	R_z		
B_1	1	−1	1	−1	1	−1			$x(x^2 - 3y^2)$
B_2	1	−1	1	−1	−1	1			$y(3x^2 - y^2)$
E_1	2	1	−1	−2	0	0	$(x, y), (R_x, R_y)$	(xz, yz)	(xz^2, yz^2)
E_2	2	−1	−1	2	0	0		$(x^2 - y^2, xy)$	$[xyz, z(x^2 - y^2)]$

The $\mathbf{C}_{nh}$ groups $\mathbf{C}_{nh} = \mathbf{C}_n \times \mathbf{C}_s$

$\mathbf{C}_{2h}$	E	C_2	i	σ_h			
A_g	1	1	1	1	R_z	x^2, y^2, z^2, xy	
B_g	1	-1	1	-1	R_x, R_y	xz, yz	
A_u	1	1	-1	-1	z		$z^3, xyz, z(x^2-y^2)$
B_u	1	-1	-1	1	x, y		$xz^2, yz^2, x(x^2-3y^2), y(3x^2-y^2)$

$\mathbf{C}_{3h}$	E	C_3	C_3^2	σ_h	S_3	S_3^5	$\varepsilon = \exp(2\pi i/3)$		
A'	1	1	1	1	1	1	R_z	x^2+y^2, z^2	$x(x^2-3y^2), y(3x^2-y^2)$
E'	1	ε	ε^*	1	ε	ε^*	(x, y)	(x^2-y^2, xy)	(xz^2, yz^2)
	1	ε^*	ε	1	ε^*	ε			
A''	1	1	1	-1	-1	-1	z		z^3
E''	1	ε	ε^*	-1	$-\varepsilon$	$-\varepsilon^*$	(R_x, R_y)	(xz, yz)	$[xyz, z(x^2-y^2)]$
	1	ε^*	ε	-1	$-\varepsilon^*$	$-\varepsilon$			

$\mathbf{C}_{4h}$	E	C_4	C_2	C_4^3	i	S_4^3	σ_h	S_4			
A_g	1	1	1	1	1	1	1	1	R_z	x^2+y^2, z^2	
B_g	1	-1	1	-1	1	-1	1	-1		x^2-y^2, xy	
E_g	1	i	-1	$-i$	1	i	-1	$-i$	(R_x, R_y)	(xz, yz)	
	1	$-i$	-1	i	1	$-i$	-1	i			
A_u	1	1	1	1	-1	-1	-1	-1	z		z^3
B_u	1	-1	1	-1	-1	1	-1	1			$xyz, z(x^2-y^2)$
E_u	1	i	-1	$-i$	-1	$-i$	1	i	(x, y)		$(xz^2, yz^2), [x(x^2-3y^2), y(3x^2-y^2)]$
	1	$-i$	-1	i	-1	i	1	$-i$			

$\mathbf{C}_{5h}$	E	C_5	C_5^2	C_5^3	C_5^4	σ_h	S_5	S_5^7	S_5^3	S_5^9	$\varepsilon = \exp(2\pi i/5)$		
A'	1	1	1	1	1	1	1	1	1	1	R_z	$x^2 + y^2, z^2$	
E'_1	1	ε	ε^2	ε^{2*}	ε^*	1	ε	ε^2	ε^{2*}	ε^*	(x, y)		(xz^2, yz^2)
	1	ε^*	ε^{2*}	ε^2	ε	1	ε^*	ε^{2*}	ε^2	ε			
E'_2	1	ε^2	ε^*	ε	ε^{2*}	1	ε^2	ε^*	ε	ε^{2*}		$(x^2 - y^2, xy)$	$[x(x^2 - 3y^2), y(3x^2 - y^2)]$
	1	ε^{2*}	ε	ε^*	ε^2	1	ε^{2*}	ε	ε^*	ε^2			
A''	1	1	1	1	1	-1	-1	-1	-1	-1	z		z^3
E''_1	1	ε	ε^2	ε^{2*}	ε^*	-1	$-\varepsilon$	$-\varepsilon^2$	$-\varepsilon^{2*}$	$-\varepsilon^*$	(R_x, R_y)	(xz, yz)	
	1	ε^*	ε^{2*}	ε^2	ε	-1	$-\varepsilon^*$	$-\varepsilon^{2*}$	$-\varepsilon^2$	$-\varepsilon$			
E''_2	1	ε^2	ε^*	ε	ε^{2*}	-1	$-\varepsilon^2$	$-\varepsilon^*$	$-\varepsilon$	$-\varepsilon^{2*}$			$[xyz, z(x^2 - y^2)]$
	1	ε^{2*}	ε	ε^*	ε^2	-1	$-\varepsilon^{2*}$	$-\varepsilon$	$-\varepsilon^*$	$-\varepsilon^2$			

$\mathbf{C}_{6h}$	E	C_6	C_3	C_2	C_3^2	C_6^5	i	S_3^5	S_6^5	σ_h	S_6	S_3	$\varepsilon = \exp(2\pi i/6)$		
A_g	1	1	1	1	1	1	1	1	1	1	1	1	R_z	$x^2 + y^2, z^2$	
B_g	1	-1	1	-1	1	-1	1	-1	1	-1	1	-1			
E_{1g}	1	ε	$-\varepsilon^*$	-1	$-\varepsilon$	ε^*	1	ε	$-\varepsilon^*$	-1	$-\varepsilon$	ε^*	(R_x, R_y)	(xz, yz)	
	1	ε^*	$-\varepsilon$	-1	$-\varepsilon^*$	ε	1	ε^*	$-\varepsilon$	-1	$-\varepsilon^*$	ε			
E_{2g}	1	$-\varepsilon^*$	$-\varepsilon$	1	$-\varepsilon^*$	$-\varepsilon$	1	$-\varepsilon^*$	$-\varepsilon$	1	$-\varepsilon^*$	$-\varepsilon$		$(x^2 - y^2, xy)$	
	1	$-\varepsilon$	$-\varepsilon^*$	1	$-\varepsilon$	$-\varepsilon^*$	1	$-\varepsilon$	$-\varepsilon^*$	1	$-\varepsilon$	$-\varepsilon^*$			
A_u	1	1	1	1	1	1	-1	-1	-1	-1	-1	-1	z		z^3
B_u	1	-1	1	-1	1	-1	-1	1	-1	1	-1	1			$x(x^2 - 3y^2), y(3x^2 - y^2)$
E_{1u}	1	ε	$-\varepsilon^*$	-1	$-\varepsilon$	ε^*	-1	$-\varepsilon$	ε^*	1	ε	$-\varepsilon^*$	(x, y)		(xz^2, yz^2)
	1	ε^*	$-\varepsilon$	-1	$-\varepsilon^*$	ε	-1	$-\varepsilon^*$	ε	1	ε^*	$-\varepsilon$			
E_{2u}	1	$-\varepsilon^*$	$-\varepsilon$	1	$-\varepsilon^*$	$-\varepsilon$	-1	ε^*	ε	-1	ε^*	ε			$(xyz, z(x^2 - y^2))$
	1	$-\varepsilon$	$-\varepsilon^*$	1	$-\varepsilon$	$-\varepsilon^*$	-1	ε	ε^*	-1	ε	ε^*			

THE DIHEDRAL GROUPS

The $\mathbf{D}_n$ groups $\mathbf{D}_n = \mathbf{C}_n \wedge \mathbf{C}_2$

$\mathbf{D}_2$	E	$C_2(z)$	$C_2(y)$	$C_2(x)$			
A	1	1	1	1		x^2, y^2, z^2	xyz
B_1	1	1	−1	−1	z, R_z	xy	$z^3, z(x^2 - y^2)$
B_2	1	−1	1	−1	y, R_y	xz	$yz^2, y(3x^2 - y^2)$
B_3	1	−1	−1	1	x, R_x	yz	$xz^2, x(x^2 - 3y^2)$

(x axis coincident with C_2)

$\mathbf{D}_3$	E	$2C_3$	$3C_2$			
A_1	1	1	1		$x^2 + y^2, z^2$	$x(x^2 - 3y^2)$
A_2	1	1	−1	z, R_z		$z^3, y(3x^2 - y^2)$
E	2	−1	0	$(x, y), (R_x, R_y)$	$(x^2 - y^2, xy), (xz, yz)$	$(xz^2, yz^2), [xyz, z(x^2 - y^2)]$

(x axis coincident with C'_2)

$\mathbf{D}_4$	E	$2C_4$	$C_2(=C_4^2)$	$2C'_2$	$2C''_2$			
A_1	1	1	1	1	1		$x^2 + y^2, z^2$	
A_2	1	1	1	−1	−1	z, R_z		z^3
B_1	1	−1	1	1	−1		$x^2 - y^2$	xyz
B_2	1	−1	1	−1	1		xy	$z(x^2 - y^2)$
E	2	0	−2	0	0	$(x, y), (R_x, R_y)$	(xz, yz)	$(xz^2, yz^2), [x(x^2 - 3y^2), y(3x^2 - y^2)]$

(x axis coincident with C_2)

$\mathbf{D}_5$	E	$2C_5$	$2C_5^2$	$5C_2$			
A_1	1	1	1	1		$x^2 + y^2, z^2$	
A_2	1	1	1	−1	z, R_z		z^3
E_1	2	$2\cos 72°$	$2\cos 144°$	0	$(x, y), (R_x, R_y)$	(xz, yz)	(xz^2, yz^2)
E_2	2	$2\cos 144°$	$2\cos 72°$	0		$(x^2 - y^2, xy)$	$[xyz, z(x^2 - y^2)], [x(x^2 - 3y^2), y(3x^2 - y^2)]$

$\mathbf{D}_6$	E	$2C_6$	$2C_3$	C_2	$3C'_2$	$3C''_2$	(x axis coincident with C'_2)		
A_1	1	1	1	1	1	1		x^2+y^2, z^2	
A_2	1	1	1	1	−1	−1	z, R_z		z^3
B_1	1	−1	1	−1	1	−1			$x(x^2-3y^2)$
B_2	1	−1	1	−1	−1	1			$y(3x^2-y^2)$
E_1	2	1	−1	−2	0	0	$(x, y), (R_x, R_y)$	(xz, yz)	(xz^2, yz^2)
E_2	2	−1	−1	2	0	0		(x^2-y^2, xy)	$[xyz, z(x^2-y^2)]$

The $\mathbf{D}_{nh}$ groups $\mathbf{D}_{nh} = \mathbf{D}_n \times \mathbf{C}_s$

$\mathbf{D}_{2h}$	E	$C_2(z)$	$C_2(y)$	$C_2(x)$	i	$\sigma(xy)$	$\sigma(xz)$	$\sigma(yz)$			
A_g	1	1	1	1	1	1	1	1		x^2, y^2, z^2	
B_{1g}	1	1	−1	−1	1	1	−1	−1	R_z	xy	
B_{2g}	1	−1	1	−1	1	−1	1	−1	R_y	xz	
B_{3g}	1	−1	−1	1	1	−1	−1	1	R_x	yz	
A_u	1	1	1	1	−1	−1	−1	−1			xyz
B_{1u}	1	1	−1	−1	−1	−1	1	1	z		$z^3, z(x^2-y^2)$
B_{2u}	1	−1	1	−1	−1	1	−1	1	y		$yz^2, y(3x^2-y^2)$
B_{3u}	1	−1	−1	1	−1	1	1	−1	x		$xz^2, x(x^2-3y^2)$

$\mathbf{D}_{3h}$	E	$2C_3$	$3C_2$	σ_h	$2S_3$	$3\sigma_v$	(x axis coincident with C_2)		
A'_1	1	1	1	1	1	1		x^2+y^2, z^2	$x(x^2-3y^2)$
A'_2	1	1	−1	1	1	−1	R_z		$y(3x^2-y^2)$
E'	2	−1	0	2	−1	0	(x, y)	(x^2-y^2, xy)	(xz^2, yz^2)
A''_1	1	1	1	−1	−1	−1			
A''_2	1	1	−1	−1	−1	1	z		z^3
E''	2	−1	0	−2	1	0	(R_x, R_y)	(xz, yz)	$[xyz, z(x^2-y^2)]$

The $\mathbf{D}_{nh}$ groups (*continued*)

$\mathbf{D}_{4h}$	E	$2C_4$	C_2	$2C'_2$	$2C''_2$	i	$2S_4$	σ_h	$2\sigma_v$	$2\sigma_d$	(x axis coincident with C'_2)		
A_{1g}	1	1	1	1	1	1	1	1	1	1		x^2+y^2, z^2	
A_{2g}	1	1	1	−1	−1	1	1	1	−1	−1	R_z		
B_{1g}	1	−1	1	1	−1	1	−1	1	1	−1		x^2-y^2	
B_{2g}	1	−1	1	−1	1	1	−1	1	−1	1		xy	
E_g	2	0	−2	0	0	2	0	−2	0	0	(R_x, R_y)	(xz, yz)	
A_{1u}	1	1	1	1	1	−1	−1	−1	−1	−1			
A_{2u}	1	1	1	−1	−1	−1	−1	−1	1	1	z		z^3
B_{1u}	1	−1	1	1	−1	−1	1	−1	−1	1			xyz
B_{2u}	1	−1	1	−1	1	−1	1	−1	1	−1			$z(x^2-y^2)$
E_u	2	0	−2	0	0	−2	0	2	0	0	(x, y)		$(xz^2, yz^2), [x(x^2-3y^2), y(3x^2-y^2)]$

$\mathbf{D}_{5h}$	E	$2C_5$	$2C_5^2$	$5C_2$	σ_h	$2S_5$	$2S_5^3$	$5\sigma_v$	(x axis coincident with C_2)		
A'_1	1	1	1	1	1	1	1	1		x^2+y^2, z^2	
A'_2	1	1	1	−1	1	1	1	−1	R_z		
E'_1	2	2 cos 72°	2 cos 144°	0	2	2 cos 72°	2 cos 144°	0	(x, y)		(xz^2, yz^2)
E'_2	2	2 cos 144°	2 cos 72°	0	2	2 cos 144°	2 cos 72°	0		(x^2-y^2, xy)	$[x(x^2-3y^2), y(3x^2-y^2)]$
A''_1	1	1	1	1	−1	−1	−1	−1			
A''_2	1	1	1	−1	−1	−1	−1	1	z		z^3
E''_1	2	2 cos 72°	2 cos 144°	0	−2	−2 cos 72°	−2 cos 144°	0	(R_x, R_y)	(xz, yz)	
E''_2	2	2 cos 144°	2 cos 72°	0	−2	−2 cos 144°	−2 cos 72°	0			$[xyz, z(x^2-y^2)]$

$\mathbf{D}_{6h}$	E	$2C_6$	$2C_3$	C_2	$3C'_2$	$3C''_2$	i	$2S_3$	$2S_6$	σ_h	$3\sigma_d$	$3\sigma_v$	(x axis coincident with C'_2)		
A_{1g}	1	1	1	1	1	1	1	1	1	1	1	1		$x^2 + y^2, z^2$	
A_{2g}	1	1	1	1	−1	−1	1	1	1	1	−1	−1	R_z		
B_{1g}	1	−1	1	−1	1	−1	1	−1	1	−1	1	−1			
B_{2g}	1	−1	1	−1	−1	1	1	−1	1	−1	−1	1			
E_{1g}	2	1	−1	−2	0	0	2	1	−1	−2	0	0	(R_x, R_y)	(xz, yz)	
E_{2g}	2	−1	−1	2	0	0	2	−1	−1	2	0	0		$(x^2 - y, xy)$	
A_{1u}	1	1	1	1	1	1	−1	−1	−1	−1	−1	−1			
A_{2u}	1	1	1	1	−1	−1	−1	−1	−1	−1	1	1	z		z^3
B_{1u}	1	−1	1	−1	1	−1	−1	1	−1	1	−1	1			$x(x^2 - 3y^2)$
B_{2u}	1	−1	1	−1	−1	1	−1	1	−1	1	1	−1			$y(3x^2 - y^2)$
E_{1u}	2	1	−1	−2	0	0	−2	−1	1	2	0	0	(x, y)		(xz^2, yz^2)
E_{2u}	2	−1	−1	2	0	0	−2	1	1	−2	0	0			$[xyz, z(x^2 - y^2)]$

$\mathbf{D}_{8h}$	E	$2C_8$	$2C_8^3$	$2C_4$	C_2	$4C'_2$	$4C''_2$	i	$2S_8^3$	$2S_8$	$2S_4$	σ_h	$4\sigma_v$	$4\sigma_d$	(x axis coincident with C'_2)		
A_{1g}	1	1	1	1	1	1	1	1	1	1	1	1	1	1		$x^2 + y^2, z^2$	
A_{2g}	1	1	1	1	1	−1	−1	1	1	1	1	1	−1	−1	R_z		
B_{1g}	1	−1	−1	1	1	1	−1	1	−1	−1	1	1	1	−1			
B_{2g}	1	−1	−1	1	1	−1	1	1	−1	−1	1	1	−1	1			
E_{1g}	2	$\sqrt{2}$	$-\sqrt{2}$	0	−2	0	0	2	$\sqrt{2}$	$-\sqrt{2}$	0	−2	0	0	(R_x, R_y)	(xz, yz)	
E_{2g}	2	0	0	−2	2	0	0	2	0	0	−2	2	0	0		$(x^2 - y^2, xy)$	
E_{3g}	2	$-\sqrt{2}$	$\sqrt{2}$	0	−2	0	0	2	$-\sqrt{2}$	$\sqrt{2}$	0	−2	0	0			
A_{1u}	1	1	1	1	1	1	1	−1	−1	−1	−1	−1	−1	−1			
A_{2u}	1	1	1	1	1	−1	−1	−1	−1	−1	−1	−1	1	1	z	z^3	
B_{1u}	1	−1	−1	1	1	1	−1	−1	1	1	−1	−1	−1	1			
B_{2u}	1	−1	−1	1	1	−1	1	−1	1	1	−1	−1	1	−1			
E_{1u}	2	$\sqrt{2}$	$-\sqrt{2}$	0	−2	0	0	−2	$-\sqrt{2}$	$\sqrt{2}$	0	2	0	0	(x, y)		(xz^2, yz^2)
E_{2u}	2	0	0	−2	2	0	0	−2	0	0	2	−2	0	0			$[xyz, z(x^2 - y^2)]$
E_{3u}	2	$-\sqrt{2}$	$\sqrt{2}$	0	−2	0	0	−2	$\sqrt{2}$	$-\sqrt{2}$	0	2	0	0			$[x(x^2 - 3y^2), y(3x^2 - y^2)]$

The $\mathbf{D}_{nd}$ groups $\mathbf{D}_{nd} = \mathbf{D}_n \wedge \mathbf{C}_2$ (n even) $= \mathbf{D}_n \times \mathbf{C}_i$ (n odd)

$\mathbf{D}_{2d}$	E	$2S_4$	C_2	$2C'_2$	$2\sigma_d$	(x axis coincident with C'_2)		
A_1	1	1	1	1	1		x^2+y^2, z^2	xyz
A_2	1	1	1	−1	−1	R_z		$z(x^2-y^2)$
B_1	1	−1	1	1	−1		x^2-y^2	
B_2	1	−1	1	−1	1	z	xy	z^3
E	2	0	−2	0	0	$(x, y), (R_x, R_y)$	(xz, yz)	$(xz^2, yz^2), [x(x^2-3y^2), y(3x^2-y^2)]$

$\mathbf{D}_{3d}$	E	$2C_3$	$3C_2$	i	$2S_6$	$3\sigma_d$	(x axis coincident with C_2)		
A_{1g}	1	1	1	1	1	1		x^2+y^2, z^2	
A_{2g}	1	1	−1	1	1	−1	R_z		
E_g	2	−1	0	2	−1	0	(R_x, R_y)	$(x^2-y^2, xy); (xz, yz)$	
A_{1u}	1	1	1	−1	−1	−1			$x(x^2-3y^2)$
A_{2u}	1	1	−1	−1	−1	1	z		$y(3x^2-y^2), z^3$
E_u	2	−1	0	−2	1	0	(x, y)		$(xz^2, yz^2), [xyz, z(x^2-y^2)]$

$\mathbf{D}_{4d}$	E	$2S_8$	$2C_4$	$2S_8^3$	C_2	$4C'_2$	$4\sigma_d$	(x axis coincident with C'_2)		
A_1	1	1	1	1	1	1	1		x^2+y^2, z^2	
A_2	1	1	1	1	1	−1	−1	R_z		
B_1	1	−1	1	−1	1	1	−1			
B_2	1	−1	1	−1	1	−1	1	z		z^3
E_1	2	$\sqrt{2}$	0	$-\sqrt{2}$	−2	0	0	(x, y)		(xz^2, yz^2)
E_2	2	0	−2	0	2	0	0		(x^2-y^2, xy)	$[xyz, z(x^2-y^2)]$
E_3	2	$-\sqrt{2}$	0	$\sqrt{2}$	−2	0	0	(R_x, R_y)	(xz, yz)	$[x(x^2-3y^2), y(3x^2-y^2)]$

$\mathbf{D}_{5d}$	E	$2C_5$	$2C_5^2$	$5C_2$	i	$2S_{10}^3$	$2S_{10}$	$5\sigma_d$	(x axis coincident with C_2)		
A_{1g}	1	1	1	1	1	1	1	1		$x^2 + y^2, z^2$	
A_{2g}	1	1	1	−1	1	1	1	−1	R_z		
E_{1g}	2	2 cos 72°	2 cos 144°	0	2	2 cos 72°	2 cos 144°	0	(R_x, R_y)	(xz, yz)	
E_{2g}	2	2 cos 144°	2 cos 72°	0	2	2 cos 144°	2 cos 72°	0		$(x^2 - y^2, xy)$	
A_{1u}	1	1	1	1	−1	−1	−1	−1			
A_{2u}	1	1	1	−1	−1	−1	−1	1	z		z^3
E_{1u}	2	2 cos 72°	2 cos 144°	0	−2	−2 cos 72°	−2 cos 144°	0	(x, y)		(xz^2, yz^2)
E_{2u}	2	2 cos 144°	2 cos 72°	0	−2	−2 cos 144°	−2 cos 72°	0			$[xyz, z(x^2 - y^2)]$, $[x(x^2 - 3y^2), y(3x^2 - y^2)]$

$\mathbf{D}_{6d}$	E	$2S_{12}$	$2C_6$	$2S_4$	$2C_3$	$2S_{12}^5$	C_2	$6C_2'$	$6\sigma_d$	(x axis coincident with C_2')		
A_1	1	1	1	1	1	1	1	1	1		$x^2 + y^2, z^2$	
A_2	1	1	1	1	1	1	1	−1	−1	R_z		
B_1	1	−1	1	−1	1	−1	1	1	−1			
B_2	1	−1	1	−1	1	−1	1	−1	1	z		z^3
E_1	2	$\sqrt{3}$	1	0	−1	$-\sqrt{3}$	−2	0	0	(x, y)		(xz^2, yz^2)
E_2	2	1	−1	−2	−1	1	2	0	0		$(x^2 - y^2, xy)$	
E_3	2	0	−2	0	2	0	−2	0	0			$[x(x^2 - 3y^2), y(3x^2 - y^2)]$
E_4	2	−1	−1	2	−1	−1	2	0	0			$[xyz, z(x^2 - y^2)]$
E_5	2	$-\sqrt{3}$	1	0	−1	$\sqrt{3}$	−2	0	0	(R_x, R_y)	(xz, yz)	

THE CUBIC GROUPS

$\mathbf{T}$	E	$4C_3$	$4C_3^2$	$3C_2$	$\mathbf{T} = \mathbf{D}_2 \wedge \mathbf{C}_3$		$\varepsilon = \exp(2\pi i/3)$
A	1	1	1	1		$x^2 + y^2 + z^2$	xyz
E	1 1	ε ε^*	ε^* ε	1 1		$(2z^2 - x^2 - y^2, x^2 - y^2)$	
T	3	0	0	−1	$(R_x, R_y, R_z), (x, y, z)$	(xy, xz, yz)	$(x^3, y^3, z^3), [x(z^2 - y^2), y(z^2 - x^2), z(x^2 - y^2)]$

$\mathbf{T}_h$	E	$4C_3$	$4C_3^2$	$3C_2$	i	$4S_6$	$4S_6^5$	$3\sigma_h$	$\mathbf{T}_h = \mathbf{T} \times \mathbf{C}_i$		$\varepsilon = \exp(2\pi i/3)$
A_g	1	1	1	1	1	1	1	1		$x^2 + y^2 + z^2$	
A_u	1	1	1	1	−1	−1	−1	−1			xyz
E_g	1 1	ε ε^*	ε^* ε	1 1	1 1	ε ε^*	ε^* ε	1 1		$(2z^2 - x^2 - y^2, x^2 - y^2)$	
E_u	1 1	ε ε^*	ε^* ε	1 1	−1 −1	$-\varepsilon$ $-\varepsilon^*$	$-\varepsilon^*$ $-\varepsilon$	−1 −1			
T_g	3	0	0	−1	3	0	0	−1	(R_x, R_y, R_z)	(xz, yz, xy)	
T_u	3	0	0	−1	−3	0	0	1	(x, y, z)		$(x^3, y^3, z^3), [x(z^2 - y^2), y(z^2 - x^2), z(x^2 - y^2)]$

$\mathbf{T}_d$	E	$8C_3$	$3C_2$	$6S_4$	$6\sigma_d$	$\mathbf{T}_d = \mathbf{D}_2 \wedge \mathbf{C}_{3v}$		
A_1	1	1	1	1	1		$x^2 + y^2 + z^2$	xyz
A_2	1	1	1	−1	−1			
E	2	−1	2	0	0		$(2z^2 - x^2 - y^2, x^2 - y^2)$	
T_1	3	0	−1	1	−1	(R_x, R_y, R_z)		$[x(z^2 - y^2), y(z^2 - x^2), z(x^2 - y^2)]$
T_2	3	0	−1	−1	1	(x, y, z)	(xy, xz, yz)	(x^3, y^3, z^3)

$\mathbf{O}$	E	$6C_4$	$3C_2(=C_4^2)$	$8C_3$	$6C_2$		$\mathbf{O} = \mathbf{D}_2 \wedge \mathbf{D}_3$	
A_1	1	1	1	1	1		$x^2 + y^2 + z^2$	
A_2	1	−1	1	1	−1			xyz
E	2	0	2	−1	0		$(2z^2 - x^2 - y^2, x^2 - y^2)$	
T_1	3	1	−1	0	−1	$(R_x, R_y, R_z), (x, y, z)$		(x^3, y^3, z^3)
T_2	3	−1	−1	0	1		(xy, xz, yz)	$[x(z^2 - y^2), y(z^2 - x^2), z(x^2 - y^2)]$

$\mathbf{O}_h$	E	$8C_3$	$6C_2$	$6C_4$	$3C_2(=C_4^2)$	i	$6S_4$	$8S_6$	$3\sigma_h$	$6\sigma_d$		$\mathbf{O}_h = \mathbf{O} \times \mathbf{C}_i$	
A_{1g}	1	1	1	1	1	1	1	1	1	1		$x^2 + y^2 + z^2$	
A_{2g}	1	1	−1	−1	1	1	−1	1	1	−1			
E_g	2	−1	0	0	2	2	0	−1	2	0		$(2z^2 - x^2 - y^2, x^2 - y^2)$	
T_{1g}	3	0	−1	1	−1	3	1	0	−1	−1	(R_x, R_y, R_z)		
T_{2g}	3	0	1	−1	−1	3	−1	0	−1	1		(xz, yz, xy)	
A_{1u}	1	1	1	1	1	−1	−1	−1	−1	−1			
A_{2u}	1	1	−1	−1	1	−1	1	−1	−1	1			xyz
E_u	2	−1	0	0	2	−2	0	1	−2	0			
T_{1u}	3	0	−1	1	−1	−3	−1	0	1	1	(x, y, z)		(x^3, y^3, z^3)
T_{2u}	3	0	1	−1	−1	−3	1	0	1	−1			$[x(z^2 - y^2), y(z^2 - x^2), z(x^2 - y^2)]$

THE ICOSAHEDRAL GROUPS[a]

$\mathbf{I}_h$	E	$12C_5$	$12C_5^2$	$20C_3$	$15C_2$	i	$12S_{10}$	$12S_{10}^3$	$20S_6$	15σ			
A_g	1	1	1	1	1	1	1	1	1	1		$x^2 + y^2 + z^2$	
T_{1g}	3	$\frac{1}{2}(1 + \sqrt{5})$	$\frac{1}{2}(1 - \sqrt{5})$	0	-1	3	$\frac{1}{2}(1 - \sqrt{5})$	$\frac{1}{2}(1 + \sqrt{5})$	0	-1	(R_x, R_y, R_z)		
T_{2g}	3	$\frac{1}{2}(1 - \sqrt{5})$	$\frac{1}{2}(1 + \sqrt{5})$	0	-1	3	$\frac{1}{2}(1 + \sqrt{5})$	$\frac{1}{2}(1 - \sqrt{5})$	0	-1			
G_g	4	-1	-1	1	0	4	-1	-1	1	0			
H_g	5	0	0	-1	1	5	0	0	-1	1		$(2z^2 - x^2 - y^2,$ $x^2 - y^2,$ $xy, yz, zx)$	
A_u	1	1	1	1	1	-1	-1	-1	-1	-1			
T_{1u}	3	$\frac{1}{2}(1 + \sqrt{5})$	$\frac{1}{2}(1 - \sqrt{5})$	0	-1	-3	$-\frac{1}{2}(1 - \sqrt{5})$	$-\frac{1}{2}(1 + \sqrt{5})$	0	1	(x, y, z)		
T_{2u}	3	$\frac{1}{2}(1 - \sqrt{5})$	$\frac{1}{2}(1 + \sqrt{5})$	0	-1	-3	$-\frac{1}{2}(1 + \sqrt{5})$	$-\frac{1}{2}(1 - \sqrt{5})$	0	1			(x^3, y^3, z^3)
G_u	4	-1	-1	1	0	-4	1	1	-1	0			$[x(z^2 - y^2), y(z^2 - x^2),$ $z(x^2 - y^2), xyz]$
H_u	5	0	0	-1	1	-5	0	0	1	-1			

[a] For the pure rotation group **I**, the outlined section in the upper left is the character table; the g subscripts should, of course, be dropped and (x, y, z) assigned to the T_1 representation.

THE GROUPS $\mathbf{C}_{\infty v}$ AND $\mathbf{D}_{\infty h}$ FOR LINEAR MOLECULES

$\mathbf{C}_{\infty v}$	E	$2C_{\infty}^{\varphi}$	$\cdots$	$\infty\sigma_v$	$\mathbf{C}_{\infty v} = \mathbf{C}_{\infty} \wedge \mathbf{C}_s$		
$A_1 \equiv \Sigma^+$	1	1	$\cdots$	1	z	$x^2 + y^2, z^2$	z^3
$A_2 \equiv \Sigma^-$	1	1	$\cdots$	-1	R_z		
$E_1 \equiv \Pi$	2	$2\cos\varphi$	$\cdots$	0	$(x, y); (R_x, R_y)$	(xz, yz)	(xz^2, yz^2)
$E_2 \equiv \Delta$	2	$2\cos 2\phi$	$\cdots$	0		$(x^2 - y^2, xy)$	$[xyz, z(x^2 - y^2)]$
$E_3 \equiv \Phi$	2	$2\cos 3\phi$	$\cdots$	0			$[x(x^2 - 3y^2), y(3x^2 - y^2)]$
$\vdots$	$\vdots$	$\vdots$	$\vdots$	$\vdots$			

$\mathbf{D}_{\infty h}$	E	$2C_{\infty}^{\phi}$	$\cdots$	$\infty\sigma_v$	i	$2S_{\infty}^{\phi}$	$\cdots$	∞C_2	$\mathbf{D}_{\infty h} = \mathbf{D}_{\infty} \times \mathbf{C}_i$		
$A_{1g} \equiv \Sigma_g^+$	1	1	$\cdots$	1	1	1	$\cdots$	1		$x^2 + y^2, z^2$	
$A_{2g} \equiv \Sigma_g^-$	1	1	$\cdots$	-1	1	1	$\cdots$	-1	R_z		
$E_{1g} \equiv \Pi_g$	2	$2\cos\phi$	$\cdots$	0	2	$-2\cos\phi$	$\cdots$	0	(R_x, R_y)	(xz, yz)	
$E_{2g} \equiv \Delta_g$	2	$2\cos 2\phi$	$\cdots$	0	2	$2\cos 2\phi$	$\cdots$	0		$(x^2 - y^2, xy)$	
$\cdots$	$\cdots$	$\cdots$	$\cdots$	$\cdots$	$\cdots$	$\cdots$	$\cdots$	$\cdots$			
$A_{1u} \equiv \Sigma_u^+$	1	1	$\cdots$	1	-1	-1	$\cdots$	-1	z		z^3
$A_{2u} \equiv \Sigma_u^-$	1	1	$\cdots$	-1	-1	-1	$\cdots$	1			
$E_{1u} \equiv \Pi_u$	2	$2\cos\phi$	$\cdots$	0	-2	$2\cos\phi$	$\cdots$	0	(x, y)		(xz^2, yz^2)
$E_{2u} \equiv \Delta_u$	2	$2\cos 2\phi$	$\cdots$	0	-2	$-2\cos 2\phi$	$\cdots$	0			$[xyz, z(x^2 - y^2)]$
$E_{3u} \equiv \Phi_u$	2	$2\cos 3\phi$	$\cdots$	0	-2	$2\cos 3\phi$	$\cdots$	0			$[x(x^2 - 3y^2), y(3x^2 - y^2)]$
$\vdots$	$\vdots$	$\vdots$	$\vdots$	$\vdots$	$\vdots$	$\vdots$	$\vdots$	$\vdots$			

Appendix **2**

Character Tables for Some Double Groups

$\mathbf{T}'_d$		E	R	$4C_3$ $4C_3^2R$	$4C_3^2$ $4C_3R$	$3C_2$ $3C_2R$	$3S_4$ $3S_4^3R$	$3S_4^3$ $3S_4R$	$6\sigma_d$ $6\sigma_dR$
$\mathbf{O}'$ $(h = 48)$		E	R	$4C_3$ $4C_3^2R$	$4C_3^2$ $4C_3R$	$3C_2$ $3C_2R$	$3C_4$ $3C_4^3R$	$3C_4^3$ $3C_4R$	$6C'_2$ $6C'_2R$
Γ_1	A'_1	1	1	1	1	1	1	1	1
Γ_2	A'_2	1	1	1	1	1	-1	-1	-1
Γ_3	E'_1	2	2	-1	-1	2	0	0	0
Γ_4	T'_1	3	3	0	0	-1	1	1	-1
Γ_5	T'_2	3	3	0	0	-1	-1	-1	1
Γ_6	E'_2	2	-2	1	-1	0	$\sqrt{2}$	$-\sqrt{2}$	0
Γ_7	E'_3	2	-2	1	-1	0	$-\sqrt{2}$	$\sqrt{2}$	0
Γ_8	G'	4	-4	-1	1	0	0	0	0

$\mathbf{D}'_{2d}$		E	R	S_4 S_4^3R	S_4^3 S_4R	C_2 C_2R	$2C'_2$ $2C'_2R$	$2\sigma_d$ $2\sigma_dR$
$\mathbf{C}'_{4v}$		E	R	C_4 C_4^3R	C_4^3 C_4R	C_2 C_2R	$2\sigma_v$ $2\sigma_vR$	$2\sigma_d$ $2\sigma_dR$
$\mathbf{D}'_4$ $(h = 16)$		E	R	C_4 C_4^3R	C_4^3 C_4R	C_2 C_2R	$2C'_2$ $2C'_2R$	$2C''_2$ $2C''_2R$
Γ_1	A'_1	1	1	1	1	1	1	1
Γ_2	A'_2	1	1	1	1	1	-1	-1
Γ_3	B'_1	1	1	-1	-1	1	1	-1
Γ_4	B'_2	1	1	-1	-1	1	-1	1
Γ_5	E'_1	2	2	0	0	-2	0	0
Γ_6	E'_2	2	-2	$\sqrt{2}$	$-\sqrt{2}$	0	0	0
Γ_7	E'_3	2	-2	$-\sqrt{2}$	$\sqrt{2}$	0	0	0

$\mathbf{C}'_{3v}$		E	R	C_3 C_3^2R	C_3^2 C_3R	$3\sigma_v$	$3\sigma_vR$
$\mathbf{D}'_3$ $(h = 12)$		E	R	C_3 C_3^2R	C_3^2 C_3R	$3C_2$	$3C_2R$
Γ_1	A'_1	1	1	1	1	1	1
Γ_2	A'_2	1	1	1	1	-1	-1
Γ_3	E'_1	2	2	-1	-1	0	0
Γ_4	E'_2	2	-2	1	-1	0	0
Γ_5	E'_3	1	-1	-1	1	i	$-i$
Γ_6		1	-1	-1	1	$-i$	i

$\mathbf{D}'_{3h}$		E	R	S_3 S_3^5R	S_3^5 S_3R	C_3 C_3^2R	C_3^2 C_3R	σ_h σ_hR	$3C'_2$ $3C'_2R$	$3\sigma_v$ $3\sigma_vR$
$\mathbf{C}'_{6v}$		E	R	C_6 C_6^5R	C_6^5 C_6R	C_3 C_3^2R	C_3^2 C_3R	C_2 C_2R	$3\sigma_v$ $3\sigma_vR$	$3\sigma_d$ $3\sigma_dR$
$\mathbf{D}'_6$ $(h = 24)$		E	R	C_6 C_6^5R	C_6^5 C_6R	C_3 C_3^2R	C_3^2 C_3R	C_2 C_2R	$3C'_2$ $3C'_2R$	$3C''_2$ $3C''_2R$
Γ_1	A'_1	1	1	1	1	1	1	1	1	1
Γ_2	A'_2	1	1	1	1	1	1	1	-1	-1
Γ_3	B'_1	1	1	-1	-1	1	1	-1	1	-1
Γ_4	B'_2	1	1	-1	-1	1	1	-1	-1	1
Γ_5	E'_1	2	2	1	1	-1	-1	-2	0	0
Γ_6	E'_2	2	2	-1	-1	-1	-1	2	0	0
Γ_7	E'_3	2	-2	$\sqrt{3}$	$-\sqrt{3}$	1	-1	0	0	0
Γ_8	E'_4	2	-2	0	0	-2	2	0	0	0
Γ_9	E'_5	2	-2	$-\sqrt{3}$	$\sqrt{3}$	1	-1	0	0	0

Appendix 3

Correlation Tables for the Species of a Group and Its Subgroups[1]

$\mathbf{O}_h$	$\mathbf{O}$	$\mathbf{T}_d$	$\mathbf{D}_{4h}$	$\mathbf{C}_{4v}$	$\mathbf{C}_{2v}$	$\mathbf{D}_3$	$\mathbf{D}_{2d}$
A_{1g}	A_1	A_1	A_{1g}	A_1	A_1	A_1	A_1
A_{2g}	A_2	A_2	B_{1g}	B_1	A_2	A_2	B_1
E_g	E	E	$A_{1g} + B_{1g}$	$A_1 + B_1$	$A_1 + A_2$	E	$A_1 + B_1$
T_{1g}	T_1	T_1	$A_{2g} + E_g$	$A_2 + E$	$A_2 + B_1 + B_2$	$A_2 + E$	$A_2 + E$
T_{2g}	T_2	T_2	$B_{2g} + E_g$	$B_2 + E$	$A_1 + B_1 + B_2$	$A_1 + E$	$B_2 + E$
A_{1u}	A_1	A_2	A_{1u}	A_2	A_2	A_1	B_1
A_{2u}	A_2	A_1	B_{1u}	B_2	A_1	A_2	A_1
E_u	E	E	$A_{1u} + B_{1u}$	$A_2 + B_2$	$A_1 + A_2$	E	$A_1 + B_1$
T_{1u}	T_1	T_2	$A_{2u} + E_u$	$A_1 + E$	$A_1 + B_1 + B_2$	$A_2 + E$	$B_2 + E$
T_{2u}	T_2	T_1	$B_{2u} + E_u$	$B_1 + E$	$A_2 + B_1 + B_2$	$A_1 + E$	$A_2 + E$

$\mathbf{T}_d$	$\mathbf{T}$	$\mathbf{D}_{2d}$	$\mathbf{C}_{2v}$	$\mathbf{S}_4$	$\mathbf{D}_2$	$\mathbf{C}_{2v}$	$\mathbf{C}_3$	$\mathbf{C}_2$
A_1	A	A_1	A_1	A	A	A_1	A	A
A_2	A	B_1	A_2	B	A	A_2	A	A
E	E	$A_1 + B_1$	E	$A + B$	$2A$	$A_1 + A_2$	E	$2A$
T_1	T	$A_2 + E$	$A_2 + E$	$A + E$	$B_1 + B_2 + B_3$	$A_2 + B_1 + B_2$	$A + E$	$A + 2B$
T_2	T	$B_2 + E$	$A_1 + E$	$B + E$	$B_1 + B_2 + B_3$	$A_1 + B_1 + B_2$	$A + E$	$A + 2B$

[1] Adapted with permission from more extensive tables in E. B. Wilson, Jr., J. C. Decius, and P. C. Cross, *Molecular Vibrations*, Dover, New York, 1980. Where necessary the symmetry element relating the groups is identified. In some cases more than one choice is shown. The correlation can be made in steps (e.g., $\mathbf{O}_h \rightarrow \mathbf{D}_{4h} \rightarrow \mathbf{C}_{4h}$).

$\mathbf{D}_{4h}$	$\mathbf{D}_4$	$C_2' \to C_2'$ $\mathbf{D}_{2d}$	$\mathbf{C}_{4v}$	$\mathbf{C}_{4h}$	C_2' $\mathbf{D}_{2h}$	C_2'' $\mathbf{D}_{2h}$	$\mathbf{C}_4$	$\mathbf{S}_4$	C_2' $\mathbf{D}_2$
A_{1g}	A_1	A_1	A_1	A_g	A_g	A_g	A	A	A
A_{2g}	A_2	A_2	A_2	A_g	B_{1g}	B_{1g}	A	A	B_1
B_{1g}	B_1	B_1	B_1	B_g	A_g	B_{1g}	B	B	A
B_{2g}	B_2	B_2	B_2	B_g	B_{1g}	A_g	B	B	B_1
E_g	E	E	E	E_g	$B_{2g} + B_{3g}$	$B_{2g} + B_{3g}$	E	E	$B_2 + B_3$
A_{1u}	A_1	B_1	A_2	A_u	A_u	A_u	A	B	A
A_{2u}	A_2	B_2	A_1	A_u	B_{1u}	B_{1u}	A	B	B_1
B_{1u}	B_1	A_1	B_2	B_u	A_u	B_{1u}	B	A	A
B_{2u}	B_2	A_2	B_1	B_u	B_{1u}	A_u	B	A	B_1
E_u	E	E	E	E_u	$B_{2u} + B_{3u}$	$B_{2u} + B_{3u}$	E	E	$B_2 + B_3$

$\mathbf{D}_{4h}$ (cont.)	C_2, σ_v $\mathbf{C}_{2v}$	C_2, σ_d $\mathbf{C}_{2v}$	C_2 $\mathbf{C}_{2h}$	C_2' $\mathbf{C}_{2h}$	C_2'' $\mathbf{C}_{2h}$	C_2 $\mathbf{C}_2$	C_2' $\mathbf{C}_2$	σ_h $\mathbf{C}_s$	σ_v $\mathbf{C}_s$	$\mathbf{C}_i$
A_{1g}	A_1	A_1	A_g	A_g	A_g	A	A	A'	A'	A_g
A_{2g}	A_2	A_2	A_g	B_g	B_g	A	B	A'	A''	A_g
B_{1g}	A_1	A_2	A_g	A_g	B_g	A	A	A'	A'	A_g
B_{2g}	A_2	A_1	A_g	B_g	A_g	A	B	A'	A''	A_g
E_g	$B_1 + B_2$	$B_1 + B_2$	$2B_g$	$A_g + B_g$	$A_g + B_g$	$2B$	$A + B$	$2A''$	$A' + A''$	$2A_g$
A_{1u}	A_2	A_2	A_u	A_u	A_u	A	A	A''	A''	A_u
A_{2u}	A_1	A_1	A_u	B_u	B_u	A	B	A''	A'	A_u
B_{1u}	A_2	A_1	A_u	A_u	B_u	A	A	A''	A''	A_u
B_{2u}	A_1	A_2	A_u	B_u	A_u	A	B	A''	A'	A_u
E_u	$B_1 + B_2$	$B_1 + B_2$	$2B_u$	$A_u + B_u$	$A_u + B_u$	$2B$	$A + B$	$2A'$	$A' + A''$	$2A_u$

$\mathbf{D}_{2h}$	$\mathbf{D}_2$	$C_2(z)$ $\mathbf{C}_{2v}$	$C_2(y)$ $\mathbf{C}_{2v}$	$C_2(x)$ $\mathbf{C}_{2v}$	$C_2(z)$ $\mathbf{C}_{2h}$	$C_2(y)$ $\mathbf{C}_{2h}$	$C_2(x)$ $\mathbf{C}_{2h}$	$C_2(z)$ $\mathbf{C}_2$	$C_2(y)$ $\mathbf{C}_2$	$C_2(x)$ $\mathbf{C}_2$	$\sigma(xy)$ $\mathbf{C}_s$	$\sigma(yz)$ $\mathbf{C}_s$
A_g	A	A_1	A_1	A_1	A_g	A_g	A_g	A	A	A	A'	A'
B_{1g}	B_1	A_2	B_2	B_1	A_g	B_g	B_g	A	B	B	A'	A''
B_{2g}	B_2	B_1	A_2	B_2	B_g	A_g	B_g	B	A	B	A''	A''
B_{3g}	B_3	B_2	B_1	A_2	B_g	B_g	A_g	B	B	A	A''	A'
A_u	A	A_2	A_2	A_2	A_u	A_u	A_u	A	A	A	A''	A''
B_{1u}	B_1	A_1	B_1	B_2	A_u	B_u	B_u	A	B	B	A''	A'
B_{2u}	B_2	B_2	A_1	B_1	B_u	A_u	B_u	B	A	B	A'	A'
B_{3u}	B_3	B_1	B_2	A_1	B_u	B_u	A_u	B	B	A	A'	A''

$\mathbf{D}_{4d}$	$\mathbf{D}_4$	$\mathbf{C}_{4v}$	$\mathbf{S}_8$	$\mathbf{C}_4$	$\mathbf{C}_{2v}$	C_2 $\mathbf{C}_2$	C_2' $\mathbf{C}_2$	$\mathbf{C}_s$
A_1	A_1	A_1	A	A	A_1	A	A	A'
A_2	A_2	A_2	A	A	A_2	A	B	A''
B_1	A_1	A_2	B	A	A_2	A	A	A''
B_2	A_2	A_1	B	A	A_1	A	B	A'
E_1	E	E	E_1	E	$B_1 + B_2$	$2B$	$A + B$	$A' + A''$
E_2	$B_1 + B_2$	$B_1 + B_2$	E_2	$2B$	$A_1 + A_2$	$2A$	$A + B$	$A' + A''$
E_3	E	E	E_3	E	$B_1 + B_2$	$2B$	$A + B$	$A' + A''$

$\mathbf{D}_{2d}$	$\mathbf{S}_4$	$C_2 \rightarrow C_2(z)$ $\mathbf{D}_2$	$\mathbf{C}_{2v}$	C_2 $\mathbf{C}_2$	C_2' $\mathbf{C}_2$	$\mathbf{C}_s$
A_1	A	A	A_1	A	A	A'
A_2	A	B_1	A_2	A	B	A''
B_1	B	A	A_2	A	A	A''
B_2	B	B_1	A_1	A	B	A'
E	E	$B_2 + B_3$	$B_1 + B_2$	$2B$	$A + B$	$A' + A''$

$\mathbf{D}_{3d}$	$\mathbf{D}_3$	$\mathbf{C}_{3v}$	$\mathbf{S}_6$	$\mathbf{C}_3$	$\mathbf{C}_{2h}$	$\mathbf{C}_2$	$\mathbf{C}_s$	$\mathbf{C}_i$
A_{1g}	A_1	A_1	A_g	A	A_g	A	A'	A_g
A_{2g}	A_2	A_2	A_g	A	B_g	B	A''	A_g
E_g	E	E	E_g	E	$A_g + B_g$	$A + B$	$A' + A''$	$2A_g$
A_{1u}	A_1	A_2	A_u	A	A_u	A	A''	A_u
A_{2u}	A_2	A_1	A_u	A	B_u	B	A'	A_u
E_u	E	E	E_u	E	$A_u + B_u$	$A + B$	$A' + A''$	$2A_u$

$\mathbf{D}_{3h}$	$\mathbf{C}_{3h}$	$\mathbf{D}_3$	$\mathbf{C}_{3v}$	$\sigma_h \rightarrow \sigma_v(zy)$ $\mathbf{C}_{2v}$	$\mathbf{C}_3$	$\mathbf{C}_2$	σ_h $\mathbf{C}_s$	σ_v $\mathbf{C}_s$
A_1'	A'	A_1	A_1	A_1	A	A	A'	A'
A_2'	A'	A_2	A_2	B_2	A	B	A'	A''
E'	E'	E	E	$A_1 + B_2$	E	$A + B$	$2A'$	$A' + A''$
A_1''	A''	A_1	A_2	A_2	A	A	A''	A''
A_2''	A''	A_2	A_1	B_1	A	B	A''	A'
E''	E''	E	E	$A_2 + B_1$	E	$A + B$	$2A''$	$A' + A''$

Appendix 4

Answers to Problems

Chapter 2

2.1 *Closure* is satisfied and addition is *associative* (i.e., $a + (b + c) = (a + b) + c$). The identity is zero. The inverse of an integer is its negative. All group postulates are satisfied. Addition is commutative.

2.2 σ_h combined with E, C_3, and C_3^2 gives positions shown in Fig. 2.1 (a) with open circles replacing solid centers. These represent new operations. The order is 12.

2.3 (a) $\sigma_h C_3^2$ or $C_3^2 \sigma_h$; (b) $\sigma_h C_3$ or $C_3 \sigma_h$ or S_3; (c) σ_h; (d) C_3^2; (e) σ_h; (f) σ_v''; (g) σ_v'.

2.4

	E	A	A^2	B	A^2B	AB
E	E	A	A^2	B	A^2B	AB
A	A	A^2	E	AB	B	A^2B
A^2	A^2	E	A	A^2B	AB	B
B	B	A^2B	AB	E	A	A^2
A^2B	A^2B	AB	B	A^2	E	A
AB	AB	B	A^2B	A	A^2	E

This table corresponds to that in Example 2.3c.

2.5 The group is cyclic of order four. The generator is i or $-i$. $(i)^4 = E$ or $(-i)^4 = E$.

2.6 σ_h plus one element from each of Figs. 2.1 (a), (b), (c).

2.7 The group (E, A, B, b', a', c') has the same multiplication table as $\mathbf{D}_3$ $(E, C_3, C_3^2, C_2, C_2', C_2'')$. One generator comes from the set (A, B) and one from the set $(b', a', c',)$. $A^3 = E$, $(b')^2 = E$, $b'Ab'A = E$.

2.8 $(abdh)(cef)$. Periods: $(abdh)$ 4, (cef) 3, $(abdh)(cef)$ 12.

2.9 Let a be an element and consider the set $(a, a^2, \ldots)$. Since these are all elements of the group, there can be only a finite number of different elements. Pick two equal elements nearest together, a^r and a^s, so that $a^s = a^r$ or $a^{(s-r)} = E$. Then $n\ (= s - r)$ is the period of a. The distinct elements are $(a, a^2, \ldots, a^{n-1}, E)$, a cyclic subgroup or the whole group.

2.10 (a) C_3; (b) C_3^2; (c) C_2''; (d) C_2.

2.11

$$\begin{aligned}(Y^{-1}X^{-1}\cdots C^{-1}B^{-1}A^{-1})&(ABC\cdots XY)\\ &= Y^{-1}X^{-1}\cdots C^{-1}B^{-1}(A^{-1}A)BC\cdots XY\\ &= Y^{-1}X^{-1}\cdots C^{-1}B^{-1}EBC\cdots XY\\ &= \cdots = Y^{-1}Y = E.\end{aligned}$$

2.12 (1) If $X^{-1}AX = X^{-1}BX$, then $XX^{-1}AXX^{-1} = XX^{-1}BXX^{-1}$ and $A = B$.

(2) Suppose A occurs in two classes and B occurs in only one of these. There must be an element Z such that $Z^{-1}AZ = B$ or $A = ZBZ^{-1}$. Substituting ZBZ^{-1} for A in the transforms of the class in which B is supposed not to occur gives $X^{-1}(ZBZ^{-1})X$. But $X = Z$ is included in these transforms, giving $Z^{-1}ZBZ^{-1}Z = B$, a contradiction. The "two" classes are the same class.

(3) Let s be the number of times that A is transformed into itself: $X_i^{-1}AX_i = A\ (i = 1, \ldots, s)$. Let Y be an element that transforms A into B: $Y^{-1}AY = B$; $Y^{-1}X_i^{-1}AX_iY = B\ (i = 1, \ldots, s)$. Thus the s elements YX_i transform A into B. Substituting $A = YBY^{-1}$ for A in the last equation gives $Y^{-1}X_i^{-1}YBY^{-1}X_iY = B$. The s elements $Y^{-1}X_iY$ transform B into itself.

Each element of a class is transformed into itself and into every other element of the class s times. Since the number of transforms of any element is h, the number of elements in a class must be h/s.

2.13 The identity is $X^{-1}EX$. The inverse of $X^{-1}QX$ is $X^{-1}Q^{-1}X$. If P and Q are elements of H then $(X^{-1}PX)(X^{-1}QX) = X^{-1}PQX$ is an element of $X^{-1}HX$ because PQ belongs to H.

2.14 The group D_2, of order four, with each element its own inverse, is isomorphic to the group with table (2.4). The relations are $A^2 = E$, $B^2 = E$, and $ABAB = E$.

2.15 $S_2 = \sigma_h C_2 = iC_2C_2 = i$

2.16 $\sigma_h C_6(z) = iC_2(z)C_6(z) = iC_3^2(z)$

2.17 (a) Even n: $S_n = \sigma_h C_n$; $S_n^2 = \sigma_h^2 C_n^2 = C_n^2$; $S_n^3 = \sigma_h^3 C_n^3 = \sigma_h C_n^3$; $S_n^k = \sigma_h C_n^k$ (k odd) and C_n^k (k even); $S_n^n = C_n^n = E$. The group is cyclic of order n, and $\mathbf{S}_n \sim \mathbf{C}_n$.

(b) Odd n: $S_n^n = \sigma_h C_n^n = \sigma_h$; $S_n^{2n} = C_n^{2n} = E$. All operators of $\mathbf{C}_{nh}$ are given by the $2n$ powers of S_n. $\mathbf{C}_{nh}$ is cyclic of order $2n$. $\mathbf{C}_{nh} \sim \mathbf{S}_{2n}$.

2.18 $\mathbf{D}_4$, $\mathbf{C}_{4h}$, $\mathbf{C}_4$, $\mathbf{S}_4$, $\mathbf{C}_{4v}$, $\mathbf{D}_{2h}$, $\mathbf{D}_2$, $\mathbf{C}_2$, $\mathbf{C}_{2v}$, $\mathbf{C}_{2h}$, $\mathbf{C}_i$, $\mathbf{C}_s$, $\mathbf{C}_1$

2.19 Same: c, e, g.

2.20 (a) $\mathbf{C}_{3h}$; (b) and (c) $\mathbf{S}_6$; (d) and (e) $\mathbf{C}_{4h}$.

2.21 (a) $\mathbf{O}$; (b) $\mathbf{D}_3$; (c), (d), and (e) $\mathbf{D}_4$.

2.22 See B. E. Douglas, D. H. McDaniel, and J. J. Alexander, *Problems for Inorganic Chemistry*, Wiley, 1983, Problems 3.7 and 3.8.

2.24 (a) $\mathbf{C}_{\infty v}$; (b) $\mathbf{C}_{2v}$; (c) $\mathbf{D}_{\infty h}$; (d) $\mathbf{D}_{3h}$; (e) $\mathbf{C}_{3v}$; (f) and (g) $\mathbf{C}_{2v}$; (h) $\mathbf{D}_{\infty h}$; (i) $\mathbf{D}_{4h}$; (j) $\mathbf{D}_{5h}$.

2.25 (a) $\mathbf{D}_{3d}$; (b) $\mathbf{D}_{3h}$; (c) and (d) $\mathbf{D}_{2d}$; (e) $\mathbf{O}_h$; (f) $\mathbf{C}_{2h}$; (g) and (h) $\mathbf{C}_{2v}$; (i) $\mathbf{C}_{2v}$; (j) $\mathbf{D}_{3d}$; (k) $\mathbf{D}_{4h}$; (l) $\mathbf{D}_{3h}$.

2.26 (a), (b) and (d) $\mathbf{D}_{3h}$; (c) $\mathbf{D}_3$.

2.27 (a) $\mathbf{D}_{\infty h}$; (b) $\mathbf{C}_{\infty}$; (c) $\mathbf{C}_{\infty v}$.

2.28 (a) $\mathbf{D}_{3h}$; (b) $\mathbf{D}_{6h}$; (c) $\mathbf{D}_{4d}$; (d) $\mathbf{D}_{8h}$; (e) $\mathbf{D}_{2d}$; (f) $\mathbf{D}_{2h}$.

2.29 (a) $\mathbf{D}_3$; (b) $\mathbf{C}_{2v}$; (c) $\mathbf{D}_{4h}$; (d) $\mathbf{C}_2$; (e) $\mathbf{D}_{2h}$; (f) $\mathbf{C}_{2v}$; (g) $\mathbf{C}_2$; (h) $\mathbf{C}_1$; (i) $\mathbf{C}_{3v}$; (j) $\mathbf{C}_{2v}$; (k) $\mathbf{C}_s$, $\mathbf{C}_1$; (l) $\mathbf{C}_{2v}$; (m) $\mathbf{C}_3$; (n) $\mathbf{C}_1$; (o) $\mathbf{D}_4$.

2.30 (a), (b), and (c) $\mathbf{T}_d$; (d) $\mathbf{D}_{2h}$; (e) $\mathbf{C}_{2v}$; (f) $\mathbf{C}_{4v}$; (g) $\mathbf{D}_{2d}$; (h) and (i) $\mathbf{D}_{4h}$; (j) and (k) $\mathbf{D}_{3h}$; (l) and (m) $\mathbf{O}_h$; (n) $\mathbf{D}_{3h}$; (o) $\mathbf{D}_{4d}$; (p) $\mathbf{D}_{3h}$; (q) $\mathbf{T}_d$.

2.31 (a) and (b) both ions $\mathbf{O}_h$; (c) and (d) both ions $\mathbf{T}_d$; (e) Ca $\mathbf{O}_h$, F $\mathbf{T}_d$; (f) Ni $\mathbf{O}_h$, As $\mathbf{D}_{3h}$; (g) Pt approximately $\mathbf{D}_{4h}$, but $\mathbf{D}_{2h}$ considering sharing of S along only one direction, S $\mathbf{T}_d$; (h) $\mathbf{O}_h$; (i) $\mathbf{D}_{3h}$.

2.32 $\mathbf{O}_h$ for both.

2.33 (a) $\mathbf{C}_3$; (b) *ccp* $4C_3$, *hcp* $1C_3$; (c) slip planes along four directions for *ccp*.

Chapter 3

3.1 The evaluation of the third-order determinants goes as follows:

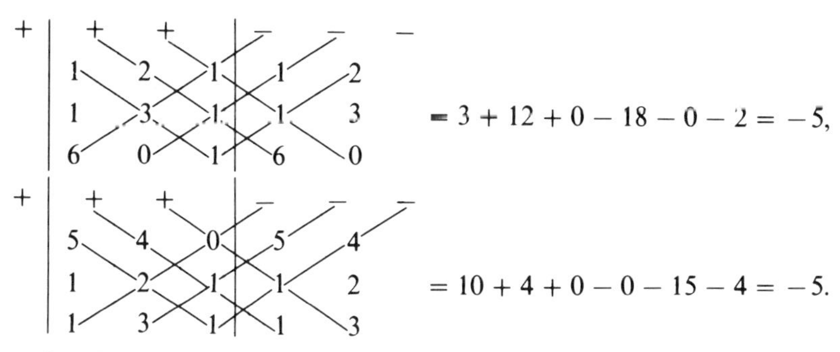

Therefore, $D = 2(-5) - 3(-5) = 5$.

3.2 The number 5 is factored from row two. Row two is multiplied by (-1) and added to row one.

3.3 Interchange rows two and four to obtain (Properties 3A.3 and 3A.10)

$$\begin{vmatrix} a & b & 0 & 0 & 0 \\ x & y & 0 & 0 & 0 \\ 2 & 5 & l & m & n \\ 1 & 4 & p & q & r \\ 7 & 6 & u & v & w \end{vmatrix} = (-1)\begin{vmatrix} a & b \\ x & y \end{vmatrix}\begin{vmatrix} l & m & n \\ p & q & r \\ u & v & w \end{vmatrix}.$$

3.4 The matrix for a two-fold rotation about the z axis is obtained by taking $\theta = \pi$ in Eq. (3.37). The result is

$$R_z(\pi) = \begin{vmatrix} -1 & 0 & 0 \\ 0 & -1 & 0 \\ 0 & 0 & 1 \end{vmatrix}.$$

Combination of this matrix with the inversion matrix gives Eq. (3.41).

3.5 (a) The result of the transformation $\mathbf{\Phi}$ on the vector (u_x, u_y, u_z) is (u_x', u_y', u_z'), where

$$u_x' = \phi_{xx}u_x + \phi_{xy}u_y + \phi_{xz}u_z,$$
$$u_y' = \phi_{yx}u_x + \phi_{yy}u_y + \phi_{yz}u_z,$$
$$u_z' = \phi_{zx}u_x + \phi_{zy}u_y + \phi_{zz}u_z,$$
$$u_x' = (1 - \cos\theta)(u_x^3 + u_xu_y^2 + u_xu_z^2) + u_x\cos\theta + u_yu_z\sin\theta - u_yu_z\sin\theta$$
$$= (1 - \cos\theta)u_x(u_x^2 + u_y^2 + u_z^2) + u_x\cos\theta.$$

But, $u_x^2 + u_y^2 + u_z^2 = 1$. Therefore, $u_x' = u_x$. Likewise, $u_y' = u_y$ and $u_z' = u_z$. It follows that since $\mathbf{u} = (u_x, u_y, u_z)$ is unchanged by the rotation, it is in the direction of the axis.

(b) For rotation about the z axis we have $u = (0, 0, 1)$ and therefore,

$$\phi_{xx} = \cos\theta, \qquad \phi_{xy} = \sin\theta, \qquad \phi_{xz} = 0,$$
$$\phi_{yx} = -\sin\theta, \qquad \phi_{yy} = \cos\theta, \qquad \phi_{yz} = 0,$$
$$\phi_{zx} = 0, \qquad \phi_{zy} = 0, \qquad \phi_{zz} = 1,$$

and

$$\mathbf{\Phi} = \begin{bmatrix} \cos\theta & \sin\theta & 0 \\ -\sin\theta & \cos\theta & 0 \\ 0 & 0 & 1 \end{bmatrix}$$

(c) The reflection is $(-1)\mathbf{\Phi}$ with $\theta = 180°$, $\cos\theta = -1$, or

$$-\phi_{xx} = -(2u_x^2 - 1); \qquad -\phi_{xy} = -2u_xu_y; \qquad -\phi_{xz} = -2u_xu_z;$$
$$-\phi_{yx} = -2u_xu_y; \qquad -\phi_{yy} = -(2u_y^2 - 1); \qquad -\phi_{yz} = -2u_yu_z;$$
$$-\phi_{zx} = -2u_xu_z; \qquad -\phi_{zy} = -2u_yu_z; \qquad -\phi_{zz} = -(2u_z^2 - 1).$$

(d) In this case $u_x = -\sin\phi$, $u_y = \cos\phi$, and $u_z = 0$, since **u** is in the xy plane and is perpendicular to the reflection plane. From Part (c) we have

$$-\phi_{xx} = -(2\sin^2\phi - 1) = \cos 2\phi; \quad -\phi_{xy} = 2\sin\phi\cos\phi = \sin 2\phi; \quad -\phi_{xz} = 0;$$
$$-\phi_{yz} = \sin 2\phi; \quad -\phi_{yy} = -(2\cos^2\phi - 1) = -\cos 2\phi; \quad -\phi_{yz} = 0;$$
$$-\phi_{zx} = 0; \quad -\phi_{zy} = 0; \quad -\phi_{zz} = 1.$$

(e) This follows directly from Part (a) because $\mathbf{\Phi u} = \mathbf{u}$ and $(-\mathbf{\Phi})\mathbf{u} = -\mathbf{u}$.

(f) $\operatorname{tr}\mathbf{\Phi} = \phi_{xx} + \phi_{yy} + \phi_{zz} = (1 - \cos\theta)(u_x^2 + u_y^2 + u_z^2) + 3\cos\theta = 1 + 2\cos\theta$.

3.6 We have

$$\begin{aligned} z_1' &= x_1'y_1' = (a_{11}x_1 + a_{12}x_2)(b_{11}y_1 + b_{12}y_2 + b_{13}y_3), \\ z_2' &= x_1'y_2' = (a_{11}x_1 + a_{12}x_2)(b_{21}y_1 + b_{22}y_2 + b_{23}y_3), \\ &\vdots \\ z_6' &= x_2'y_3' = (a_{21}x_1 + a_{22}x_2)(b_{31}y_1 + b_{32}y_2 + b_{33}y_3). \end{aligned}$$

This gives

$$\begin{aligned} z_1' &= a_{11}b_{11}z_1 + a_{11}b_{12}z_2 + a_{11}b_{13}z_3 + a_{12}b_{11}z_4 + a_{12}b_{12}z_5 \\ &\quad + a_{12}b_{13}z_6, \\ z_2' &= a_{11}b_{21}z_1 + a_{11}b_{22}z_2 + a_{11}b_{23}z_3 + a_{12}b_{21}z_4 + a_{12}b_{22}z_5 \\ &\quad + a_{12}b_{23}z_6, \\ &\vdots \\ z_6' &= a_{21}b_{31}z_1 + a_{21}b_{32}z_2 + a_{21}b_{33}z_3 + a_{22}b_{31}z_4 + a_{22}b_{32}z_5 \\ &\quad + a_{22}b_{33}z_6. \end{aligned}$$

The matrix for this transformation is exactly that given in Equation (3.31).

3.7 Let $\left[\begin{smallmatrix} a' & b' \\ b' & a' \end{smallmatrix}\right]$ be a second matrix of the same form. Then

$$\begin{bmatrix} a & b \\ b & a \end{bmatrix}\begin{bmatrix} a' & b' \\ b' & a' \end{bmatrix} = \begin{bmatrix} aa' + bb' & ab' + ba' \\ ba' + ab' & bb' + aa' \end{bmatrix},$$
$$\begin{bmatrix} a' & b' \\ b' & a' \end{bmatrix}\begin{bmatrix} a & b \\ b & a \end{bmatrix} = \begin{bmatrix} a'a + b'b & a'b + b'a \\ b'a + a'b & b'b + a'a \end{bmatrix}.$$

Since a, a', b, and b' are scalars, the two product matrices have all their elements equal.

3.8 We have (see Example 3.16)

$$P_{E,22}F = \tfrac{2}{6}[E - \tfrac{1}{2}C_3 - \tfrac{1}{2}C_3^2 + \sigma_v - \tfrac{1}{2}\sigma_v' - \tfrac{1}{2}\sigma_v'']F = by.$$

The partner of by can be obtained as follows:

$$T_{E,21}by = \frac{2}{6}\left[-\frac{\sqrt{3}}{2}C_3 + \frac{\sqrt{3}}{2}C_3^2 - \frac{\sqrt{3}}{2}\sigma_v' + \frac{\sqrt{3}}{2}\sigma_v''\right]by = bx.$$

3.9

$$\int \Psi_i^* \Psi_j \, d\tau = \int \left(\sum_k a_{ik}^* \phi_k^* \right) \left(\sum_l a_{jl} \phi_l \right) d\tau = \sum_{k,l} \left| a_{ik}^* a_{jl} \int \phi_k^* \phi_l \, d\tau \right|$$
$$= \sum_{k,l} a_{ik}^* a_{jl} \delta_{kl} = \sum_k a_{ik}^* a_{jk}.$$

If $\int \Psi_i^* \Psi_j \, d\tau = 0$ for $i \neq j$, we have $\sum_k a_{ik}^* a_{jk} = \delta_{ij}$.

This means that the rows of $[a_{ij}]$ from an orthonormal set so that $[a_{ij}]$ is unitary. On the other hand, if $[a_{ij}]$ is unitary, then

$$\sum_k a_{ik}^* a_{jk} = \delta_{ij} \qquad \text{and} \qquad \int \Psi_i^* \Psi_j \, d\tau = \delta_{ij}.$$

3.10 Let $\sigma_v{}^{\theta_0}$ represent the vertical plane at angle $\theta = \theta_0$. Reflection through this plane has the following effects:

$$\theta \xrightarrow{\sigma_v{}^{\theta_0}} \theta_0 + \theta_0 - \theta = 2\theta_0 - \theta,$$
$$u_k(\theta) \xrightarrow{\sigma_v{}^{\theta_0}} e^{-i2k\theta_0} u_k(-\theta) = e^{-i2k\theta_0} u_k(\theta)^*.$$

The matrix representations are

$$\sigma_v{}^{\theta_0} = \begin{bmatrix} 0 & e^{-i2k\theta_0} \\ e^{i2k\theta_0} & 0 \end{bmatrix} \qquad \text{and} \qquad C_\infty^\phi = \begin{bmatrix} e^{ik\phi} & 0 \\ 0 & e^{-ik\phi} \end{bmatrix}$$

The characters are zero for $\sigma_v{}^{\theta_0}$ and $2 \cos k\phi$ for C_∞^ϕ. These are the characters for E_k.

3.11 This follows directly from the results obtained in Problem 3.10.

3.12 We take the axis of rotation to be the z axis.

$$Y_{lm}(\theta, \phi) = P_{lm}(\cos \theta) e^{im\phi}, \qquad m = 0, \pm 1, \ldots, \pm l,$$
$$C_\infty^\omega Y_{lm} = e^{im\omega} Y_{lm}.$$

Equation (3.126) follows directly.

3.13 From Eq. (3.126) we have

$$\chi^{21+1}(\omega) = \sum_{m=-l}^{l} e^{im\omega}.$$

Letting $x = e^{i\omega}$, we obtain

$$\chi^{2l+1}(\omega) = 1 + x + x^2 + \cdots + x^l + x^{-1} + \cdots + x^{-l}$$
$$= x^{-l}(1 + x + \cdots + x^{2l})$$
$$= \frac{x^{-l}(1 - x^{2l+1})}{1 - x}$$
$$= \frac{x^{1/2}(x^{-(l+1/2)} - x^{l+1/2})}{1 - x} = \frac{x^{-(l+1/2)} - x^{l+1/2}}{x^{-1/2} - x^{1/2}}$$
$$= \frac{\exp[-(l + \frac{1}{2})i\omega] - \exp[(l + \frac{1}{2})i\omega]}{\exp(-\frac{1}{2}i\omega) - \exp(\frac{1}{2}i\omega)} = \frac{\sin[(l + \frac{1}{2})\omega}{\sin \frac{1}{2}\omega}.$$

3.14 Let $\boldsymbol{\Psi}$ be one of the common eigenvectors; then we have

$$\mathbf{A}\boldsymbol{\Psi} = a\boldsymbol{\Psi} \qquad \text{and} \qquad \mathbf{B}\boldsymbol{\Psi} = b\boldsymbol{\Psi},$$

where a and b are the eigenvalues. Operating on the first equation with **B** and on the second with **A** gives

$$\mathbf{BA}\boldsymbol{\Psi} = a\mathbf{B}\boldsymbol{\Psi} = ab\boldsymbol{\Psi} \qquad \text{and} \qquad \mathbf{AB}\boldsymbol{\Psi} = b\mathbf{A}\boldsymbol{\Psi} = ba\boldsymbol{\Psi} = ab\boldsymbol{\Psi}.$$

Thus, when operating on $\boldsymbol{\Psi}$, **A** and **B** commute.

3.15 From Eq. (3.136) we have

$$[\boldsymbol{\Theta}, \mathbf{t}][\boldsymbol{\Theta}^{-1}, -\boldsymbol{\Theta}^{-1}\mathbf{t}] = [\boldsymbol{\Theta}^{-1}\boldsymbol{\Theta}, \boldsymbol{\Theta}^{-1}\mathbf{t} + (-\boldsymbol{\Theta}^{-1}\mathbf{t})] = [\mathbf{I}, \mathbf{0}].$$

3.16 Suppose that $\boldsymbol{\Theta}$ is a rotation, and its axis is $\mathbf{t}_a$, $\mathbf{t}_a = \alpha\tau_1 + \beta\tau_2 + \gamma\tau_3$, where α, β, and γ are constants. $\mathbf{t}_a$ is invariant under $\boldsymbol{\Theta}$, $\boldsymbol{\Theta}\mathbf{t}_a = \mathbf{t}_a$. (We can say that the axis is an eigenvector with eigenvalue unity.) Written out, this equation becomes

$$\begin{aligned}(\theta_{11} - 1)\alpha + \theta_{12}\beta + \theta_{13}\gamma &= 0,\\ \theta_{21}\alpha + (\theta_{22} - 1)\beta + \theta_{23}\gamma &= 0,\\ \theta_{31}\alpha + \theta_{32}\beta + (\theta_{33} - 1)\gamma &= 0.\end{aligned}$$

We can solve these equations for α/γ and β/γ by the use of Cramer's rule (Chapter 3, Appendix 3A.12) on the first two equations.

$$\frac{\alpha}{\gamma} = -\frac{\begin{vmatrix}\theta_{13} & \theta_{12}\\ \theta_{23} & (\theta_{22} - 1)\end{vmatrix}}{\begin{vmatrix}(\theta_{11} - 1) & \theta_{12}\\ \theta_{21} & \theta_{22} - 1\end{vmatrix}} \qquad \text{and} \qquad \frac{\beta}{\gamma} = -\frac{\begin{vmatrix}(\theta_{11} - 1) & \theta_{13}\\ \theta_{21} & \theta_{23}\end{vmatrix}}{\begin{vmatrix}(\theta_{11} - 1) & \theta_{12}\\ \theta_{21} & (\theta_{22} - 1)\end{vmatrix}}.$$

The right-hand sides of these equations are ratios of integers. Therefore, α/γ and β/γ are ratios of integers. This means that $\mathbf{t}_a$ must have the same form as Eq. (3.128).

3.17 Let the two-fold rotation be about the x axis. The matrices are

$$R_z = \begin{bmatrix}\cos\theta & \sin\theta & 0\\ -\sin\theta & \cos\theta & 0\\ 0 & 0 & 1\end{bmatrix} \qquad \text{and} \qquad C_2 = C_2^{-1} = \begin{bmatrix}1 & 0 & 0\\ 0 & -1 & 0\\ 0 & 0 & -1\end{bmatrix}.$$

The similarity transformation is $C_2^{-1} R_z C_2$. We use the blocked forms for multiplication

$$\begin{bmatrix}1 & 0\\ 0 & -1\end{bmatrix}\begin{bmatrix}\cos\theta & \sin\theta\\ -\sin\theta & \cos\theta\end{bmatrix}\begin{bmatrix}1 & 0\\ 0 & -1\end{bmatrix}$$

$$= \begin{bmatrix}1 & 0\\ 0 & -1\end{bmatrix}\begin{bmatrix}\cos\theta & -\sin\theta\\ -\sin\theta & -\cos\theta\end{bmatrix} = \begin{bmatrix}\cos\theta & -\sin\theta\\ \sin\theta & \cos\theta\end{bmatrix}$$

and $[-1][1][-1] = [1]$.

These give the result

$$\begin{bmatrix} \cos\theta & -\sin\theta & 0 \\ \sin\theta & \cos\theta & 0 \\ 0 & 0 & 1 \end{bmatrix},$$

which is the same as one obtains by changing the sign of θ in R_z. It is, therefore, R_z^{-1}.

3.18 $\mathbf{C}_{4v}$ p_z A_1, (p_x, p_y) E, d_{z^2} A_1, $d_{x^2-y^2}$ B_1, d_{xy} B_2, (d_{xz}, d_{yz}) E. The pairs (p_x and p_y) and (d_{xz}, d_{yz}) are interchanged (mixed) by C_4 so they belong to E. Note that the choice of σ_v and σ_d makes the distinction between the B_1 and B_2 representations. $\mathbf{D}_{3h}$ p_z A_2'', (p_x, p_y) E', d_{z^2} A_1', $(d_{x^2-y^2}, d_{xy})$ E', (d_{xz}, d_{xy}) E''. C_3 mixes the pairs of E representations, but for only (d_{xz}, d_{yz}) does σ_h change the sign, making the representation E''.

3.19

$\mathbf{D}_{4h}$	E	C_4	C_2	$(C_2')_x$	C_2''	i	S_4	σ_h	$(\sigma_v)_x$	σ_d	
x'	x	$-y$	$-x$	x	y	$-x$	$-y$	x	x	y	
y'	y	x	$-y$	$-y$	x	$-y$	x	y	$-y$	x	
z'	z	z	z	$-z$	$-z$	$-z$	$-z$	$-z$	z	z	
χ_{xy}	1	-1	1	-1	1	1	-1	1	-1	1	B_{2g}
χ_{xyz}	1	-1	1	1	-1	-1	1	1	-1	1	B_{1u}
$\chi_{x^2-y^2}$	1	-1	1	1	-1	1	-1	1	1	-1	B_{1g}
$\chi_{z(x^2-y^2)}$	1	-1	1	-1	1	-1	1	-1	1	-1	B_{2u}

We see that the xz function is interchanged with yz by some operations. Since x and y are interchanged, they must be considered together, so xz and yz belong to an E representation. These centrosymmetric d orbitals belong to E_g.

3.20 Any C_n along z other than C_1 and C_2 mix x and y, resulting in an E representation. The C_3 axes not coincident with x, y, or z of the cubic and icosahedral groups mix x, y, and z, resulting in a T representation. A C_4 axis along z mixes p_x and p_y, but not $d_{x^2-y^2}$ and d_{xy}. These orbitals are doubly degenerate under $\mathbf{D}_{4d}$ because of the S_8 axis.

Chapter 4

4.1 We have

$$H\psi_m = E_m\psi_m, \qquad \int \psi_{m'}^* H\psi_m \, d\tau = E_m \int \psi_{m'}^* \psi_m \, d\tau;$$

$$H_{m'm} = E_m S_{m'm}, \qquad \int \psi_m^* H\psi_{m'} \, d\tau = E_{m'} \int \psi_m^* \psi_{m'} \, d\tau;$$

$$H_{mm'} = E_{m'} S_{mm'}, \qquad H_{mm'}^* = E_m^* S_{mm'}^*.$$

Thus, we obtain

$$H_{m'm} - H^*_{mm'} = E_m S_{m'm} - E^*_{m'} S^*_{mm'}.$$

But $S^*_{m'm} = S_{mm'}$, and because H is Hermitian, $H_{m'm} - H^*_{mm'} = 0$. Therefore, $0 = (E_m - E^*_{m'})S_{m'm}$. Since $(E_m - E^*_{m'}) \neq 0$, we must have $S_{m'm} = \int \psi^*_{m'}\psi_m \,\mathrm{d}\tau = 0$.

4.2 In the solution to Problem 4.1 take $m = m'$. Then we obtain $(E_m - E^*_m)S_{mm} = 0$. Since $S_{mm} \neq 0$, we must have $E_m - E^*_m = 0$, or E_m is real.

4.3 (a) We have given that

$$H\psi_1 = E\psi_1 \qquad \text{and} \qquad H\psi_2 = E\psi_2.$$

Multiply these equations by the arbitrary constants a and b, respectively, and add the two resulting equations to obtain

$$aH\psi_1 + bH\psi_2 = aE\psi_1 + bE\psi_2.$$

Because H is linear, this is the same as

$$H(a\psi_1 + b\psi_2) = E(a\psi_1 + b\psi_2) \qquad \text{or} \qquad H\psi = E\psi,$$

where ψ is the arbitrary linear combination $\psi = a\psi_1 + b\psi_2$ and is an eigenfunction of H with eigenvalue E.

(b) $SHS^{-1} = H$. Multiplying both sides of this equation on the right with S gives $SH = HS$.

(c) Let the degeneracy of the level E be g, and let the set $\psi_i \, (i = 1, \ldots, g)$ be a set of the linearly independent eigenfunctions with energy E. Let S be a symmetry operator for H. We have $H\psi_i = E\psi_i \; (i = 1, \ldots g)$. Operating from the left on both sides of this equation with S gives $SH\psi_i = SE\psi_i$ But by the result of Part (b) we have $HS\psi_i = ES\psi_i$ or $H\psi'_i = E\psi'_i$, where $\psi'_i = S\psi_i$. ψ'_i is an eigenfunction with eigenvalue E. Since ψ_i $(i = 1, \ldots, g)$ are independent, $S\psi_i$ must be a linear combination of the ψ_i

$$S\psi_i = \sum_{j=1}^{g} a_{ij}\psi_j, \qquad i = 1, \ldots, g.$$

The matrix $[a_{ij}]$ is a g-dimensional representation of S.

(d) Suppose that $\mathbf{A}$ and $\mathbf{B}$ are two matrices that commute with $\mathbf{H}$; then $\mathbf{AH} = \mathbf{HA}$ and $\mathbf{AHA}^{-1} = \mathbf{H}$. Likewise, $\mathbf{BHB}^{-1} = \mathbf{H}$.

Multiplying both sides of this last equation on the right with $\mathbf{A}^{-1}$ and on the left with $\mathbf{A}$, we obtain $\mathbf{ABHB}^{-1}\mathbf{A}^{-1} = \mathbf{AHA}^{-1}$ or $\mathbf{(AB)H(AB)}^{-1} = \mathbf{H}$.

This establishes the *closure* property.

The *associative* law holds because matrix multiplication is associative.

The *identity* is present because the unit matrix commutes with $\mathbf{H}$.

The *inverse* of each of these elements commutes with $\mathbf{H}$, because we have $\mathbf{AHA}^{-1} = \mathbf{H}$, $\mathbf{A}^{-1}\mathbf{AHA}^{-1} = \mathbf{A}^{-1}\mathbf{H}$, and $\mathbf{HA}^{-1} = \mathbf{A}^{-1}\mathbf{H}$.

All the group postulates hold.

4.4 (a) For ψ_i to form an orthonormal set, we must have

$$\int \psi_k^* \psi_l \, d\tau = \int \left(\sum_i a_{ki}\phi_i\right)^* \left(\sum_j a_{lj}\phi_j\right) d\tau = \sum_{i,j} a_{ki}^* a_{lj} \int \phi_i^* \phi_j \, d\tau$$
$$= \sum_{i,j} a_{ki}^* a_{lj} S_{ij} = \delta_{kl}.$$

(b) For ψ_1 and ψ_3 of Example 4.1(b) to be orthogonal, the following expression must be zero:

$$\begin{aligned} &a_{11}a_{31}S_{11} + a_{11}a_{32}S_{12} + a_{11}a_{33}S_{13} \\ &+ a_{12}a_{31}S_{21} + a_{12}a_{32}S_{22} + a_{12}a_{33}S_{23} \\ &+ a_{13}a_{31}S_{31} + a_{13}a_{32}S_{32} + a_{13}a_{33}S_{33} = 0. \end{aligned}$$

If we substitute in this expression the values of the S_{ij} and the a_{ij} found in Example 4.1(b), we obtain the value 1.0×10^{-6}, which is zero within the accuracy of our calculations.

4.5 $\mathbf{C}_{3h}$ $2A'$, A'', E', E''; $\mathbf{D}_{4h}$ A_{2u}, B_{1u}, B_{2u}, $2E_u$; $\mathbf{O}_h$ A_{2u}, T_{1u}, T_{2u}.

Chapter 5

5.1 N: p^3, 3 unpaired electrons, $M_L = 0$, $L = 0$, $M_S = 3/2$, $S = 3/2$, ${}^4S_{3/2}$;
Fe^{2+}: d^6 $\underset{2}{\underline{\uparrow\downarrow}}$ $\underset{1}{\underline{1}}$ $\underset{0}{\underline{1}}$ $\underset{-1}{\underline{1}}$ $\underset{-2}{\underline{1}}$ $M_L = 2$, $M_S = 2$, ${}^5D_4$;
Cr^{3+}: d^3, $M_L = 3$, $M_S = 3/2$, ${}^4F_{3/2}$.

5.2 d^4 210 microstates;
$e_\alpha^4 \rightarrow {}^5D$ (for all interchange α and β);
$e_\alpha^3 e_\beta^1 \rightarrow {}^3[D(F+P)] \rightarrow {}^3P(2), {}^3D, {}^3F(2), {}^3G, {}^3H$;
$e_\alpha^2 e_\beta^2 \rightarrow {}^1[(F+P)(F+P)] \rightarrow {}^1S(2), {}^1D(2), {}^1G(2), {}^1F, {}^1I$;
Ground level ${}^5D_0$ (replicated terms dropped).
f^{12} 49 microstates;
$e_\alpha^5 e_\beta^7 \rightarrow {}^3[S(P+F+H)] \rightarrow {}^3P, {}^3F, {}^3H$;
$e_\alpha^6 e_\beta^6 \rightarrow {}^1[F \times F)] \rightarrow {}^1I, {}^1\not{H}, {}^1G, {}^1\not{F}, {}^1D, {}^1\not{P}, {}^1S$;
ground level ${}^3H_6$.

5.3 Considering d^1–d^5 cases, we see that the maximum M_L values are 2, 3, and 0.

5.4 $T_{2u} + G_u$ (use **I** group and add u subscripts).

5.5 For $\mathbf{D}_{3h}$ and $\mathbf{C}_{4v}$, $\chi(\sigma) = +1$ for $l = 1$, $\chi(\sigma) = -1$ for $L = 1$.

	p	P
$\mathbf{D}_{3h}$	$A_2'' + E'$	$A_2' + E''$
$\mathbf{C}_{4v}$	$A_1 + E$	$A_2 + E$

5.6 C_{4v}: $\chi(\sigma) = +1$ ($l = 2,3^-$), $\chi(\sigma) = -1$ ($L = 3^+$).

$$l = 2: \quad A_1 + B_1 + B_2 + E;$$
$$l = 3: \quad A_1 + B_1 + B_2 + 2E;$$
$$L = 3^+: \quad A_2 + B_1 + B_2 + 2E.$$

5.7 (b)

	D_{4h}	C_{4v}		D_{4h}	C_{4v}
F	A_{2g}	A_2	f	A_{2u}	A_1
	B_{1g}	B_1		B_{1u}	B_2
	B_{2g}	B_2		B_{2u}	B_1
	$2E_g$	$2E$		$2E_u$	$2E$

5.8 $g(O_h)$: $A_{1g} + E_g + T_{1g} + T_{2g}$;
D_{4h}: $2A_{1g} + B_{1g} + A_{2g} + B_{2g} + 2E_g$.

5.9

	Rotational group	Order	
C_{3v}	C_3	3	$\frac{4!}{3} = 8$
D_{2d}	D_2	4	$\frac{4!}{4} = 6$

5.10 Rotational group is C_2; $4!/2 = 12$ stereoisomers. Since there are no planes of symmetry with the ligands added, there are six pairs of optical isomers.

5.11 $360/6 = 60$ stereoisomers (D_3); $360/12 = 30$ geometrical isomers (D_{3h}).

5.12 $60/6 = 10$ stereoisomers; $72/12 = 6$ geometrical isomers ($3P_2^2P_1^2$ for $3\sigma_v$). The 2 inactive isomers have vertical planes.

5.13 $8/2 = 4$ stereoisomers (C_2); $12/4 = 3$ geometrical isomers (C_{2v}). The optical isomers have *a*'s *cis* and *b*'s *cis*.

5.14 D_2 $8!/4 = 10080$ stereoisomers. There are no planes of symmetry, so there are $10080/2 = 5040$ pairs of enantiomers.

Chapter 6

6.1

	CO^+	CO	CO^-	NO^+	NO	NO^-	
Bond energy	804	1070	784	1047	627	488	kJ/mole

CO: $\sigma_g^2\ \sigma_u^2\ \pi_x^2\ \pi_y^2\ \sigma_g^2$. No unpaired electrons. An electron is removed from a somewhat nonbonding σ_g orbital (the orbital containing the lone pair donated to metals) to form CO^+ (one unpaired electron). Nevertheless, the screening of the positive nuclei decreases, and the bond energy decreases. An antibonding electron is added to π^* to form CO^-, weakening the bond. NO has one more electron than CO in a π^* orbital. The antibonding electron is removed to form NO^+ (diamagnetic), increasing the bond energy. Another antibonding electron is added to form NO^-, weakening the bond and giving two unpaired electrons.

6.2 The bond order of $Cl_2{}^+$ is 1.5; an antibonding π^* electron is removed.

6.3 $\sigma_g^2\ \pi_u^4\ \delta_g^4$. Bond order 5.

$$\text{Nb } 5s^2 4d^3 \rightarrow 5s^1 4d^4 \quad s + s \rightarrow \sigma \quad (\text{or } s\text{-}d_{z^2} \text{ hybrid})$$
$$2(xz, yz) \rightarrow 2\pi$$
$$2(x^2 - y^2, xy) \rightarrow 2\delta$$

6.4 Bond order 6 with two π bonds, two δ bonds, one strongly bonding σ bond, and one weakly bonding σ bond from the s-d_{z^2} hybrids.

6.5 $NF_4{}^+F^-$. $NF_4{}^+$ is tetrahedral (sp^3).

6.6 NO_2 (17 electrons) is angular. $NO_2{}^+$ has 16 valence electrons and has the linear CO_2 structure. $NO_2{}^-$ has an unshared electron pair on N, instead of a single electron as for NO_2, so a smaller bond angle is expected for $NO_2{}^-$.

6.7 (a) Planar; (b) pyramidal; (c) bent **T**.

6.8 $ClOF_2{}^+$ sp^3 hybrid with one lone pair—pyramidal with $\angle$OClF greater than $\angle$FClF because of higher bond order for Cl—O.
$ClOF_3$ dsp^3 hybrid with one lone pair—the lone pair and the oxygen (with some degree of double bonding) will be in the equatorial plane of a TBP. The $\angle$FClF should be less than 90° and the $\angle$OClF(axial) should be greater than 90°.
$ClOF_4{}^-$ d^2sp^3 hybrid—the lone pair should be trans to the oxygen in an octahedral arrangement. The structure of the ion (atom positions) is SP.
$ClO_2F_2{}^-$ dsp^3 hybrid—the lone pair and oxygens are in equatorial positions of a TBP, giving a saw-horse shaped ion.

6.9 C_{2v}: $\Gamma_\sigma = A_1 + B_2$, $\Gamma_\pi = A_2 + B_1$.
Orbitals on N: a_1, s, p_z; b_1, p_x; b_2, p_y.
The sp hybrids on N can combine with the oxygen a_1 group orbital to give bonding (σ), nonbonding, and antibonding MOs. The second σ bond is formed by combination of p_y with the b_2 group orbital. The π a_2 orbital is nonbonding (no N a_2 orbital). The p_x on N combines with the b_1 group orbital to form the π bond. See B. Douglas, D. H. McDaniel,

and J. J. Alexander, *Problems for Inorganic Chemistry*, Wiley, New York, 1983, Problems 4.11 and 4.14 for sketches.

6.10 ClO_3^- had $\mathbf{C}_{3v}$ symmetry; use sp^3 hybridization. $\Gamma_\sigma = 2A_1 + E$. The lone pair is in an a_1 orbital along C_3.

$$\phi_1 = \sigma_2 + \sigma_3 + \sigma_4 \quad (a_1)$$
$$\phi_2 = 2\sigma_2 - \sigma_3 - \sigma_4 \quad (e)$$
$$\phi_3 = \sigma_3 - \sigma_4$$

(sketches not shown)

— a_1^*
— — e^*
:: :: e
:: a_1
:: a_1 (nonbonding)

6.11 $\mathbf{C}_{3v}$ d orbitals: z^2, A_1; $(x^2 - y^2, xy)$, E; (xz, yz), E; p_x, $A_2 + E$; p_y, $A_1 + E$ (combinations of oxygen orbitals). The a_2 group orbital is nonbonding. The p_y group orbitals combine with z^2 and (xz, yz). The p_x (e) group orbitals combine with $(x^2 - y^2, xy)$. Sketches are not shown.

6.12 SF_4 has $\mathbf{C}_{2v}$ symmetry. $\Gamma_\sigma = 2A_1 + B_1 + B_2$. The nonbonding orbital (lone pair) is a_1 ($s + p_z + d_{z^2}$ hybrid). One a_1 orbital ($s + p_z - d_{z^2}$ hybrid) is used to bond the axial F atoms. The other σ a_1 orbital ($s - p_z + d_{z^2}$ hybrid) is used to bond the equatorial F atoms. The S p_x orbital (b_1) is used for the axial F atoms and p_y (b_2) is used for the equatorial F atoms. Sketches are not shown.

6.13

	$\mathbf{T}_d$	$\mathbf{D}_2$
$\phi_1 = \frac{1}{2}(s + p_z + p_y + p_x)$	A_1	$A(s)$
$\phi_2 = \frac{1}{2}(s + p_z - p_y - p_x)$		$B_1(z)$
$\phi_3 = \frac{1}{2}(s - p_z - p_y + p_x)$	T_2	$B_2(y)$
$\phi_4 = \frac{1}{2}(s - p_z + p_y - p_x)$		$B_3(x)$

6.14

$$\phi_1 = \frac{1}{\sqrt{6}}s + \frac{1}{\sqrt{3}}d_{z^2} + \frac{1}{\sqrt{2}}p_z$$

$$\phi_2 = \frac{1}{\sqrt{6}}s - \frac{1}{\sqrt{12}}d_{z^2} + \frac{1}{2}d_{x^2-y^2} + \frac{1}{\sqrt{2}}p_x$$

$$\phi_3 = \frac{1}{\sqrt{6}}s - \frac{1}{\sqrt{12}}d_{z^2} - \frac{1}{2}d_{x^2-y^2} + \frac{1}{\sqrt{2}}p_y$$

$$\phi_4 = \frac{1}{\sqrt{6}}s - \frac{1}{\sqrt{12}}d_{z^2} + \frac{1}{2}d_{x^2-y^2} - \frac{1}{\sqrt{2}}p_x$$

$$\phi_5 = \frac{1}{\sqrt{6}}s - \frac{1}{\sqrt{12}}d_{z^2} - \frac{1}{2}d_{x^2-y^2} - \frac{1}{\sqrt{2}}p_y$$

$$\phi_6 = \frac{1}{\sqrt{6}}s + \frac{1}{\sqrt{3}}d_{z^2} - \frac{1}{\sqrt{2}}p_z$$

Chapter 7

7.1 (a) The molecule should distort to remove the degeneracy of the singly occupied e' orbitals, giving a linear arrangement.
(b) If triangular, H_3^- would be a paramagnetic diradical. With paired electrons the same distortion as in (a) should occur to give a linear ion.

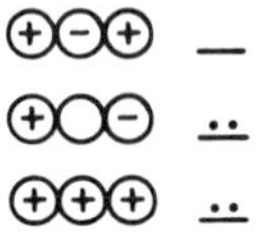

⊕⊖⊕ —
⊕○⊖ ··
⊕⊕⊕ ··

7.2

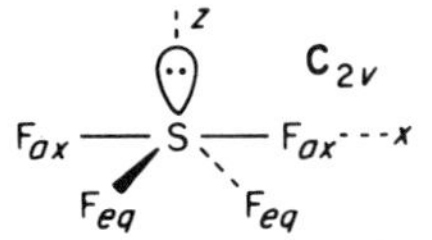

C_{2v}

F_{ax} — S — F_{ax} - - - x ; F_{eq}, F_{eq}; z

5 orbitals about TBP on its side

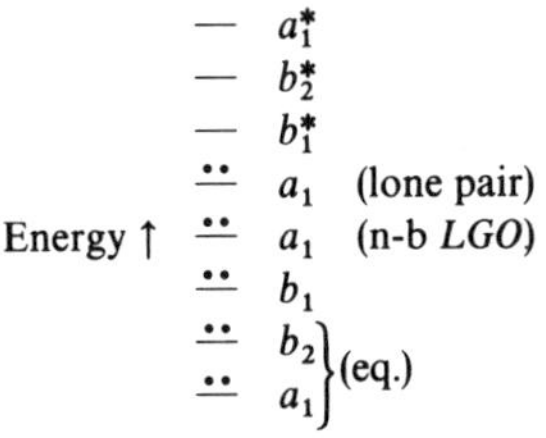

Energy ↑

— a_1^*
— b_2^*
— b_1^*
·· a_1 (lone pair)
·· a_1 (n-b *LGO*)
·· b_1
·· b_2 } (eq.)
·· a_1

See Problem 6.12 for a bonding description using d orbitals.

If d orbitals are excluded (too high in energy), we can use sp^2 hybrid orbitals (using s, p_y and p_z) for bonding to the two equatorial F atoms. The symmetry-adapted bonding MOs are a_1 and b_2 with a_1 (along z) for the lone pair. The p_x (b_1) orbital is used for 3-c, 4-e bonding. The b_1 $(+, -)$ LGO (orbitals of two axial F atoms) combines with p_x to give bonding and antibonding MOs. Combination of three AOs (p_x of S and two F orbitals) gives three MOs. The third MO consists of a nonbonding a_1 *LGO* $(+, +)$ for the two F orbitals. Actually the s orbital of S can participate here and in equatorial bonding, but this only alters relative energies.

7.3

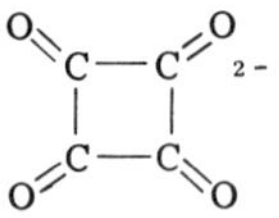

[Structure: squarate ion $C_4O_4^{2-}$]

Each C has 3 σ bonds. We can use an sp^2 basis set, leaving 1 p_z orbital per C for π bonding within the ring and with the O atoms. The C p_z orbitals transform as do those of C_4H_4 (Section 7.3.2) and the O p_z orbitals transform in the same way:

$$\Gamma_\pi(C) = A_{2u} + B_{2u} + E_g, \qquad \Gamma_\pi(O) = A_{2u} + B_{2u} + E_g.$$

The MOs are

$$a_{2u} \pm a_{2u} \rightarrow a_{2u} + a_{2u}^*, \qquad e_g \pm e_g \rightarrow e_g + e_g^*,$$
$$b_{2u} \pm b_{2u} \rightarrow b_{2u} + b_{2u}^*.$$

The MOs correspond to those in Fig. 7.1 For the bonding combinations, the signs are the same for the C and O sets and opposite for the antibonding combinations.

7.4 The representations for the π MOs for benzene are given in Fig. 7.2. The energies ($E = \int \Psi_i H \Psi_i \, d\tau$) are

$$E_{a_{2u}} = \tfrac{1}{6}(6\alpha + 12\beta) = \alpha + 2\beta, \qquad E_{b_{2g}} = \tfrac{1}{6}(6\alpha - 12\beta) = \alpha - 2\beta,$$
$$E_{(e_{1g})_a} = \tfrac{1}{4}(4\alpha + 4\beta) = \alpha + \beta \quad [\text{same for } (e_{1g})_b],$$
$$E_{(e_{2u})_a} = \tfrac{1}{4}(4\alpha - 4\beta) = \alpha - \beta \quad [\text{same for } (e_{2u})_b].$$

7.5 The ligand π MOs are the sums and differences of the π MOs for C_6H_6.

$a_{1u} \pm a_{1u} \rightarrow a_{1g} + a_{2u}$

$e_{1g} \pm e_{1g} \rightarrow e_{1u} + e_{1g}$

$e_{2u} \pm e_{2u} \rightarrow e_{2u} + e_{2g}$

$b_{2g} \pm b_{2g} \rightarrow b_{1u} + b_{2g}$

Metal orbitals	($\mathbf{D}_{6h}$)
s and z^2	A_{1g}
(xz, yz)	E_{1g}
$(x^2 - y^2, xy)$	E_{2g}
z	A_{2u}
(x, y)	E_{1u}

Metal and ligand orbitals belonging to the same representation combine to give bonding and nonbonding combinations. The e_{2u}, b_{1u}, and b_{2g} LGOs are nonbonding (no Cr orbitals). See B. Douglas, D. H. McDaniel, and J. J. Alexander, *Problems for Inorganic Chemistry*, Wiley, New York, 1983, Problem 7.22 for sketches.

7.6 $\mathbf{D}_{4h}$

$a_{2u} \pm a_{2u} \rightarrow a_{1g} + a_{2u}$

$e_g \pm e_g \rightarrow e_u + e_g$

$b_{2u} \pm b_{2u} \rightarrow b_{1g} + b_{2u}$

Metal orbitals			
s, z^2	A_{1g}	z	A_{2u}
$x^2 - y^2$	B_{1g}	(x, y)	E_u
(xz, yz)	E_g	xy	B_{2g}

Nonbonding: b_{2u} LGO (no M orbital),
b_{2g} (xy) (no LGO).

From s, d_{z^2}, and the a_{1g} (LGO) there are bonding, nonbonding, and antibonding combinations. Sketches are not shown.

Ni would be most favorable. The Ni^{IV} is a formalism counting $C_4H_4^{2-}$ as a six π-electron system. The complex could be

$$Ni^{IV}(C_4H_4^{2-})_2 \qquad Ni^{0}(C_4H_4)_2$$
$$6 + 2(6) = 18, \qquad 10 + 2(4) = 18.$$

7.7

B_5H_9

$\mathbf{C}_{4v}$ B (apex): sp hybrid a_1
(p_x, p_y) e

4 B (base)—orbitals directed toward apex.

C_{4v}	E	C_4	C_2	σ_v	σ_d	
Γ	4	0	0	2	0	$= A_1 + B_1 + E$

Using the projection operator we get

$$a_1 = \frac{1}{2}(\phi_1 + \phi_2 + \phi_3 + \phi_4), \qquad b_1 = \frac{1}{2}(\phi_1 - \phi_2 + \phi_3 - \phi_4),$$

$$e_1 = \frac{1}{\sqrt{2}}(\phi_1 - \phi_3), \qquad e_2 = \frac{1}{\sqrt{2}}(\phi_2 - \phi_4).$$

$(a_1)_{\text{apex}} \pm (a_1)_{\text{base}} \to a_1 + a_1^*$

$(b_1)_{\text{base}}$ is nonbonding

$(e)_{\text{apex}} \pm (e)_{\text{base}} \to e + e^*$

} Draw sketches using LCAO signs

Bridge bonding

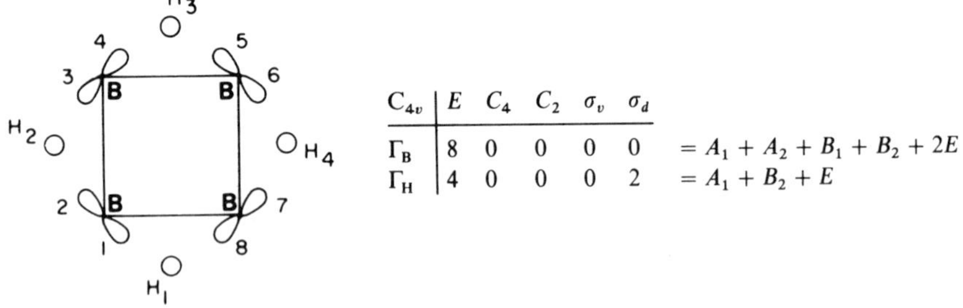

C_{4v}	E	C_4	C_2	σ_v	σ_d	
Γ_B	8	0	0	0	0	$= A_1 + A_2 + B_1 + B_2 + 2E$
Γ_H	4	0	0	0	2	$= A_1 + B_2 + E$

Using the projection operator we get

$$a_1 = \frac{1}{2\sqrt{2}}(\phi_1 + \phi_2 + \phi_3 + \phi_4 + \phi_5 + \phi_6 + \phi_7 + \phi_8),$$

$$\left[a_1 = \frac{1}{2}(\mathrm{H}_1 + \mathrm{H}_2 + \mathrm{H}_3 + \mathrm{H}_4)\right],$$

$$a_2 = \frac{1}{2\sqrt{2}}(\phi_1 - \phi_2 + \phi_3 - \phi_4 + \phi_5 - \phi_6 + \phi_7 - \phi_8),$$

[nonbonding—no $(a_2)_{\mathrm{H}}$],

$$b_1 = \frac{1}{2\sqrt{2}}(\phi_1 + \phi_2 - \phi_3 - \phi_4 + \phi_5 + \phi_6 - \phi_7 - \phi_8),$$

[nonbonding—no $(b_1)_{\mathrm{H}}$],

$$b_2 = \frac{1}{2\sqrt{2}}(\phi_1 - \phi_2 - \phi_3 + \phi_4 + \phi_5 - \phi_6 - \phi_7 + \phi_8),$$

$$\left[b_2 = \frac{1}{2}(\mathrm{H}_1 - \mathrm{H}_2 + \mathrm{H}_3 - \mathrm{H}_4)\right].$$

New linear combinations to involve all boron orbitals:

$$e_1 = \frac{1}{\sqrt{2}}(\phi_1 - \phi_5)$$

$$(e_a)_1 = e_1 + e_2 + e_3 - e_4 = \frac{1}{2\sqrt{2}}(\phi_1 + \phi_2 + \phi_3 - \phi_4 - \phi_5 - \phi_6 - \phi_7 + \phi_8),$$

$$e_2 = \frac{1}{\sqrt{2}}(\phi_2 - \phi_6)$$

$$(e_a)_2 = e_1 - e_2 - e_3 - e_4 = \frac{1}{2\sqrt{2}}(\phi_1 - \phi_2 - \phi_3 - \phi_4 - \phi_5 + \phi_6 + \phi_7 + \phi_8),$$

$$e_3 = \frac{1}{\sqrt{2}}(\phi_3 - \phi_7)$$

$$(e_b)_1 = e_1 - e_2 + e_3 + e_4 = \frac{1}{2\sqrt{2}}(\phi_1 - \phi_2 + \phi_3 + \phi_4 - \phi_5 + \phi_6 - \phi_7 - \phi_8),$$

$$e_4 = \frac{1}{\sqrt{2}}(\phi_4 - \phi_8)$$

$$(e_b)_2 = e_1 + e_2 - e_3 + e_4 = \frac{1}{2\sqrt{2}}(\phi_1 + \phi_2 - \phi_3 + \phi_4 - \phi_5 - \phi_6 + \phi_7 - \phi_8).$$

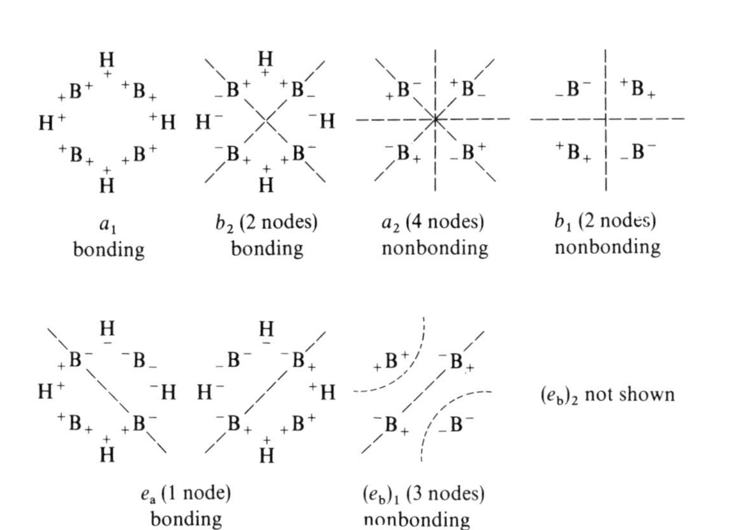

$3 \times 5(\text{B}) + 1 \times 9(\text{H}) = 24$ electrons
-10 (5 B–H)
Framework 14 electrons

$4 \times 5(\text{B}) + 1 \times 9(\text{H}) = 29$ orbitals
-10 (5 B–H bonds)
Framework 19 orbitals

Energy ↑

Level	Occupancy	Label	
—		a_1 (base)*	
—		a_1	
— —		e (base)	
— —		e	
—		b_2 (base)	
—		a_2 (base) 4 nodes	nonbonding
— —		e (base) 3 nodes	nonbonding
—		b_1 2 nodes	nonbonding
—		b_1 (base) 2 nodes	nonbonding
—	··	b_2 (base) 2 nodes	bonding
— —	·· ··	e 1 node	bonding
— —	·· ··	e (base) 1 node	bonding
—	··	a_1	bonding
—	··	a_1 (base)	bonding

7.8 $B_3H_8^-$ $(2 \times 3) + 6$ framework electrons, *arachno*, C_{2v}. Total number of electrons $= 3 \times 3 + 8 \times 1 = 18$. The BH_2 groups are arranged in a triangle with two bridging H atoms and one B—B two-center bond.

6 B—H	1 B—B	2 B–H–B (bridge)	
12 +	2 +	4	= 18 electrons

7.9 (a) *closo*; (b) *nido*; (c) *arachno*; (d) *hypho*; (e) *nido*.

7.10 (a) $[Rh(CO)_2]_6(CO)_4$, $(6 \times 1) + (4 \times 2) = 14$ e $(2 \times 6 + 2)$, *closo*;
(b) $[Os(CO)_3]_5(CO)$, $(5 \times 2) + (2) = 12$ e $(2 \times 5 + 2)$, *closo*;
(c) $[Rh(CO)_4]_2(FeCp)_2$, $(2 \times 5) + (2 \times 1) = 12$ e $(2 \times 4 + 4)$, *nido*;
(d) $(CoCp)_2(BH)_6(CH)_2$, $(2 \times 2) + (6 \times 2) + (2 \times 3) = 22$ e $(2 \times 10 + 2)$, *closo*;
(e) $(CoCp)(BH)_9(CH)_2$, $(2) + (9 \times 2) + (2 \times 3) = 26$ e $(2 \times 12 + 2)$, *closo*.

7.11

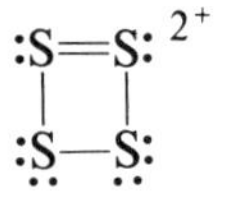

:S=S: $^{2+}$
| |
:S—S:

VB structure

S_4^{2+} — square of S atoms with σ vectors numbered 1–8: S(4)→ ←S(5); 3↓ ↓6; 2↑ ↑7; S→(1) ←S(8)

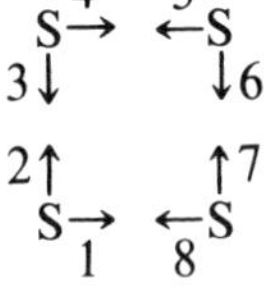

$(6 \times 4) - 2 = 22$ electrons

Consider the s orbitals as nonbonding. Use p_x and p_y for σ bonding (vectors shown). The p_z orbitals are used for π bonding. The π MOs are the same as for C_4H_4 (Section 7.3.2). $\Gamma_\pi = a_{2u} + e_g + b_{2u}^*$.

$\mathbf{D}_{4h}$	E	C_4	C_2	C_2'	C_2''	i	σ_h	σ_v	σ_d	
Γ_σ	8	0	0	0	0	0	8	0	0	$= A_{1g} + B_{2g} + A_{2g} + B_{1g} + 2E_u$

The LCAOs are the same as for the B orbitals for bridge bonding in the base of B_5H_9 (Problem 7.7) without g or u subscripts. The energy sequence is $a_{1g}^2 b_{2g}^2 e_u^4 a_{2u}^2 e_g^4 b_{2u}^* e_u^* b_{1g}^* a_{2g}^*$. Since e_g is π nonbonding, there are 10 bonding electrons, giving a bond order of 1.25. See J. D. Corbett, *Inorg. Nucl. Chem. Lett.* 1969, **5**, 81. The symmetry-adapted orbitals for the *s* lone pairs not included above are $a_{1g} + b_{1g} + e_u$.

a_{1g} b_{2g} $(e_u)_a$ b_{1g}^*

a_{2g}^* $(e_u^*)_b$

Chapter 8

8.1 Octahedral: low spin d^6, LFSE $24Dq$.
Tetrahedral: low-spin d^4, LFSE $24Dq$.
The low-spin tetrahedral case is not realized because of the weaker tetrahedral field.

8.2 No distortion is expected for d^8, so d^9, with one electron in the highest-energy orbital, is most favorable.

8.3 Compression along *z*, with one orbital of highest energy:

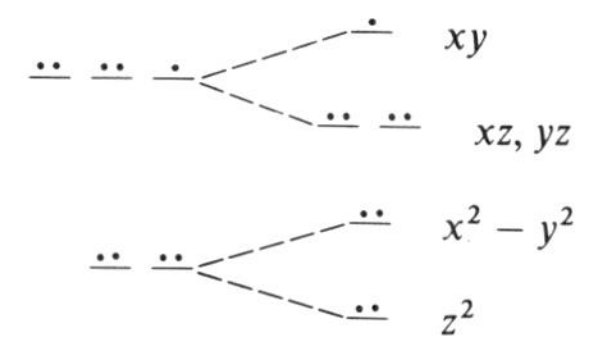

8.4 (a) Cube, ML_8. The cube consists of two tetrahedra, or 2 × Case III.
(b) Square pyramid, ML_5. Octahedral ML_6 minus L along *z*.
(c) Pentagonal pyramid, ML_6. $\frac{5}{2} \times ML_2\ (xy)$ + L along *z*.

	z^2	$(x^2 - y^2)$	xy	xz	yz	
(a)	−5.34	−5.34	3.56	3.56	3.56	Dq(L)
(b)	0.86	9.14	−0.86	−4.57	−4.57	Dq(L)
(c)	−0.21	12.21	−0.29	−5.86	−5.86	Dq(L)
average		5.96	5.96			

8.5

	z^2	x^2-y^2	xy	xz	yz	
(a) MX_6	6.00	6.00	−4.00	−4.00	−4.00	$Dq(X)$
(b) MY_6	4.20	4.20	−2.80	−2.80	−2.80	$Dq(X)$
(c) MX_5Y						
$MX_6 - X(z) + Y(z)$						
$\mathbf{C}_{4v}$	4.46	6.94	−3.06	−4.17	−4.17	$Dq(X)$
(d) *trans*-$[MX_4Y_2]$						
$MX_6 - 2X(z) + 2Y(z)$						
$\mathbf{D}_{4h}$	2.92	7.88	−2.12	−4.34	−4.34	$Dq(X)$
(e) *cis*-$[MX_4Y_2]$						
$MX_6 - 2X(xy) + 2Y(xy)$						
$\mathbf{C}_{2v}$ (but $x = y$)	6.64	4.16	−4.34	−3.23	−3.23	$Dq(X)$
(f) *fac*-$[MX_3Y_3]$						
$MX_6 - X(z) - 2X(xy) + Y(z) + 2Y(xy)$						
$\mathbf{C}_{3v}$ (but the field is cubic, $x = y = z$)	5.10	5.10	−3.40	−3.40	−3.40	$Dq(X)$

8.6 LCAOs for σ bonding: $a_1' + a_2''$ (axial) and $a_1' + e'$ (equatorial). Sketches are the same as those for PF_5 in Section 6.3.3.

LCAOs for π bonding: equatorial $\pi(\perp)\ a_2'' + e''$, $\pi(\|)\ a_2' + e'$.

Axial

$\mathbf{D}_{3h}$	E	C_3	C_2	σ_h	S_3	σ_v	
Γ_π	4	−2	0	0	0	0	$= E' + E''$

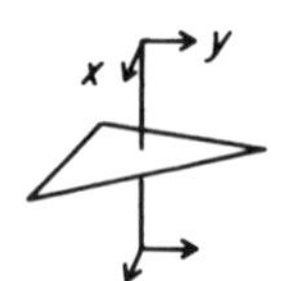

For C_3, x and y are mixed:

$$\begin{bmatrix} x \\ y \end{bmatrix} \begin{bmatrix} -\frac{1}{2} & \frac{\sqrt{3}}{2} \\ -\frac{\sqrt{3}}{2} & -\frac{1}{2} \end{bmatrix}.$$

$\chi = -1$ for each x,y pair. If we choose σ_v through y, $\chi = +2(y) - 2(x) = 0$. Sketches are not shown; see vector combinations in Section 10.6 as a guide.

8.7

$$\Gamma_\sigma = a_1' + e' + a_2'' + e''$$

$$(\Gamma_\pi)_x = a_1' + a_2'' + e' + e'' \qquad (\Gamma_\pi)_y = a_1'' + a_2' + e' + e''$$

Sketches are not shown.

8.8 XeF_4. Two p orbitals (p_x and p_y) are used for bonding to F atoms. The two lone pairs are accommodated in the s and p_z orbitals (or a hybrid). Each of the p_x and p_y orbitals is used for a 4e, 3c bond to two F atoms, giving a square planar molecule with lone pairs above and below. The 4e, 3c bonds are weaker than 2e, 2c bonds.

$$\begin{array}{l} \qquad\qquad\qquad\qquad\qquad\qquad -\,-\;\sigma^* \\ \text{Xe}\;p\;\;\underline{\cdot\cdot}\;\underline{\cdot\cdot}\;\underline{\cdot\cdot}\;\cdots\cdots\;\;\underline{\cdot\cdot}\;\;p_z \\ \varepsilon = 2e - 4f \;\updownarrow\; \underline{\cdot\cdot}\;\underline{\cdot\cdot}\;\cdots\cdots\;\underline{\cdot}\;\underline{\cdot}\;\underline{\cdot}\;\underline{\cdot}\;-\;\text{F}\,\sigma \\ \qquad\qquad\qquad\qquad\quad \underline{\cdot\cdot}\;\underline{\cdot\cdot}\;\;\sigma \\ s\;\underline{\cdot\cdot}\qquad\qquad\qquad\quad \underline{\cdot\cdot} \end{array}$$

8.9 Using the s and p orbitals we could form three two-center bonds using sp^2 hybrid orbitals and then use p_z for 3c bonding to the two axial F atoms. This is probably better than a p-only AOM description. However, using p orbitals only, we can form the 3c bonds to two axial F atoms (1 and 5) and then use p_x and p_y for bonding to the three equatorial F atoms (2, 3, and 4). The three equatorial ligands combine with the two p orbitals using the following LCAOs:

$$\psi_{e'}(1) = \frac{1}{\sqrt{6}}(2\phi_2 - \phi_3 - \phi_4) \quad \text{combines with } p_y;$$

$$\psi_{e'}(2) = \frac{1}{\sqrt{2}}(\phi_3 - \phi_4) \qquad\qquad \text{combines with } p_x.$$

For $\psi_{e'}(1)$, F_2 along x: $\theta = 0°$, $(1/\sqrt{6}) \times 2 \times (\cos\theta)S_\sigma = (2/\sqrt{6})S_\sigma$.

$$F_3\colon \theta = 120°,\ \cos 120° = -\frac{1}{2}; \qquad -\frac{1}{\sqrt{6}}\left(-\frac{1}{2}\right)S_\sigma = \frac{1}{2\sqrt{6}}S_\sigma.$$

Same for F_4.

$$\text{Total: } \left(\frac{2}{\sqrt{6}} + \frac{1}{2\sqrt{6}} + \frac{1}{2\sqrt{6}}\right)S_\sigma = \frac{\sqrt{3}}{\sqrt{2}}S_\sigma.$$

For $\psi_{e'}(2)$, F_2: $\theta = 90°$, $\cos 90° = 0$.

$$F_3\colon \theta = 30°,\ \cos 30° = \frac{\sqrt{3}}{2}; \qquad \frac{1}{\sqrt{2}}(1)\frac{\sqrt{3}}{2}S_\sigma = \frac{\sqrt{3}}{2\sqrt{2}}S_\sigma.$$

Same for F_4.

$$\text{Total: } 0 + \frac{\sqrt{3}}{2\sqrt{2}}S_\sigma + \frac{\sqrt{3}}{2\sqrt{2}}S_\sigma = \frac{\sqrt{3}}{\sqrt{2}}S_\sigma$$

The stabilization of each orbital for equatorial bonding is

$$\left(\frac{3}{\sqrt{2}}\right)^2 e_\sigma - \left(\frac{\sqrt{3}}{\sqrt{2}}\right)^4 f_\sigma = \frac{3}{2}e_\sigma - \frac{9}{4}f_\sigma.$$

For axial 3c bonding, the stabilization of the bonding orbital is $2e_\sigma - 4f_\sigma$. Total stabilization is as follows.

Equatorial: four electrons	Axial: two electrons
$4\left(\frac{3}{2}e_\sigma - \frac{9}{4}f_\sigma\right)$	$2(2e_\sigma - 4f_\sigma)$
Per bond: $\frac{4}{3}\left(\frac{3}{2}e_\sigma - \frac{9}{4}f_\sigma\right) = 2e_\sigma - 3f_\sigma$	Per bond: $\frac{2}{2}(2e_\sigma - 4f_\sigma) = 2e_\sigma - 4f_\sigma$

Because of the larger negative f_σ term, the axial bonds are weaker.

$2a_2''$

$2e'$

$2a_1' + a_2'' + e'$

$\frac{3}{2}e_\sigma - \frac{9}{4}f_\sigma$

$1e'$

$1a_2''$

$2e_\sigma - 4f_\sigma$

8.10 $:\ddot{S}F_2$ obeys the octet rule (sp^3). We can add two F atoms using one of the p orbitals (p_z) to form a three-center bond. For simplicity, let us assume that the other two F atoms and the unshared electron pair are in a plane perpendicular to p_z, using sp^2 hybrid orbitals. The linear 4e, 3c bonding to the two F atoms using p_z is the same as for AL_2 described in Figure 8.11.

8.11

—	a_1'	$2.75e_\sigma$
— —	e'	$\frac{9}{8}e_\sigma + 1.5e_\pi$
$\cdot$ $\cdot$	e''	$3.5e_\pi$

The d^2 configuration would not involve any antibonding σ interaction and should give maximum AOMSE, or for π donor ligands, minimal π antibonding interaction.

8.12 Using Table 8.8 and Figure 8.15,

$8e_\sigma - 2e_\sigma(a_{1g}^*) + 2(4e_\pi) + 4(2e_\pi) = 6e_\sigma + 16e_\pi$.

8.13 See J. M. Burdett, *Inorg. Chem.* 1975, **14**, 375. Consider the ligands in octahedral positions 1, 2, and 3. Low-spin d^6 (or high-spin d^3) would give maximum AOMSE.

— —	$1.5e_\sigma$
$\cdot\cdot$ $\cdot\cdot$ $\cdot\cdot$	$2e_\pi$

8.14 Since $d_{x^2-y^2}$ is stabilized relative to e (T_d) and d_{xy} is stabilized relative to t_2 (T_d), the d^1 and d^3 configurations should be stabilized by the distortion. d_{z^2} becomes σ bonding at the expense of d_{xy} and d_{z^2} becomes more strongly π bonding and $d_{x^2-y^2}$ less so. The totals are still $4e_\sigma + 8e_\pi$.

$\frac{3}{2}e_\sigma + e_\pi$	— —	(xz, yz)
$\frac{3}{4}e_\sigma + e_\pi$	—	xy
$\frac{1}{4}e_\sigma + 3e_\pi$	—	z^2
$2e_\pi$	—	$x^2 - y^2$

Chapter 9

9.1 (a) d^8 $\mathbf{O}_h$ $^3A_{2g} \rightarrow {}^3T_{2g}, {}^3T_{1g}$ $\mathbf{D}_3$ $A_2 \rightarrow A_1, E(T_{2g}); A_2, E(T_{1g})$.

(b) All $d \rightarrow d$ transitions are electric-dipole forbidden in $\mathbf{O}_h$.

$\mathbf{D}_3$ $A_2 \times A_1 = A_2$ allowed, z polarization

$A_2 \times E = E$ allowed, x,y polarization

$A_2 \times A_2 = A_1$ forbidden

(c) Since R_z belongs to A_2 and (R_x, R_y) belong to E, the selection rules for magnetic-dipole transitions apply in the same way as (b).

9.2 (a) There are no spin-allowed transitions for high-spin Fe^{3+} so only very weak peaks are expected. The spin-allowed transitions for low-spin complexes have greater intensity.

(b) $^2T_{2g} \rightarrow {}^2A_{2g}, {}^2T_{1g}, {}^2E_g, {}^2A_{1g}$. For $\mathbf{O}_h$, all $d \rightarrow d$ transitions are forbidden.

9.3 For $\mathbf{O}_h$, $g \rightarrow g$ transitions are forbidden. For $\mathbf{T}_d$, there are no g or u subscripts, and some mixing of d and p orbitals can increase intensities.

9.4 x, y, and z transform as T_{1u}.

$$A_{2g} \times T_{2g} = T_{1g} \qquad T_{1g} \times T_{1u} = A_{1u} + E_u + T_{1u} + T_{2u}$$

$$A_{2g} \times T_{1g} = T_{2g} \qquad T_{2g} \times T_{1u} = A_{2u} + E_u + T_{1u} + T_{2u}$$

Both transitions are vibronically allowed since the products contain T_{1u} and T_{2u}.

9.5 (a) $^3A_{2g} \rightarrow {}^1T_{1g}$ should be less intense since it is spin-forbidden.

(b) $[Co(H_2O)_6]^{2+}$ bands should be less intense. The symmetry selection rule does not apply as strictly to $\mathbf{T}_d$ complexes. See Problem 9.3.

(c) For $[Cr(NH_3)_6]^{3+}$ ($\mathbf{O}_h$), all $d \rightarrow d$ bands are symmetry-forbidden. An allowed charge-transfer band is expected for $[CrCl(NH_3)_5]^{2+}$, and even the $d \rightarrow d$ bands might be more intense for the noncentrosymmetric $\mathbf{C}_{4v}$ group.

(d) $^3T_{1g} \rightarrow {}^3A_{2g}$ should be less intense since it corresponds to a two-electron transition, $t_{2g}^2 \rightarrow e_g^2$.

9.6 We might expect greater intensities for the noncentrosymmetric $[Cr(en)_3]^{3+}$. However, the effective symmetry is $\mathbf{O}_h$ for the CrN_6 chromophore in each case.

9.7 The point groups are those for which there is no representation to which at least one of the p orbitals and one of the d orbitals belongs:

$$\mathbf{C}_{5h}, \mathbf{D}_{5h}, \mathbf{D}_{4d}, \mathbf{D}_{6d}, \mathbf{S}_8, \mathbf{O}, \mathbf{I}.$$

9.8 There are no representations in common for d and f orbitals for $\mathbf{D}_{7h}$, $\mathbf{D}_{6d}$, $\mathbf{D}_{8d}$, and $\mathbf{I}$. The greater energy separation between $3d$ and $4f$ orbitals makes mixing of little importance.

9.9 ν_1 $^3T_{1g} \rightarrow {}^3T_{2g}$ 14,800 = $8Dq$ + bending; (d^2)

$Dq \cong 1800\ \text{cm}^{-1}$, but this is too high because of bending.

Use $Dq \approx 1500\ \text{cm}^{-1}$ and estimate $B' \approx 800\ \text{cm}^{-1}$, $Dq/B' \cong 1.9$. From the Tanabe–Sugano diagram we see that $^3T_2$ is a bit too high, Dq must be even lower than $1500\ \text{cm}^{-1}$, $^3T_1 \approx 24{,}000$, and $^3A_2 \approx 29{,}000\ \text{cm}^{-1}$. The second band is $^3T_{1g} \rightarrow {}^3T_{1g}$. $^3T_{1g} \rightarrow {}^3A_{2g}$ is too high in energy and is less likely to be observed (two-electron transition). We cannot calculate the bending parameter without the A_{2g} band.

9.10 Expect the following.

ν_1	$^3A_2 \rightarrow {}^3T_2$	$8{,}600\ \text{cm}^{-1} = 10Dq$	$Dq = 860\ \text{cm}^{-1}$
ν_2	$^3T_1$	$13{,}700 = 18Dq - c$	$c = 1800$
ν_3	$^3T_1$	$24{,}500 = 12\text{Dq} + 15B' + c$	$B' = 830$

$Dq/B' = 1.04$. From the Tanabe–Sugano diagram $^1E(D)$ has about the same energy as $^3T_1(F)$. The spin-forbidden band at $21{,}500\ \text{cm}^{-1}$ could be $^1A_1(G)$ or $^1T_2(D)$.

9.11 Spin-allowed transitions (symmetry of μ_e is T_2):

	$A_2 \rightarrow T_2\ \nu_1$	$A_2 \times T_2 = T_1$	forbidden, and no T_1 vibrations
$5460\ \text{cm}^{-1}$	$\rightarrow T_1(F)\ \nu_2$	$A_2 \times T_1 = T_2$	allowed
14,700	$\rightarrow T_1(P)\ \nu_3$	T_2	allowed

Since ν_2 is in the near IR, ν_1, at still lower energy, would be difficult to observe and it should have low intensity.

9.12 Weak field states: $^4[(P+F)(S)]$ and $^2[(P+F)(D)] \rightarrow$ Same free ion states as for d^3 (Figure 9.10), use the same energy sequence on the left. Strong field configurations (all g) are as follows

$-8Dq$ $\quad t_2^5e^2 \rightarrow {}^4T_1,\ {}^2T_1(2),\ {}^2T_2(2)$

$+2Dq$ $\quad t_2^4e^3 \rightarrow {}^4T_1,\ {}^4T_2,\ {}^2T_1(2),\ {}^2T_2(2),\ {}^2E(2),\ {}^2A_1,\ {}^2A_2$

$+12Dq$ $\quad t_2^3e^4 \rightarrow {}^4A_2,\ {}^2E,\ {}^2T_1,\ {}^2T_2$

$-18Dq + P$ $\quad t_2^6e^1 \rightarrow {}^2E$

The diagram is similar to that for d^3 except for the energies of strong-field configurations.

9.13 Weak field states—same as in Figure 9.12. Orbital energies—see Table 9.4. The diagram is similar to Figure 9.12.

$-18.28 + P$ $\quad e^4b_2^2a_1^2 \rightarrow {}^1A_1$

-10.0 $\quad e^4b_2^2a_1^1b_1^1 \rightarrow {}^3B_1,\ {}^1B_1$

-8.28 $\quad e^4b_2^1a_1^2b_1^1 \rightarrow {}^3A_2,\ {}^1A_2$

-4.57 $\quad e^3b_2^2a_1^2b_1^1 \rightarrow {}^3E,\ {}^1E$

$-1.72 + P$	$e^4 b_2^2 b_1^2$	$\rightarrow {}^1A_1$
0	$e^4 b_2^1 a_1^1 b_1^2$	$\rightarrow {}^3B_2, {}^1B_2$
$+1.72 + P$	$e^4 b_2^2 b_1^2$	$\rightarrow {}^1A_1$
$+3.71$	$e^3 b_2^2 a_1^1 b_1^2$	$\rightarrow {}^3E, {}^1E$
$+5.43$	$e^3 b_2^1 a_1^2 b_1^2$	$\rightarrow {}^3E, {}^1E$
$+9.14$	$e^2 b_1^2 a_1^2 b_1^2$	$\rightarrow {}^3A_2, {}^1A_1, {}^1B_1, {}^1B_2$

9.14 Ground state ${}^3B_1$; μ_z, A_1; $\mu_{x,y}$, E

All ${}^3B_1 \rightarrow$ singlets are spin-forbidden.

$${}^3B_1 \rightarrow {}^3A_2 \qquad B_1 \times A_2 = B_2$$
$${}^3B_1 \rightarrow {}^3E \qquad B_1 \times E = E$$
$${}^3B_1 \rightarrow {}^3B_2 \qquad B_1 \times B_2 = A_2$$

Only $B_1 \rightarrow E$ is electric-dipole allowed (without vibronic coupling).

9.15 For the weak-field d^2 states, see the left side of Figure 9.9 For the strong-field $\mathbf{D}_{3h}$ states, see McDaniel, *J. Chem. Educ.* 1977, **54,** 149.

9.16 $[Cr(NH_3)_6]^{3+}$, *trans*-$[CrCl_2(NH_3)_4]^+$, *cis*-$[CrCl_2(NH_3)_4]^+$: $\mathbf{C}_{2v}$,

$\mathbf{O}_h$	$\mathbf{D}_{4h}$
${}^4A_{2g} \rightarrow {}^4T_{2g}$	${}^4B_{1g} \rightarrow B_{2g} + E_g$
${}^4A_{2g} \rightarrow {}^4T_{1g}$	${}^4B_{1g} \rightarrow A_{2g} + E_g$

but because of the averaging of fields along x and y it can be treated as effectively $\mathbf{D}_{4h}$. The splitting is only 1/2 as great as for the *trans* isomer and with reversed sign. *fac*-$[CrCl_3(NH_3)_3]$: $\mathbf{C}_{3v}$, but effectively cubic without splitting—same field along x, y, and z.

9.17 ${}^1A_1 \rightarrow {}^1T_1$ ($\mathbf{O}_h$) for $[Co(en)_3]^{3+}$ $\qquad 21.47 = 10Dq - C$

${}^1A_1 \rightarrow {}^1E_a$ for $[Co(en)_2(H_2O)_2]^{3+}$ $\qquad 18.2 = 10Dq + \frac{35}{4}Dt - C$

$$18.2 - 21.5 = \tfrac{35}{4}Dt \qquad Dt = -0.38$$

$${}^1A_1 \rightarrow {}^3T_1 \qquad 13.8 = 10Dq - 3C \qquad 21.5 - 13.8 = 2C \qquad C = 3.8$$

$$21.47 = 10Dq - C \qquad Dq(\text{en}) = 2.53$$

$$Dt = -\tfrac{4}{7}[Dq(\text{en}) - Dq(H_2O)] \qquad Dq(H_2O) = 1.86$$

9.18 Since H_2O is the weaker field ligand, Dt is negative and ${}^4E_g$ (${}^4T_{2g}$) is the lower-energy component. Since only $B_{1g} \rightarrow E_g$ transitions are allowed with z polarization, the assignment are as follows.

${}^4B_{1g} \rightarrow {}^4E_g({}^4T_{2g})$	20.0 kK	(z pol.)	
$\rightarrow {}^4B_{2g}$	23.5	$= 10Dq(\text{en})$	$Dq(\text{en}) = 2.35$ kK
$\rightarrow {}^4E_g({}^4T_{1g})$	27.5	(z pol.)	
$\rightarrow {}^4A_{2g}$	29.3		

Separation of T_{2g} components $= 3.50 = -\frac{35}{4}Dt \quad Dt = -0.40$ kK

$$Dt = -\tfrac{4}{7}[Dq(\mathrm{en}) - Dq(\mathrm{H_2O})] \qquad Dq(\mathrm{H_2O}) = 1.65$$

$$^4B_{1g} \rightarrow {}^4E_g(T_{1g}) \qquad 18Dq - c - 2Ds - \tfrac{3}{4}Dt = 27.7$$

$$\rightarrow {}^4A_{2g} \qquad 18Dq - c + 4Ds - 2Dt = 29.3$$

$$29.3 - 27.5 = +6Ds - 1.25Dt \qquad Ds = 0.22 \qquad \kappa = Ds/Dt = -0.55$$

9.19 d^9 $\mathbf{T}_d$ $\mu_z(B_2)$ $\mu_{x,y}(\mathrm{E})$ $\mathbf{T}_d$ $\mathbf{D}_{2d}$

$$^2E \begin{cases} ^2A_1 \\ ^2B_1 \end{cases}$$

$$^2T_2 \begin{cases} ^2E \\ ^2B_2 \end{cases}$$

			Product with		
	Transition	Symmetry	B_2	E	Polarization
5,000 cm^{-1}	$B_2 \rightarrow E$	E	E	A_1, A_2, B_1, B_2	(x, y)
8,000	$\rightarrow B_1$	A_2	B_1	E	forbidden
9,000	$\rightarrow A_1$	B_2	A_1	E	z

The weak forbidden $B_2 \rightarrow B_1$ band might appear because of distortion by the lower site symmetry or spin–orbit coupling.

9.20

$$16.94 \text{ kK} \quad {}^4B_{1g} \rightarrow {}^4E_g$$

$$21.11 \quad \rightarrow {}^4B_{2g} \quad \Delta_\mathrm{N} = 21.11 \text{ kK}$$

$$24.88 \quad \rightarrow {}^4A_{2g}$$

$$25.66 \quad \rightarrow {}^4E_g$$

$$\Delta(t_{2g}) = -3.30 = -\tfrac{1}{2}\Delta'_{\pi\mathrm{Cl}} \qquad \Delta'_{\pi\mathrm{Cl}} = 6.60$$

$$\Delta(e_g) = -0.66 = -\tfrac{2}{3}\Delta'_{\sigma\mathrm{Cl}} + \tfrac{2}{3}(21.11) \qquad \Delta'_{\sigma\mathrm{Cl}} = 22.10$$

$$\Delta(d) = \tfrac{1}{3}(22.10) - \tfrac{1}{3}(6.6) + \tfrac{2}{3}(21.11) = 19.24 \text{ kK}$$

9.21 $\mathbf{C}_n$, $\mathbf{D}_n$, **T**, **O**, **I**.

9.22 d^8 $\mathbf{O}_h$ $\qquad \mathbf{D}_3 \qquad \mu_\mathrm{e} \qquad \mu_\mathrm{m}$

$$^3A_{2g} \rightarrow {}^3T_{2g} \qquad {}^3A_2 \rightarrow {}^3A_1 + {}^3E \qquad z \quad A_2 \quad \mathrm{R}_z$$

$$\rightarrow {}^3T_{1g}(F) \qquad \rightarrow {}^3A_2 + {}^3E \qquad (x, y) \quad E \quad (\mathrm{R}_x, \mathrm{R}_y)$$

$$\rightarrow {}^3T_{1g}(P) \qquad \rightarrow {}^3A_2 + {}^3E$$

$$A_2 \times A_1 = A_2 \qquad A_2 \times A_2 = A_1 \qquad A_2 \times E = E$$

$A_2 \rightarrow A_1$ (z pol.) and $A_2 \rightarrow E$ (x, y pol.) are electric- and magnetic-dipole-allowed and should appear in CD.

9.23

	$\mathbf{O}_h$	$\mathbf{C}_{4v}$	d^9 $J = \frac{3}{2}, \frac{5}{2}$
^2D	$^2T_{2g}$	$^2E + {}^2B_2$	
	^2E_g	$^2B_1 + {}^2A_1$	($^2B_1$ ground state)
	$S = \frac{1}{2}$	$\Gamma_{\frac{1}{2}} = E_2(\Gamma_6)\ \mathbf{C}'_{4v}$	

$A_1 \times E_2 = E_2(\Gamma_6)$ $\quad J = \frac{3}{2} \quad \Gamma_{3/2} = E_2(\Gamma_6) + E_3(\Gamma_7)$

$B_1 \times E_2 = E_3(\Gamma_7)$ $\quad J = \frac{5}{2} \quad \Gamma_{5/2} = E_2(\Gamma_6) + 2E_3(\Gamma_7)$

$E_1 \times E_2 = E_2(\Gamma_6) + E_3(\Gamma_7)$

$B_2 \times E_2 = E_3(\Gamma_7)$

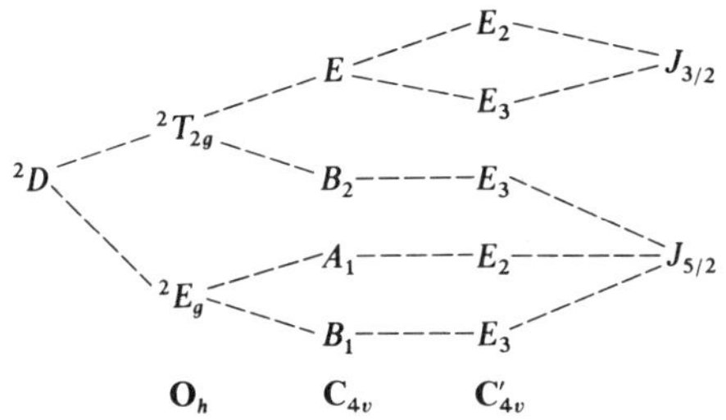

Chapter 10

10.1 (a) $H_2 > N_2 > O_2 > F_2 > Mg_2$.
(b) v varies greatly with isotopic mass, but k does not.

10.2 Omit C_∞ and S_∞, reduce the representation by inspection.

$2A_{1g}, A_{1u}, E_{1g}, E_{1u}$

←H—C→←C—H→ A_{1g}

←H←C—C→H→ A_{1g}

H→←C←C—H→ A_{1u}

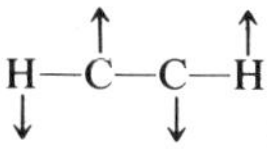

H—C—C—H (↑ on C, ↑ on H; ↓ on H, ↓ on C) E_{1g} (one shown)

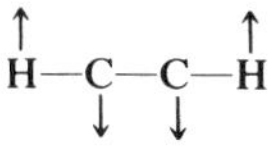

H—C—C—H (↑ on H, ↑ on H; ↓ on C, ↓ on C) E_{1u} (one shown)

10.3

	Number	Symmetry species	Stretching
(a)	6	$2A_1, 2E$	A_1, E
(b)	6	$A'_1, A''_2, 2E'$	A'_1, E'
(c)	9	$A_1, E, 2T_2$	A_1, T_2
(d)	9	$4A_1, A_2, 2B_1, 2B_2$	$2A_1, B_1, B_2$
(e)	12	$3A_1, 2B_1, B_2, 3E$	$2A_1, B_1, E$
(f)	15	$A_{1g}, E_g, 2T_{1u}, T_{2g}, T_{2u}$	A_{1g}, E_g, T_{1u}

10.4

	IR active	Raman active
(a)	A_1, E	A_1, E
(b)	A''_2, E'	A'_1, E'
(c)	T_2	A_1, E, T_2
(d)	A_1, B_1, B_2	All
(e)	A_1, E	A_1, B_1, B_2, E
(f)	T_{1u}	A_{1g}, E_g, T_{2g}

10.5 (a) $3A_1$, $2B_1$, B_2; (b) $2A_1$, B_1 (stretches); (c) all IR and Raman allowed.

10.6 $3A_g, 2B_{1g}, B_{2g}, A_u, B_{1u}, 2B_{2u}, 2B_{3u}$.

10.7 (a) $3A_g$, A_u, $2B_u$; (b) IR active: A_u, B_u; Raman active: A_g; (c) A_g and B_u ($+1$ for σ_h); (d) not shown.

10.8 (a) All first overtones are A_g and Raman active only.
(b) $A_g \times A_u = A_u$ IR $\quad A_g \times B_u = B_u$ IR
$A_u \times B_u = B_g$ Raman

10.9 ν_1 and ν_2 A_1; ν_3 and ν_4 E.

10.10 (a) A_{1g}, B_{1g}, E_u; (b) IR: E_u; Raman: A_{1g}, B_{1g}; (c) A_1, T_2.

10.11 (a) Terminal ($\sim 2050\ \text{cm}^{-1}$) and bridging ($\sim 1850\ \text{cm}^{-1}$) COs are present. (b) $\mathbf{D}_{3h}$
(c) $2A'_1, 2E', A''_2, E''$. In-plane: A'_1, E' (bridging CO).
(d) Yes; bridging E', terminal A''_2, E'.
(e) A'_1, E', E''.

10.12 (a) $2A'_1, A'_2, A''_2, 3E', E''$; (b) In plane: A'_1, A'_2, E' ($+$ for σ_h); (c) IR: A''_2, E'; Raman: A'_1, E', E''; (d) A'_1; sketches for X_3Y_3 are given in G. Herzberg, *Molecular Spectra and Molecular Structure II. Infrared and Raman Spectra of Polyatomic Molecules*, Van Nostrand, Princeton, New Jersey, 1945, p. 91.

10.13 Sketches of the normal modes are given by Herzberg, p. 118 of reference in problem 10.12. (a) $2A_{1g}, A_{2g}, A_{2u}, 2B_{1u}, 2B_{2g}, 2B_{2u}, E_{1g}, 3E_{1u}, 4E_{2g}, 2E_{2u}$. (b) In plane: $2A_{1g}, A_{2g}, 2B_{1u}, 2B_{2u}, 3E_{1u}, 4E_{2g}$, all those with positive characters for σ_h. (c) IR: A_{2u}, E_{1u}; Raman: A_{1g}, E_{1g}, E_{2g}.

10.14 (a) $\Gamma^2 = A_{1g}$, so all first overtones are Raman active only.
(b) $\Gamma^3 = \Gamma$, so the IR and Raman activity is the same as for the fundamentals.
(c) $\left.\begin{array}{l} A_{1g} \times A_{2u} = A_{2u} \\ B_{1g} \times B_{2u} = A_{2u} \end{array}\right\}$ IR active
All $\Gamma_g \times \Gamma_g$ are Raman active except $B_{2g} \times B_{1g} = A_{2g}$.
(d) A_{1g}

10.15 (a) $3A_1$, $3E$; all are IR and Raman active.
(b) ν_1, ν_2, and ν_3 are A_1 (Raman polarized), others are E.
(c)

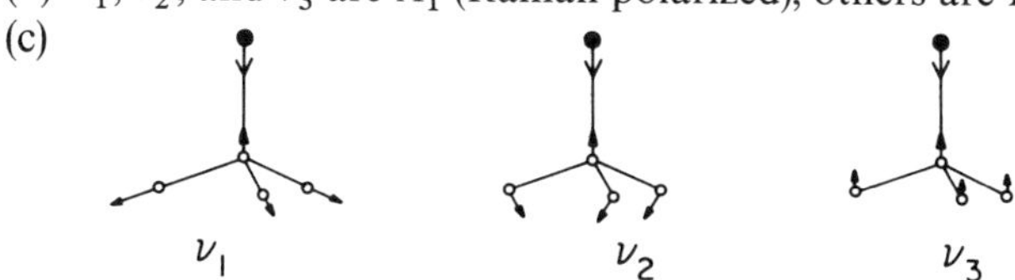

ν_1 occurs at highest frequency and involves least displacement of Cl and greater displacement of H—there is a large shift upon deuteration. ν_3 occurs at the lowest frequency of the A_1 bands and involves the greatest movement of the heavy Cl atoms.

Chapter 11

11.1 A triangular arrangement would result from attack of 1s of H^- on σ of H_2, but both orbitals are filled. H^- attacks one H atom with electron transfer from $1s \rightarrow \sigma^*$, leading to a linear arrangement.

11.2 (a) Cl—Cl—F^+ is preferred. LUMO of Cl^+ combines with HOMO of ClF (a Cl p orbital).
(b) Cl—Cl—F^- is preferred. HOMO of Cl^- (filled p orbital) attacks the LUMO (σ^*) of ClF at the more positive site, Cl. The linear ion involves a four-electron, three-center bond. The nonbonding electron pair is concentrated on the terminal atoms, including the electronegative F.

11.3 The reaction is allowed with conrotatory motion (**C**$_2$). Lowest-energy π MO (b) correlates with HOMO (b) of the cyclic cation. The HOMO (a) correlates with the σ (a) orbital of the product.

11.4 Process is symmetry-allowed. See R. D. Woodward and R. Hoffmann, *Angew. Chem. Int. Edit.* 1969, **8,** 781, Fig. 13.

11.5 The conrotatory process is symmetry-allowed.

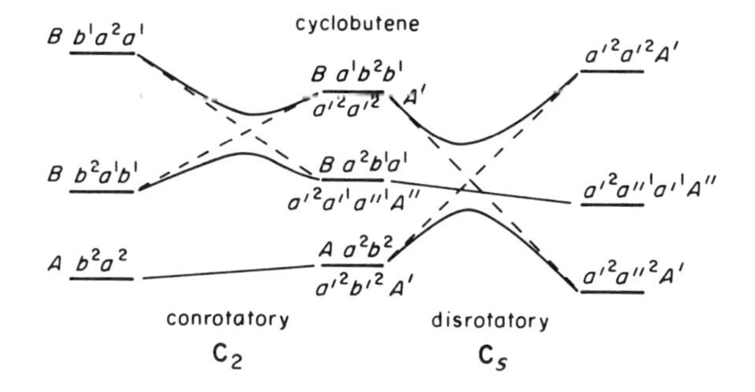

11.6 The conrotatory mode for ring closure of the allyl anion is symmetry-allowed, as predicted from orbital correlations. See H. C. Longuet-Higgins and E. W. Abrahamson, *J. Am. Chem. Soc.* 1965, **87,** 2045. Note that the high activation energy of the disrotatory process is predicted even if $\sigma^1\pi^2\sigma^{*2}$ is assumed to be lower in energy than $\sigma^2\sigma^{*2}$.

11.7 a) $\mathbf{C}_s$ $(a' + a'') + a' \rightarrow a' + (a' + a'')$ Terminal C atoms are substituted. The reaction is allowed.

b) The correlation diagram is similar to Fig. 11.7 except for using σ (a') and σ^* (a'') of Cl_2. The reaction is symmetry-allowed. The butadiene a' and a'' MOs and Cl_2 $\sigma\,(a')$ correlate with bonding orbitals of the product.

11.8 a) $\mathbf{C}_s$ $a' + a' \rightarrow a' + a''$. Forbidden.

b) $\mathbf{C}_s$ $(a' + a'') + a' \rightarrow a' + a'' + a'$ (lone pair). Allowed.

11.9 $\mathbf{C}_s$ $a' + (a' + a'') \rightarrow 2a' + a''$. Allowed.

11.10 a) $\mathbf{C}_2$ $a_1 + b_2 \rightarrow 2a_1$. Forbidden.

b) $\mathbf{C}_s$ $2a' + a'' \rightarrow 2a' + a''$. Allowed.

11.11 a) $\mathbf{C}_s$ $a' + a'' \rightarrow 2a'$ (π bond and lone pair). Forbidden.

b) $\mathbf{C}_s$ $2a' + a'' \rightarrow 2a' + a''$. Allowed.

11.12 *Trans* loss is allowed. *Cis* loss is forbidden since a lone pair appears in a p orbital (a_1). See R. W. Pearson, *Symmetry Rules for Chemical Reactions*, Wiley (Interscience), New York, 1976, p. 268.

11.13 $\mathbf{C}_{2v}$ $a_1(F_2) + (a_1 + b_1 + b_2) \rightarrow a_1 + b_2$ (bonds) $+ a_1 + b_1$ lone pairs.

$\quad\quad\quad\quad\quad\quad\quad\quad\quad s \quad p_x \quad p_y$

Reaction is allowed.

11.14 The reaction is forbidden for any d^n configuration.

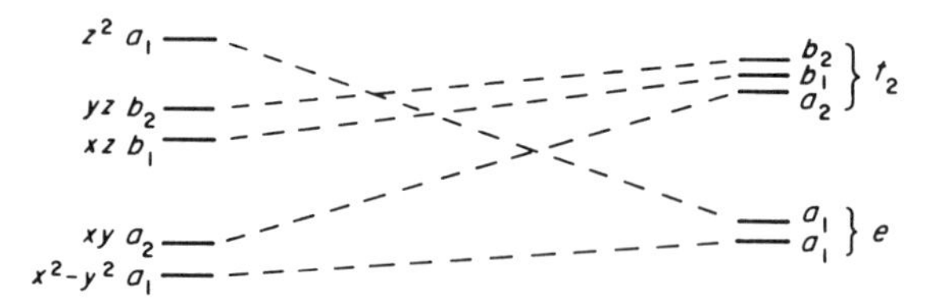

Index

D